Separation Science and Technology
Volume 15

Separations of Water Pollutants with Nanotechnology

A reference series edited by

Satinder Ahuja

Separation Science and Technology
Volume 15

Separations of Water Pollutants with Nanotechnology

Edited by

Satinder Ahuja
Ahuja Consulting, Myrtle Beach, SC, United States

ELSEVIER
ACADEMIC PRESS
An imprint of Elsevier

Academic Press is an imprint of Elsevier
125 London Wall, London EC2Y 5AS, United Kingdom
525 B Street, Suite 1650, San Diego, CA 92101, United States
50 Hampshire Street, 5th Floor, Cambridge, MA 02139, United States
The Boulevard, Langford Lane, Kidlington, Oxford OX5 1GB, United Kingdom

Notices
Knowledge and best practice in this field are constantly changing. As new research and experience broaden our understanding, changes in research methods, professional practices, or medical treatment may become necessary.

Practitioners and researchers must always rely on their own experience and knowledge in evaluating and using any information, methods, compounds, or experiments described herein. In using such information or methods they should be mindful of their own safety and the safety of others, including parties for whom they have a professional responsibility.

To the fullest extent of the law, neither the Publisher nor the authors, contributors, or editors, assume any liability for any injury and/or damage to persons or property as a matter of products liability, negligence or otherwise, or from any use or operation of any methods, products, instructions, or ideas contained in the material herein.

ISBN: 978-0-323-90763-7
ISSN: 1877-1718

Publisher: Susan Dennis
Acquisitions Editor: Kathryn Eryilmaz/Charlotte Rowley
Editorial Project Manager: Mica Ella Ortega
Production Project Manager: Sruthi Satheesh
Cover Designer: Victoria Pearson

Typeset by STRAIVE, India

Contents

Contributors

Numbers in parentheses indicate the pages on which the authors' contributions begin.

Satinder Ahuja (1, 13, 209), Ahuja Consulting, Myrtle Beach, SC, United States

Iftaykhairul Alam (281), Department of Civil and Environmental Engineering, Washington State University, Pullman, WA, United States

Mian Muhammad Ahson Aslam (177), Department of Environmental Science and Engineering, Tunghai University, Taichung City, Taiwan, R.O.C

Muhammad Bilal (177), Department of Agricultural Engineering, Bahauddin Zakariya University, Multan, Pakistan

Chartanay D.J. Bonner (55), Joint School of Nanoscience and Nanoengineering, University of North Carolina Greensboro and North Carolina A&T, Greensboro, NC, United States

William Borges (87), Department of Chemistry, Stony Brook University, Stony Brook, NY, United States

Isha Brahmbhatt (87), Department of Chemistry, Stony Brook University, Stony Brook, NY, United States

George Cai (87), Department of Chemistry, Stony Brook University, Stony Brook, NY, United States

James Chapman (287), Advanced Institute of Science and Technology, The University of Da Nang, Da Nang, Viet Nam

Indranil Chowdhury (281), Department of Civil and Environmental Engineering, Washington State University, Pullman, WA, United States

Daniel Cozzolino (287), QAAFI, University of Queensland, QLD, Australia

Yen T.H. Dang (287), Advanced Institute of Science and Technology, The University of Da Nang, Da Nang, Viet Nam; School of Applied Chemistry and Environmental Sciences, RMIT University, Melbourne, VIC, Australia

Johannes A. DeMessie (163), Environmental Engineering and Science, College of Engineering and Applied Sciences, University of Cincinnati, Cincinnati; William Mason High School, Mason, OH, United States

Walter Den (177), Institute for Water Resources Science and Technology, Department of Mathematical, Physical, and Engineering Sciences, Texas A&M University-San Antonio, San Antonio, TX, United States

Ranjana Dixit (97), Ramjas College, Department of Chemistry, University of Delhi, Delhi, India

Sriparna Dutta (97), Green Chemistry Network Centre, Department of Chemistry, University of Delhi, Delhi, India

A. Figoli (255), Institute on Membrane Technology, Rende, CS, Italy

Sheeana Gangadoo (287), School of Applied Chemistry and Environmental Sciences, RMIT University, Melbourne, VIC, Australia

Reetesh K. Gangwar (139), Department of Physics and CAMOST, Indian Institute of Technology Tirupati, Tirupati, India

Tanvi Gupte (37), DST Unit of Nanoscience (DST UNS) and Thematic Unit of Excellence, Department of Chemistry, Indian Institute of Technology Madras; Department of Metallurgical and Materials Engineering, Indian Institute of Technology, Chennai, India

Afrah Harafan (139), Department of Civil and Environmental Engineering, Indian Institute of Technology Tirupati, Tirupati, India

Benjamin S. Hsiao (69, 87), Department of Chemistry, Stony Brook University, Stony Brook, NY, United States

Ken I. Johnson (87), Department of Chemistry, Stony Brook University, Stony Brook, NY, United States

Sana Khalid (223), College of Earth and Environmental Sciences, University of the Punjab, Lahore, Pakistan

Hsion-Wen Kuo (177), Department of Environmental Science and Engineering, Tunghai University, Taichung City, Taiwan, R.O.C

Songtao Li (87), Department of Chemistry, Stony Brook University, Stony Brook, NY, United States

Yogita Madan (87), Amity School of Applied Sciences, Amity University, Jaipur, Rajasthan, India

Shihabudheen M. Maliyekkal (139), Department of Civil and Environmental Engineering, Indian Institute of Technology Tirupati, Tirupati, India

Rabia Nazir (223), Pakistan Council of Scientific and Industrial Research Laboratories Complex, Lahore, Pakistan

Sherine O. Obare (55), Joint School of Nanoscience and Nanoengineering, University of North Carolina Greensboro and North Carolina A&T, Greensboro, NC, United States

Sukanchan Palit (243), Department of Chemical Engineering, University of Petroleum and Energy Studies, Dehradun, India

Thalappil Pradeep (37), DST Unit of Nanoscience (DST UNS) and Thematic Unit of Excellence, Department of Chemistry, Indian Institute of Technology Madras, Chennai, India

Kashif Rasool (177), Qatar Environment and Energy Research Institute, Hamad Bin Khalifa University (HBKU), Qatar Foundation, Doha, Qatar

Endalkachew Sahle-Demessie (163), Department of Chemistry, Missouri University of Science & Technology, Rolla, MO, United States

Swagotom Sarkar (13, 209), Bangladesh Reference Institute for Chemical Measurements (BRiCM), Dhaka, Bangladesh

Priyanka R. Sharma (69, 87), Department of Chemistry, Stony Brook University, Stony Brook, NY, United States

R.K. Sharma (97), Green Chemistry Network Centre, Department of Chemistry, University of Delhi, Delhi, India

Sunil K. Sharma (69, 87), Department of Chemistry, Stony Brook University, Stony Brook, NY, United States

George A. Sorial (163), Environmental Engineering and Science, College of Engineering and Applied Sciences, University of Cincinnati, Cincinnati, OH, United States

Muhammad Sultan (177), Department of Agricultural Engineering, Bahauddin Zakariya University, Multan, Pakistan

Vi Khanh Truong (287), School of Applied Chemistry and Environmental Sciences, RMIT University, Melbourne, VIC, Australia

C. Ursino (255), Institute on Membrane Technology, Rende, CS, Italy

Preface

Over 80 countries in the world suffer from water shortages. As a result, more than one billion people in the world do not have access to clean drinking water. According to the United Nations (UN), water shortages will be encountered by an estimated 2.7 billion people by 2025. This means we need to find more reliable sources of water and we have to reclaim wastewater. To achieve water sustainability, we must improve water quality by using various purification techniques. This book addresses water purification issues with nanotechnology. A large number of applications are covered, and the limitations of this technique are also discussed.

Chapter 1 provides an overview of water sources and its potential pollutants. It describes the importance of using ultra-trace analysis to monitor contaminants. These contaminants can be removed by using nanomaterials and nanotechnology. Nanomaterials are efficient in that only small amounts are required and they are multifunctional. They can purify water by removing water pollutants such as heavy metals, pathogens, organic and inorganic compounds, and drugs.

Chapter 2 describes the importance of nanomaterials being used for water purification. It covers how nanophotocatalysts, nanomembranes, nanoreactors, and nanoadsorbents can decontaminate polluted water sources. In addition, this chapter includes a discussion on the removal of five selected pollutants (arsenic, lead, pharmaceuticals, per- and poly-fluoroalkyl substances, and corona viruses) from water by various nanomaterials.

Chapter 3 highlights the potential of nanosensor technologies and recent advances in their use for water quality monitoring and analysis. Researchers are focused on nanomaterial-enabled nanosensors for water quality monitoring because they have an inherent capacity for sensitive, selective, rapid, and efficient on-site detection. Various nanosensors are highlighted in this chapter.

Chapter 4 examines the interaction of heavy metal lead ions (Pb^{+2}) on Ag nanoparticles (Ag NPs) stability and antimicrobial response. Ag NPs change the availability of Pb^{+2} to bacterial cells and alter individual antimicrobial toxicity. The results heighten the need to study nanoparticles' interaction in combination with other contaminant stability and biological response as they are seldom found in the environment independently.

Chapter 5 describes the advanced applications of NC (nanocellulose)-based membranes in tackling water impurities. Nanocellulose possesses many unique features including high surface area, tunable porosity, immense functionality, flexibility, transparency, and excellent compatibility with various compounds. This chapter discusses NC's implication in microfiltration, ultrafiltration, nanofiltration, reverse osmosis, forward osmosis, and membrane distillation to tackle a variety of impurities such as biological (viruses, bacteriophages, antifouling agents), organic dyes, oil/water, and solvent/water mixtures, metal ions, and salts.

Chapter 6 describes how carboxycellulose nanofibers (CNF) can be used for removing lead from water. CNFs were prepared from raw jute fibers using a one-step nitro-oxidation approach based on nitric acid/sodium nitrite. The main characteristics of extracted CNF with low crystallinity, good carboxylate content, and highly negative surface charge make them suitable nanomaterials for removal of Pb(II) ions from water.

Chapters 7 and 8 cover remediation of heavy metals and pesticides with nanomaterials. Chapter 9 discusses applications and limitations of graphene oxide for remediating contaminants of emerging concern in wastewater. Chapter 10 describes removal of fluoride, a major problem of water contamination in the world.

Chapters 11 and 12 cover chitosan composites and carbon nanotubes (CNTs) with significant and tunable physical–chemical properties for remarkable Cr(VI) removal ability from an aqueous medium. Major influencing factors are highlighted, including operational pH range, adsorbent dosage, contact time, and interference. The adsorption equilibrium, kinetics, and thermodynamics are also reviewed to identify the adsorption mechanisms and nature of the adsorption process. The important considerations such as stability and reusability of composite adsorbents and their recovery are also discussed for economic and sustainability concerns.

Chapter 13 elucidates the success of nanofiltration and nanotechnology. It covers the vast domain of environmental sustainability and its interface with environmental pollution control. Recent advances in the field of nanofiltration, nanotechnology, novel separation processes, and environmental sustainability are included in this chapter.

Chapter 14 describes why the application of polymeric membranes containing nanomaterials (NMs) represents an interesting and advanced approach to the preparation of tailored membranes and how to affect overcome biofouling problems. The presence of the NMs in the polymeric matrix allows avoiding some membrane limitations as fouling, microbial growth, flux reduction, and a reduced lifetime. The most common NPs for the preparation of these nanocomposite membranes, such as CNTs, titanium oxide (TiO_2), silver (Ag), and graphene oxide (GO), are discussed in the chapter.

Two-dimensional nanomaterials are of great interest because of their potential characteristics such as increased surface area, mechanical strength, thermal conductivity, etc. (Chapter 15). With proper synthesis and modification, these high-aspect ratio sheet-like 2D layer solids can lead to novel and improved technologies to meet the global environmental challenges. Even though many water filtration and membrane technologies are in use, fouling remains one of the main problems. Fouling can seriously hamper the performance of the membranes. Two-dimensional nanomaterials like GO, chalcogenides, and oxides have shown great potential in water filtration technology. Nanosheets show both antifouling and antibacterial properties that are technologically useful properties.

Water purification techniques frequently rely on univariate approaches in experimental design and synthesis and for the analyses of the results of the materials or strategies used in the water depollution process (Chapter 16). Access to safe and clean water is one of the most pressing issues that humanity will be confronted with over the next 50 years. This chapter focuses on chemometric investigations that generate a multimodal approach to the design and the use of nanomaterials for water cleanup.

September 13, 2021
Satinder Ahuja

Chapter 1

Overview of separations of water pollutants with nanotechnology

Satinder Ahuja*
Ahuja Consulting, Myrtle Beach, SC, United States
**Corresponding author: e-mail address: sutahuja@atmc.net*

Introduction

More than 1 billion people in the world have no access to clean drinking water, and over 80 countries in the world suffer from water shortages. This situation is not likely to improve much, according to the United Nations (UN): Water shortages will be encountered by an estimated 2.7 billion people by 2025. These shortages could lead to major conflicts over water. To achieve water sustainability, we need to improve water quality by various purification techniques [1–8]. To attain this objective, we must address technical, economic, and social issues [4] relating to purification of water for drinking; this must include reclamation of wastewater. Water reclamation (the act or process of recovering), is an absolute necessity because we have managed to pollute our surface water, and even groundwater in some cases, to a point that water needs to be purified for drinking [1].

Drinking water comes mainly from rivers, lakes, wells, and natural springs. These sources are exposed to a variety of situations that can lead to water contamination.

The failure of safety measures related to production, utilization, and disposal of thousands of inorganic and organic compounds is causing pollution of our water supplies. Water contaminants that occur from materials that we frequently use to improve the quality of life are as follows [8]:

- Combustion of coal and oil
- Detergents
- Disinfectants
- Fertilizers
- Gasoline combustion products and additives
- Herbicides
- Insecticides and pesticides
- Per- and polyfluoroalkyl substances
- Pharmaceuticals (including endocrine disruptors)
- Phthalates
- Radionuclides

In 2002, the US Geological Survey (USGS) found pharmaceuticals (hormones and other drugs) in 80% of streams sampled in 30 states of the United States. This situation has not improved significantly. An enormous quantity of antimicrobials and antibiotics is administered to healthy animals on US farms each year. These compounds end up in our water supplies when the animal waste is not handled properly There are numerous potential sources of water contamination: volatile and semi-volatile compounds; improperly disposed chemicals; animal wastes; introduction of toxic agents by terrorists; wastes injected underground; and naturally occurring substances. Furthermore, some emerging contaminants can also impact water quality. Drinking water that is not properly treated or disinfected, or which travels through an improperly maintained distribution system (infrastructure), may also pose a health risk. It is important to recognize that a large number of inorganic/organic

Separation Science and Technology, Volume 15. https://doi.org/10.1016/B978-0-323-90763-7.00015-9

compounds that cover the entire range of the alphabet, from A to Z (arsenic to zinc), can cause contamination of our water supplies [9]. Not all chemicals are necessarily harmful; e.g., zinc in small amounts is desirable, whereas, arsenic even at concentrations above 10 parts per billion (ppb) is quite harmful. A variety of pollutants can come from wastewater because it is generally recycled to surface water or groundwater after some processing. Wastewater can be defined as water that has been used in the home, in a business, or as part of an industrial process. So, wastewater can originate from many places: households, industries, commercial developments, road runoff, etc. In addition, we have to contend with nonpoint sources of pollution. As diverse as the sources of wastewater are, so too are their potential constituents.

Monitoring ultratrace pollutants

Over 700 different chemicals have been found in US drinking water—as it comes out of the tap! The EPA classifies 129 of these chemicals as being particularly dangerous (see List of Priority Pollutants in Appendix), and it sets standards for approximately 90 contaminants in drinking water. Those standards, along with each contaminant's likely source and its health effects, are available at www.epa.gov/safewater/mcl.html. Even though consuming tap water can be risky, there is no assurance that bottled water is any safer.

A better understanding of wastewater constituents and their abundance at different stages is a first step in recognizing appropriate opportunities for pretreatment. The composition of wastewater affects not only the treatment processes applied but also their source recovery opportunities. It is clear that wastewater contains pollutants/contaminants that have to be removed and/or reduced to safe levels before it is directed to a surface water source (river, ocean, bay, lake, etc.) or to groundwater.

In the 1978 Metrochem meeting, I presented a paper "In Search of Femtogram." A femtogram is 10^{-15} g, or 1 part per quadrillion—a phantom quantity at that time. I pointed out that it is essential to analyze very small quantities of various contaminants to fully understand the impact of an assortment of chemicals on our body [9–11,22]. For example, dioxin (2,3,7,8-tetrachloro-dibenzodioxin) can cause abortion in monkeys at the 200 ppt (parts per trillion) level, and PCBs at 0.43 ppb level can weaken the backbones of trout. It has been known for some time now that water that we call potable may contain many trace and ultratrace contaminants [11]. For more detailed discussion, please see Refs. [1–8].

A large variety of methods are available for monitoring point-source pollutants because we can reasonably assess which pollutants may be present. By contrast, this task becomes much more difficult with nonpoint source pollution. The overwhelming majority of water quality problems are now caused by diffuse nonpoint sources of pollution from agricultural land, urban development, tree harvesting, and the atmospheric pollutants. The nonpoint-source contaminants are more difficult to effectively monitor, evaluate, and control than those from point sources (such as discharges of sewage and industrial waste).

It is important to assure that the sample used for ultratrace analyses (at or below ppb level) should be representative of the "bulk material." The major considerations are [12]: The analytical uncertainty should be reduced to one third or less of sampling uncertainty [13]. Poor analytical results can be obtained because of reagent contamination, operator errors in procedure or data handling, biased methods, and so on. These errors can be controlled by the proper use of blanks, standards, and reference samples. It is also important to determine the extraction efficiency of the method.

Preconcentration of the analyte is frequently necessary because the detector used for quantification does not have the necessary detectability, selectivity, or freedom from matrix interferences [14]. Significant losses can occur during this step because of small volume losses to glass walls of recovery flasks or disposable glass pipets and other glassware. However, with suitable precautions, preconcentration of metals at concentrations down to 10^{-12} g/g for copper, lead, and zinc, and 10^{-13} g/g for cadmium have been successfully demonstrated in a typical polar snow matrix [15]. It is crystal clear that ultratrace analyses can help us detect potential contaminants; however, when these elements do get into water, purification becomes essential.

Ideally, we should look for all pollutants at trace or ultratrace levels (at or below the ppb level). This means that our methodologies should separate all contaminants based on their molecular weight, using techniques such as size-exclusion chromatography. And the contaminants should be characterized by IR, NMR, MS (mass spectrometry), or various hyphenated techniques. Alternatively, we can separate the pollutants on the basis of whether they are organic or inorganic, and also do additional tests to determine if they are polymers. It is important to remember that water has to meet certain requirements to be called potable (see Section "Quality of drinking water"). Real-time monitoring can enable a quick response to water quality concerns that arise from natural, accidental, or malicious contamination and allow the greatest protection of public health (see chapter 7 in Ref. [4]).

Pollutants arising from water disasters

Recent water disasters in Flint, MI, and in Wilmington and Fayetteville, NC, are distressing reminders of what can go terribly wrong with our water supplies [16]. In Flint, instead of buying Lake Huron water from Detroit, the city began drawing its water from the local river in April 2014 in an effort to save money. Residents started complaining about burning skin, hand tremors, hair loss, and even seizures. In 2015, high levels of lead were found in the water supply of Flint. Unsafe levels had turned up earlier in tap water in Washington, DC in 2001, in Columbia, SC, in 2005; in Durham and Greenville, NC, in 2006; and in Jackson, MS, and Sebring, OH, in 2015.

DuPont (Chemours) introduced GenX in 2009 to replace PFOA (perfluorooctanoic acid), a compound it used to manufacture Teflon and coatings for stain-resistant carpeting and waterproof clothes. GenX has been detected in the drinking water in Fayetteville, Wilmington, and Brunswick County NC. In Fayetteville, soil and groundwater have also been affected. Levels of GenX in the drinking water of the Cape Fear Public Utility Authority in Wilmington averaged 631 ppt (parts per trillion), according to a study published by Knappe and coworkers in *Environmental Science & Technology Letters* in 2016. The EPA has set drinking water standards for PFOA at 70 ppt; GenX limit has been calculated at 140 ppt based on some assumptions. Water utilities are trying to find a reliable and economic means to remove GenX.

Another looming disaster may arise from harmful cyanobacterial algal blooms. In aquatic environments, they are likely to be a major worldwide problem, causing a significant adverse impact on public health. Microcystins are hepatotoxins produced by cyanobacteria. They also produce a wide variety of secondary metabolites, and most important, potent toxins called cyanobacterial toxins or cyanotoxins. They are known to affect a wide range of living organisms including humans. Cylindrospermopsin (CYN) is one of the most widely distributed cyanotoxins in bodies of water. CYN is produced by large groups of cyanobacteria that are highly adaptive and invasive. They have been detected in tropical, subtropical, and even in temperate areas. Compared with microcystins, CYN can be actively released into the environment, leading to higher extracellular toxin concentrations than the intracellular levels. Unregulated Contaminant Monitoring Rule 4 (UCMR4) that covers cyanotoxins went into effect in 2016.

There is still another impending water disaster—microplastics—are a major problem. Microplastics can stunt fish growth and alter their behavior. The extensive use of plastics and their careless disposal has led to the pollution of various water bodies. Large parts of the Pacific Ocean are referred to as "plastic oceans," where enormous gyres about the size of Texas are covered with plastic debris. The Pacific is the largest ocean realm on our planet, approximately the size of Africa—over 10 million square miles—and it is the home of two very large gyres of plastics. The Atlantic Ocean contains two or more gyres, and other plastic oceans exist in other bodies of water.

Quality of drinking water

A simple definition of drinking water (potable water) is as follows: Any water that is safe and clean. The National Primary Drinking Water Regulations control water quality in the United States. The Federal Water Pollution Control Act of 1948 was the first major US law that was passed to address water pollution. In response to public concern about degraded water quality and a widespread view that pollution of our rivers and lakes was unacceptable, the Water Act became law in 1972. Control of point-source contamination, traced to specific "end of pipe" points of discharge or outfalls such as those at factories and combined sewers, was the primary focus of the Clean Water Act (CWA). The 1972 amendments to the act

- Established the basic structure for regulating pollutant discharges into waters of the United States
- Provided the EPA the authority to implement pollution control programs such as setting wastewater standards for industry
- Maintained existing requirements to set water quality standards for all contaminants in surface waters
- Made it unlawful for any person to discharge any pollutant from a point source into navigable waters, unless a permit is obtained under its provisions
- Funded the construction of sewage treatment plants under the construction grants program
- Recognized the need for planning to address the critical problems posed by nonpoint-source pollution

The Safe Drinking Water Act (SDWA) was passed in the United States in 1974, giving the EPA the responsibility for monitoring and enforcing the public's drinking water safety rules. The EPA sets standards for drinking water quality, and with its partners, implements various technical and financial programs to ensure drinking water safety. In 1996, SDWA was amended to assure that the EPA will continually seek, identify, and monitor potentially harmful or relevant contaminants that may be currently unregulated. Other nations adopted similar measures and have seen improvement in point-source contamination as well. In 2010, a United Nations resolution declared the human right to "safe and clean drinking water and sanitation."

As mentioned earlier, a large number of aquatic microorganisms can infect or parasitize humans, and these pathogens are responsible for considerable morbidity and mortality worldwide. The strategies and methods for studying these organisms are discussed in chapter 8 in Ref. [1].

Table 1 shows how one region in the United States (Brunswick County, NC) monitors the quality of potable water. It should be noted that a number of unregulated substances such as 1,4 dioxane and manganese are also monitored. After GenX was revealed in the Cape Fear River, Brunswick County started monitoring per- and polyfluoroalkyl substances (PFASs) including GenX. Some of the contaminants such as arsenic, perchlorate, pesticides, and pharmaceuticals are not analyzed routinely. Among pharmaceutical contaminants, the problem of endocrine disruptors in water supplies demands greater vigilance (see chapter 5 in Ref. [4] and chapter 13 in vol. 1 of Ref. [5]).

To purify water, it is necessary to remove or separate water pollutants. Separate is defined by Webster's dictionary as set or kept apart and separation is an act or process of separating. Pollutant is defined as something that pollutes or makes impure. Nanotechnology is an advanced technique that can be helpful in the purification of water. It can be defined as methods that create materials or structures with designed features in the 1–100 nm-size range [17].

Nanomaterials are useful because they are efficient in that only small amounts are required and they are multifunctional [18]. They can purify water by removing water pollutants such as heavy metals, pathogens, organic and, inorganic harmful metals, and drugs, etc. [19].

TABLE 1 Water quality results of potable water in Brunswick County, NC in 2020.

Substances	EPAs MCL	Amount detected	Source of contaminant
Turbidity	1.0 ntu	Ave. 0.050 ntu	Soil runoff
Raw water TOC	Min. 45% removal	*Removal efficiency* Ave. 55.1%	Naturally present in the environment
Finished water TOC	N/A	Ave. 1.215 ppm	Naturally present in the environment
pH	6.8–8.5	7.6	By-product of caustic addition
Regulated inorganic chemicals			
Chlorite	1.0 ppm	Ave. 0.57 ppm	By-product of disinfection
Chlorine dioxide	0.8 ppm	0.02 ppm	Water additive for microbial control
Fluoride	4 ppm	0.63 ppm	Water additive to promote strong teeth
Orthophosphate	17 ppm	1.76 ppm	Water additive to control corrosion
Total chlorine	4 ppm	2.96 ppm	Water additive to control microbes
Monochloramine, residual	4 ppm	2.73 ppm	Water additive to control microbes
Unregulated substances			
1,4 Dioxane	N/R	0.80 ppb	Purifying agent in pharmaceuticals and by-product of PET plastic production
Hardness	N/R	22.5 ppm	Part of treatment process Erosion of natural deposits
Iron	N/R	0.02–0.10 ppm	Part of treatment process Erosion of natural deposits
Manganese	Secondary 0.05 ppm	0.014 ppm	Part of treatment process Erosion of natural deposits
Free ammonia	N/R	0.109 ppm	Water additive to control microbes
Sodium	N/R	20.749 ppm	Part of treatment process Erosion of natural deposits

Advanced research in this area suggests that existing water treatment problems can be solved by applying the following nanotechnology-based processes: antimicrobial nanotechnology, zero-valent iron nanoparticles, nanoadsorbents, nano-enhanced membranes, nanometal oxides, and nanophotocatalysts [20]. This book covers in detail how nanomaterials can help purify water. Applications and limitations are also provided. New developments in nanotechnology pertaining to water purification are discussed as well.

Importance of nanomaterials in water purification

Pure drinking water is absolutely essential for our survival (Chapter 2). Unfortunately, drinking water resources are being polluted in a variety of ways. Conventional water purification technologies are generally not efficient. To mitigate this problem, scientists around the world are developing new cost-effective, environment-friendly technologies. Nanomaterials with extraordinary physical, chemical, mechanical, and morphological properties offer a potential solution to water pollution issues. This chapter points out some basic information on different contaminant sources and their effects on human beings and environment, differences between wastewater and drinking water treatment processes, and limitations of conventional water treatment technologies. It describes how nanophotocatalysts, nanomembranes, nanoreactors, and nanoadsorbents can decontaminate polluted water sources. In addition, it includes discussion on removal of five selected pollutants (arsenic, lead, pharmaceuticals, per- and polyfluoroalkyl substances, and corona viruses) from water by various nanomaterials. Some limitations of using nanomaterials for water treatment are also described.

Nanosensors for water quality monitoring

Lack of safe water is a global problem that is caused by water pollution. The development of reliable sensors for the detection of contaminants is a necessity and an urgent need (Chapter 3). This chapter highlights the potential of nanosensor technologies and recent advances in them for water quality monitoring and analysis. Contaminants such as organic pollutants, pharmaceuticals, heavy metals, pathogens, and pesticides are toxic and hazardous to the ecosystem. Researchers are focused on nanomaterial-enabled nanosensors for water quality monitoring as they have an inherent capacity for sensitive, selective, rapid, and efficient on-site detection. Various nanosensors are discussed in this chapter based on the signal transduction mechanism (electrochemical, optical, and magnetic), along with multiple nanomaterial platforms employed in the sensing. A few promising examples of novel (biomimetic and hybrid) nanosensing technologies are also reviewed.

Assessing the environmental impact of anthropogenic nanoparticles

Silver nanoparticles' (Ag NPs) interaction with other anthropogenic contaminants influences their surface properties, aggregation state, stability, and, consequently, their toxic response toward microbial pathogens (Chapter 4). This study examines the interaction of heavy metal lead ion (Pb^{+2}) on Ag NPS's stability and antimicrobial response. The effect of the metal ions on the toxicity of Ag NPs was investigated using *Escherichia coli* and *Staphylococcus aureus* bacterial strains. Ag NPs changed the availability of Pb^{+2} to bacterial cells and altered individual antimicrobial toxicity. Exposure of Pb^{+2} with Ag NPs displayed higher toxicity toward the bacterial strain. In the presence of humic acid, the Pb^{+2} reduced the adsorption of Ag NPs, thus making them more available to bacterial cells. The results heighten the need to study nanoparticles' interaction in combination with other contaminant stability and biological responses, as they are seldom in the environment independently.

Nanocellulose in membrane technology for water purification

Chapter 5 describes the advanced applications of NC (nanocellulose)-based membranes in tackling water impurities. Nanocellulose possesses many unique features including high surface area, tunable porosity, immense functionality, flexibility, transparency, and excellent compatibility with other organic and inorganic molecules. The focus of this chapter is to compile all the studies focused on NC-based membranes derived from cellulose nanofibers (CNF), cellulose nanocrystals (CNC), bacterial nanocellulose (BNC). This chapter also discusses their implication as microfiltration (MF), ultrafiltration (UF), nanofiltration (NF), reverse osmosis (RO), forward osmosis (FO), and membrane distillation (MD) to tackle a variety of impurities such as biological (e.g., viruses, bacteriophages, antifouling agents), organic dyes, oil/water and solvent/water mixtures, metal ions, and salts. The properties of NC membranes, their separation performance coupled with the competency of flux, and their structural–physical properties are also covered in this chapter.

Plant-derived carboxycellulose: Highly efficient bionanomaterials for removal of toxic lead from contaminated water

Lead (Pb) toxicity in drinking water has become a serious problem worldwide, particularly in developing nations (Chapter 6). This has encouraged development of various Pb removal methods from polluted water. Carboxycellulose nanofibers (CNF) were prepared from raw jute fibers using a one-step nitro-oxidation approach based on nitric acid/sodium nitrite. The main characteristics of extracted CNF with low crystallinity (~35%), good carboxylate content (1.15 mmol/g) and highly negative surface charge (−70 mV) made them suitable nanomaterial for removal of Pb(II) ions from water. Experimental results showed that 0.2 wt% concentration of CNF suspension is capable of removing a wide range of Pb(II) ions from 50 to 5000 ppm in a very short time, where the adsorption proficiency of CNF was found to be around 2200 mg/g based on the Langmuir isotherm. The high removal efficiency of CNF was due to the combined effects of adsorption and mineralization of lead crystals. Finally, nanoscale lead crystallization on nanocellulose scaffolds were further investigated by FTIR (Fourier Transform infrared spectroscopy), UV–visible spectroscopy, SEM-EDS (scanning electron microscopy–energy dispersive X–ray spectroscopy), and TEM (transmission electron microscopy) analysis.

Remediation of heavy metals with nanomaterials

The advent of nanomaterials with remarkable advancements in nanoscience has opened new horizons in the field of water purification, especially heavy metal remediation because of their fascinating attributes arising from their nanodimensions (Chapter 7). The possibility to tune their properties at the nanoscale via rational surface engineering has further enhanced the potential use of these materials through increased selectivity toward the target contaminant, thus producing safer drinkable water. The chapter provides desirable information to the researchers interested in designing and developing promising nanomaterials for removing absolutely noxious contaminants such as mercury, lead, cadmium, arsenic and chromium. With the compilation of latest research accomplished so far in metal remediation through development of nano-based technologies (carbon, silica, iron, other metal oxides, and bionano-based materials as either filters or adsorbents, the chapter intends to equip the reader with the knowledge to critically analyze the benefits of such innovative techniques over the traditional methods. There is also a significant discussion on the 2030 agenda of U.N. Sustainable Developments goals that have prioritized providing cleaner water to all. This chapter sheds some light on how the proficient incorporation of nanomaterials can contribute substantially in this regard.

Abatement of pesticides in drinking water by nanoscale carbon materials

Pesticides are a group of chemicals widely detected in water bodies, mainly because of their indiscriminate use by the agricultural sector (Chapter 8). Because of their possible entry into human and animal food chains, health hazards posed by pesticides have become a considerable area of concern worldwide. Nevertheless, the production and use of pesticides are increasing, and the global pesticide market is expected to reach 130.7 billion USD $ by 2022–2023. Given the widespread occurrence and potential toxicity, many treatment technologies are in place to treat pesticide-contaminated water. However, the diverse chemical nature of the pesticides and the stringent regulations in drinking water standards, 0.1 μg/L for single pesticide and 0.5 μg/L for the sum of all pesticides, limit the use of many existing treatment systems. The advancement in nanoscience and nanotechnology suggests that nanoscale materials, especially nanocarbon and its derivatives, are promising candidates for scavenging pesticides in water. This chapter reviews the literature relating to various nanoscale carbons with exciting properties and their applications for treating pesticide-laden drinking water. The mechanism of removal, challenges, prospects, and future development in the area are discussed. This chapter also covers the origin, occurrence of pesticides in water bodies, and their human, and ecological health impacts.

Applications and limitations of graphene oxide for remediating contaminants of emerging concern in wastewater

Water resources are being polluted with a variety of contaminants which include contaminants of emerging concern (CEC). More research is needed to develop new, cost-effective, and environment-friendly technologies for water purification because conventional water purification technologies such as sedimentation, filtration, and chemical or biological degradation are unable to effectively remove some of CECs such as antibiotics, pesticides, pharmaceutical and personal care products, and hormones (Chapter 9). Graphene oxide (GO) nanomaterials with specific physical, chemical, mechanical, and optical properties can be excellent candidates for removing CECs efficiently. This chapter covers water pollution by CECs and their effects, limitations of current technologies in removing CECs, applications of pure and modified GO nanomaterials to remove different types of CECs which include antibiotic-resistant bacteria and genes, endocrine disrupting chemicals (pesticides and pharmaceutical, antibiotic, and perfluoroalkyl substances), disinfectant by-products, dioxin, UV filter chemicals, algal toxins, and microplastics from wastewater. Finally, limitations of graphene oxide nanomaterials for water purification are also discussed.

Nanoadsorption: An innovative and compelling approach for fluoride remediation of water

Various human activities pose severe environmental threats to air, land, and water quality (Chapter 10). Fluoride contamination is one of the biggest environmental challenges being faced by numerous countries of the world. With the emergence of awareness of deadly health problems associated with fluoride, it is imperative to design efficacious purification technologies for its decontamination of water. A multitude of conventional methods involving the basic principles of precipitation, coagulation, flocculation, ion exchange, membrane filtrations, and reverse osmosis have been tested to combat fluoride pollution, but among these methods, adsorption is considered effective in terms of a cost-cutting and convenient method of operation. Nanotechnology has attracted paramount attention and is being extensively used in water treatment by devising efficient nanoadsorbents with interesting attributes such as large surface area, short diffusion rate, and enhanced optical, chemical, electronic, and magnetic characteristics. In this chapter, several metal elements, geopolymers, biopolymers, and carbon-based nanomaterials are discussed with special emphasis on their potential applications in defluoridation of aqueous solutions.

Removing chromium (VI) from contaminated water using a nano-chitosan–coated diatomaceous earth

Access to clean water is a major challenge in many developing counties due to municipal and industrial pollution (Chapter 11). Therefore, developing low-cost technologies to provide clean water is vital. Hexavalent chromium, Cr(Vi), is a toxic substance that is not biodegradable and has adverse effects on the ecosystem and human health. In this work, chitosan-coated diatomaceous earth (CCDE) granules were synthesized at 20 and 30 wt% and characterized by Fourier transfer infrared, scanning electron microscopy, and thermogravimetric analysis. CCDE is used for Cr(VI) ion removal from aqueous solution in batch processes. Several parameters have been studied such as solution-pH, initial Cr(VI) ion concentration, the mass of adsorbent, and contact time to investigate the efficiency of Cr(VI) removal. In these studies, CCDE achieved 99% removal efficiency of Cr(VI) and adsorption capacity was 18.69 mg Cr(VI)/g. The adsorption followed pseudo–second-order and was well fitted to the Langmuir isotherm model, indicating monolayer adsorption behavior. Anions in the solution had an insignificant effect on Cr (VI) ion uptake by CCDE powder achieving clean water standard from initial Cr(VI) concentrations of 1 mg/L. The removal efficiency increased at lower pH and the optimal pH was 4. Adsorption equilibrium was reached after approximately 60 min, and the adsorption kinetics fitted second-order model for Cr(VI). This study demonstrates that nature-based adsorbents can be employed as a low-cost, sustainable, and excellent alternative material for removing heavy metals from contaminated water.

Recent trends of carbon nanotubes and chitosan composites for hexavalent chromium removal from aqueous samples

Traditional adsorbents have been studied extensively to lower the level of Cr in an aqueous media, but their efficiencies are low (Chapter 12). Carbon nanotubes (CNTs) and chitosan composites with significant and tunable physical–chemical properties have been developed to exhibit remarkable Cr(VI) removal ability from an aqueous media. However, there remain certain challenges that need be addressed to develop CNTs and chitosan-based adsorbent treatment systems from lab-scale to practical applications. This chapter provides coverage of recent developments and application of CNTs and chitosan composites for the adsorptive removal of Cr(VI) to an acceptable level. Major influencing factors are highlighted, including the operational pH range, adsorbent dosage, contact time, and interference. The adsorption equilibrium, kinetics, and thermodynamics are also reviewed to identify the adsorption mechanisms and nature of the adsorption process. The important considerations such as stability and reusability of composite adsorbents and their recovery are also discussed for economic and sustainability concerns.

Nanofiltration and environmental sustainability

Chapter 13 elucidates the success of nanofiltration and nanotechnology. It covers the vast domain of environmental sustainability and its interface with environmental pollution control. Recent advances in the field of nanofiltration, nanotechnology, novel separation processes, and environmental sustainability are included in this chapter.

Nanomaterials in polymeric membranes for water treatment applications

The application of polymeric membranes containing nanomaterials (NMs) represents an interesting and advanced approach to prepare tailored membranes and overcome biofouling problems (Chapter 14). The presence of the NMs in the polymeric matrix allows getting around some membrane limitations as fouling, microbial growth, flux reduction, and reduced lifetime. The most common nanoparticles (NPs) for the preparation of these nanocomposite membranes, such as carbon nanotubes (CNTs), titanium oxide (TiO_2), silver (Ag) and graphene oxide (GO), have been discussed in the chapter. In particular, the chapter highlights the NMs properties, membranes production, and applications in different fields of water treatment, including wastewater treatment.

Applications of two-dimensional nanostructures for water filtration

Two dimensional nanomaterials have become a major focus in modern research worldwide because of their potential characteristics like increased surface area, mechanical strength, thermal conductivity, etc. (Chapter 15) With proper synthesis and modification these high-aspect ratio sheet-like 2D layer solids can lead to novel and improved technologies to meet the global environmental challenges. Supplying safe drinking water is a major problem we face in the modern era. Even though many water filtration and membrane technologies are in use, fouling remains one of the main problems. Fouling can seriously hamper the performance of the membranes. Two-dimensional nanomaterials like graphene oxide, chalcogenides, and oxides have shown great potential in water filtration technology to modify the membrane. The nanosheets show both antifouling and antibacterial properties that are certainly fascinating and technologically useful properties. This chapter assesses the recent development of 2D nanomaterials as building blocks for upcoming generation's water treatment, desalination, antifouling, and antibacterial membranes. The most promising application areas are highlighted to show how these materials or combination and modification of these materials can lead to new cleanup technologies in the field of water filtration.

New nanomaterials for wastewater depollution: Methods using chemometric approaches

Water purification techniques frequently rely on univariate approaches in experimental design and synthesis, and for the analyses of the results of the materials or strategies used in the water depollution process (Chapter 16). Access to safe and clean water is one of the most pressing issues that humanity will be confronted with over the next 50 years. Traditional wastewater treatment and water-depollution nanomaterials used for water pollution applications are reviewed. Additionally, the chapter focuses on chemometric investigations that generate a multimodal approach to the design and usage of nanomaterials for water cleanup.

Appendix

Priority pollutants

Priority pollutants are a set of chemical pollutants which should get important consideration in developing national discharge standards (such as Effluent Guidelines) or in national permitting programs (such as NPDES).

1. Acenaphthene
2. Acrolein
3. Acrylonitrile
4. Benzene
5. Benzidine
6. Carbon tetrachloride
7. Chlorobenzene
8. 1,2,4-Trichlorobenzene
9. Hexachlorobenzene
10. 1,2-Dichloroethane
11. 1,1,1-Trichloreothane
12. Hexachloroethane
13. 1,1-Dichloroethane
14. 1,1,2-Trichloroethane

15. 1,1,2,2-Tetrachloroethane
16. Chloroethane
17. (Removed)
18. Bis(2-chloroethyl) ether
19. 2-Chloroethyl vinyl ethers
20. 2-chloronaphthalene
21. 2,4,6-trichlorophenol
22. Parachlorometa cresol
23. Chloroform
24. 2-chlorophenol
25. 1,2-dichlorobenzene
26. 1,3-dichlorobenzene
27. 1,4-dichlorobenzene
28. 3,3-dichlorobenzidine
29. 1,1-dichloroethylene
30. 1,2-trans-dichloroethylene
31. 2,4-dichlorophenol
32. 1,2-dichloropropane
33. 1,3-dichloropropylene
34. 2,4-dimethylphenol
35. 2,4-dinitrotoluene
36. 2,6-dinitrotoluene
37. 1,2-diphenylhydrazine
38. Ethylbenzene
39. Fluoranthene
40. 4-chlorophenyl phenyl ether
41. 4-bromophenyl phenyl ether
42. Bis(2-chloroisopropyl) ether
43. Bis(2-chloroethoxy) methane
44. Methylene chloride
45. Methyl chloride
46. Methyl bromide
47. Bromoform
48. Dichlorobromomethane
49. (Removed)
50. (Removed)
51. Chlorodibromomethane
52. Hexachlorobutadiene
53. Hexachlorocyclopentadiene
54. Isophorone
55. Naphthalene
56. Nitrobenzene
57. 2-nitrophenol
58. 4-nitrophenol
59. 2,4-dinitrophenol
60. 4,6-dinitro-*o*-cresol
61. *N*-nitrosodimethylamine
62. *N*-nitrosodiphenylamine
63. *N*-nitrosodi-*N*-propylamine
64. Pentachlorophenol
65. Phenol
66. Bis(2-ethylhexyl) phthalate
67. Butyl benzyl phthalate
68. Di-*n*-butyl phthalate

69. Di-*n*-octyl phthalate
70. Diethyl phthalate
71. Dimethyl phthalate
72. Benzo (α) anthracene
73. Benzo (α) pyrene
74. Benzo (β) fluoranthene
75. Benzo (α) fluoranthene
76. Chrysene
77. Acenaphthylene
78. Anthracene
79. Benzo (ghi) perylene
80. Fluorene
81. Phenanthrene
82. Dibenzo (a,h) anthracene
83. Indeno (1,2,3-cd) pyrene
84. Pyrene
85. Tetrachloroethylene
86. Toluene
87. Trichloroethylene
88. Vinyl chloride
89. Aldrin
90. Dieldrin
91. Chlordane
92. 4,4-DDT
93. 4,4-DDE
94. 4,4-DDD
95. α-Endosulfan
96. β-Endosulfan
97. Endosulfan sulfate
98. Endrin
99. Endrin aldehyde
100. Heptachlor
101. Heptachlor epoxide
102. α-BHC
103. β-BHC
104. γ-BHC
105. δ-BHC
106. PCB-1242 (Arochlor 1242)
107. PCB-1254 (Arochlor 1254)
108. PCB-1221 (Arochlor 1221)
109. PCB-1232 (Arochlor 1232)
110. PCB-1248 (Arochlor 1248)
111. PCB-1260 (Arochlor 1260)
112. PCB-1016 (Arochlor 1016)
113. Toxaphene
114. Antimony
115. Arsenic
116. Asbestos
117. Beryllium
118. Cadmium
119. Chromium
120. Copper
121. Cyanide, Total

122. Lead
123. Mercury
124. Nickel
125. Selenium
126. Silver
127. Thallium
128. Zinc
129. 2,3,7,8-TCDD

References

[1] S. Ahuja, Handbook of Water Purity and Quality, second ed., Elsevier, Amsterdam, 2021.
[2] S. Ahuja, Advances in Water Purification Techniques, Elsevier, Amsterdam, 2019.
[3] S. Ahuja, K. Hristovski, Novel Solutions to Water Pollution, American Chemical Society, Washington, DC, 2013.
[4] S. Ahuja, Monitoring Water Quality: Pollution Assessment, Analysis, and Remediation, Elsevier, Waltham, MA, 2013.
[5] S. Ahuja, Comprehensive Water Quality and Purification, vols. 1-4, Elsevier, Kidlington, Oxford, 2013.
[6] S. Ahuja, J.B. de Andrade, D. Dionysiou, K.D. Hristovski, B. Loganathan, Water Challenges and Solutions on a Global Scale, American Chemical Society, Washington, DC, 2015.
[7] S. Ahuja, Chemistry and Water: The Science Behind Sustaining World's Most Crucial Resource, Elsevier, New York, 2017.
[8] S. Ahuja, Water Reclamation and Sustainability, Elsevier, New York, 2014.
[9] S. Ahuja, Assuring water purity by monitoring water contaminants from arsenic to zinc, in: American Chemical Society Meeting, Atlanta, March 26–30, 2006, 2006.
[10] S. Ahuja, Ultratrace Analysis of Pharmaceuticals and Other Compounds of Interest, Wiley, New York, 1986.
[11] E.E. McNeil, R. Otson, W.F. Miles, F.J.M. Rahabalee, J. Chromatogr. 132 (1977) 277.
[12] B. Kratochvil, J.K. Taylor, Anal. Chem. 53 (1981) 924A.
[13] W.J. Youden, J. Assoc. Off. Anal. Chem. 50 (1967) 1007.
[14] F.W. Karasek, R.E. Clement, J.A. Sweetman, J. Assoc. Off. Anal. Chem. 53 (1981) 1050.
[15] E.W. Wolff, M. Landy, D.A. Peel, J. Assoc. Off. Anal. Chem. 53 (1981) 1566.
[16] S. Ahuja, Evaluating Water Quality to Prevent Future Disasters, Elsevier, 2019.
[17] R. Suthar, B. Gao, Nanotechnology for drinking water purification, in: Water Purification, 2017, pp. 75–118.
[18] X. Qu, P. Alvarez, Q. Li, Applications of nanotechnology in water and wastewater treatment, Water Res. 47 (12) (2013) 3931–3946.
[19] A. Yaqoob, T. Parveen, K. Umar, M. Mohamad Ibrahim, Role of nanomaterials in the treatment of wastewater: a review, Water 12 (2) (2020) 495.
[20] A. Figoli, M. Dorraji, A. Amani-Ghadim, Application of nanotechnology in drinking water purification, Water Purif. (2017) 119–167.
[21] S. Ahuja, Water sustainability and reclamation, in: American Chemical Society Meeting, San Diego, March 25–29, 2012, 2012.

Further reading

[22] G.H. Brundtland, Our Common Future, the World Commission on Environment and Development, United Nations, 1987.

Chapter 2

Importance of nanomaterials in water purification

Swagotom Sarkar[a] and Satinder Ahuja[b,*]
[a]*Bangladesh Reference Institute for Chemical Measurements (BRiCM), Dhaka, Bangladesh,* [b]*Ahuja Consulting, Myrtle Beach, SC, United States*
**Corresponding author: e-mail address: sutahuja@atmc.net*

Introduction

The importance of pure drinking water for our well-being cannot be overemphasized [1–3]. Unfortunately, over 2 billion people live in countries that are experiencing high water stresses [4]. As a result, approximately 1.2 billion people have limited access to fresh, potable water, and 2.6 billion people have issues relating to poor sanitation that seriously affects water quality. Over 1 million people die every year from sicknesses communicated through the use of unsafe water [5].

Drinking water pollution is caused by the extensive use and disposal of chemicals, fertilizers, and other hazardous wastes. This results in contamination of drinking water sources with heavy metals, radioactive substances, plastics, organic nutrients, inorganic pollutants, pharmaceuticals, etc. [1,2]. Polluted drinking water is a major cause for many different diseases, e.g., cancers, and reproductive and digestive problems. Polluted surface water affects the environment and aquatic lives as well [2].

To address these critical issues, it is necessary to purify contaminated water. Conventional water treatment technology suffers from a variety of issues: Coagulation and flocculation produce toxic sludge. The least expensive technique, using activated carbon, is able to remove odor, color, and some organics; it is not efficient in removing heavy metals, etc. Reverse osmosis removes most contaminants; however, it is expensive, and the pores of nanofiltration membranes get clogged after several uses. One of the most powerful approaches to water purification might be the use of nanomaterials and nanotechnology [2,3]. Nanomaterials offer high surface areas, photocatalytic activity, and good filtration capacity (pore sizes down to 0.1 nm) for water treatment processes [2]. In this chapter, we will discuss the latest developments in water and wastewater treatment based on nanomaterials and nanobiotechnologies.

Sources of water pollution

Water can be harmful to our health if it is contaminated by bacteria, viruses, and other pathogens. Fecal *Escherichia coli* is a significant cause of diarrhea. The impact of Covid-19, responsible for the pandemic, that started in December of 2019 and its role in water pollution is being currently assessed. There are other contamination sources of pollutants: agriculture, heavy metal in groundwater, chemical pollutants from industry, sewage water, and wastewater, etc. [2]. Table 1 summarizes possible water pollution sources and the effect on human beings, aquatic life and environment.

EPA primary drinking water standards

Table A1 in the Appendix shows a variety of pollutants that can be encountered in water, and the maximum concentration level (MCL) allowed for these contaminants in drinking water [17]. Secondary drinking water regulations are given in Table A2 in the Appendix) [17]. These contaminants may produce skin or tooth discoloration or changes in taste, odor, or color of drinking water. The US Environmental Protection Agency (EPA) recommends secondary standards to water systems but does not require systems to comply. These regulations vary from country to country. This means we need

Separation Science and Technology, Volume 15. https://doi.org/10.1016/B978-0-323-90763-7.00002-0

TABLE 1 Sources and effects of water pollutants.

Water pollutants	Source of pollutants	Effects of pollutants	References
Pathogens	Viruses and bacteria	Waterborne diseases	[6]
Agricultural pollutants	Agricultural chemicals	Directly affect freshwater resources	[7]
Sediments and suspended solids	Land cultivation, demolition, and mining operations	Damaging fish spawning, affecting aquatic environment of insects and fishes	[8]
Inorganic pollutants	Metal compounds, trace elements, inorganic salts, heavy metals, and mineral acids	Aquatic flora and fauna public health problems	[9]
Organic pollutants	Detergents, insecticides, and herbicides	Aquatic life problems, carcinogenic	[10]
Industrial pollutants	Municipal polluted water	Water and air pollution	[11]
Radioactive pollutants	Different isotopes	Possibly carcinogenic	[12]
Nutrients pollutants	Plant debris, fertilizer	Effect on eutrophication processes	[13]
Macroscopic pollutants	Marine debris	Microplastics	[14]
Sewage, contaminated water, SARS-CoV-2	Domestic wastewater	Waterborne diseases, potential impact of Covid-19	[15,16]

Adapted from A. Yaqoob, T. Parveen, K. Umar, M. Mohamad Ibrahim, Role of nanomaterials in the treatment of wastewater: a review, Water 12 (2) (2020) 495.

to look at each country's drinking water standards. Table A3 [18] in the Appendix demonstrates that wastewater from industry, power plants, etc. should meet certain quality requirements. If treated water complies with the requirements, then it can be reused for different purposes such as gardens, agricultural fields, replenishing surface water and groundwater (i.e., groundwater recharge), residences (e.g., toilet flushing), businesses, and industry, and could even be treated to reach drinking water standards.

Wastewater and drinking water treatment processes

There are fundamental differences between wastewater and drinking water treatments (see Table 2). Wastewater contains toxic pollutants as well as other pollutants which need to be removed to a purity level of natural water resources. This is done to minimize the pollution of rivers, lakes, ponds, oceans, etc. We know that the dissolved oxygen level of water

TABLE 2 Comparison of wastewater and drinking water treatment processes.

	Wastewater treatment	Drinking water treatment
Purification process	A process to purify industrial waste or sewage should include rainwater runoff so that it cannot pollute our natural water resources (rivers, ponds, oceans)	A process to purify groundwater or surface water so that it can be consumed without any ill effects on humans
Purification steps	Pretreatment: screening Primary treatment: gravity clarification, coagulation, precipitation, flocculation, sedimentation Secondary treatment: biodegradation, filtration, adsorption Tertiary treatment: oxidation, membrane filtration, disinfection, and treatment of sludge: supervised tipping, recycling, incineration	Screening, coagulation, flocculation, sedimentation, filtration, disinfection, and fluoride application
Characteristics of water	Effluent water which can be released into environment	Water that can be consumed safely

Adapted from C. Gerba, I. Pepper, Drinking water treatment. Environ. Pollut. Sci. (2019) 435–454; A. Englande, P. Krenkel, J. Shamas, Wastewater treatment &water reclamation. Reference Module in Earth Systems and Environmental Sciences, 2015.

decreases because of pollution. This leads to an increase in biochemical oxygen demand (BOD) and chemical oxygen demand (COD) levels. Therefore, it is very important to keep water bodies pollution-free to assure safety of aquatic life.

Providing safe drinking water (potable water) is one of the most important issues for mankind. Drinking water needs to be free from toxic chemicals and microbial contamination; however, it should contain essential minerals necessary for our health. Consequently, drinking water treatment agencies need to take adequate steps to achieve this goal.

Conventional water purification methods

Conventional water treatment processes include screening, coagulation/flocculation, sedimentation, filtration, disinfection, adsorption, advanced oxidation, distillation, ion exchange, distillation, biological treatment, ultraviolet (UV) light, reverse osmosis, carbon filter, etc. Table 3 provides the main application and limitations of some conventional water treatment processes.

TABLE 3 Applications and limitations of conventional water treatment processes.

Conventional water treatment process	Applications	Limitations	References
Screening	Debris removal	Cannot remove pathogens and other organic and inorganic pollutants	[19]
Coagulation/ flocculation	Formation of floc and large particles	Toxicity and health hazard of inorganic contaminants, production of toxic sludge, ineffective for removing heavy metals	[20]
Sedimentation	Particles/floc	Time-consuming, cannot remove viruses and bacteria effectively	[21]
Filtration	Remove algae, sediment, insoluble organic/inorganic compounds	Membrane clogging and fouling	[19]
Disinfection	Inactivate microorganisms	Cannot remove organic and inorganic pollutants, create carcinogenic by-products in the presence of organic compounds	[22]
Adsorption	Improve taste and remove odor-producing compounds, synthetic organic compounds, disinfection by-product precursors, and inorganic chemicals	Lack of selectivity and produces waste products	[19,21]
Advanced oxidation	Degrade and potentially remove organic chemicals	Cannot remove inorganic contaminants	[19,21]
Ion exchange	Remove dissolved ionic constituents	Cannot remove organic compounds, viruses, and bacteria	[19,21]
Distillation	Remove dissolved material and biological contamination	Volatile and semi-volatile organic compounds cannot be removed	[21,23]
Biological treatment	Remove organic matter	Expensive, time-consuming, and cannot handle microorganisms properly	[21,23]
UV treatment	Remove viruses and bacteria	Expensive, cannot work properly if water is highly turbid	[21,23]
Reverse osmosis	Remove virus, bacteria, salt, and other low molecular weight organic molecules	Not energy-efficient. Water is wasted extensively. Process should not be used with water that has total dissolved solids below 500 ppm. Filters out essential minerals like Ca, Mg, etc., that are important to our health	[21,23]
Carbon filter	Remove organic and total dissolved solids	This method is not efficient for removing nitrates, fluoride, metals, e.g., sodium. Clogging and mold growth occur unless filter is changed regularly.	[21,23]

Nanotechnology-based water purification processes

Nanotechnology creates materials or structures with features in the 1–100-nm size range [24]. Nanomaterials have been defined by the International Organization for Standardization (ISO) as materials whose diameter ranges from 1 to 100 nm [25]. According to the European Commission, nanomaterials are natural, incidental, or manufactured materials containing particles in an unbound state or as an aggregate or an agglomerate and where, for 50% or more of the particles in the number size distribution, one or more external dimensions is in the size range from 1 to 100 nm [26]. Nanomaterials can be classified based on their origin, dimensions, chemical composition, and toxicity, as shown in Table 4 [27].

Nanotechnology plays an important role in different sectors of our economy [28–32]. Recently, this technology has been applied for water purification. It can purify water by removing pollutants such as pathogens, organic compounds, inorganic compounds including harmful metals, and pharmaceuticals/drugs, etc. [33]. Advanced research in this area suggests that existing water treatment problems can be solved by applying the following nanotechnology-based processes: nanoadsorbents, nanoenhanced membranes, nano/microreactors, and nanophotocatalysts [2,33]. Nanomaterials are very useful because they are efficient; only a small amount is required, and they are multifunctional [34].

Water treatment with various nanomaterials

There is a significant need for advanced water purification technique such as nanotechnology to ensure high quality drinking water by eliminating water pollutants. Nanoengineered materials, such as nanoadsorbents, nanometal oxides, nanomembranes, and nanophotocatalysts, offer great potential for novel water purification technologies that can be easily integrated into customer-specific applications [24]. Most of them can be compatible with existing treatment technologies and can be integrated simply in conventional modules. Retention and reuse of nanomaterials is a key aspect of nanotechnology-enabled device design because of both cost and public health concerns.

As mentioned before, nanotechnology offers promising solutions for water supplies contaminated by organic and inorganic compounds, heavy metal ions, and microorganisms. Recent advances strongly suggest that many of the current problems involving water quality can be addressed and potentially resolved using nanoadsorbents, nanoenhanced membranes, nanometal oxides, and nanophotocatalysts. Table 5 shows applications of nanomaterials. Among them, three categories show most promising applications, based on current research and development, commercial availability, cost of nanomaterials involved, and compatibility with the existing infrastructure: nanoadsorbents, nanotechnology-enabled membranes, and nanophotocatalysts [23,24].

TABLE 4 Nanomaterial classification based on origin, dimension, composition and toxicity.

Nanomaterials				
Origin		Dimensions	Chemical composition	Toxicity
Natural nanomaterials		Zero-dimension nanomaterials	Metal and metal alloy	Fiber-like nanomaterials
Anthropogenic nanomaterials		One-dimension nanomaterials	Metal oxide	Bio-persistent granular nanoparticles
Incidental nanomaterials	Engineered nanomaterials	Two-dimension nanomaterials	Semiconductor	Carcinogenic, mutagenic, asthmagenic, reproductive toxin nanoparticles (CMAR)
		Three-dimension nanomaterials	Silicate, carbonate, and nitride polymer	
			Carbon nano object	

TABLE 5 Applications of nanomaterials.

Nanomaterial	Pollutant	Type of pollutant	Removal efficiency or adsorption capacity	References
Carbon nanotube (CNT) membrane	*Escherichia coli*	Bacteria	Complete reduction	[35]
Nano silver-impregnated membranes	*E. coli*	Bacteria	99%	[36]
Ag/TiO_2 nanofiber membrane	*E. coli*	Bacteria	99.9%	[37]
Silver-doped titanium dioxide nanoparticles	Bacteriophage MS2	Virus	Inactivation rate $0.089\,s^{-1}$	[38]
Nano WO_3	*E. coli*	Bacteria	Bacteria decay constant $= 0.94\,min^{-1}$	[39]
CNT ferrihydrite composite	As^{5+}	Heavy metal	44.1 mg/g	[40]
Dithiocarbamate functionalized and silica-coated F_3O_4 magnetic nanoparticle (Fe_3O_4/SiO_2/NH/CS_2)	Hg^{2+}	Heavy metal	74%	[41]
Fe_3O_4 nanoparticles functionalized with dithiocarbamate group	Hg^{2+}	Heavy metal	206 mg/g	[42]
Magnetic Fe_3O_4 nanospheres	Cr^{6+}; Pb^{2+}	Heavy metal	9; 19 mg/g	[43]
Nanotubes Bi_2O_3	Cr^{6+}	Heavy metal	79 mg/g	[44]
Alumina membranes functionalized with poly (styrene sulfonate) or poly (allylamine hydrochloride)	$MgSO_4$	Salt	96%	[45]
CeO_2–CNT	As^{5+}	Heavy metal	Regenerated efficiency: 94%	[46]
Carbon nanoparticles conjugated with polyethylenimine	Zn^{2+}, Cd^{2+}, Cu^{2+}, Hg^{2+}, Ni^{2+}, Cr^{6+}	Heavy metal	4–5.7 mmol/g	[47]
Oxidized carbon nanotube sheet	Cd^{2+}, Pb^{2+}, Co^{2+}, Zn^{2+}, Cu^{2+}	Heavy metal	101.05, 75.84, 69.63, 58.00, 50.37 mg/g	[48]
Fe_3O_4 particles encapsulated in thiol-containing polymers (Fe_3O_4–PEDOT)	Ag^{+}, Hg^{2+}, Pb^{2+}	Heavy metal	>95%	[49]
Fe_3O_4 magnetic nanoparticles coated with humic acid	Hg^{2+}, Pb^{2+}; Cu^{2+}, Cd^{2+}	Heavy metal	>99%; >95%	[50]
Nanocrystalline TiO_2	As^{5+}, As^{3+}	Heavy metal	>0.5 mmol/g	[51]
Nanoscale FeMn binary oxides loaded on Zeolite	As^{3+}	Heavy metal	Initial conc. = 2 mg/L; after adsorption, conc. = 0.01 mg/L	[52]
Fe-MWNT-CH_2COONa	Benzene, toluene, dimethylbenzene and styrene	Organic	79%, 81%, 83% and 88%	[53]
Alkylated hyperbranched polymers	Phenanthrene	Organic	Loading capacity 18–54 mg/g	[54]

Continued

TABLE 5 Applications of nanomaterials.—cont'd

Nanomaterial	Pollutant	Type of pollutant	Removal efficiency or adsorption capacity	References
Ceramic membranes impregnated with cross-linked silylated dendritic and cyclodextrin polymers	Polycyclic aromatic hydrocarbon	Organic	99%	[55]
Graphitized CNT	1,2-Dichlorobenzene	Organic	28.8 mg/g	[56]
Cyclodextrin-polyurethanes containing carbon nanotubes	*p*-Nitrophenol	Organic	99%	[57]
Cadmium sulfide/hexagonal mesoporous silica (HMS) sphere-poly(diallyl dimethyl ammonium) chloride (PDDA) (CdS/HMS–PDDA) photocatalyst	Phenol, *p*-chlorophenol, *p*-nitrophenol,2,4,6-trichlorophenol, and hydroquinone	Organic	Complete degradation within 240 min	[58]
TiO_2/Al_2O_3 composite membrane + UV light	Direct black 168 dye	Dye	82% degradation after 300 min	[59]
ZnS nanoporous nanoparticles + UV light	Eosin B	Dye	Complete degradation after about 40 min	[60]
Manganese-doped ZnO NPs + visible light	Methylene blue	Dye	50% degradation within 5 min	[61]
Au/Bi_2O_3 nanorods + ozone + visible light	Orange II dye	Dye	Degradation rate constant $= 3.92 \times 10^{-3}\ s^{-1}$	[62]
Nanocrystalline CeO_2	Acidic black 10 B dye	Dye	97%	[63]
Plasmonic photocatalyst Ag@Ag(Cl, Br)	Methyl orange	Dye	Complete degradation after 15 min	[64]
Magnetic graphene nanoplates	Amoxicillin	Pharmaceutical residue	40–90%	[65]
Nano-structured In_2O_3	Poly- and per-fluoroalkyl substances (PFAS)	Endocrine disrupting chemical	71%	[66]

Adapted from R. Suthar, B. Gao, Nanotechnology for drinking water purification, Water Purif. (2017) 75–118.

One of the most important advantages of nanomaterials compared to conventional water treatment technologies is their ability to integrate various properties, resulting in multifunctional systems, such as nanocomposite membranes that enable both particle retention and elimination of contaminants. Furthermore, nanomaterials enable higher processing efficiency because of their unique characteristics such as a high reaction rate. Covered below are more details of the following: nanophotocatalysts, nanoadsorbents, nanoenhanced membranes, and nanoreactors.

Nanophotocatalysts

Photocatalysis entails the breaking down or decomposition of compounds in the presence of light. Nanophotocatalysts can change the rate of chemical reactions [2,67–71]. Their higher surface area makes them excellent candidates for water and wastewater purification. There are two types of processes in nanophotocatalysis: Mineralization (complete destruction) and degradation (decomposition of organic pollutants).

Working mechanism: A semiconductor like TiO_2 or ZnO absorbs the light which is greater than or equal to their band gap width, and carries it to electron–hole pairs (e^- h^+) (through Ref. [2]). If the separation of charge is continued, the electron–hole may travel to the surface of the catalyst where they contribute with sorbed species in redox reactions. Particularly, $h^+{}_{vb}$ reacts with water (surface-bound) to generate the hydroxyl radicals and simultaneously $e^-{}_{cb}$ selected by oxygen to produce the radical anion (superoxide radicals) as designated below in Eqs. (1)–(3).

$$TiO_2 + h\nu \rightarrow e^-{}_{cb} + h^+{}_{vb} \tag{1}$$

$$O_2 + e^-{}_{cb} \rightarrow O^{2-\cdot} \tag{2}$$

$$H_2O + h^+{}_{vb} \rightarrow OH^{\cdot} + H^+ \tag{3}$$

After pollutant degradation with nanomaterials in the presence of light, HPLC-MS or GC-MS is employed to analyze the produced degraded products. It is necessary to confirm, using a toxicity test, whether these degradation products are more toxic or less toxic as compared to the parent compound. It should be noted that nanophotocatalysts can undergo electron–hole pair recombination and have stability issues. Table 6 exhibits applications of nanophotocatalysts. These applications include removal of sulfamethoxazole, *E. coli*, and some dyes.

Nanomembranes

Filtration by membranes is mostly automated technology that can be used for conventional water purification [2,4–78]. Nanomembranes offer an excellent choice for water and wastewater purification and desalination. However, they are expensive, have poor mechanical stability, and can have fouling problems. Table 7 shows applications of nanomembranes which include removal of bacteria, heavy metals, salt, and oil.

Working mechanism: A membrane is a porous thin-layered material that allows water molecules to pass through it; however, it simultaneously restricts the passage of bacteria, viruses, salts, and metals [79]. Membranes use either pressure-driven forces or electrical. Pressure-driven membrane technology provides a suitable method for water purification to the desired quality.

TABLE 6 Applications of nanophotocatalysts.

Nanomaterial	Pollutant	Type of pollutant	Removal efficiency	References
Pt/TiO_2	*Escherichia coli*	Bacteria	Complete removal after 3 h	[66]
(SMTZ) P-Fe_2O_3/TiO_2	Sulfamethoxazole (SMTZ)	Organic compound	Complete degradation	[67]
Mo and Mn doped ZnO	Anthraquinone dye	Dye	84%	[68]
TiO_2/ZnO/GO composite	Methyl orange dye	Dye	57.7%	[69]
ZnO/In_2S_3 core/shell nanorods arrays	RhB dye	Dye	83.7%	[70]

Adapted from A. Yaqoob, T. Parveen, K. Umar, M. Mohamad Ibrahim, Role of nanomaterials in the treatment of wastewater: a review, Water 12 (2) (2020) 495.

TABLE 7 Applications of nanomembranes.

Nanomembrane	Pollutants	Type of pollutant	Removal efficiency	References
Cellulose nitrate UF membrane reinforced with poly(N-vinylcarbazole)-graphene oxide (PVK-GO)	*Escherichia coli* and *Bacillus subtilis*	Bacteria	95% and 97%, respectively	[74]
Chlorinated poly (vinyl chloride) nanofiber membrane	Cu^{2+}, Pb^{2+} & Cd^{2+}	Heavy metals	73%, 82%, and 90%, respectively	[75]
rGO/TiO_2 Reverse osmosis membrane	NaCl	Salt	99.45%	[76]
PVA electrospun nanofibrous scaffold support and a PVA hydrogel coating membrane	Oil	Organic	>99.5%	[77]
A thin layer of a hydrophilic but water-resistant chitosan coating, an asymmetric electrospun PAN nanofibrous mid-layer support, and a nonwoven PET substrate containing a three-layer membrane	Oil	Organic	>99.9%	[78]

Adapted from A. Yaqoob, T. Parveen, K. Umar, M. Mohamad Ibrahim, Role of nanomaterials in the treatment of wastewater: a review, Water 12 (2) (2020) 495.

Membrane separation processes are being increasingly used for the treatment of water and wastewater. Membranes separate substances depending on pore and molecule size. They offer reliable and automated processes for wastewater treatment [23,24]. Microscopic openings in the membrane allow water molecules to pass through, but not compounds that are larger than the opening. Membrane filtration can be functionalized either as dead-end filtration or cross-flow filtration. In dead-end filtration, the feed water flows perpendicular to the membrane surface where all solids will amass onto the membrane surface during filtration, and removed via backwashing. In cross-flow filtration, the feed water is parallel to the membrane surface. Since the majority of the solid contaminants will pass with the water instead of collecting on the membrane surface, the system can function at higher flux [24].

Nanoreactors

Nanoreactor/motors can be run with or without fuel [2]. Acoustics, magnetic fields, and electric fields can be the source of energy during a nonfuel nanoreactor run. Using this technology, water and wastewater can be treated by different mechanisms. This process is fast and controllable.

Working mechanism: There are two major mechanisms used by nano- and micromotors [80]: They either adsorb the pollutants or they degrade them by different routes (H_2O_2-assisted, $NaBH_4$-assisted, photocatalytic, or biocatalytic).

Adsorption: This is a widely used process for the removal of contaminants in water [2]. It is a multistep procedure that requires, first, the transport of target pollutants from the aqueous phase to the surface of porous solid particles and, second, the diffusion of these target pollutants into the interior of the reactor.

Degradation: In this case, organic pollutants are directly degraded with the help of H_2O_2, $NaBH_4$, and photocatalysts or biocatalysts [2].

Table 8 describes applications of nano/micromotors. These include removal of bacteria, organic compounds, and pharmaceutical residues.

Nanoadsorbents

During adsorption, molecules of a substance (adsorbate) become attached onto the surface of a solid/substrate (adsorbent) [24]. This method is used to remove organic and inorganic contaminants from polluted water. The rate of adsorption is affected by the properties of adsorbate and adsorbent: surface area and pore size, temperature, foreign particles, particle size, chemistry of surrounding environment, and duration of contact. In general, the smaller the pore size, the greater the surface area. The greater the surface area, the greater the number of adsorption sites. It should be noted that adsorption can be hindered by steric effects of adsorbate, smaller surface area, less selectivity, and adsorption kinetics as well. It is difficult to recover used nanoadsorbents. Table 9 shows applications of nanosorbents. Interesting applications include remediation of inorganic and organic compounds and pathogen contaminants of wastewater.

TABLE 8 Applications of nano/micromotors.

Nano/micro reactor	Pollutants	Type of pollutant	Removal efficiency	References
Glucose oxidase (GOX) and myeloper oxidase (MPO) within dendrimers DSs (GOX-MPO-DSs) nanoreactor	*Staphylococcus aureus* and *Pseudomonas aeruginosa*	Bacteria	99.9%	[81]
UV/TiO_2/Electroflotation microreactors	COD removal	Organic	97.5%	[82]
Zinc oxide nano-enabled microfluidic reactor	VOC (benzene, toluene, ethylbenzene, *m–p* xylenes, *o*-xylene (BTEX)) removal	Organic	95%	[83]
Yolk–shell Co_3O_4@ metal–organic frame-works (MOFs) nanoreactor	4-Cholorophenol	Organic	Almost 100%	[84]
Magnetic hierarchical porous Cd^{2+} imprinted photocatalytic nanoreactors	Tetracycline	Pharmaceutical residues	75.32%	[85]

Adapted from A. Yaqoob, T. Parveen, K. Umar, M. Mohamad Ibrahim, Role of nanomaterials in the treatment of wastewater: a review, Water 12 (2) (2020) 495.

TABLE 9 Applications of the most commonly used nanosorbents.

Nanosorbent	Pollutant	Type of pollutant	Removal efficiency	References
Fe_3O_4/graphene	*Escherichia coli*	Bacteria	93.09%	[86]
MWCNT (multiwalled carbon nanotubes)	Hg (II)	Heavy metal	151.51 mg/g	[87]
GO and GO-COOH (carboxylate graphene)	Cu^{2+}	Heavy metal	277.77 and 347.14 mg/g	[88]
Cobalt/manganese mixed ferrite with polydimethylsiloxane	Toluene	Organic	325 mg/g	[89]
MoS_2/CuS nanosheet	RhB, MB, RhB6G	Dyes	93.3%, 100%, 4.7%, respectively	[90]

Adapted from A. Yaqoob, T. Parveen, K. Umar, M. Mohamad Ibrahim, Role of nanomaterials in the treatment of wastewater: a review, Water 12 (2) (2020) 495.

Choosing the best water purification technology to purify water

From the nanotechnologies discussed above, nanoadsorbents offer the best option for water purification in terms of selectivity, efficiency, and cost-effectiveness [91–93]. Nanoadsorbents can remove various types of contaminants that include bacteria and a large number of organic compounds. Furthermore, adsorbent materials are easily available, and their low cost makes them suitable for developing countries. Nanomembranes are selective and efficient in removing various types of contaminants from water; however, they are very expensive and require the application of extensive pressure to purify water. Nanophotocatalysts are selective and efficient only for particular contaminants, e.g., destruction of viruses and bacteria, +dyes present in contaminated water. Finally, nanoreactor technology is still evolving; they are able to remove mostly organic pollutants. The order of preference for the use of these technologies is as follows: nanoadsorbents > nanomembranes > nanophotocatalysts > nano/micro reactors.

Effects of nanomaterials on humans and the environment

Risks associated with nanomaterials were first noticed in 1990 [94]. Although nanomaterials offer promising technology in terms of waste and water treatment, recent toxicological studies show that they can be harmful to humans and the environment because of their size, shape, length, surface area, distribution ratio, aggregation degree, concentration, reactivity, contact with the environment and humans [23]. If nanoparticles enter our bloodstream through skin, inhalation, or ingestion, they might destroy protein, membrane, DNA, and enzymes [23,95–106]. For example, carbon nanotubes are associated with pulmonary inflammation, oxidative stress, granuloma in lungs, basic inflammation, apoptosis, and fibrosis [107]. So it is necessary to conduct extensive research to evaluate risks associated with nanomaterial use in water and wastewater treatment.

Select applications

Contamination from arsenic, lead, pharmaceuticals, and PFASs have been significant problems in the water treatment sector. Dr. Satinder Ahuja has been involved for a long time in resolution of these issues [108–117]. Applications of nanomaterials for some select applications including SARS-CoV-2 virus have been discussed here at some length.

Arsenic removal from water

Arsenic contamination of groundwater has been described as the world's worst catastrophic problem. More than 100 million people were affected in Bangladesh alone. Contamination of natural waters with arsenic is widespread in southeast Asian countries like Bangladesh, India, China, Thailand, Cambodia, etc. [118]. This problem is also encountered in advanced countries like USA. Groundwater arsenic, arsenic mining and smelting, coal combustion, and industrial production with arsenic as a raw material are the major sources of arsenic contamination of water and the environment [119]. Arsenic can cause melanosis, keratosis, hyperpigmentation, immunotoxicity, genotoxicity through generation of reactive oxygen species and lipid peroxidation, modulation of coreceptor expression, toxicological, and carcinogenic effects [120]. Considering its toxicity and carcinogenicity, WHO determined the maximum permissible limit for arsenic as 0.01 mg/L in drinking water [121]. In Bangladesh, maximum permissible limit for arsenic is 0.05 mg/L based on national standards [122].

Various technologies have been employed for arsenic removal from water: coagulation-filtration [123], membrane separation [124], ion exchange [125], and adsorption [126]. Some of the arsenic removal technologies which are currently being used in arsenic affected countries like Bangladesh, India, Nepal, Argentina, Mexico, Peru, etc. are as follows: Small-scale column test, Alufloc, natural geological materials, RO desalinator, passive sedimentation, in situ oxidation, SORAS, bucket treatment units, Sono filter, subterranean arsenic removal technology, DRDO arsenic removal technology, Neosepta AX 01 membrane, multiwall carbon nanotubes, magnetite-reduced graphene oxide, and Kanchan arsenic filter [127]. Among different technologies, adsorption, specifically nanoadsorption by nanoparticles, i.e., magnetite (Fe_3O_4), maghemite (c-Fe_2O_3), nickel ferrite ($NiFe_2O_4$), cobalt ferrite ($CoFe_2O_4$), magnetic mobile crystalline material functionalized by chlorosulfonic acid (MMCM-41-SO_3H), etc., offers many advantages because of their unique size and structure, as well as magnetic properties [128], including simple and stable operation, easy handling of waste, absence of added reagents, compact facilities, and generally lower operation cost [129]. Table 10 shows methodology used for various nanoadsorbents.

TABLE 10 Arsenic removal with various nanoadsorbents.

Nanomaterial	Adsorption methodology	References
Fe_3O_4 Surface area: 178.48 m^2/g	Contact time: 24 h Temp: 298 K pH: 2–13 Removal efficiency: 45%	[130]
Nanostructured cerium incorporated manganese oxide (NCMO) Surface area: 116.96 m^2/g	pH: 7 Contact time: 3 h Temp: 303 K Removal efficiency: 38%	[131]
Fe_3O_4-NCs (Nanocrystals) Surface area: 300 m^2/g	Contact time: 24 h Removal efficiency: 99.2%	[132]
M-RGO (Magnetite-reduced graphite oxide) Surface area: 137–148 m^2/g	pH: 7 Contact time: 120 min Removal efficiency: 99.9%	[133]

Lead removal from water

Lead is second on the list of most hazardous metals with significant health risks for living beings. The presence of lead in even trace amounts is a major concern [134]. The permissible lead limit in drinking water is 0.05 mg/L, whereas in the case of surface water and discharge in public sewers, the quantum of lead should be below 0.10 mg/L as per ISO 105 001992. Globally, EPA allows only 0.015 mg/L Pb in the drinking water; wheras, the limit is farther lowered to 0.01 mg/L by WHO [135]. Water resources can be polluted from discharge from pipe corrosion, faucets, household plumbing, paints, smelting, metallurgy, mining, tannery, chemical manufacturing, battery manufacturing, and plastics industries [136]. The accumulation of lead severely affects the central nervous system and may lead to short-term memory loss, neurological, gastrointestinal, renal, and cardiovascular disorders, along with a low IQ, learning disabilities, attention deficit disorders, behavioral problems, impaired hearing, growth, kidney damage, coma, mental disorders, cancer, high blood pressure, infertility, and muscle disorders [137–139]. All of these effects clearly require development of efficient methods for removal of lead from water and wastewater. Lead can be removed from water using nanoadsorbents as shown in Table 11.

TABLE 11 Lead removal technologies with nanoadsorbents

Concentration of Pb	Nanomaterial for destruction	Adsorption methodology	References
100 mg/L	CuO Commercially available Size: 75 nm, spheroid Surface area: 12.1 m^2/g Adsorbent dose: 2 g	pH = 6.0 Temperature: 25°C Contact time: 120–180 min Removal %: >80	[140]
100 mg/L	Fe_2O_3 Commercially available Size = 37 nm Surface area: NA Adsorbent dose: 0.5 g	pH: 8.0 Temperature: 45°C Contact time: 120 min Removal %: 100	[141]
100 mg/L	ZnO Commercially available Size: 25 nm, rod-shaped Surface area: 31.2 m^2/g Adsorbent dose: 2 g	pH: 6 Temperature: 25°C Contact time: 120–180 min Removal % = >80	[140]
100 mg/L	MgO Commercially available Size: 24 nm, rod-shaped Surface area: 20.1 m^2/g Adsorbent dose: 2 g	pH: 5.0 Room temperature Contact time: 60–90 min Removal %: >90	[142]
100 mg/L	TiO_2 Commercially available Aggregated particles Surface area: 185.5 m^2/g Adsorbent dose: 0.5 g	pH: 8.0 Temperature: 25°C Contact time: 120 min Removal %: 100	[143]
10 mg/L	MWCNT/MnO_2 Preparation: mixing MWCNT in MnO_2. Manganese sulfate and NaOH heated and stirred, added dropwise to MnO_2 and stirred at 70°C Size: 50–250 nm diameter (long entangled tube), rod-shaped structure, diameter (25–110 nm) Surface area: 6.4 m^2/g Adsorbent dose: 10 g	pH: 7.0 Temperature: 25°C Contact time: 120 min Removal %: 98	[144]

Removal of pharmaceuticals from water

Pharmaceutical contaminants (PCs) are contaminants of emerging concern (CECs), because of their potential adverse effects to humans and the ecosystem [145–148]. For instance, PCs such as antibiotics can give rise to antibiotic-resistant bacteria, which can cause irreparable harm to humans and the ecosystem [149]. Other pharmaceutical contaminants can cause problems such as development and reproduction disorders, plant growth, birth defects, severe bleeding, cancer, organ damage, endocrine disorder, toxic to aquatic organism, etc. [150–160]. Adsorption and photocatalysis technologies can be used for the decontamination of pharmaceutical-based emerging contaminants, e.g., antibiotics [161]. Table 12 shows remediation of pharmaceuticals with nanomaterials.

Per- and poly-fluoroalkyl substances removal from water

Perfluoroalkyl and polyfluoroalkyl substances (PFASs), because of their excellent thermal stability, have been widely used in various applications, such as fire-fighting foams, nonstick cookware, stain-resistant carpets, paints, adhesives, waxes,

TABLE 12 Remediation of pharmaceuticals with nanomaterial [160].

Pharmaceutical contaminants	Nanomaterials	Methodology	References
Antibiotic (Amoxicillin) Concentration: 10ppm	Magnetic graphene nanoplatelets (M-GNPs) Surface area: 543.2 m^2/g	Nanoadsorption Adsorption capacity: 14.1 mg/g Removal %: 40–90	[65]
Penicillin antibiotic (Ampicillin) WO_3/ZrO_2 $Ru/WO_3/ZrO_2$ Concentration: 10ppm	WO_3/ZrO_2 $Ru/WO_3/ZrO_2$	Photocatalysis Reaction source: 150W solar light Removal %: 96–100	[162]
Analgesic (Aspirin) Concentration: 20ppm	Graphene Surface area: NA	Nanoadsorption Adsorption capacity: 17 mg/g Removal %: 81	[163]
Analgesic (Acetaminophen) Concentration: 10ppm	$BaTiO_3/TiO_2$ composite	Photocatalysis Reaction source: UV–visible light, 500W Removal %: 82.8	[164]
Antihistamine (Diphenhydramine) Concentration: 1000ppm	Reduced graphene oxide TiO_2 composite Surface area: 148 m^2/g	Nanoadsorption Adsorption capacity: NA Removal %: ~100	[165]
Antihistamine (Diphenhydramine) Concentration: 100ppm	ZnO/Fe_2O_3/Zeolite TiO_2/Fe_2O_3/Zeolite	Photocatalysis Reaction source: UV light (254 nm), 6W Removal %: 100	[166]
Antidiabetic (Metformin) Concentration: 10ppm	Graphene oxide Surface area: 108.7 m^2/g	Nanoadsorption Adsorption capacity: 47.1 mg/g Removal %: 80	[167]
Anticoagulant (Warfarin) Concentration: 0.2–0.4ppm	PVDF membrane with TiO_2	Photocatalysis Reaction source: low pressure, UV light (254 nm), 40W Removal %: NA	[168]
Anticonvulsant (Carbamazepine) Concentration: 0.168ppm	Graphene-P25 (Gr-P25) nanocomposites Surface area: 45.0–48.1 m^2/g	Nanoadsorption Adsorption capacity: NA Removal %: 100	[169]
Anticonvulsant (Carbamazepine) Concentration: 0.050ppm	$C\text{-}TiO_2$	Photocatalysis Reaction source: visible light (>400 nm), 150W Removal %: 100	[170]

TABLE 13 Remediation of PFAS in wastewater with nanomaterials.

PFAS	Nanomaterials for PFAS destruction	Methodology	References
PFOA (100 μg/L)	Carbon nanotube Usefulness: PFOA is adsorbed on the CNT, increasing the concentration in anode interface	Electrochemical oxidation and hydrolysis. Degradation performance: 3 h, >90%	[177]
PFOA (30 mg/L)	Nanoporous In_2O_3 nanosphere Usefulness: better performance because of larger specific surface area and nanoporous structure	Photocatalytic decomposition Degradation performance: 3 h, 71% at pH 3.9	[66]
PFOA (30 mg/L)	Titanium dioxide with multiple wall carbon nanotubes (TiO_2-MWCNT) Usefulness: adsorption ability and electron transport capacity of MWCNT-enhanced photocatalytic degradation of PFOA	Photocatalytic degradation Degradation performance: 8 h, ~100 in acid medium	[178]
PFOS (0.019 mM)	Graphene quantum dots (GQDs) attached to SiC nanoparticles (SiC/GQDs) Usefulness: efficient electron transitions of GQDs and band gap of SiC/GQDs provide strong reductive potential to degrade PFOS	Photocatalytic decomposition Degradation performance: 20 h, 88.5% at pH 7	[179]
12 PFCs (0.03–6.37 ng L^{-1})	Nano-ZnO Usefulness: nano-ZnO-coated anode showed higher efficiency than Ti-anode	Electrochemical Oxidation Degradation performance: 40 min, 38.95–65.84% at pH 7	[180]
PFOA, PFOS, PFNA, PFDA (0.2 mg/L)	Nanoscale zero-valent iron (nZVI) Usefulness: nanoparticle with large surface area, and reduction via dehydrohalogenation are main factors	Reductive degradation Degradation performance: 1 h, 38–96% at pH 3	[181]
PFBA, PFPeA, PFHxA, PFHpA, PFOA (100 mg/L)	Ce-doped modified porous nanocrystalline PbO_2 Usefulness: porous, nanostructures can improve the decomposition performance	Electrochemical mineralization Degradation performance: 1.5 h 49–95%	[182]

and surface-active agents in waterproof clothing over the last five decades [170–174]. Because of their unique amphiphilic properties and stability, PFASs are recognized as highly persistent, toxic, and environmentally bioaccumulative, endocrine disrupters, and possible carcinogens [175]. In 2016, the EPA issued health advisory levels (HAL) for drinking water of 70 ng L^{-1} for PFOA and PFOS, and 100–7000 ng L^{-1} for C4–C7 PFASs [176]. Among several physicochemical technologies, nanoadsorption and nano-photocatalysis have been extensively used as effective methods for removing PFASs from an aqueous environment (Table 13).

SARS-CoV-2 virus (Covid-19) removal from water

The novel coronavirus (SARS-CoV-2) that caused numerous deaths worldwide in recent pandemic was first identified in China in December 2019 [183]. On March 11, 2020, the WHO declared the COVID-19 outbreak a pandemic [183]. As of April 23, 2021, SARS-CoV-2 caused around 3 million deaths worldwide [184] and the incubation period is reported to be between 5 and 14 days before disease onset [185]. Recent reports indicate that some COVID-19 patients exhibit damage in organs such as the heart, kidney, eye (conjunctivitis), and brain (encephalitis) [186]. The COVID-19 virus spreads primarily through droplets of saliva or discharge from the nose when an infected person coughs or sneezes [187]. Recent report suggests that this virus could spread through water, if wastewater and drinking water are not treated with great care [188–199].

To make water free from Covid-19 and other viruses, different strategies could be employed, such as applying temperature, disinfectants (chlorine or chlorine dioxide), and chitosan-based coagulation-flocculation [200,201]. Inactivation of SARS-CoV using nanophotocatalysts could be a good strategy in the future. Photocatalysts coupled to UV light could lead to a decreased infectious capacity of the virus. For example, photocatalytic titanium apatite filters (PTAFs) showed effective inactivation of SARS-CoV when exposed to UV light for 6 h [202]. Nanophotocatalyst Ag and Cu nanoparticles, pure and doped TiO_2, $BiVO_4$, $CuFeO_2$, $CuY_xFe_{2-x}O_4$, $LaFeO_8$, $CuMn_2O_4$, $ZnMn_2O_4$, $BaCr_2O_4$, $SrCr_2O_4$, $NiCo_2O_4$, $CuCo_2O_4$, $LaCoO_3$, and $La_{0.9}Sr_{0.1}CoO_3$ may be used to inactivate SARS-CoV-2 in water treatment processes [203–212].

Conclusions

The importance of nanomaterials in water purification cannot be overemphasized. There is a need for water purification technologies that can provide high-quality drinking water and are cost-effective, eco-friendly, and easy to carry out. Nanomaterials can help us meet these goals. This technology can be applied to remove a large variety of contaminants that may be present in water. However, there is a need to overcome some challenges mentioned in this chapter to optimize these technologies for removing various types of contaminants.

Acknowledgment

Swagotom Sarkar acknowledges the valuable advice provided by Dr. Satinder Ahuja.

Appendix

TABLE A1 National primary drinking water regulations [17]

S/N	Pollutants	MCL[a]
1.	Total coliform	See footnote[b]
2.	Fecal coliform/*Escherichia coli*	Zero
3.	Fecal indicators (GWR): 1. *E. coli* 2. Enterococci 3. Coliphage	*Escherichia*-1. TT 2. TT 3. TT
4.	Groundwater rule (GWR) TT violations	TT
5.	Turbidity (MCL)[c]	1 NTU[d]/5 NTU
6.	Turbidity (SWTRTT)[e]	TT[f]
7.	Turbidity (IESWTR TT and LT1ESWTR TT)[g]	TT
8.	*Giardia lamblia* (SWTR/IESWTR/LT1ESWTR)	TT[h]
9.	Viruses (SWTR/IESWTR/LT1ESWTR)	TT[h]
10.	Heterotrophic plate count (HPC) bacteria[i] (SWTR/IESWTR/LT1ESWTR)	TT[h]
11.	Legionella (SWTR/IESWTR/LT1ESWTR)	TT[h]
12.	Legionella (SWTR/IESWTR/LT1ESWTR)	TT[h]
13.	Cryptosporidium (IESWTR/FBRR/LT1ESWTR)	TT[h]
14.	Antimony	0.006
15.	Arsenic	0.010
16.	Asbestos	7 MFL
17.	Barium	2

TABLE A1 National primary drinking water regulations [17]—cont'd

18.	Beryllium	0.004
19.	Cadmium	0.005
20.	Chromium (total)	0.1
21.	Cyanide	0.2
22.	Fluoride	4.0
23.	Mercury (inorganic)	0.002
24.	Nitrate	10
25.	Nitrite	1
26.	Total nitrate and nitrite	10
27.	Selenium	0.05
28.	Thallium	0.002
29.	Lead	TT[j]
30.	Copper	TT[k]
31.	2,4–D	0.07
32.	2,4,5–TP (Silvex)	0.05
33.	Alachlor	0.002
34.	Atrazine	0.003
35.	Benzo(α)pyrene (PAHs)	0.0002
36.	Carbofuran	0.04
37.	Chlordane	0.002
38.	Dalapon	0.2
39.	Di(2-ethylhexyl) adipate	0.4
40.	Di(2-ethylhexyl) phthalate	0.006
41.	Dibromochloropropane (DBCP)	0.0002
42.	Dinoseb	0.007
43.	Dioxin	3×10^{-8}
44.	Diquat	0.02
45.	Endothall	0.1
46.	Endrin	0.002
47.	Ethylene dibromide	0.00005
48.	Glyphosate	0.7
49.	Heptachlor	0.0004
50.	Heptachlor epoxide	0.0002
51.	Hexachlorobenzene	0.001
52.	Hexachlorocyclo-pentadiene	0.05
53.	Lindane	0.0002
54.	Methoxychlor	0.04
55.	Oxamyl (Vydate)	0.2

Continued

TABLE A1 National primary drinking water regulations [17]—cont'd

56.	Pentachlorophenol	0.001
57.	Picloram	0.5
58.	Polychlorinated biphenyls (PCBs)	0.0005
59.	Simazine	0.004
60.	Toxaphene	0.003
61.	Benzene	0.005
62.	Carbon tetrachloride	0.005
63.	Chlorobenzene (monochloro-benzene)	0.1
64.	*o*-Dichlorobenzene	0.6
65.	*p*-Dichlorobenzene	0.075
66.	1,2-Dichloroethane	0.005
67.	Dichloroethylene	0.007
68.	*cis*-1,2-Dichloroethylene	0.07
69.	*trans*-1,2-Dichloroethylene	0.1
70.	Dichloromethane	0.005
71.	1,2-Dichloropropane	0.005
72.	Ethylbenzene	0.7
73.	Styrene	0.1
74.	Tetrachloroethylene	0.005
75.	Toluene	1
76.	1,2,4-Trichlorobenzene	0.07
77.	1,1,1-Trichloroethane	0.2
78.	1,1,2-Trichloroethane	0.005
79.	Trichloroethylene	0.005
80.	Vinyl chloride	0.002
81.	Xylenes (total)	10
82.	Beta photon emitters	4 mrem/year[l]
83.	Alpha emitters	17 pCi/L[m]
84.	Combined radium (226 & 228)	5 pCi/L
85.	Uranium[n]	30 μg/L
86.	Total trihalomethanes (TTHMs)	0.080[o,p]
87.	Haloacetic Acids (HAA)	0.060[q]
88.	Bromate	0.010
89.	Chlorite	1.0
90.	Chlorine	4.0 (MRDL)[r]s
91.	Chloramines	4.0 (MRDL)
92.	Chlorine dioxide	0.8 (MRDL)
93.	Chlorine dioxide	0.8 (MRDL)
94.	Control of DBP precursors (TOC)	TT

TABLE A1 National primary drinking water regulations [17]—cont'd

95.	Acrylamide	TT
96.	Epichlorohydrin	TT

[a]MCL, maximum contaminant level.
[b]For water systems analyzing at least 40 samples per month, no more than 5.0% of the monthly samples may be positive for total coliforms. For systems analyzing fewer than 40 samples per month, no more than one sample per month may be positive for total coliforms.
[c]There are various regulations that set turbidity standards for different types of systems, including 40 CFR 141.13, and the 1989 Surface Water Treatment Rule, the 1998 Interim Enhanced Surface Water Treatment Rule and the 2002 Long Term 1 Enhanced Surface Water Treatment Rule. The MCL for the monthly turbidity average is 1 NTU; the MCL for the 2-day average is 5 NTU for systems that are required to filter but have not yet installed filtration (40 CFR 141.13).
[d]NTU, nephelometric turbidity unit.
[e]There are various regulations that set turbidity standards for different types of systems, including 40 CFR 141.13, and the 1989 Surface Water Treatment Rule, the 1998 Interim Enhanced Surface Water Treatment Rule and the 2001 Long Term 1 Enhanced Surface Water Treatment Rule. Systems subject to the Surface Water Treatment Rule (both filtered and unfiltered) may not exceed 5 NTU. In addition, in filtered systems, 95% of samples each month must not exceed 0.5 NTU in systems using conventional or direct filtration and must not exceed 1 NTU in systems using slow sand or diatomaceous earth filtration or other filtration technologies approved by the primacy agency.
[f]TT—treatment technique on DSK5CLS3C1PROD with CFR.
[g]There are various regulations that set turbidity standards for different types of systems, including 40 CFR 141.13, the 1989 Surface Water Treatment Rule (SWTR), the 1998 Interim Enhanced Surface Water Treatment Rule (IESWTR) and the 2002 Long Term 1 Enhanced Surface Water Treatment Rule (LT1ESWTR). For systems subject to the IESWTR (systems serving at least 10,000 people, using surface water or groundwater under the direct influence of surface water), that use conventional filtration or direct filtration, after January 1, 2002, the turbidity level of a system's combined filter effluent may not exceed 0.3 NTU in at least 95% of monthly measurements, and the turbidity level of a system's combined filter effluent must not exceed 1 NTU at any time. Systems subject to the IESWTR using technologies other than conventional, direct, slow sand, or diatomaceous earth filtration must meet turbidity limits set by the primacy agency. For systems subject to the LT1ESWTR (systems serving fewer than 10,000 people, using surface water or ground Pt. 141, Subpt. Q, App. B water under the direct influence of surface water) that use conventional filtration or direct filtration, after January 1, 2005, the turbidity level of a system's combined filter effluent may not exceed 0.3 NTU in at least 95% of monthly measurements, and the turbidity level of a system's combined filter effluent must not exceed 1 NTU at any time. Systems subject to the LT1ESWTR using technologies other than conventional, direct, slow sand, or diatomaceous earth filtration must meet turbidity limits set by the primacy agency.
[h]SWTR, IESWTR, and LT1ESWTR treatment technique violations that involve turbidity exceedances may use the health effects language for turbidity instead.
[i]The bacteria detected by heterotrophic plate count (HPC) are not necessarily harmful. HPC is simply an alternative method of determining disinfectant residual levels. The number of such bacteria is an indicator of whether there is enough disinfectant in the distribution system.
[j]Action level = 0.015 mg/L.
[k]Action level = 1.3 mg/L.
[l]Millirems per year.
[m]Picocuries per liter.
[n]The uranium MCL is effective December 8, 2003, for all community water systems.
[o]Community and nontransient noncommunity systems must comply with subpart V TTHM and HAA5 MCLs of 0.080 and 0.060 mg/L, respectively (with compliance calculated as a locational running annual average) on the schedule in §141.620.
[p]The MCL for total trihalomethanes is the sum of the concentrations of the individual trihalomethanes.
[q]The MCL for haloacetic acids is the sum of the concentrations of the individual haloacetic acids.
[r]MRDL, maximum residual disinfectant level.
[s]These arsenic values are effective January 23, 2006. Until then, the MCL is 0.05 mg/L and there is no MCL.

TABLE A2 National secondary drinking water contaminants [17]

S/N	Contaminants	MCL
1.	Aluminum	0.05–0.2 mg/L
2.	Chloride	250 mg/L
3.	Color	15 Color units
4.	Copper	1 mg/L
5.	Corrosivity	Noncorrosive
6.	Fluoride	250 mg/L
7.	Foaming agent	0.5 mg/L
8.	Iron	0.3 mg/L
9.	Manganese	0.05 mg/L
10.	Odor	3 Threshold odor number
11.	pH	6.5–8.5
12.	Silver	0.1 mg/L
13.	Sulfate	250 mg/L
14.	Total dissolved solids (TDS)	500 mg/L
15.	Zinc	5 mg/L

TABLE A3 Typical reclaimed water quality parameters [18]

Parameters	Value
1. Typical range alkalinity, total as $CaCO_3$	–150 mg/L
2. Ammonia as N	0–0.4 mg/L
3. Boron	0.2–0.4 mg/L
4. Calcium	40–70 mg/L
5. Chloride	65–170 mg/L
6. Chlorine, total	1.5–4.0 mg/L
7. Iron	0.05–0.6 mg/L
8. Magnesium	5–20 mg/L
9. Manganese	0.003–0.08 mg/L
10. Nitrate + Nitrite as N	5–30 mg/L
11. Nitrate as N	5–20 mg/L
12. Nitrite as N	0.01–0.05 mg/L
13. Ortho phosphorus, dissolved as P	0.04–0.3 mg/L
14. pH	6–8
15. Phosphorus, total as P	0.04–0.4 mg/L
16. Potassium	10–20 mg/L
17. Sodium	90–200 mg/L
18. Specific conductance	360–1250 μS/cm
19. Sulfate	80–250 mg/L
20. Temperature	10–30°C
21. Total coliform (MPN/100 mL)	<1.0
22. Total Kjeldahl nitrogen	0.2–2 mg/L
23. Total organic carbon	4–8 mg/L

References

[1] S. Ahuja, Green chemistry and other novel solutions to water pollution: overview, in: ACS Symposium Series, 2013, pp. 1–14.

[2] A. Yaqoob, T. Parveen, K. Umar, M. Mohamad Ibrahim, Role of nanomaterials in the treatment of wastewater: a review, Water 12 (2) (2020) 495.

[3] X. Qu, J. Brame, Q. Li, P. Alvarez, Nanotechnology for a safe and sustainable water supply: enabling integrated water treatment and reuse, Acc. Chem. Res. 46 (3) (2012) 834–843.

[4] Unwater, 2020. Access date: 06/05/2020 https://www.unwater.org/water-facts/scarcity/.

[5] M. Montgomery, M. Elimelech, Water and sanitation in developing countries: including health in the equation, Environ. Sci. Technol. 41 (1) (2007) 17–24.

[6] K. Umar, T. Parveen, M. Khan, M. Ibrahim, A. Ahmad, M. Rafatullah, Degradation of organic pollutants using metal-doped TiO_2 photocatalysts under visible light: a comparative study, Desalin. Water Treat. 161 (2019) 275–282.

[7] K. Tang, C. Gong, D. Wang, Reduction potential, shadow prices, and pollution costs of agricultural pollutants in China, Sci. Total Environ. 541 (2016) 42–50.

[8] K. Richter, J. Ayers, An approach to predicting sediment microbial fuel cell performance in shallow and deep water, Appl. Sci. 8 (12) (2018) 2628.

[9] T. Sizmur, T. Fresno, G. Akgül, H. Frost, E. Moreno-Jiménez, Biochar modification to enhance sorption of inorganics from water, Bioresour. Technol. 246 (2017) 34–47.

[10] J. Wang, Z. Wang, C. Vieira, J. Wolfson, G. Pingtian, S. Huang, Review on the treatment of organic pollutants in water by ultrasonic technology, Ultrason. Sonochem. 55 (2019) 273–278.

[11] C. Liu, T. Hong, H. Li, L. Wang, From club convergence of per capita industrial pollutant emissions to industrial transfer effects: an empirical study across 285 cities in China, Energy Policy 121 (2018) 300–313.

[12] T. Bayoumi, H. Saleh, Characterization of biological waste stabilized by cement during immersion in aqueous media to develop disposal strategies for phytomediated radioactive waste, Prog. Nucl. Energy 107 (2018) 83–89.

[13] H. Ma, Y. Guo, Y. Qin, Y. Li, Nutrient recovery technologies integrated with energy recovery by waste biomass anaerobic digestion, Bioresour. Technol. 269 (2018) 520–531.
[14] G. Hlongwane, P. Sekoai, M. Meyyappan, K. Moothi, Simultaneous removal of pollutants from water using nanoparticles: a shift from single pollutant control to multiple pollutant control, Sci. Total Environ. 656 (2019) 808–833.
[15] P. Rajasulochana, V. Preethy, Comparison on efficiency of various techniques in treatment of waste and sewage water—a comprehensive review, Resour. Eff. Technol. 2 (4) (2016) 175–184.
[16] CDC, 2020, Access date:05/06/2020; https://www.cdc.gov/coronavirus/2019ncov/community/sanitation-wastewater-workers.html.
[17] https://www.epa.gov/sdwa/how-epa-regulates-drinking-water-contaminants-documents.
[18] https://www.epa.gov/waterreuse/guidelines-water-reuse.
[19] R. Jain, M. Camarillo, W.T. Stringfellow, Drinking Water Security for Engineers, Planners, and Managers, Butterworth-Heinemann, 2014, pp. 45–67.
[20] C. Teh, P. Budiman, K. Shak, T. Wu, Recent advancement of coagulation–flocculation and its application in wastewater treatment, Ind. Eng. Chem. Res. 55 (16) (2016) 4363–4389.
[21] G. Crini, E. Lichtfouse, Advantages and disadvantages of techniques used for wastewater treatment, Environ. Chem. Lett. 17 (1) (2018) 145–155.
[22] M.S. Ishaq, Z. Afsheen, A. Khan, A. Khan, in: S.B. Khan, K. Akhtar (Eds.), Disinfection Methods, Photocatalysts—Applications and Attributes, IntechOpen, 2018.
[23] K. Kunduru, M. Nazarkovsky, S. Farah, R. Pawar, A. Basu, A. Domb, Nanotechnology for water purification: applications of nanotechnology methods in wastewater treatment, Water Purif. (2017) 33–74.
[24] R. Suthar, B. Gao, Nanotechnology for drinking water purification, Water Purif. (2017) 75–118.
[25] https://www.iso.org/standard/44278.html.
[26] European Commission, Commission recommendation of 18 October 2011 on the definition of nanomaterial, Off. J. Eur. Union 275 (2011) 38–40.
[27] P. Dolez, Nanomaterials definitions, classifications, and applications, Nanoengineering (2015) 3–40.
[28] V.M. Anischik, V.E. Borisenko, S.A. Jdanok, N.K. Tolochko, V.M. Fedosyuk, Nanomaterialy i nanotehnologii, National Academy of Science, Belorussia Republic, 2008, ISBN: 978-985-476-618-8.
[29] S. Read, A. Jiménez, B. Ross, R. Aitken, M. van Tongeren, Nanotechnology and exposure scenarios, in: Handbook of Nanosafety, 2014, pp. 17–58.
[30] S. Kaluza, J. Kleine Balderhaar, B. Orthen, B. Honnert, E. Jankowska, P. Pietrowski, et al., in: J. Kosk-Bienko (Ed.), Workplace Exposure to Nanoparticles, European Agency for Safety and Health at Work, 2009.
[31] D. Hristozov, I. Malsch, Hazards and risks of engineered nanoparticles for the environment and human health, Sustainability 1 (4) (2009) 1161–1194.
[32] M. Hedmer, M. Kåredal, P. Gustavsson, J. Rissler, Carbon Nanotubes, Arbets-och miljömedicin, Göteborgs universitet, 2013. 47 (5).
[33] A. Figoli, M. Dorraji, A. Amani-Ghadim, Application of nanotechnology in drinking water purification, Water Purif. (2017) 119–167.
[34] X. Qu, P. Alvarez, Q. Li, Applications of nanotechnology in water and wastewater treatment, Water Res. 47 (12) (2013) 3931–3946.
[35] A. Srivastava, O. Srivastava, S. Talapatra, R. Vajtai, P. Ajayan, Carbon nanotube filters, Nat. Mater. 3 (9) (2004) 610–614.
[36] K. Zodrow, L. Brunet, S. Mahendra, D. Li, A. Zhang, Q. Li, P. Alvarez, Polysulfone ultrafiltration membranes impregnated with silver nanoparticles show improved biofouling resistance and virus removal, Water Res. 43 (3) (2009) 715–723.
[37] L. Liu, Z. Liu, H. Bai, D. Sun, Concurrent filtration and solar photocatalytic disinfection/degradation using high-performance ag/TiO_2 nanofiber membrane, Water Res. 46 (4) (2012) 1101–1112.
[38] M. Liga, E. Bryant, V. Colvin, Q. Li, Virus inactivation by silver doped titanium dioxide nanoparticles for drinking water treatment, Water Res. 45 (2) (2011) 535–544.
[39] M. Gondal, M. Dastageer, A. Khalil, Synthesis of nano-WO_3 and its catalytic activity for enhanced antimicrobial process for water purification using laser induced photo-catalysis, Catal. Commun. 11 (3) (2009) 214–219.
[40] G. Poinern, D. Parsonage, T. Issa, M. Ghosh, E. Paling, P. Singh, Preparation, characterization and as(V) adsorption behaviour of CNT-ferrihydrite composites, Int. J. Eng. Sci. Technol. 2 (8) (2011).
[41] P. Girginova, A. Daniel-da-Silva, C. Lopes, P. Figueira, M. Otero, V. Amaral, E. Pereira, T. Trindade, Silica coated magnetite particles for magnetic removal of Hg^{2+} from water, J. Colloid Interface Sci. 345 (2) (2010) 234–240.
[42] P. Figueira, C. Lopes, A. Daniel-da-Silva, E. Pereira, A. Duarte, T. Trindade, Removal of mercury (II) by dithiocarbamate surface functionalized magnetite particles: application to synthetic and natural spiked waters, Water Res. 45 (17) (2011) 5773–5784.
[43] M. Kumari, C. Pittman, D. Mohan, Heavy metals [Chromium (VI) and Lead (II)] removal from water using mesoporous magnetite (Fe_3O_4) nanospheres, J. Colloid Interface Sci. 442 (2015) 120–132.
[44] F. Qin, G. Li, R. Wang, J. Wu, H. Sun, R. Chen, Template-free fabrication of Bi_2O_3 and $(BiO)_2CO_3$ nanotubes and their application in water treatment, Chem. A Eur. J. 18 (51) (2012) 16491–16497.
[45] B. Stanton, J. Harris, M. Miller, M. Bruening, Ultrathin, multilayered polyelectrolyte films as nanofiltration membranes, Langmuir 19 (17) (2003) 7038–7042.
[46] X. Peng, Z. Luan, J. Ding, Z. Di, Y. Li, B. Tian, Ceria nanoparticles supported on carbon nanotubes for the removal of arsenate from water, Mater. Lett. 59 (4) (2005) 399–403.
[47] R. Khaydarov, R. Khaydarov, O. Gapurova, Water purification from metal ions using carbon nanoparticle-conjugated polymer nanocomposites, Water Res. 44 (6) (2010) 1927–1933.
[48] M. Tofighy, T. Mohammadi, Adsorption of divalent heavy metal ions from water using carbon nanotube sheets, J. Hazard. Mater. 185 (1) (2011) 140–147.
[49] S. Shin, J. Jang, Thiol containing polymer encapsulated magnetic nanoparticles as reusable and efficiently separable adsorbent for heavy metal ions, Chem. Commun. 41 (2007) 4230.
[50] J. Liu, Z. Zhao, G. Jiang, Coating Fe_3O_4 magnetic nanoparticles with humic acid for high efficient removal of heavy metals in water, Environ. Sci. Technol. 42 (18) (2008) 6949–6954.

[51] M. Pena, G. Korfiatis, M. Patel, L. Lippincott, X. Meng, Adsorption of As(V) and As(III) by nanocrystalline titanium dioxide, Water Res. 39 (11) (2005) 2327–2337.

[52] S. Kong, Y. Wang, Q. Hu, A. Olusegun, Magnetic nanoscale Fe–Mn binary oxides loaded zeolite for arsenic removal from synthetic groundwater, Colloids Surf. A Physicochem. Eng. Asp. 457 (2014) 220–227.

[53] J. Jin, R. Li, H. Wang, H. Chen, K. Liang, J. Ma, Magnetic Fe nanoparticle functionalized water-soluble multi-walled carbon nanotubes towards the preparation of sorbent for aromatic compounds removal, Chem. Commun. 4 (2007) 386–388.

[54] M. Arkas, L. Eleades, C. Paleos, D. Tsiourvas, Alkylated hyperbranched polymers as molecular nanosponges for the purification of water from polycyclic aromatic hydrocarbons, J. Appl. Polym. Sci. 97 (6) (2005) 2299–2305.

[55] R. Allabashi, M. Arkas, G. Hörmann, D. Tsiourvas, Removal of some organic pollutants in water employing ceramic membranes impregnated with cross-linked silylated dendritic and cyclodextrin polymers, Water Res. 41 (2) (2007) 476–486.

[56] X. Peng, Y. Li, Z. Luan, Z. Di, H. Wang, B. Tian, Z. Jia, Adsorption of 1,2-dichlorobenzene from water to carbon nanotubes, Chem. Phys. Lett. 376 (1–2) (2003) 154–158.

[57] K. Salipira, B. Mamba, R. Krause, T. Malefetse, S. Durbach, Carbon nanotubes and cyclodextrin polymers for removing organic pollutants from water, Environ. Chem. Lett. 5 (1) (2006) 13–17.

[58] Y. Yang, N. Ren, Y. Zhang, Y. Tang, Nanosized cadmium sulfide in polyelectrolyte protected mesoporous sphere: a stable and regeneratable photocatalyst for visible-light-induced removal of organic pollutants, J. Photochem. Photobiol. A Chem. 201 (2–3) (2009) 111–120.

[59] H. Zhang, X. Quan, S. Chen, H. Zhao, Y. Zhao, Fabrication of photocatalytic membrane and evaluation its efficiency in removal of organic pollutants from water, Sep. Purif. Technol. 50 (2) (2006) 147–155.

[60] J. Hu, L. Ren, Y. Guo, H. Liang, A. Cao, L. Wan, C. Bai, Mass production and high photocatalytic activity of ZnS nanoporous nanoparticles, Angew. Chem. Int. Ed. 44 (8) (2005) 1269–1273.

[61] R. Ullah, J. Dutta, Photocatalytic degradation of organic dyes with manganese-doped ZnO nanoparticles, J. Hazard. Mater. 156 (1–3) (2008) 194–200.

[62] S. Anandan, G. Lee, P. Chen, C. Fan, J. Wu, Removal of Orange II dye in water by visible light assisted photocatalytic ozonation using Bi_2O_3 and au/Bi_2O_3 nanorods, Ind. Eng. Chem. Res. 49 (20) (2010) 9729–9737.

[63] Y. Zhai, S. Zhang, H. Pang, Preparation, characterization and photocatalytic activity of CeO_2 nanocrystalline using ammonium bicarbonate as precipitant, Mater. Lett. 61 (8–9) (2007) 1863–1866.

[64] H. Wang, L. Yang, H. Yu, F. Peng, A highly efficient and stable visible-light plasmonic photocatalyst Ag-AgCl/CeO_2, World J. Nano Sci. Eng. 01 (04) (2011) 129–136.

[65] Ö. Kerkez-Kuyumcu, Ş. Bayazit, M. Salam, Antibiotic amoxicillin removal from aqueous solution using magnetically modified graphene nanoplatelets, J. Ind. Eng. Chem. 36 (2016) 198–205.

[66] Z. Li, P. Zhang, T. Shao, J. Wang, L. Jin, X. Li, Different nanostructured In_2O_3 for photocatalytic decomposition of perfluorooctanoic acid (PFOA), J. Hazard. Mater. 260 (2013) 40–46.

[67] J. He, A. Kumar, M. Khan, I. Lo, Critical review of photocatalytic disinfection of bacteria: from Noble metals- and carbon nanomaterials-TiO_2 composites to challenges of water characteristics and strategic solutions, Sci. Total Environ. 758 (2021) 143953.

[68] S. Mendiola-Alvarez, A. Hernández-Ramírez, J. Guzmán-Mar, M. Maya-Treviño, A. Caballero-Quintero, L. Hinojosa-Reyes, A novel P-doped Fe_2O_3-TiO_2 mixed oxide: synthesis, characterization and photocatalytic activity under visible radiation, Catal. Today 328 (2019) 91–98.

[69] K. Umar, A. Aris, T. Parveen, J. Jaafar, Z. Abdul Majid, A. Vijaya Bhaskar Reddy, J. Talib, Synthesis, characterization of Mo and Mn doped ZnO and their photocatalytic activity for the decolorization of two different chromophoric dyes, Appl. Catal. Gen. 505 (2015) 507–514.

[70] R. Raliya, C. Avery, S. Chakrabarti, P. Biswas, Photocatalytic degradation of methyl orange dye by pristine titanium dioxide, zinc oxide, and graphene oxide nanostructures and their composites under visible light irradiation, Appl. Nanosci. 7 (5) (2017) 253–259.

[71] S. Khanchandani, S. Kundu, A. Patra, A. Ganguli, Band gap tuning of ZnO/In_2S_3 core/shell nanorod arrays for enhanced visible-light-driven photocatalysis, J. Phys. Chem. C 117 (11) (2013) 5558–5567.

[72] C. Gerba, I. Pepper, Drinking water treatment, Environ. Pollut. Sci. (2019) 435–454.

[73] A. Englande, P. Krenkel, J. Shamas, Wastewater treatment &water reclamation, in: Reference Module in Earth Systems and Environmental Sciences, 2015.

[74] Y. Musico, C. Santos, M. Dalida, D. Rodrigues, Surface modification of membrane filters using graphene and graphene oxide-based nanomaterials for bacterial inactivation and removal, ACS Sustain. Chem. Eng. 2 (7) (2014) 1559–1565.

[75] Y. Sang, F. Li, Q. Gu, C. Liang, J. Chen, Heavy metal-contaminated groundwater treatment by a novel nanofiber membrane, Desalination 223 (1–3) (2008) 349–360.

[76] M. Safarpour, A. Khataee, V. Vatanpour, Thin film nanocomposite reverse osmosis membrane modified by reduced graphene oxide/TiO_2 with improved desalination performance, J. Membr. Sci. 489 (2015) 43–54.

[77] X. Wang, D. Fang, K. Yoon, B. Hsiao, B. Chu, High performance ultrafiltration composite membranes based on poly(vinyl alcohol) hydrogel coating on crosslinked nanofibrous poly(vinyl alcohol) scaffold, J. Membr. Sci. 278 (1–2) (2006) 261–268.

[78] K. Yoon, K. Kim, X. Wang, D. Fang, B. Hsiao, B. Chu, High flux ultrafiltration membranes based on electrospun nanofibrous PAN scaffolds and chitosan coating, Polymer 47 (7) (2006) 2434–2441.

[79] L. Madhura, S. Singh, A review on the advancements of nanomembranes for water treatment, in: C.M. Hussain, A.K. Mishra (Eds.), Nanotechnology in Environmental Science, 2018.

[80] M. Esfahani, S. Aktij, Z. Dabaghian, M. Firouzjaei, A. Rahimpour, J. Eke, I. Escobar, M. Abolhassani, L. Greenlee, A. Esfahani, A. Sadmani, N. Koutahzadeh, Nanocomposite membranes for water separation and purification: fabrication, modification, and applications, Sep. Purif. Technol. 213 (2019) 465–499.

[81] M. Potter, A. Najer, A. Klöckner, S. Zhang, M. Holme, V. Nele, J. Che, L. Massi, J. Penders, C. Saunders, J. Doutch, A. Edwards, O. Ces, M. Stevens, Controlled dendrimersome nanoreactor system for localized hypochlorite-induced killing of bacteria, ACS Nano 14 (12) (2020) 17333–17353.

[82] A. Talaiekhozani, M. Reza Mosayebi, M. Fulazzaky, Z. Eskandari, R. Sanayee, Combination of TiO_2 microreactor and electroflotation for organic pollutant removal from textile dyeing industry wastewater, Alex. Eng. J. 59 (2) (2020) 549–563.

[83] I. Azzouz, Y. Habba, M. Capochichi-Gnambodoe, F. Marty, J. Vial, Y. Leprince-Wang, T. Bourouina, Zinc oxide nano-enabled microfluidic reactor for water purification and its applicability to volatile organic compounds, Microsyst. Nanoeng. 4 (1) (2018).

[84] T. Zeng, X. Zhang, S. Wang, H. Niu, Y. Cai, Spatial confinement of a Co_3O_4 catalyst in hollow metal–organic frameworks as a nanoreactor for improved degradation of organic pollutants, Environ. Sci. Technol. 49 (4) (2015) 2350–2357.

[85] Z. Lu, F. He, C. Hsieh, X. Wu, M. Song, X. Liu, Y. Liu, S. Yuan, H. Dong, S. Han, P. Du, G. Xing, Magnetic hierarchical photocatalytic nanoreactors: toward highly selective Cd^{2+} removal with secondary pollution free tetracycline degradation, ACS Appl. Nano Mater. 2 (3) (2019) 1664–1674.

[86] S. Zhan, D. Zhu, S. Ma, W. Yu, Y. Jia, Y. Li, H. Yu, Z. Shen, Highly efficient removal of pathogenic bacteria with magnetic graphene composite, ACS Appl. Mater. Interfaces 7 (7) (2015) 4290–4298.

[87] A. Gupta, S. Vidyarthi, N. Sankararamakrishnan, Enhanced sorption of mercury from compact fluorescent bulbs and contaminated water streams using functionalized multiwalled carbon nanotubes, J. Hazard. Mater. 274 (2014) 132–144.

[88] R. White, C. White, H. Turgut, A. Massoud, Z. Tian, Comparative studies on copper adsorption by graphene oxide and functionalized graphene oxide nanoparticles, J. Taiwan Inst. Chem. Eng. 85 (2018) 18–28.

[89] F. Rodovalho, G. Capistrano, J. Gomes, F. Sodré, J. Chaker, A. Campos, A. Bakuzis, M. Sousa, Elaboration of magneto-thermally recyclable nanosorbents for remote removal of toluene in contaminated water using magnetic hyperthermia, Chem. Eng. J. 302 (2016) 725–732.

[90] Z. Wang, B. Mi, Environmental applications of 2D molybdenum disulfide (MoS_2) nanosheets, Environ. Sci. Technol. 51 (15) (2017) 8229–8244.

[91] X. Qu, J. Brame, Q. Li, P. Alvarez, Nanotechnology for a safe and sustainable water supply: enabling integrated water treatment and reuse, Acc. Chem. Res. 46 (3) (2012) 834–843.

[92] N. McGuinness, M. Garvey, A. Whelan, H. John, C. Zhao, G. Zhang, D. Dionysiou, J. Byrne, S. Pillai, Nanotechnology solutions for global water challenges, in: ACS Symposium Series, 2015, pp. 375–411.

[93] R. Li, L. Zhang, P. Wang, Rational design of nanomaterials for water treatment, Nanoscale 7 (41) (2015) 17167–17194.

[94] G. Oberdörster, V. Stone, K. Donaldson, Toxicology of nanoparticles: a historical perspective, Nanotoxicology 1 (1) (2007) 2–25.

[95] M.M. Roberto, C.A. Christofoletti, How to assess nanomaterial toxicity? An environmental and human health approach, in: Nanomaterials Toxicity, Human Health and Environment, 2020.

[96] M. Croteau, S. Misra, S. Luoma, E. Valsami-Jones, Silver bioaccumulation dynamics in a freshwater invertebrate after aqueous and dietary exposures to nanosized and ionic Ag, Environ. Sci. Technol. 45 (15) (2011) 6600–6607.

[97] R. Miller, H. Lenihan, E. Muller, N. Tseng, S. Hanna, A. Keller, Impacts of metal oxide nanoparticles on marine phytoplankton, Environ. Sci. Technol. 44 (19) (2010) 7329–7334.

[98] D. Chatterjee, L. Fong, Y. Zhang, Nanoparticles in photodynamic therapy: an emerging paradigm, Adv. Drug Deliv. Rev. 60 (15) (2008) 1627–1637.

[99] G. Bielmyer-Fraser, T. Jarvis, H. Lenihan, R. Miller, Cellular partitioning of nanoparticulate versus dissolved metals in marine phytoplankton, Environ. Sci. Technol. 48 (22) (2014) 13443–13450.

[100] X. Zhu, L. Zhu, Z. Duan, R. Qi, Y. Li, Y. Lang, Comparative toxicity of several metal oxide nanoparticle aqueous suspensions to zebrafish (danio rerio) early developmental stage, J. Environ. Sci. Health A 43 (3) (2008) 278–284.

[101] B. Jovanović, L. Anastasova, E. Rowe, Y. Zhang, A. Clapp, D. Palić, Effects of nanosized titanium dioxide on innate immune system of fathead minnow (*Pimephales Promelas* Rafinesque, 1820), Ecotoxicol. Environ. Saf. 74 (4) (2011) 675–683.

[102] J. Zhang, M. Wages, S. Cox, J. Maul, Y. Li, M. Barnes, L. Hope-Weeks, G. Cobb, Effect of titanium dioxide nanomaterials and ultraviolet light coexposure on African clawed frogs (*Xenopus Laevis*), Environ. Toxicol. Chem. 31 (1) (2011) 176–183.

[103] G. Marcone, Á. Oliveira, G. Almeida, G. Umbuzeiro, W. Jardim, Ecotoxicity of TiO_2 to *Daphnia Similis* under irradiation, J. Hazard. Mater. 211-212 (2012) 436–442.

[104] M. Heinlaan, A. Ivask, I. Blinova, H. Dubourguier, A. Kahru, Toxicity of nanosized and bulk ZnO, CuO and TiO_2 to bacteria *Vibrio Fischeri* and crustaceans daphnia magna and *Thamnocephalus Platyurus*, Chemosphere 71 (7) (2008) 1308–1316.

[105] R. Griffitt, J. Luo, J. Gao, J. Bonzongo, D. Barber, Effects of particle composition and species on toxicity of metallic nanomaterials in aquatic organisms, Environ. Toxicol. Chem. 27 (9) (2008) 1972.

[106] K. Hund-Rinke, K. Schlich, D. Kühnel, B. Hellack, H. Kaminski, C. Nickel, Grouping concept for metal and metal oxide nanomaterials with regard to their ecotoxicological effects on algae, daphnids and fish embryos, NanoImpact 9 (2018) 52–60.

[107] A. Hubbs, R. Mercer, S. Benkovic, J. Harkema, K. Sriram, D. Schwegler-Berry, M. Goravanahally, T. Nurkiewicz, V. Castranova, L. Sargent, Nanotoxicology—a Pathologist'S perspective, Toxicol. Pathol. 39 (2) (2010) 301–324.

[108] S. Ahuja, Handbook of Water Purity and Quality, second ed., Elsevier, Amsterdam, 2021.

[109] S. Ahuja, Evaluating Water Quality to Prevent Future Disasters, Elsevier, Amsterdam, 2019.

[110] S. Ahuja, Advances in Water Purification, Elsevier, Amsterdam, 2019.

[111] S. Ahuja, Arsenic Contamination of Groundwater; Mechanism, Analysis, and Remediation, Wiley, New York, 2008.

[112] S. Ahuja, K. Hristovski, Novel Solutions to Water Pollution, American Chemical Society, Washington, DC, 2013.

[113] S. Ahuja, Monitoring Water Quality: Pollution Assessment, Analysis, and Remediation, Elsevier, Waltham, MA, 2013.

[114] S. Ahuja, Comprehensive Water Quality and Purification, vols. 1-4, Elsevier, Kidlington, Oxford, 2013.

[115] S. Ahuja, Chemistry and Water: The Science Behind Sustaining the World's Most Crucial Resource, Elsevier, Amsterdam, 2017.

[116] S. Ahuja, Monitoring Water Quality and Infrastructure to Prevent Future Flints, American Chemical Society Meeting, Washington DC, 2017.
[117] S. Ahuja, Water Reclamation and Sustainability, Elsevier, Amsterdam, 2014.
[118] M. Habuda-Stanić, M. Nujić, Arsenic removal by nanoparticles: a review, Environ. Sci. Pollut. Res. 22 (11) (2015) 8094–8123.
[119] T. Viraraghavan, K. Subramanian, J. Aruldoss, Arsenic in drinking water-problems and solutions, Water Sci. Technol. 40 (2) (1999) 69–76.
[120] L. Liu, X. Luo, L. Ding, S. Luo, Application of nanotechnology in the removal of heavy metal from water, in: Nanomaterials for the Removal of Pollutants and Resource Reutilization, 2019, pp. 83–147.
[121] S. Lata, S. Samadder, Removal of arsenic from water using Nano adsorbents and challenges: a review, J. Environ. Manage. 166 (2016) 387–406.
[122] WHO, 2020, Access date: 16/06/2020; https://www.who.int/news-room/fact-sheets/detail/arsenic.
[123] S. Wickramasinghe, B. Han, J. Zimbron, Z. Shen, M. Karim, Arsenic removal by coagulation and filtration: comparison of groundwaters from the United States and Bangladesh, Desalination 169 (3) (2004) 231–244.
[124] R. Ning, Arsenic removal by reverse osmosis, Desalination 143 (3) (2002) 237–241.
[125] W. Ficklin, Separation of arsenic(III) and arsenic(V) in ground waters by ion-exchange, Talanta 30 (5) (1983) 371–373.
[126] T. Lin, J. Wu, Adsorption of arsenite and arsenate within activated alumina grains: equilibrium and kinetics, Water Res. 35 (8) (2001) 2049–2057.
[127] S. Palit, K. Misra, J. Mishra, Arsenic contamination in south Asian regions: the difficulties, challenges and vision for the future, Sep. Sci. Technol. (2019) 113–123.
[128] S. Dutta, R. Sharma, Sustainable magnetically retrievable nanoadsorbents for selective removal of heavy metal ions from different charged wastewaters, Sep. Sci. Technol. (2019) 371–416.
[129] I. Akin, G. Arslan, A. Tor, M. Ersoz, Y. Cengeloglu, Arsenic(V) removal from underground water by magnetic nanoparticles synthesized from waste red mud, J. Hazard. Mater. 235-236 (2012) 62–68.
[130] L. Feng, M. Cao, X. Ma, Y. Zhu, C. Hu, Superparamagnetic high-surface-area Fe_3O_4 nanoparticles as adsorbents for arsenic removal, J. Hazard. Mater. 217-218 (2012) 439–446.
[131] K. Gupta, S. Bhattacharya, D. Chattopadhyay, A. Mukhopadhyay, H. Biswas, J. Dutta, N. Ray, U. Ghosh, Ceria associated manganese oxide nanoparticles: synthesis, characterization and arsenic(V) sorption behavior, Chem. Eng. J. 172 (1) (2011) 219–229.
[132] C. Yavuz, J. Mayo, W. Yu, A. Prakash, J. Falkner, S. Yean, L. Cong, H. Shipley, A. Kan, M. Tomson, D. Natelson, V. Colvin, Low-field magnetic separation of monodisperse Fe_3O_4 nanocrystals, Science 314 (5801) (2006) 964–967.
[133] V. Chandra, J. Park, Y. Chun, J. Lee, I. Hwang, K. Kim, Water-dispersible magnetite-reduced graphene oxide composites for arsenic removal, ACS Nano 4 (7) (2010) 3979–3986.
[134] Z. Özlem Kocabaş-Ataklı, Y. Yürüm, Synthesis and characterization of anatase nanoadsorbent and application in removal of Lead, copper and arsenic from water, Chem. Eng. J. 225 (2013) 625–635.
[135] A. Farghali, M. Bahgat, A. Enaiet Allah, M. Khedr, Adsorption of pb(II) ions from aqueous solutions using copper oxide nanostructures, Beni-Suef Univ. J. Basic Appl. Sci. 2 (2) (2013) 61–71.
[136] N. Mubarak, J. Sahu, E. Abdullah, N. Jayakumar, Removal of heavy metals from wastewater using carbon nanotubes, Sep. Purif. Rev. 43 (4) (2013) 311–338.
[137] W. Mortada, M. Sobh, M. El-Defrawy, S. Farahat, Study of Lead exposure from automobile exhaust as a risk for nephrotoxicity among traffic policemen, Am. J. Nephrol. 21 (4) (2001) 274–279.
[138] T. Lidsky, J. Schneider, Lead neurotoxicity in children: basic mechanisms and clinical correlates, Brain 126 (1) (2003) 5–19.
[139] K. Steenland, P. Boffetta, Lead and cancer in humans: where are we now? Am. J. Ind. Med. 38 (3) (2000) 295–299.
[140] S. Mahdavi, M. Jalali, A. Afkhami, Removal of heavy metals from aqueous solutions using Fe_3O_4, ZnO, and CuO nanoparticles, J. Nanopart. Res. 14 (8) (2012).
[141] H. Shipley, K. Engates, V. Grover, Removal of Pb(II), Cd(II), Cu(II), and Zn(II) by hematite nanoparticles: effect of sorbent concentration, Ph, temperature, and exhaustion, Environ. Sci. Pollut. Res. 20 (3) (2012) 1727–1736.
[142] S. Mahdavi, M. Jalali, A. Afkhami, Heavy metal removal from aqueous solution using TiO_2, MgO, and Al_2O_3 nanoparticles, Chem. Eng. Commun. 200 (3) (2013) 448–470.
[143] K. Engates, H. Shipley, Adsorption of Pb, Cd, Cu, Zn, and Ni to titanium dioxide nanoparticles: effect of particle size, solid concentration, and exhaustion, Environ. Sci. Pollut. Res. 18 (3) (2010) 386–395.
[144] M. Salam, Coating carbon nanotubes with crystalline manganese dioxide nanoparticles and their application for lead ions removal from model and real water, Colloids Surf. A Physicochem. Eng. Asp. 419 (2013) 69–79.
[145] M. Scheurer, H. Brauch, F. Lange, Transformation products of artificial sweeteners, in: Transformation Products of Emerging Contaminants in the Environment, 2014, pp. 525–544.
[146] M. Mauter, I. Zucker, F. Perreault, J. Werber, J. Kim, M. Elimelech, The role of nanotechnology in tackling global water challenges, Nat. Sustain. 1 (4) (2018) 166–175.
[147] S. Richardson, T. Ternes, Water analysis: emerging contaminants and current issues, Anal. Chem. 90 (1) (2017) 398–428.
[148] R. Bennett, et al., White Paper: Aquatic Life Criteria for Contaminants on Emerging Concern. Part I General Challenges and Recommendations, vol. 46, US EPA, 2008.
[149] F. Baquero, J. Martínez, R. Cantón, Antibiotics and antibiotic resistance in water environments, Curr. Opin. Biotechnol. 19 (3) (2008) 260–265.
[150] G. Bruce, R. Pleus, S. Snyder, Toxicological relevance of pharmaceuticals in drinking water, Environ. Sci. Technol. 44 (14) (2010) 5619–5626.
[151] O. Jones, J. Lester, N. Voulvoulis, Pharmaceuticals: a threat to drinking water? Trends Biotechnol. 23 (4) (2005) 163–167.
[152] B. Escher, N. Bramaz, R. Eggen, M. Richter, In vitro assessment of modes of toxic action of pharmaceuticals in aquatic life, Environ. Sci. Technol. 39 (9) (2005) 3090–3100.

[153] K.S. Crider, M.A. Cleves, J. Reefhuis, R.J. Berry, C.A. Hobbs, D.J. Hu, Antibacterial medication use during pregnancy and risk of birth defects: National Birth Defects Prevention Study, Arch. Pediatr. Adolesc. Med. 163 (11) (2009) 978–985.

[154] WHO, IPCS International Programme on Chemical Safety, Healthy and Safety Guide No. 96, 1989.

[155] R. Bull, J. Crook, M. Whittaker, J. Cotruvo, Therapeutic dose as the point of departure in assessing potential health hazards from drugs in drinking water and recycled municipal wastewater, Regul. Toxicol. Pharmacol. 60 (1) (2011) 1–19.

[156] B. Lopez, P. Ollivier, A. Togola, N. Baran, J. Ghestem, Screening of French groundwater for regulated and emerging contaminants, Sci. Total Environ. 518-519 (2015) 562–573.

[157] B. Du, A. Price, W. Scott, L. Kristofco, A. Ramirez, C. Chambliss, J. Yelderman, B. Brooks, Comparison of contaminants of emerging concern removal, discharge, and water quality hazards among centralized and on-site wastewater treatment system effluents receiving common wastewater influent, Sci. Total Environ. 466-467 (2014) 976–984.

[158] J. Reichert, D. Souza, A. Martins, Antipsychotic drugs in hospital wastewater and a preliminary risk assessment, Ecotoxicol. Environ. Saf. 170 (2019) 559–567.

[159] C. Tan, N. Gao, Y. Deng, Y. Zhang, M. Sui, J. Deng, S. Zhou, Degradation of antipyrine by UV, UV/H_2O_2 and UV/PS, J. Hazard. Mater. 260 (2013) 1008–1016.

[160] K. Fent, A. Weston, D. Caminada, Ecotoxicology of human pharmaceuticals, Aquat. Toxicol. 76 (2) (2006) 122–159.

[161] S. Fanourakis, J. Peña-Bahamonde, P. Bandara, D. Rodrigues, Nano-based adsorbent and photocatalyst use for pharmaceutical contaminant removal during indirect potable water reuse, NPJ clean Water 3 (1) (2020).

[162] M. Gar Alalm, S. Ookawara, D. Fukushi, A. Sato, A. Tawfik, Improved WO_3 photocatalytic efficiency using ZrO_2 and Ru for the degradation of carbofuran and ampicillin, J. Hazard. Mater. 302 (2016) 225–231.

[163] L. Al-Khateeb, S. Almotiry, M. Salam, Adsorption of pharmaceutical pollutants onto graphene nanoplatelets, Chem. Eng. J. 248 (2014) 191–199.

[164] T. Kurniawan, L. Yanyan, T. Ouyang, A. Albadarin, G. Walker, $BaTiO_3$/TiO_2 composite-assisted photocatalytic degradation for removal of acetaminophen from synthetic wastewater under UV–vis irradiation, Mater. Sci. Semicond. Process. 73 (2018) 42–50.

[165] L. Pastrana-Martínez, S. Morales-Torres, V. Likodimos, J. Figueiredo, J. Faria, P. Falaras, A. Silva, Advanced nanostructured photocatalysts based on reduced graphene oxide–TiO_2 composites for degradation of diphenhydramine pharmaceutical and methyl orange dye, Appl. Catal. Environ. 123-124 (2012) 241–256.

[166] N. Davari, M. Farhadian, A. Nazar, M. Homayoonfal, Degradation of diphenhydramine by the photocatalysts of ZnO/Fe_2O_3 and TiO_2/Fe_2O_3 based on clinoptilolite: structural and operational comparison, J. Environ. Chem. Eng. 5 (6) (2017) 5707–5720.

[167] S. Zhu, Y. Liu, S. Liu, G. Zeng, L. Jiang, X. Tan, L. Zhou, W. Zeng, T. Li, C. Yang, Adsorption of emerging contaminant metformin using graphene oxide, Chemosphere 179 (2017) 20–28.

[168] L. Paredes, S. Murgolo, H. Dzinun, M. Dzarfan Othman, A. Ismail, M. Carballa, G. Mascolo, Application of immobilized Tio_2 on PVDF dual layer hollow fibre membrane to improve the photocatalytic removal of pharmaceuticals in different water matrices, Appl. Catal. Environ. 240 (2019) 9–18.

[169] I. Amalraj Appavoo, J. Hu, Y. Huang, S. Li, S. Ong, Response surface modeling of carbamazepine (CBZ) removal by graphene-P25 nanocomposites/UVA process using central composite design, Water Res. 57 (2014) 270–279.

[170] A. Surenjan, B. Sambandam, T. Pradeep, L. Philip, Synthesis, characterization and performance of visible light active C-TiO_2 for pharmaceutical photodegradation, J. Environ. Chem. Eng. 5 (1) (2017) 757–767.

[171] C. Moody, J. Field, Perfluorinated surfactants and the environmental implications of their use in fire-fighting foams, Environ. Sci. Technol. 34 (18) (2000) 3864–3870.

[172] A. Paul, K. Jones, A. Sweetman, A first global production, emission, and environmental inventory for perfluorooctane sulfonate, Environ. Sci. Technol. 43 (2) (2009) 386–392.

[173] K. Prevedouros, I. Cousins, R. Buck, S. Korzeniowski, Sources, fate and transport of perfluorocarboxylates, ChemInform 37 (11) (2006).

[174] E. Houtz, C. Higgins, J. Field, D. Sedlak, Persistence of perfluoroalkyl acid precursors in AFFF-impacted groundwater and soil, Environ. Sci. Technol. 47 (15) (2013) 8187–8195.

[175] E. Sunderland, X. Hu, C. Dassuncao, A. Tokranov, C. Wagner, J. Allen, A review of the pathways of human exposure to poly-and perfluoroalkyl substances (PFASs) and present understanding of health effects, J. Expo. Sci. Environ. Epidemiol. 29 (2) (2018) 131–147.

[176] USEPA, Lifetime Health Advisories and Health Effects Support Documents for Perfluorooctanoic Acid and Perfluorooctane Sulfonate, Environmental Protection Agency, Washington, DC, 2016.

[177] A. Xue, Z. Yuan, Y. Sun, A. Cao, H. Zhao, Electro-oxidation of perfluorooctanoic acid by carbon nanotube sponge anode and the mechanism, Chemosphere 141 (2015) 120–126.

[178] C. Song, P. Chen, C. Wang, L. Zhu, Photodegradation of perfluorooctanoic acid by synthesized TiO_2–MWCNT composites under 365 nm UV irradiation, Chemosphere 86 (8) (2012) 853–859.

[179] D. Huang, L. Yin, X. Lu, S. Lin, Z. Niu, J. Niu, Directional electron transfer mechanisms with graphene quantum dots as the electron donor for photodecomposition of Perfluorooctane sulfonate, Chem. Eng. J. 323 (2017) 406–414.

[180] C. Zhang, J. Tang, C. Peng, M. Jin, Degradation of Perfluorinated compounds in wastewater treatment plant effluents by electrochemical oxidation with Nano-ZnO coated electrodes, J. Mol. Liq. 221 (2016) 1145–1150.

[181] O. Arvaniti, Y. Hwang, H. Andersen, A. Stasinakis, N. Thomaidis, M. Aloupi, Reductive degradation of Perfluorinated compounds in water using mg-aminoclay coated nanoscale zero valent iron, Chem. Eng. J. 262 (2015) 133–139.

[182] J. Niu, H. Lin, J. Xu, H. Wu, Y. Li, Correction to electrochemical mineralization of perfluorocarboxylic acids (PFCAs) by Ce-doped modified porous nanocrystalline Pbo2 film electrode, Environ. Sci. Technol. 47 (9) (2013) 4959.

[183] WHO, 2020, Access date: 19/06/2020; https://www.who.int/emergencies/diseases/novel-coronavirus-2019/events-as-they-happen.

[184] WHO, 2021. Access date: 23/04/2021; https://covid19.who.int/.

[185] S. Lauer, K. Grantz, Q. Bi, F. Jones, Q. Zheng, H. Meredith, A. Azman, N. Reich, J. Lessler, The incubation period of coronavirus disease 2019 (COVID-19) from publicly reported confirmed cases: estimation and application, Ann. Intern. Med. 172 (9) (2020) 577–582.

[186] C. Weiss, M. Carriere, L. Fusco, I. Capua, J. Regla-Nava, M. Pasquali, J. Scott, F. Vitale, M. Unal, C. Mattevi, D. Bedognetti, A. Merkoçi, E. Tasciotti, A. Yilmazer, Y. Gogotsi, F. Stellacci, L. Delogu, Toward nanotechnology-enabled approaches against the COVID-19 pandemic, ACS Nano 14 (6) (2020) 6383–6406.

[187] WHO, 2020, Access date: 19/06/2020, https://www.who.int/healthtopics/coronavirus#tab=tab_1.

[188] M. Holshue, C. DeBolt, S. Lindquist, K. Lofy, J. Wiesman, H. Bruce, C. Spitters, K. Ericson, S. Wilkerson, A. Tural, G. Diaz, A. Cohn, L. Fox, A. Patel, S. Gerber, L. Kim, S. Tong, X. Lu, S. Lindstrom, M. Pallansch, W. Weldon, H. Biggs, T. Uyeki, S. Pillai, First case of 2019 novel coronavirus in the United States, N. Engl. J. Med. 382 (10) (2020) 929–936.

[189] W. Lodder, A. de Roda Husman, SARS-CoV-2 in wastewater: potential health risk, but also data source, Lancet Gastroenterol. Hepatol. 5 (6) (2020) 533–534.

[190] G. Medema, L. Heijnen, G. Elsinga, R. Italiaander, A. Brouwer, Presence of SARS-Coronavirus-2 in Sewage, 2020.

[191] A. Tang, Z. Tong, H. Wang, Y. Dai, K. Li, J. Liu, W. Wu, C. Yuan, M. Yu, P. Li, J. Yan, Detection of novel coronavirus by RT-PCR in stool specimen from asymptomatic child, China, Emerg. Infect. Dis. 26 (6) (2020) 1337–1339.

[192] World Health Organization & United Nations Children's Fund (UNICEF), Water, Sanitation, Hygiene, and Waste Management for the COVID-19 Virus: Interim Guidance, World Health Organization, 2020.

[193] F. Wu, J. Zhang, A. Xiao, X. Gu, W. Lee, F. Armas, K. Kauffman, W. Hanage, M. Matus, N. Ghaeli, N. Endo, C. Duvallet, M. Poyet, K. Moniz, A. Washburne, T. Erickson, P. Chai, J. Thompson, E. Alm, SARS-CoV-2 titers in wastewater are higher than expected from clinically confirmed cases, mSystems 5 (4) (2020).

[194] F. Wu, S. Zhao, B. Yu, Y. Chen, W. Wang, Z. Song, Y. Hu, Z. Tao, J. Tian, Y. Pei, M. Yuan, Y. Zhang, F. Dai, Y. Liu, Q. Wang, J. Zheng, L. Xu, E. Holmes, Y. Zhang, A new coronavirus associated with human respiratory disease in China, Nature 579 (7798) (2020) 265–269.

[195] Y. Wu, C. Guo, L. Tang, Z. Hong, J. Zhou, X. Dong, H. Yin, Q. Xiao, Y. Tang, X. Qu, L. Kuang, X. Fang, N. Mishra, J. Lu, H. Shan, G. Jiang, X. Huang, Prolonged presence of SARS-Cov-2 viral RNA in faecal samples, Lancet Gastroenterol. Hepatol. 5 (5) (2020) 434–435.

[196] Y. Ye, R. Ellenberg, K. Graham, K. Wigginton, Survivability, partitioning, and recovery of enveloped viruses in untreated municipal wastewater, Environ. Sci. Technol. 50 (10) (2016) 5077–5085.

[197] W. Ahmed, N. Angel, J. Edson, K. Bibby, A. Bivins, J. O'Brien, P. Choi, M. Kitajima, S. Simpson, J. Li, B. Tscharke, R. Verhagen, W. Smith, J. Zaugg, L. Dierens, P. Hugenholtz, K. Thomas, J. Mueller, First confirmed detection of SARS-Cov-2 in untreated wastewater in Australia: a proof of concept for the wastewater surveillance of COVID-19 in the community, Sci. Total Environ. 728 (2020) 138764.

[198] S. Wurtzer, V. Marechal, J. Mouchel, Y. Maday, R. Teyssou, E. Richard, J. Almayrac, L. Moulin, Evaluation of Lockdown Impact on SARS-CoV-2 Dynamics Through Viral Genome Quantification in Paris Wastewaters, 2020.

[199] A. Carducci, I. Federigi, D. Liu, J. Thompson, M. Verani, Making waves: coronavirus detection, presence and persistence in the water environment: state of the art and knowledge needs for public health, Water Res. 179 (2020) 115907.

[200] X. Wang, J. Li, T. Guo, B. Zhen, Q. Kong, B. Yi, Z. Li, N. Song, M. Jin, W. Xiao, X. Zhu, C. Gu, J. Yin, W. Wei, W. Yao, C. Liu, J. Li, G. Ou, M. Wang, T. Fang, G. Wang, Y. Qiu, H. Wu, F. Chao, J. Li, Concentration and detection of SARS coronavirus in sewage from Xiao Tang Shan hospital and the 309Th hospital of the Chinese People's Liberation Army, Water Sci. Technol. 52 (8) (2005) 213–221.

[201] E. Christensen, M. Myrmel, Coagulant residues' influence on virus enumeration as shown in a study on virus removal using aluminium, zirconium and chitosan, J. Water Health 16 (4) (2018) 600–613.

[202] H. Wei, Z. Banhe, C. Wuchun, et al., The inactivation effect of photocatalytic titanium apatite filter on SARS virus, Prog. Biochem. Biophys. 31 (11) (2004) 982–985.

[203] J. Han, L. Chen, S.M. Duan, Q.X. Yang, M. Yang, C. Gao, B.Y. Zhang, H. He, X.P. Dong, Efficient and quick inactivation of SARS coronavirus and other microbes exposed to the surfaces of some metal catalysts, Biomed. Environ. Sci. 18 (3) (2005) 176–180.

[204] S. Loeb, C. Li, J. Kim, Solar photothermal disinfection using broadband-light absorbing gold nanoparticles and carbon black, Environ. Sci. Technol. 52 (1) (2017) 205–213.

[205] C. Dunnill, I. Parkin, Nitrogen-doped Tio2 thin films: photocatalytic applications for healthcare environments, Dalton Trans. 40 (8) (2011) 1635–1640.

[206] J. Byrne, P. Dunlop, J. Hamilton, P. Fernández-Ibáñez, I. Polo-López, P. Sharma, A. Vennard, A review of heterogeneous photocatalysis for water and surface disinfection, Molecules 20 (4) (2015) 5574–5615.

[207] J. Bogdan, J. Zarzyńska, J. Pławińska-Czarnak, Comparison of infectious agents susceptibility to photocatalytic effects of nanosized titanium and zinc oxides: a practical approach, Nanoscale Res. Lett. 10 (1) (2015).

[208] M. Liga, E. Bryant, V. Colvin, Q. Li, Virus inactivation by silver doped titanium dioxide nanoparticles for drinking water treatment, Water Res. 45 (2) (2011) 535–544.

[209] M. Behnajady, H. Eskandarloo, Silver and copper co-impregnated onto TiO_2-P25 nanoparticles and its photocatalytic activity, Chem. Eng. J. 228 (2013) 1207–1213.

[210] G. Rao, K. Brastad, Q. Zhang, R. Robinson, Z. He, Y. Li, Enhanced disinfection of *Escherichia coli* and bacteriophage MS2 in water using a copper and silver loaded titanium dioxide nanowire membrane, Front. Environ. Sci. Eng. 10 (4) (2016).

[211] M. Liga, S. Maguire-Boyle, H. Jafry, A. Barron, Q. Li, Silica decorated TiO_2 for virus inactivation in drinking water—simple synthesis method and mechanisms of enhanced inactivation kinetics, Environ. Sci. Technol. 47 (12) (2013) 6463–6470.

[212] E. Kozlova, A. Safatov, S. Kiselev, V. Marchenko, A. Sergeev, M. Skarnovich, E. Emelyanova, M. Smetannikova, G. Buryak, A. Vorontsov, Inactivation and mineralization of aerosol deposited model pathogenic microorganisms over TiO_2 and Pt/TiO_2, Environ. Sci. Technol. 44 (13) (2010) 5121–5126.

Chapter 3

Nanosensors for water quality monitoring

Tanvi Gupte[a,b] and Thalappil Pradeep[a,*]
[a]*DST Unit of Nanoscience (DST UNS) and Thematic Unit of Excellence, Department of Chemistry, Indian Institute of Technology Madras, Chennai, India,*
[b]*Department of Metallurgical and Materials Engineering, Indian Institute of Technology, Chennai, India*
**Corresponding author: e-mail address: pradeep@iitm.ac.in*

Significance of water quality monitoring

Water is the essence of life on earth. Yet, the availability of water of appropriate quality in adequate quantity is an ever-increasing concern due to a number of issues, including pollution of water bodies. Robust technologies are being developed for water quality monitoring and purification. Identification and quantitation of contaminant(s) are essential aspects of controlling water pollution. According to UNICEF and WHO, the proportion of population in the least developed countries using limited water services doubled between 2000 and 2017 [1]. Organic pollutants, pharmaceuticals, heavy metals, pathogens, explosives, pesticides, and other such water contaminants have posed an acute threat to the ecosystem worldwide. Key findings of the WHO report attribute 3.3% of deaths and 4.6% of disability-adjusted life years to the effects of inadequate water, sanitation, and hygiene for the year 2016 at a global scale [2].

Hence, one of the global targets for UN Sustainable Development Goals (SDGs) is the availability of safe and affordable drinking water for all by 2030. Some of the significant challenges for achieving these goals are increasing population, climate change, expanding cities, rapid industrial development, improper waste management, eutrophication of water bodies, etc. In these emerging global development conditions, prompt pollutant recognition has become the need of the hour.

Water quality plays a significant role in human health and economic growth, from household to global scale. Hence, the accurate and timely evaluation of its quality is essential for initiating remedial strategies to ensure that the contaminants are at acceptable levels and prevent them from causing environmental catastrophes [3]. Water quality monitoring provides information for regulatory measures and surveillance. Water is expected to contain small amounts of contaminants, and in fact, some of them such as essential and trace minerals are important for life. However, certain toxic substances pose a threat to human health even at low concentrations. The maximum acceptable level of each contaminant is different. Therefore, the development of sensitive water quality sensors for the detection of trace pollutants is essential.

Water quality sensors

A sensor is broadly defined as a device that detects events or changes in its physical environment. Information is relayed as data to other devices, such as a computer. Water quality sensors are essential for ensuring availability and accessibility for quality water for human activities ranging from drinking to agricultural and industrial purposes. The sensors can be located at the point of use, at the water treatment facilities, or within the water distribution system. The information from such sensors is vital for water management issues such as meeting the required regulations for water quality, industrial and municipal wastewater treatment (e.g., monitoring specific target compounds), agricultural irrigation, identification of non-compliant water bodies, and establishing an early warning system (e.g., leakage detection).

Monitoring water quality requires qualitative and quantitative measurement of various pollutants. However, developing such a sensor is a challenging task because of the diversity and complexity of the water samples and trace amounts of contaminants that need to be detected. In addition, several aspects of the water such as temperature, pH, dissolved solids and gases, etc. can have an impact on the chemical state and concentration of the contaminants. Hence, physical, chemical, and biological parameters need to be taken into consideration for water quality monitoring. Physical parameters include color, suspended particles, turbidity, density, conductivity, temperature, and total dissolved solids. Chemical parameters

Separation Science and Technology, Volume 15. https://doi.org/10.1016/B978-0-323-90763-7.00010-X

may be organic and inorganic (pH, dissolved oxygen, hardness, disinfectants, phosphates, nitrogen content, micronutrients, and other metal contaminants). Biological parameters include various microorganisms such as algae, bacteria, and other pathogens [4]. Current sensing technologies are mainly based on colorimetry, electrochemistry, or lab-based spectroscopic methods. Traditional methods for water quality monitoring are largely dependent on laboratory analysis, complex instrumentation, and computing infrastructure. They require frequent calibration and are often not suitable for continuous sampling [4]. Thus, there is a high demand for sensitive, simple, cost-effective, and portable *in situ* operating sensors for real-time monitoring of water quality. Incorporation of nanotechnology and innovative techniques in fabrication of sensors is proving to be a promising tool. Sensors for monitoring water quality are an essential emerging field in nanosensor technology research. Nanosensors with the capabilities to provide quantitative information comparable to conventional analysis methods along with the advantage of miniaturization of devices are an effective alternative.

Nanosensors

Nanosensors are sensing devices that utilize nanomaterials as recognition elements to detect changes. In recent years, nanomaterials-based sensors have gained popularity. They provide high sensitivity, specificity, a large surface area to volume ratio, quantum confinement effects, rapid detection, and tunable magnetic, electrical, and optical properties [5]. They have wide applications in detecting toxic gases, food packaging, healthcare diagnostics, and water quality monitoring. This chapter focuses on the recent advances and techniques in nanosensing technology, emphasizing research toward real-life applications that can replace traditional water quality evaluation methods.

Types of water quality nanosensors based on the signal transduction mechanism employed for sensing are discussed in this chapter. These include electrochemical, optical, and magnetic nanosensors. Biomimetic and hybrid nanosensing technologies are also discussed briefly. Various suitable nanomaterial platforms such as noble metal nanostructures, quantum dots, magnetic nanoparticles (MNPs), metal–organic frameworks (MOFs), clusters, enzymes, molecularly imprinted polymers, aptamers, carbon nanomaterials, etc., are discussed concurrently. Table 1 enlists applications of various nanosensors covered in the chapter for monitoring water quality.

Electrochemical nanosensors

Electrochemical nanosensors are sensing devices that measure the change in current, voltage, conductance or impedance resulting from a reaction, using nanomaterials as recognition elements that convert the signal to digital information. Typically, an electrochemical sensor consists of three electrodes—a working electrode, a reference electrode, and a counter (auxiliary) electrode. The working working (WE) serves as the surface where the reaction/process of interest occurs. Different electrochemical techniques are amperometry, voltammetry, potentiometry, impedance spectroscopy, and conductometry (capacitive and resistive). Because of the different detection mechanisms involved, information on various aspects of an electrochemical reaction is gained from different techniques [6]. Electrochemical sensing methods are sensitive and reliable because of because of their advantages such as low cost, lower limit of detection, quick response, and portability. They are promising candidates for making real-time and in-process monitoring and detection of trace amounts of water pollutants feasible [7,8].

The geometry, design, and configuration of the working electrode play an important role in electrochemical sensors [9]. A combination of nanotechnology and electrochemistry can produce robust and reliable electrical instruments for monitoring environmental contaminants [10]. Usually, the WE is miniaturized to provide portability for on-field sensing. A nanoelectrode (NE) or a WE electrode that is modified using nanomaterial platforms and surface modifications provides improved conductivity, sensitivity, and surface area to detect target contaminant(s) [7]. Screen printing is a fabrication technique that promotes miniaturization of electrochemical sensing devices. A screen printed electrode (SPE) integrates the three electrodes on an insulated solid substrate made of polyvinyl chloride, polyethylene terephthalate, ceramic, etc.) [11]. SPE is typically fabricated using layer-by-layer deposition of ink onto the substrate. The technique involves an ink-blocking stencil supported on a woven mesh. The ink and other printable modifiers are pumped across the stencil by a roller or squeegee. The ink is usually allowed to solidify by thermal treatment between printing of successive ink layers. Commonly used inks include carbon, platinum, gold, and silver [12]. The thickness of an SPE is usually a few microns which is easily controlled by the number of ink layers and the amount of printed ink [11]. Suitable modifiers such as catalysts, nanomaterials, metallic nanocomposites, organic compounds, etc. may be used along with the ink (bulk modified recognition systems) or included on the electrode surface. This increases the effective surface area, absorption capability, electrical conductivity, selectivity and sensitivity [9,13]. The modifications for SPEs may also be in the form of specific electrode design or a suitable solid substrate [9]. Fig. 1A shows the typical design of an SPE used in electrochemical sensing.

TABLE 1 Applications of various nanosensors for monitoring water quality.

Sr. No	Transduction mechanism	Sensor material	Analyte	Limit of detection	References
1.	Electrochemical	Choline oxidase enzyme and MWCNT	Pb^{2+}	0.0083 ppb	[16]
		Graphene conjugated with CdSe QDs, HRP-labeled Au NPs, and specific monoclonal antibody on a GCE	Hg^{2+}	0.06 ppb	[17]
		$HS-(CH_2)_2-[OCH_2CH_2]_6-OCH$ self-assembled monolayer on an aptasensor	Ampicillin and sulfadimethoxine drugs	10 pM and 1 nM, respectively	[18]
		Immobilized aptamers and IC-DNA-labeled Au NPs on rGO and Au dendritic nanostructures	Hg^{2+}, Cd^{2+}, and As^{3+}	2.0, 0.62, and 0.17 pM, respectively	[37]
		DNA-modified Fe_3O_4@Au MNPs on a magnetic GCE	Ag^+ and Hg^{2+}	3.4 nM (0.37 ppb) and 1.7 nM (0.34 ppb), respectively	[41]
		Polypyrrole on pencil graphite electrode	Dibutyl phthalate	4.5 nM	[21]
		$CuO-TiO_2$ hybrid nanocomposite on a GCE	Methyl parathion	1.21 ppb	[22]
		Poly(3,4-ethylene-dioxythiophene) nanorods/GO nanocomposite on a GCE	Hg^{2+}	2.78 nM (0.55 ppb)	[23]
		Amorphous carbon and Bi NPs on an SPE	Pb^{2+} and Cd^{2+}	3.8 and 2.2 ppb, respectively	[45]
		N-Methyl-2-pyrrolidone (NMP) exfoliated graphene nanosheets on a GCE	Hydroquinone, catechol, 4-chlorophenol, and 4-nitrophenol	0.012, 0.015, 0.01, and 0.04 μM, respectively	[47]
		Vertically aligned two-dimensional (2D) MoS_2	Pb^{2+}	0.3 ppb	[50]
		Fe_3O_4 NPs hybridized with S-MWCNTs that were combined with poly (2,6-pyridine dicarboxylic acid) on a GCE	Omeprazole	15 nM	[24]
		Au NPs/α-MnO_2	As^{3+}	0.019 ppb	[55]
2.	Optical	44-Mercaptopyridine functionalized Au NPs	Hg^{2+}	8 nM (1.60 ppb)	[60]
		Monoclonal cholera toxin antibody-conjugated Au NPs modified with mercaptopropionic acid	Cholera toxin	10 nM	[58]
		Silver dendrite SERS substrate	Ferbam pesticide	0.861 ppm	[64]
		Tetraphenylporphyrin tetrasulfonic acid (TPPS) and black phosphorus QDs	Hg^{2+}	0.39 nM (0.078 ppb)	[67]
		Nitrogen-doped carbon QDs	Humic substances	0.2 mg/L	[69]

Continued

TABLE 1 Applications of various nanosensors for monitoring water quality.—cont'd

Sr. No	Transduction mechanism	Sensor material	Analyte	Limit of detection	References
		Fe_3O_4 microspheres and Au NPs using SERS	16 polycyclic aromatic hydrocarbons (PAHs)	100–5 nM	[73]
		Gold-coated magnetic nanosphere–anti-*Escherichia coli* antibody complex and Raman label	*E. coli*	8 cfu/mL	[71]
		Core–shell MOF@Ag NPs on screen-printed carbon electrodes (SPCEs)	PAHs	pM	[88]
3.	Magnetic nanosensors	Schiff base/silica/magnetite as preconcentration phase	Pb^{2+}, Cd^{2+} and Cu^{2+},	0.14, 0.19 and 0.12 $\mu g L^{-1}$, respectively	[89]
4.	Biomimetic and hybrid nanosensing	Plant with lead-sensitive DNA and CNTs in leaves	Pb^{2+}	—	[90]
		Nanoswimmer: PEDOT/Au bilayer microtubes with catalase enzyme	Mercury, copper, herbicide-aminotriazole, and pesticide-NaN_3	Micromolar levels	[93]

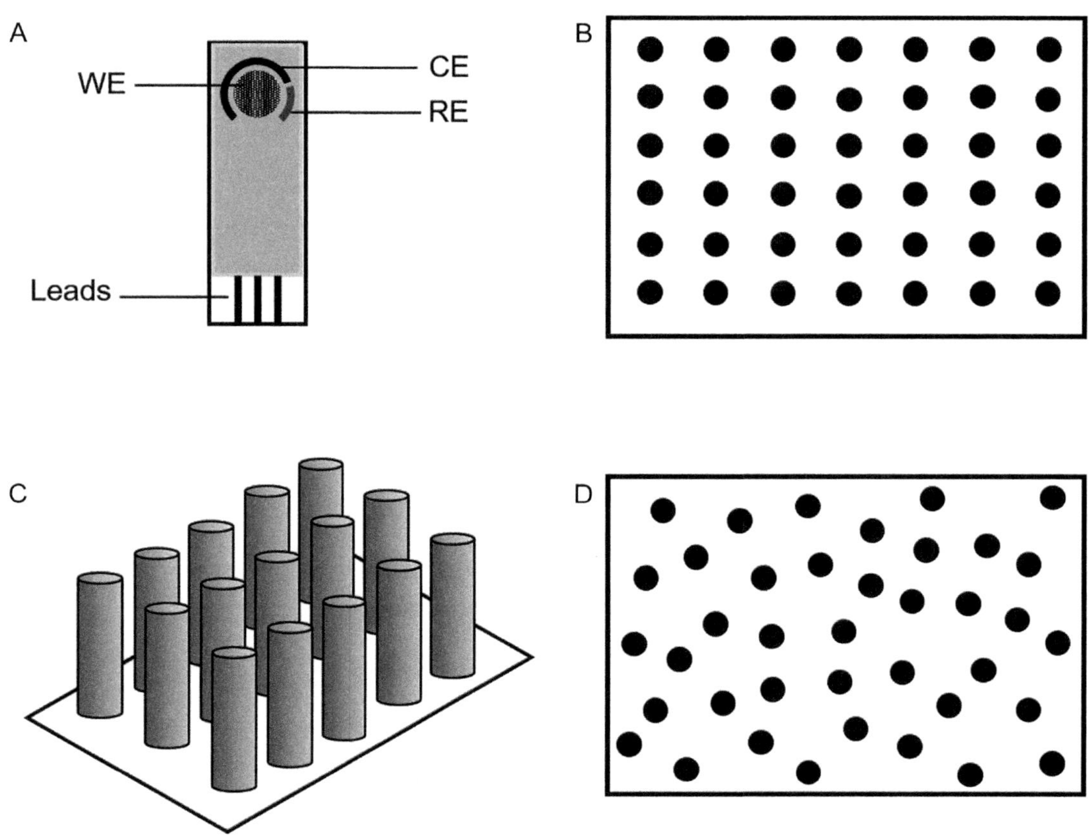

FIG. 1 Examples of typical geometries of a few electrochemical electrodes: (A) screen-printed electrode (SPE), (B) square nanoelectrode array (NEA), (C) three-dimensional nanowire NEA, (D) randomly arranged NE ensemble (NEE). WE: working electrode, CE: counter electrode, and RE: reference electrode.

This fabrication technique produces mechanically stable, cost-effective, and disposable electrodes. SPEs can be used easily with miniaturized, portable electrochemical sensing instruments for on-field detection of water contaminants [12].

Electrochemical nanosensors can also be in the form of an individual NE, an electrode with at least one dimension (e.g., radius for discs or nanowires and thickness for lines) of approximately 50 nm. Typical geometries of an NE are NE arrays (NEA) and NE ensemble (NEE), as shown in Fig. 1B–D. NEAs are perfectly ordered sets of NEs that have single as well as multianalyte measurement capabilities (Fig. 1B and C). NEEs are randomly arranged groups of NEs that can be used to obtain spatial distribution measurements for a single analyte (Fig. 1D) [6,14]. Multiplexed designs offer the possibility of differently modifying individual electrodes in an array to detect different target analytes [9].

Different nanomaterial platforms used for electrode modification include biorecognition elements (e.g., enzymes [15,16], nucleic acids, cells, antibodies [17], aptamers [18], microorganisms, etc.) [19], molecularly imprinted polymers (MIPs) [20,21], nanocomposites [22,23], nanoparticles [24], nanostructures [25], carbon nanomaterials such as carbon nanotubes [15,26–28], carbon nanofibers, graphene, etc. Using nanomaterials in electrochemical sensing provides advantages such as greater mass transport, high current densities, low background charging current, and an enhanced surface-to-volume ratio [14,20]. This results in higher sensitivity and lower detection limits [14].

The electrode surface modification can be in the form of adsorption, covalent bonding, electrochemical deposition, and electrochemical polymerization. In the adsorption type of modification, noncovalent interactions exist between the electrode surface and the modifier, mainly as coatings, chemical adsorption, or self-assembled monolayers. Electrochemical deposition and polymerization surface modification methods involve applying a voltage to the WE in an electrolyte containing the preferred modifier. This results in the formation of polymeric or metallic films on the surface. In the case of SPEs, the preferred material is added in the ink [9].

Electrochemical nanosensors have been used for detection of various water contaminants such as heavy metals (e.g., Pb^{2+}, Cd^{2+}, As^{3+}, etc.) [7,25,28], pharmaceuticals [20,24], pesticides [22,29], etc. Moreover, portable and low-cost potentiostats are now widely available commercially that are useful for *in situ*/online/point-of-care analysis [9]. Some of the widely used nanomaterial platforms employed in electrochemical nanosensors are described in brief in the following subsections.

Electrochemical biosensors

Electrochemical biosensors are a class of nanosensors that use various biorecognition elements such as enzymes, antibodies, aptamers, nucleic acids, cells, etc. to detect analytes electrochemically. Fig. 2 shows the schematic of a typical electrochemical biosensor comprising a biorecognition element (which allows selective detection of specific analyte(s)) and other components like transducers, amplifiers, processors, and displays. Enzyme, antibody, aptamer, and DNA-based electrochemical biosensors to detect water contaminants with suitable examples are briefly described.

Enzymes

Enzymes have been widely used for the detection of heavy metal ions, as the interaction causes a change in the structure and properties of the enzyme. This leads to an activation or inhibition effect on enzyme activity in the presence of specific ions. Commonly used enzymes in electrochemical metal ion sensors include oxidase, peroxidase, and urease [30]. An amperometric choline biosensor has been developed to detect Pb^{2+} using choline oxidase enzyme (ChOx) and multiwalled carbon nanotubes (MWCNT). In the presence of Pb^{2+} ions, the enzyme activity (oxidization of choline into betaine) was inhibited, causing a decrease in the choline oxidation current. The detection limit was 0.04 nM (0.0083 ppb) with a high selectivity for Pb^{2+} ions, and the detection was successful in tap water samples [16].

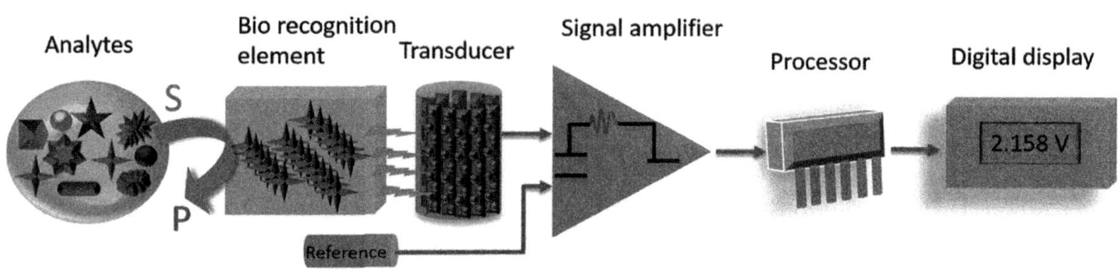

FIG. 2 Schematic illustration showing the key components of an electrochemical biosensor. Here, the biorecognition element converts the substrate (S) to the product (P). The transducer converts the biological response into an electrical signal. The output is then amplified and processed by a transistor processor, and, finally, the output is displayed on a computer. *From H.S. Magar, M.E. Ghica, M.N. Abbas, C.M.A. Brett, Highly sensitive choline oxidase enzyme inhibition biosensor for Lead ions based on multiwalled carbon nanotube modified glassy carbon electrodes, Electroanalysis 29 (7) (2017) 1741–1748. doi:10.1002/elan.201700111.*

Antibodies

Antibody-based electrochemical nanosensors are centered around the conformational modifications produced by the bio-recognition interaction between the antibody and the antigen which are detected by a transducer. They are also termed electrochemical immunosensors. The immunochemical reactions are converted into an electrical signal such as current, potential difference, or resistivity change. The antibody is usually immobilized onto the electrode surface. An antibody (Ab) monomer is a globular plasma protein made of four polypeptide chains—two light chains (~25 kDa each) and two heavy chains (~50 kDa each) which are connected by disulfide bonds. Each of these chains has a constant and a variable region and forms a characteristic "Y" shape. The variable regions determine the selectivity and binding to specific antigens or analytes. Various immobilization techniques include microcontact printing, direct spotting, covalent binding, and adsorption to a conductive polymer matrix [31]. The performance of these sensors is usually enhanced by modifying the working electrode with electrically conductive nanomaterials. This helps in the amplification of their signal and immobilization of more antibodies [32]. In the case of heavy metal ion detection, the change in the current on the electrode surface due to oxidation and reduction of selectively adsorbed species is quantified. An ultrasensitive electrochemiluminescent (ECL) immunoassay for mercury detection has been developed based on graphene conjugated with CdSe quantum dots, horseradish peroxidase (HRP)-labeled gold nanoparticles (Au NPs), and specific monoclonal antibodies (mAbs) against Hg^{2+}. The detection was performed using a GCE and cyclic voltammetry. The immunosensor exhibited good stability, accuracy, and acceptable reproducibility in environmental samples with a detection limit of 0.06 ppb [17]. The high specificity and sensitivity aspects of antibody-based electrochemical sensors have been helpful in the detection of contaminants such as pharmaceuticals [33,34], toxic microorganisms such as microalgae [35], pesticides, etc. However, immunosensors have certain drawbacks, including the high production costs of antibodies, instability, and difficulty with multiplexing [32].

Aptamers

Electrochemical aptamer-based nanosensor platforms have received significant attention in the past decade. Aptamers are single-stranded oligonucleotides or peptides that show high affinity and specificity to their targets [36]. These are usually immobilized on an electrode substrate. Aptamer-based nanosensors interact in two ways with the target analyte. The interaction can result in structural modification or a change in both flexibility and conformation of the immobilized aptamer. In the first type of interaction, the electron transfer resistance between the electrode surface and the redox-active species in the solution is altered. In the second case, the sensing is based on either an increase or a decrease in the redox current. Aptamers can be labeled with electrochemical and fluorescent reporter tags and are stable in a wide range of experimental conditions [36]. Yang et al. reported a signaling-probe displacement electrochemical aptasensor (EAS). The sensing mechanism in signaling-probe displacement sensor is based on the theoretical assumption that the target analyte will cause displacement and release of the signaling probe (ferrocene in this experiment) from the electrode surface, leading to a decrease in the current signal. The researchers passivated the aptasensor surface with $HS\text{-}(CH_2)_2\text{-}[OCH_2CH_2]_6\text{-}OCH$ self-assembled monolayer which enabled picomolar detection of antibacterial drugs (ampicillin and sulfadimethoxine). This is the first reported EAS that works well for both hydrophilic and lipophilic targets. They also demonstrated the efficacy in lake water samples [18]. Tang et al. developed an aptasensor array using a multichannel disposable screen-printed carbon electrode (SPCE) for the simultaneous detection of three heavy metal ions (Hg^{2+}, Cd^{2+}, and As^{3+}) by open circuit potential (OCP) technology. The sensor had an internal calibration DNA sequence (IC-DNA) to correct the background signal generated by the detection system. The array was first modified with reduced graphene oxide (rGO) and dendritic gold nanostructures followed by immobilization of aptamers and IC-DNA-labeled Au NPs. The aptasensor array showed high sensitivity toward Hg^{2+}, Cd^{2+}, and As^{3+} with detection limits of 2.0, 0.62, and 0.17 pM, respectively, without mutual interference or interference from other ions. Moreover, the array was successful in simultaneous detection of the three ions in real water samples, and the results accurately matched with the data obtained from inductively coupled plasma mass spectrometry [37].

Deoxyribonucleic acid (DNA)

Deoxyribonucleic acid (DNA) as a bio-recognition element for water quality assessment has vast potential because of its properties such as selectivity, sensitivity, biodegradability, cost-effectiveness, portability, and easy manipulation [38,39]. Also, for certain analytes, electrochemical DNA nanosensors produce direct electronic signals overcoming the need for expensive signal converting instruments. Moreover, these sensors work efficiently in turbid or colored samples too [39]. Also, various strategies exist to amplify signals of electrochemical DNA biosensors to achieve ultrasensitivity, e.g., nicking enzyme-assisted amplification by using various endonucleases. This strategy allows rapid exponential amplification of a target nucleic acid. Other techniques utilize core-shell nanoparticles and nanocomposites (e.g., Au NPs, graphene, carbon nanotubes (CNTs)) for surface modification and signal amplification [38]. Electrochemical DNA sensors are utilized for sensing heavy metal ions, pharmaceuticals, and microorganisms (e.g., coliform bacteria, *E. coli*). Detection of heavy metal ions depends on selective and specific binding of the ions with distinct DNA bases, resulting in a stable DNA-duplex [39]; for example, thymine-Hg^{2+}-thymine (T-Hg^{2+}-T) mismatch, cytosine-Ag^{+}-cytosine (C-Ag^{+}-C) mismatch, and Pb^{2+}-stabilized G-quadruplex [38]. Heavy metal ions may also facilitate the cleavage of deoxyribozymes (DNAzymes) [39]. DNAzymes are artificial single-stranded DNA sequences that can cleave RNA. They possess a substrate and an enzyme strand. The substrate strand has a ribonucleoside adenosine at the cleavage site, while the enzyme strand shows catalytic activity in the presence of the analyte [40]. An electrochemical DNA sensor has been developed based on DNA-modified Fe_3O_4@Au MNPs on a magnetic glassy carbon electrode for detection of heavy metal ions. Two electrochemically active species, ferrocene (Fc) and methylene blue (MB), were used as labels for the simultaneous detection of Ag^{+} and Hg^{2+} by square wave voltammetry. The detection limit was in the nanomolar range for both the ions, and the sensor worked efficiently in lake and drinking water. The developed biosensor can be translated into a universal platform for the detection of other heavy metal ions [41].

Electrochemical molecularly imprinted polymer (MIP) nanosensors

Molecularly imprinted polymers (MIPs) are also termed plastic antibodies or bio-mimetic receptors. They are produced by the template-induced formation of specific recognition sites in a polymer. The synthesis of MIPs includes components such as monomers, cross-linkers, initiators, and solvents. The monomers polymerize in the presence of the target/imprint molecule, which acts as a template to form recognition sites. Removal of the imprint molecule yields a highly cross-linked polymeric scaffold with selective, non–covalent-binding sites/cavities for the target analyte [9]. MIPs can be synthesized in various ways, such as polymerization (bulk, suspension, or precipitation), self-assembled electropolymerization, soft lithography, and spin coating [21]. The affinity and selectivity properties of synthesized receptors are comparable to the natural receptors. They possess several advantageous properties such as enhanced stability, robustness, low-cost and straightforward production, scalability, and ease of adapting the template-target molecule system to practical applications in assays and sensors, thus making them a viable alternative to natural receptors [9,42]. These are also compatible with SPEs [42]. MIP sensors are applied to the determination of heavy metals, biomolecules, pharmaceuticals, explosives, etc. [9,42]. Currently, nanoparticles (NPs) and nanostructured coatings are being widely explored for MIP sensing technology [42]. Bolat et al. developed a MIP nanosensor for the quantification of dibutyl phthalate (DBP). DBP is one of the widely used phthalate esters (PEs), which are emergent pollutants for water, soil, and atmosphere. The synthesis of the sensor was a single-step process involving electrochemical polymerization of pyrrole on the pencil graphite electrode (PGE) surface. Electrochemical impedance spectroscopy (EIS) was used for the detection of DBP with a detection limit of 4.5 nM [21]. Despite widespread research and several advantageous features of MIP nanosensors, there are a few fundamental problems that need to be addressed before commercial applications are feasible. Some of the issues include improved stability, cost-effectiveness, repeatability, accuracy, calibration, portability, and multianalyte detection in complex real water samples.

Nanomaterials-based electrochemical nanosensors

Significant advancements have been made in nanomaterials-based electrochemical sensor platforms, resulting in new technologies to ensure environmental and human safety. To develop highly sensitive electrochemical nanosensors, researchers have utilized various nanomaterials such as nanocomposites, nanoparticles, metal-oxides, two-dimensional (2D), and carbon nanomaterials to modify the surfaces of the electrodes for the detection of environmental pollutants.

Nanocomposites

Nanocomposites are multiphase solid materials with at least one component having nanoscale dimensions (termed nanofillers). The constituent materials possess significantly different physical or chemical properties; however, their combination yields a material with distinct characteristics, which differ from those of individual components. Commonly used nanofillers are 0D particles (fullerene), 1D nanofibers (carbon nanotubes), and 2D nanosheets (graphene and its oxygenated derivatives, boron nitride) [43]. A diverse range of nanocomposites can be fabricated by combining nanofillers with semiconducting nanoparticles, metals, metal oxides, polymers, quantum dots, hydrogels, and biomolecules. Nanocomposites constitute emerging electrochemical and fluorescent sensing platforms [44]. Nanocomposite-based electrochemical sensors are robust, highly sensitive, low cost, portable, decentralized, reliable, rapid, and can exhibit detection limits in the picomolar range. An example of such a sensor involves CuO-TiO_2 hybrid nanocomposites on the GCE for the sensitive and selective detection of methyl parathion (an organophosphorus pesticide) with a limit of detection of 1.21 ppb.

The sensor showed efficient sensing ability in the presence of interfering materials and groundwater samples [22]. Conducting polymers have also been extensively investigated for their use in nanocomposite electrochemical sensors. Polymer nanocomposites are comprised of copolymer/polymer matrix or nanofillers dispersed in the polymer matrix. They offer distinct properties such as high electrical conductivity, rapid electrode kinetics, high chemical stability in aqueous solutions, and environmental stability. They have a wide range of applications in analyzing several pollutants (e.g., pesticides, heavy metals, nitrite, nitroaromatic compounds, phenolic compounds, hydrazine, etc.). The most commonly used methods for synthesizing polymer nanocomposites are *in situ* polymerization, direct mixing of polymer and nanofillers (melt-mixing and solution-mixing), sol–gel, and electrochemical synthesis [10]. Researchers have developed an electrochemical Hg^{2+} sensor using a poly(3,4-ethylene-dioxythiophene) nanorods/GO nanocomposite-modified GCE (PEDOT/GO/GCE). PEDOT polymer with nanorod-like structures attached to the surface of GO nanosheets helped enhance the nanocomposite's electroactive sites. Differential pulse stripping voltammetry (DPSV) was used for the analysis. The detection limit reported was 2.78 nM (0.55 ppb) at a signal-to-noise ratio of 3. The applicability of the developed sensor was successfully demonstrated in tap water samples [23]. Major challenges with polymeric nanocomposite-based electrochemical sensors are dispersing nanofillers in the polymer matrix uniformly and enabling a strong interaction at their interface. The dispersion issue can be addressed by surfactants, physical treatment, and chemical functionalization of the surface of nanofillers [23]. Researchers have also used disposable, miniaturized SPEs for nanocomposite -based electrochemical water sensors. Niu et al. synthesized a nanocomposite made of bismuth nanoparticles and amorphous carbon. Ink was prepared by milling the prepared nanocomposite to a specific particle-size distribution suitable for screen printing. High surface area imparted by porous carbon and Bi NPs enabled the detection of trace concentrations of Pb^{2+} and Cd^{2+} ions (3.8 and 2.2 ppb, respectively) using square wave anodic stripping voltammetry technique (SWASV). The performance of these SPEs was successfully tested in tap water and wastewater from an urban wastewater treatment plant [45].

Two-dimensional (2D) nanomaterials

Two-dimensional (2D) nanomaterials such as graphite-carbon nitride (g-C_3N_4) nanosheet, boron nitride (BN) nanosheet, graphene, and various transition metal dichalcogenides (TMDs) (MoS_2, $MoSe_2$, WS_2, etc.) possess unique chemical and physical properties, which have been explored to monitor the environmental pollutants [46]. Two-dimensional nanomaterials have high surface area, chemical stability, good electrical conductivity, easy functionalization, and mechanical strength. In several instances, nanohybrids are fabricated by introducing metal oxides, noble metal NPs, or organic compounds on the surface of 2D nanomaterials to obtain sensors with better synergistic properties. Several 2D nanomaterials-based electrochemical sensors have been developed for the detection of heavy metal ions (e.g., Pb^{2+}, Cd^{2+}), pharmaceuticals (e.g., metronidazole, kanamycin), pesticides (e.g., chlorpyrifos, carbofuran), organic compounds (e.g., 4-chlorophenol, 4-nitrophenol, bisphenol A), bacteria (e.g., *Staphylococcus aureus*, *Staphylococcus arlettae*, *E. coli, Shewanella oneidensis, Salmonella*), etc. [46]. Wu et al. developed a sensor to detect phenols using single-step *N*-methyl-2-pyrrolidone (NMP) exfoliated graphene nanosheets on GCE as a sensing platform. Such NMP-exfoliated graphene nanosheets exhibited enhanced electrochemical activity compared to multistep chemical oxidation synthesis. Moreover, the electrochemical performance was further improved after anodization because of increased oxygen-related active defect sites and amplified effective surface area resulting from increased surface roughness. The developed sensor showed detection limits of 0.012, 0.015, 0.01, and 0.04 μM for hydroquinone, catechol, 4-chlorophenol, and 4-nitrophenol, respectively [47]. Similarly, many MoS_2-based nanomaterials have also been used as sensors to analyze water contaminants, e.g., MoS_2/polyaniline nanocomposite to determine chloramphenicol [48], Ag nanoprism@MoS_2 for sensing disinfection by-product trichloroacetamide in drinking water [49], vertically aligned 2D MoS_2 (edge exposed) nanofilm for lead ion (Pb^{2+}) detection [50], etc.

Carbon nanomaterials

Electrochemical sensors have been fabricated with carbon-based nanomaterials such as carbon nanotubes, graphene, graphene oxide, carbon nanofibers, and nanocrystalline diamond for highly sensitive and selective detection of various water contaminants. Carbon nanomaterials offer unique physical, chemical and electrical properties like large surface area, high carrier mobility, low residual current, flexibility, readily renewable surface, and ambipolar electric field effect [51]. Geometrical advantage of these materials ensures the exposure of maximum surface atoms, which facilitates binding of a substantial fraction of analytes. Carbon nanomaterials are usually coupled with biorecognition elements, metal oxides, nanoparticles, noble metals, etc. to enhance selectivity, specificity, and sensitivities. Their applications include sensing various water contaminants such as heavy metals, pesticides, pathogens, nitroaromatic explosives, and antibiotics [51]. An electrochemical omeprazole sensor based on shortened multiwalled carbon nanotubes (S-MWCNT) has been developed. Fe_3O_4 nanoparticles were hybridized with S-MWCNTs, which were combined with electropolymerized poly (2,6-pyridine dicarboxylic acid) on GCE for nanomolar detection of omeprazole [24]. Recently, a rapid, fully integrated, inexpensive electrochemical sensor array was reported for *in situ* water quality monitoring. Simultaneous monitoring of pH (57.5 mV/pH), free chlorine (186 nA/ppm), temperature (16.9 mV/°C), heavy metal (<10 ppb), and on-demand monitoring of acetaminophen and 17β-estradiol (<10 nM) was successfully achieved. The researchers merged nanomaterials and CNT-based sensors on microscopic glass slides. The sensor had a custom-made readout circuit, a potentiostat, and an Android app. The integrated system was efficient in monitoring tap, swimming pool, and lake water. Moreover, it can be easily modified to integrate other sensors, enabling multiplexing capability to monitor a range of water quality parameters [52].

In general, carbon nanomaterial-based sensors enable highly selective and sensitive detection of pollutants without the need for labeling or amplification techniques [51].

Nanoparticles and nanostructures

Electrochemical sensors made using nanoparticle-modified electrodes possess high surface area, improved mass-transport and electron transfer rates, high signal-to-noise ratio, and low solution resistance [53]. Metal, metal oxide, polymeric, and noble metal nitrophenols (NPs) can improve binding of analytes and performance of electrochemical sensor for the quantification of environmental samples. NPs also exhibit notably different characteristics from bulk materials (e.g., bulk Au vs Au NP). Particle size and size distribution of NPs are essential for their performance as an electrochemical sensor as it determines the degree of diffusion-layer overlap [54]. NPs are often used in conjunction with other materials such as polymers, carbon-based nanomaterials, biorecognition elements, etc. Au NPs have been widely used to modify bulk electrodes. Au NP-modified electrodes eliminate the memory effect and interferences of other ions and improve the sensitivity and detection limits [53]. An efficient and anti-interference electrochemical sensor for As^{3+} detection has been developed with Au NPs/α-MnO_2. The reported limit of detection was 0.019 ppb, and the sensor was successfully employed for detection of As^{3+} in a real water sample [55].

Optical nanosensors

Optical sensors are dependent on the changes in the optical properties of water. Such sensors are generally preferred for sensing applications as they provide low detection limits, rapid response, selectivity, are unaffected by electromagnetic noise, and have multiplexing potential [56,57]. They can be colorimetric (visible color change), fluorescent, chemiluminescent, surface plasmon resonance sensors, fiber-optic sensors, etc. Various nanomaterial platforms employed in optical sensing include noble metals (e.g., gold [58–62] and silver nanostructures [63–66]), quantum dots [67–70], MNPs [71–73], metal–organic frameworks (MOFs) [74–80], core – shell nanocages [81], and clusters [57].

Noble metals

Noble metal nanostructures possess plasmonic behavior because of their distinct localized surface plasmon resonance (LSPR), which provides the basis for the fabrication of visual and spectroscopic optical sensors. Plasmonic gold (Au NPs) and silver nanoparticles (Ag NPs) possess high extinction coefficients leading to distinct color when dispersed in solution [82]. The LSPR of Au NPs and Ag NPs is sensitive to the shape, composition, size, refractive index, pH of the surrounding environment, and aggregation properties of these plasmonic NPs [3,83,84]. Two types of optical signal transduction methods utilize LSPR: surface-enhanced Raman spectroscopy (SERS) and colorimetric (absorption) methods. The LSPR plays a role in enhancing the electromagnetic field near the NPs when coupled with the wavelength of incident light. This enables surface-enhanced Raman spectroscopy (SERS) based detection. It is possible to achieve multiplex analysis and sensitivity to the level of single molecule and single pathogen detection by SERS. When a molecule is in close proximity or adsorbed on Au NPs or Ag NPs, its Raman cross section is enhanced several folds due to SERS [3]. For colorimetric detection, the change in refractive index of solution upon introduction of analyte molecule results in a shift in LSPR peak. This effect on the refractive index of the medium could be attributed to aggregation or conjugation of NPs. The interaction between the analyte and NPs exists in the form of hydrogen bonding, donor-acceptor reaction, and electrostatic interaction. Aggregation-induced decrease in the interparticle distance leads to overlap between the plasmon fields of neighboring NPs, prompting a red shift of the LSPR band and a strong visible color change. The change in color is from red to blue for Au NPs and yellow to brown for Ag NPs. This property has enabled visual detection and label-free plasmonic sensing [82].

Depending on the application of the sensor, the NPs can be modified with suitable functional groups. Usually, Au NPs are functionalized using thiol, phosphine, disulfide, or isocyanide. Thiol groups can also be used to functionalize Ag NPs. Noble metal nanoparticle sensors have their application to detect bacteria, fungi, viruses, inorganic anions, heavy metal ions, and pesticides [82]. A portable and cost-effective fiber-optic sensor has been developed for Hg^{2+} detection using 4-mercaptopyridine (4-MPY)-functionalized Au nanoparticles (Au NPs/4-MPY). A plasmonic coupling structure was generated by self-assembly of 4-MPY on Au film surface to capture Hg^{2+} followed by introduction of Au NPs/4-MPY. This coupling produced changes in LSPR wavelength leading to highly sensitive and selective detection of Hg^{2+} with a detection limit of 8 nM (1.60 ppb). It was tested for Hg^{2+} detection in tap water and has practical potentials for on-site monitoring [60]. A colorimetric assay has been developed for sensitive detection of cholera toxin using monoclonal cholera toxin antibody-conjugated Au NPs. The Au NP surface was modified with mercaptopropionic acid (MPA) before conjugating with antibody. The toxin binds with antibody-conjugated Au NPs resulting in a network of aggregated NPs. This leads to discernible color change, and the red shift can be detected by UV–Visible spectroscopy. The assay showed high specificity and a detection limit of 10 nM. They also demonstrated the feasibility of the assay to detect cholera toxin in tap and lake water. This method can be extended to any target toxin for which specific antibodies are available [58]. Integrating a SERS sensor into a portable Raman device could facilitate economic, on-site pollutant detection [3,64]. Such an instrument quantifying SERS was fabricated using a silver dendrite SERS substrate for ferbam (pesticide) detection. It showed results in semi-quantitative, layman's format (without showing the spectra), and its use could also be extended to other target molecules [64].

Quantum dots (QDs)

Quantum dots (QDs) are nanoscale semiconductor particles with distinct optical and electronic properties having diameters ranging between 1 and 6 nm. Although they are crystals, they behave like atoms because of their bound, discrete electronic states and show quantum confinement effects. Symmetric emission profiles with narrow bandwidths and size tunability of their emission color contribute to their application as excellent fluorophores for sensors [85]. Their optoelectronic properties depend on composition, size, shape, and surface chemistry [86]. A ratiometric fluorescence sensor has been developed for the selective detection of Hg^{2+} using tetraphenylporphyrin tetrasulfonic acid (TPPS) and black phosphorus quantum dots (BP QDs). Ratiometric fluorescence assays depend on the measurement of the ratio of two signals, hence minimizing the influence of instrument and environmental conditions. The sensing was based on two phenomenons. Inhibition of the inner filter effect originating between BP QDs and TPPS in the presence of Hg^{2+} results in restoration of fluorescence of BP QDs. At the same time, Hg^{2+} causes quenching of the red fluorescence of TPPS due to its coordination with Mn^{2+}. They also tested tap and river water samples with a detection limit of 0.39 nM (0.078 ppb) [67]. Yu et al. developed a sensor to detect humic substances (HS), a key component of dissolved organic matter (DOM) using nitrogen-doped carbon quantum dots. HS are present in the aqueous environment (0.2–20 mg/L) and can influence the dissolution, transformation, and bioavailability of heavy metals. Hence their quantification is essential. The sensor was effective in a wide pH range of 3–12 with a detection range of 0–100 mg/L and detection limit of 0.2 mg/L (more efficient than the frequently used modified Lowry method). They validated the feasibility of the sensor in natural surface water samples. They also fabricated a carbon dot-based portable test paper for successful quantification of HS, which, when coupled with a light-emitting diode lamp, can be used for on-site detection [69].

Magnetic nanoparticles (MNPs)

Fig. 3 shows the applications of MNP-based SERS analysis of water pollutants. Several advantages are high and controllable enhancement factor, extraction and preconcentration of pollutants (due to adsorption and magnetism properties), effective magnetic separation, high selectivity (with suitable coating or specific capture molecules). MNPs can be used as substrates or Raman tags (MNPs labeled with specific Raman reporters) for SERS-based detection [87]. A magnetic substrate has been developed comprising Fe_3O_4 microspheres and Au NPs to detect the 16 EPA priority polycyclic aromatic hydrocarbons (PAHs) using SERS. The 16 PAHs were easily differentiated based on their characteristic SERS peaks with a 100–5 nM detection limit. Moreover, the substrate was successfully tested for sensing PAHs in river water. Additionally, it can be used for on-site monitoring of PAHs as the quantitative detection was conducted using a portable Raman spectrometer. The method can be used to identify chemically similar hydrophobic contaminant molecules whose effective separation may be difficult [73]. MNPs have also been utilized in the detection of pathogens in water. A method to enumerate *E. coli* has been developed using a combination of immunomagnetic separation (IMS) and SERS. Upon interaction with *E. coli*, a sandwich structure was formed comprised of gold-coated magnetic nanosphere–anti-*E. coli* antibody complex and Raman label (gold nanorods coated with 5,5-dithiobis-(2-nitrobenzoic acid) (DTNB)). The assay was selective, and the limit of detection (LOD) was 8 cfu/mL. *E. coli* detection was successful in real water sample analysis, and the results were comparable with the classic counting methods [71].

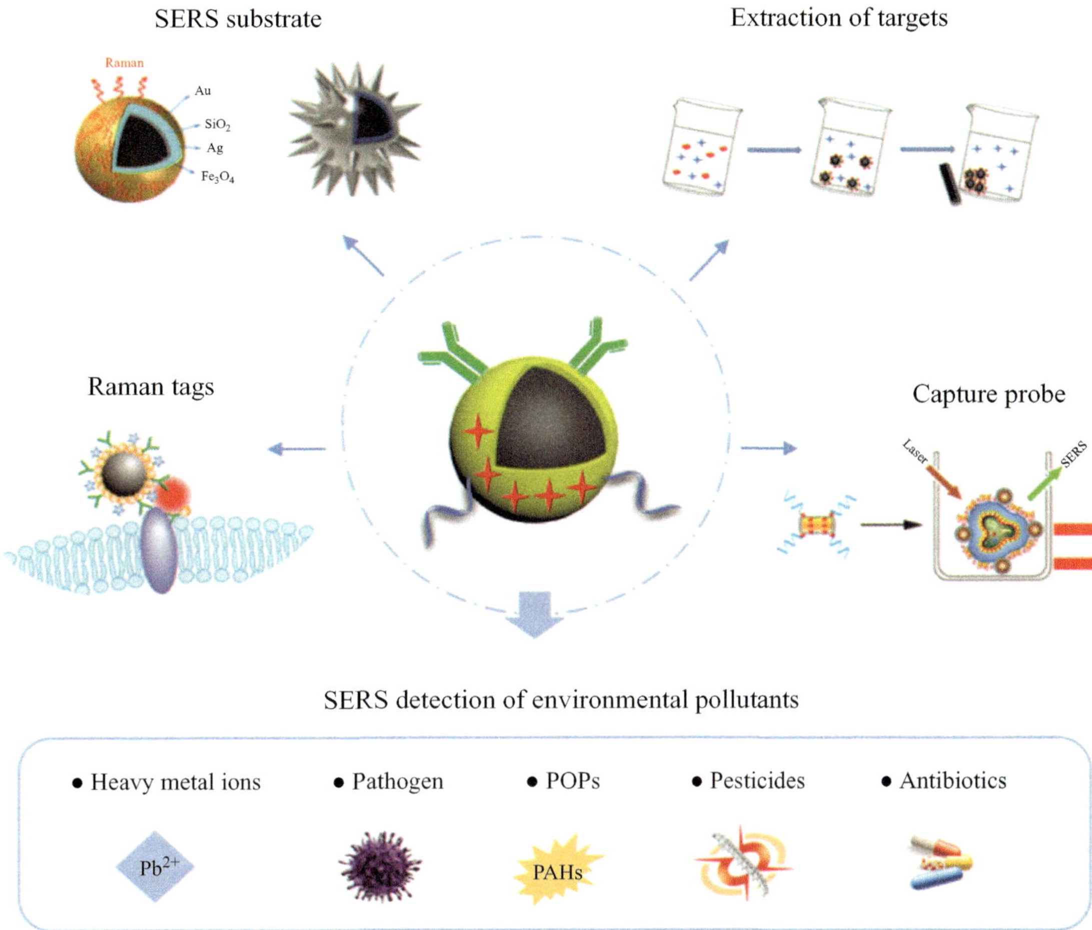

FIG. 3 Schematic of the applications of MNPs in SERS-based detection of environmental pollutants. SERS: surface-enhanced Raman scattering, POPs: persistent organic pollutants, PAHs: polycyclic aromatic hydrocarbons. *From H. Wei, K. Rodriguez, S. Renneckar, W. Leng, P.J. Vikesland, Preparation and evaluation of nanocellulose–gold nanoparticle nanocomposites for SERS applications, Analyst 140 (2015) 5640–5649. doi:10.1039/c5an00606f.*

Metal–organic frameworks (MOFs)

MOFs are a class of porous structures comprising metal ions or metal clusters and organic ligands, which are considered a promising platform for sensing and adsorption applications. They have unique optoelectronic properties, high porosity, long fluorescence lifetime, large surface area, and structural diversity. Fluorophores can be easily introduced into the pores of MOFs by ligand modifications, making them excellent candidates for optical detection [75]. Their luminescent properties are influenced by their structure, pore surfaces, their host–guest chemistry, coordination environment, etc. [80]. Fluorescent MOFs have demonstrated their quick and sensitive sensing capability to detect heavy-metal ions [80], pharmaceuticals (e.g., tetracycline [76], sulfamethazine (SMZ) [77], nitrofurantoin (NFT) [75], etc.), explosives [75,78] (e.g., 2,4,6-trinitrophenol (TNP), 4-nitrophenol (4-NP), etc.), pathogens (*Bacillus anthracis* [74], *Staphylococcus arlettae* [79], etc.), PAHs [88], etc. Several MOFs demonstrate excellent adsorption property, which plays a vital role in the preconcentration of the analyte. Hence, MOFs not only have potential in selective detection of contaminants and monitoring water quality but also in contaminant removal and wastewater treatment. Wang et al. synthesized two highly porous, stable isostructural Zr(IV)-based MOFs with excellent fluorescent and adsorption properties. Both the MOFs efficiently and selectively detected ppb levels of nitrofurazone (NZF) and nitrofurantoin (NFT) antibiotics as well as explosives such as 2,4,6-trinitrophenol (TNP) and 4-nitrophenol (4-NP). The electron and energy transfers in the host–guest system played a vital role in their exceptional quenching efficiency [75]. Another concept for MOF-based highly sensitive SERS detection includes encapsulation of NPs within MOFs. The NPs serve as SERS "hot spots" and the adsorption property of the MOF enables preconcentration of target molecules near these "hot spots" resulting in enhancement of SERS sensitivity. Utilizing this principle, Li et al. synthesized core–shell MOF@Ag NPs composite on a screen-printed carbon electrode via *in situ* electrodeposition. The prepared substrate was reusable and highly sensitivity for detection of PAHs at concentrations as low as hundreds of [88].

Magnetic nanosensors

Sensors based on magnetic nanoparticles (MNPs) have applications in environmental water containant detection because of their capability to magnetize under an applied magnetic field [56]. Their general formula is MFe_2O_4 (M = Fe, Co, Cu, Mn, etc.), such as maghemite (γ-Fe_2O_3), magnetite (Fe_3O_4), and cobalt oxide (Co_3O_4) [89]. They offer several advantages, such as extraction and preconcentration of the target analyte from complex water samples [56,87]. Second, MNPs are reusable as they can be isolated by the application of an external magnetic field [89]. Additionally, their biocompatibility permits immobilization of biorecognition elements (e.g., protein, antibody, enzyme, aptamer) on their surfaces, which can enhance their specificity toward environmental pollutants [87]. MNPs show a finite-size effect or high surface-to-volume ratio and are nonporous (no diffusion limitations), enabling higher surface adsorption capacity [89]. However, pure magnetic particles tend to form aggregates. Hence, the magnetic core is often coated with a suitable protective coating. Silica, gold, and water-soluble polymers are commonly used for protection and rendering versatility in surface modification [56,89]. Once the target analyte is magnetically isolated, it is usually detected using various optical or standard spectroscopic tools [56]. Applications of MNPs using optical nanosensor-based detection of environmental pollutants have already been discussed in this chapter. A magnetic nanocomposite based on Schiff base/silica/magnetite as a preconcentration phase has been developed for the trace determination of Pb^{2+}, Cd^{2+}, and Cu^{2+}. Iron–oxide-silica magnetic particles were modified with a Schiff base ($Fe_3O_4/SiO_2/L$). The developed magnetic solid phase is easily retrievable from large volumes of aqueous solutions. The separated metal ions were subsequently analyzed by flame atomic absorption spectrometry (FAAS). The detection limits were 0.14, 0.19 and 0.12 $\mu g L^{-1}$ for Pb^{2+}, Cd^{2+} and Cu^{2+}, respectively. The established method was successfully applied to analyze seven different natural water samples, and satisfactory results were obtained [89]. Currently, magnetic nanosensors are limited by portability and instrument availability issues [56].

Novel nanosensing technologies

Novel nanosensors which do not classify under conventional nanosensing technologies based on electrochemical, optical, or magnetic signal transduction are described here briefly. This section includes nanosensors that are hybrid (whose function and signal transduction involves interaction between synthetic and nonsynthetic components) or biomimicking (in terms of design and physical or chemical properties). Researchers are exploring alternative techniques that are not entirely dependent on complex materials and devices to assess water quality.

The hybridization approach that involves combining synthetic procedures or materials with natural systems is an emerging field. Researchers from the Parson school of design have developed an *in vivo*, plant-based novel water monitoring device named Argus. It is a living plant with DNA nanosensors (lead-sensitive DNA and CNTs) in the leaves capable of real-time detection of Pb^{2+} in water. The nanosensors are injected inside the leaf of a plant and stay within the intercellular space. It shows fluorescence when excited at 535 nm. Due to plants' natural property of water uptake, the impurities in water reach the leaf. The working principle involves breaking double-stranded DNA into single-stranded DNA (ssDNA) in the presence of Pb^{2+}. The ssDNA binds strongly to the walls of single-walled CNTs that cause quenching of fluorescence. The plants provide an optical readout of lead in water, with a concentration-dependent reduction in fluorescence. Such sensors can be modified appropriately for the detection of other elements for continuous water monitoring. It can also be used in a domestic set-up apart from the industrial environment. This could be as simple as watering a houseplant regularly to determine the chemical levels in the water [90].

Another novel nanosensor technology that has attracted significant attention is nanorobots, also termed nanomotors or nanoswimmers. Nanorobots are locomotive machines of nanoscale dimensions that are designed to perform directed tasks via self or an externally controlled propulsion mechanism. Their movement can be influenced by chemical, magnet, light, ultrasound, or electrical agents [91]. They possess immense potential in environmental detection and remediation applications. This concept is equivalent to the use of live aquatic organisms for water quality testing. Various materials have been investigated for the fabrication of nanorobots. One of the most unique and incomparable advantage over other sensing technologies is their movement ability in complex mediums. This property enables the active attachment of nanorobots to the target and enhances the mass transfer in solution [92]. Also, the stirring action that is due to the motion of nanorobots can improve the detection sensitivity and amplify signals [91,92]. The sensing mechanism could be based on chemical changes that are due to the nanorobot's motion, electrochemical, fluorescent, electrochemiluminescent (ECL), or colorimetric characteristics [92]. Based on the principle of nanorobots, researchers have developed an enzyme-powered nanoswimmer for water-quality testing. It was fabricated by template-based electrodeposition of PEDOT/Au bilayer microtubes. A catalase enzyme was cross-linked to the Au layer. The biocatalytic decomposition of the hydrogen peroxide fuel at the enzymatic layer generates the oxygen bubbles at the tail end of the fish. The thrust thus generated leads to a swimming motion. Exposure of the nanoswimmers to chemical stress (e.g., heavy metals—mercury and copper, herbicide—aminotriazole, and pesticide—NaN_3) causes inhibition of the activity of catalase, leading to a lower bubble frequency. Thus, the biocatalytic bubble propulsion mechanism gets impaired. Time-dependent optical tracking of changes in the speed and lifetime (survival) of the nanoswimmer, offers a simple and real-time visualization of pollutant-induced changes. Thus, pollutants that are known to inhibit the catalase activity can be efficiently detected down to the micromolar levels by this novel sensing technology [93].

Challenges and concerns

Nanomaterials allow sensors to achieve very low limits of detection with enhanced analytical performance. However, the advancement in the field of nanosensors for water quality monitoring comes with certain challenges. Water is a highly variable and complex matrix. The global water quality can vary substantially due to differences in hardness, alkalinity, and inorganic and organic contaminants [94]. Hence, the sensors must function efficiently, and the detection limits must be tested across this broad range of water quality conditions.

Moreover, strategic planning and design of nanosensor research are essential for minimizing health risks associated with unintended release of nanomaterials used in nanosensors to prevent hazardous environmental implications. Conventionally, researchers focus on performance efficiency and price for material selection. However, the choice for optimum material must be made considering the environmental and ecological health impacts as well [95]. The toxicity of nanoscale materials is often different from their bulk counterparts and is dependent on their size, shape, and charge. Teratogenicity, DNA damage, and decreased cell mitochondrial membrane potential are a few toxic effects caused by nanomaterials. Hence, they must be used with caution and their exposure to ecosystems must be minimum. Also, the materials must meet the quality criteria and be compatible with mass manufacturing methods. This helps in ensuring robustness and translation from laboratory-scale prototypes to commercial applications. Ideally, the developed nanosensor should be sustainable, scalable, and nontoxic and still elicit the desired functionality.

Real-time, on-site water analysis preferably requires simultaneous monitoring of multiple contaminants using a single sensor. Developing nanosensors with multiplexing (multianalyte) determination capabilities along with resistance to various interferences are significant challenges preventing commercial applications.

Future outlook

The global nanosensors market generated revenue worth USD 432 million in 2019, with electrochemical nanosensors accounting for the largest market share [96]. Significant progress has been made in the design and development of nanotechnology-enabled sensors across different industries. However, extensive research is awaited for the fabrication of automated, compact, on-site water monitoring devices that are preferably integrated with a mobile phone. Such devices would enable *in situ* operations and online, remote, real-time analysis. This advancement will lead to immediate notification and initiation of quick remedial action for averting potential large-scale environmental damage. Also, disposable testing strip-based sensors would enable residential water quality testing.

The field of nanosensors for water quality monitoring is still at its infancy, and very few such sensor devices have made it to market (e.g., graphene nanosheets-based sensors by NanoAffix Science LLC). Extensive collaborative efforts between multidisciplinary science and engineering domains are need-of-the-hour for the design and fabrication of easy-to-use environmental sensors. The potential impact of nanosensors in water quality monitoring is expected to be profound.

References

[1] UNICEF & WHO, Progress on Household Drinking Water, Sanitation and Hygiene 2000–2017. Special Focus on Inequalities, 2019. https://washdata.org/sites/default/files/documents/reports/2019-07/jmp-2019-wash-households.pdf.

[2] WHO, Safer Water, Better Health, World Health Organization, Geneva, 2019. Update https://apps.who.int/iris/bitstream/handle/10665/329905/9789241516891-eng.pdf.

[3] H. Wei, S.M.H. Abtahi, P.J. Vikesland, Plasmonic colorimetric and SERS sensors for environmental analysis, Environ. Sci. Nano 2 (2015) 120–135, https://doi.org/10.1039/c4en00211c.

[4] P. Kruse, Review on water quality sensors, J. Phys. D Appl. Phys. 51 (2018) 203002, https://doi.org/10.1088/1361-6463/aabb93.

[5] N. Bhardwaj, S.K. Bhardwaj, D. Bhatt, D.K. Lim, K.H. Kim, A. Deep, Optical detection of waterborne pathogens using nanomaterials, TrAC - Trends Anal. Chem. 113 (2019) 280–300, https://doi.org/10.1016/j.trac.2019.02.019.

[6] D. Wei, M.J.A. Bailey, P. Andrew, T. Ryhänen, Electrochemical biosensors at the nanoscale, Lab Chip 9 (15) (2009) 2123–2131, https://doi.org/10.1039/b903118a.

[7] B. Maleki, M. Baghayeri, M. Ghanei-Motlagh, et al., Polyamidoamine dendrimer functionalized iron oxide nanoparticles for simultaneous electrochemical detection of pb 2+ and cd 2+ ions in environmental waters, Meas J. Int. Meas Confed. 140 (2019) 81–88, https://doi.org/10.1016/j.measurement.2019.03.052.

[8] Z. Keivani, M. Shabani-Nooshabadi, H. Karimi-Maleh, An electrochemical strategy to determine thiosulfate, 4-chlorophenol and nitrite as three important pollutants in water samples via a nanostructure modified sensor, J. Colloid Interface Sci. 507 (2017) 11–17, https://doi.org/10.1016/j.jcis.2017.07.098.

[9] A. Garciá-Miranda Ferrari, P. Carrington, S.J. Rowley-Neale, C.E. Banks, Recent advances in portable heavy metal electrochemical sensing platforms, Environ. Sci. Water Res. Technol. 6 (10) (2020) 2676–2690, https://doi.org/10.1039/d0ew00407c.

[10] S. Tajik, H. Beitollahi, F.G. Nejad, et al., Recent developments in polymer nanocomposite-based electrochemical sensors for detecting environmental pollutants, Ind. Eng. Chem. Res. 60 (3) (2021) 1112–1136, https://doi.org/10.1021/acs.iecr.0c04952.

[11] X. Liu, Y. Yao, Y. Ying, J. Ping, Recent advances in nanomaterial-enabled screen-printed electrochemical sensors for heavy metal detection, TrAC Trends Anal. Chem. 115 (2019) 187–202, https://doi.org/10.1016/J.TRAC.2019.03.021.

[12] J. Barton, M.B.G. García, D.H. Santos, et al., Screen-printed electrodes for environmental monitoring of heavy metal ions: a review, Microchim. Acta 183 (2) (2015) 503–517, https://doi.org/10.1007/S00604-015-1651-0.

[13] H. Beitollahi, S.Z. Mohammadi, M. Safaei, S. Tajik, Applications of electrochemical sensors and biosensors based on modified screen-printed electrodes: a review, Anal. Methods 12 (12) (2020) 1547–1560, https://doi.org/10.1039/c9ay02598g.

[14] N. Karimian, P. Ugo, Recent advances in sensing and biosensing with arrays of nanoelectrodes, Curr. Opin. Electrochem. 16 (2019) 106–116, https://doi.org/10.1016/j.coelec.2019.04.026.

[15] K.B. Male, S. Hrapovic, J.M. Santini, J.H.T. Luong, Biosensor for Arsenite using Arsenite oxidase and multiwalled carbon nanotube modified electrodes, Anal. Chem. 79 (20) (2007) 7831–7837, https://doi.org/10.1021/AC070766I.

[16] H.S. Magar, M.E. Ghica, M.N. Abbas, C.M.A. Brett, Highly sensitive choline oxidase enzyme inhibition biosensor for Lead ions based on multiwalled carbon nanotube modified glassy carbon electrodes, Electroanalysis. 29 (7) (2017) 1741–1748, https://doi.org/10.1002/elan.201700111.

[17] F. Cai, Q. Zhu, K. Zhao, A. Deng, J. Li, Multiple signal amplified electrochemiluminescent immunoassay for Hg2+ using graphene-coupled quantum dots and gold nanoparticles-labeled horseradish peroxidase, Environ. Sci. Technol. 49 (8) (2015) 5013–5020, https://doi.org/10.1021/acs.est.5b00690.

[18] Z. Yang, X. Ding, Q. Guo, et al., Second generation of signaling-probe displacement electrochemical aptasensor for detection of picomolar ampicillin and sulfadimethoxine, Sens. Actuators B 253 (2017) 1129–1136, https://doi.org/10.1016/j.snb.2017.07.119.

[19] K. Mondal, A. Sharma, Recent advances in electrospun metal-oxide nanofiber based interfaces for electrochemical biosensing, RSC Adv. 6 (97) (2016) 94595–94616, https://doi.org/10.1039/c6ra21477k.

[20] G. Ozcelikay, S. Kurbanoglu, A. Yarman, F.W. Scheller, S.A. Ozkan, Au-Pt nanoparticles based molecularly imprinted nanosensor for electrochemical detection of the lipopeptide antibiotic drug daptomycin, Sens. Actuators B 320 (2020) 128285, https://doi.org/10.1016/j.snb.2020.128285.

[21] G. Bolat, Y.T. Yaman, S. Abaci, Molecularly imprinted electrochemical impedance sensor for sensitive dibutyl phthalate (DBP) determination, Sens. Actuators B 299 (2019) 127000, https://doi.org/10.1016/j.snb.2019.127000.

[22] X. Tian, L. Liu, Y. Li, et al., Nonenzymatic electrochemical sensor based on CuO-TiO2 for sensitive and selective detection of methyl parathion pesticide in ground water, Sens. Actuators B 256 (2018) 135–142, https://doi.org/10.1016/j.snb.2017.10.066.

[23] Y. Zuo, J. Xu, X. Zhu, et al., Poly(3,4-ethylenedioxythiophene) nanorods/graphene oxide nanocomposite as a new electrode material for the selective electrochemical detection of mercury (II), Synth. Met. 220 (2016) 14–19, https://doi.org/10.1016/j.synthmet.2016.05.022.

[24] K. Deng, X. Liu, C. Li, Z. Hou, H. Huang, An electrochemical omeprazole sensor based on shortened multi-walled carbon nanotubes-Fe3O4 nanoparticles and poly(2, 6-pyridinedicarboxylic acid), Sens. Actuators B 253 (2017) 1–9, https://doi.org/10.1016/j.snb.2017.06.117.

[25] T. Gupte, S.K. Jana, J.S. Mohanty, et al., Highly sensitive As3+ detection using electrodeposited nanostructured MnO x and phase evolution of the active material during sensing, ACS Appl. Mater. Interfaces 11 (31) (2019) 28154–28163, https://doi.org/10.1021/acsami.9b06023.

[26] L. Xiao, G.G. Wildgoose, R.G. Compton, Sensitive electrochemical detection of arsenic (III) using gold nanoparticle modified carbon nanotubes via anodic stripping voltammetry, Anal. Chim. Acta 620 (1–2) (2008) 44–49, https://doi.org/10.1016/j.aca.2008.05.015.

[27] X. Li, H. Zhou, C. Fu, F. Wang, Y. Ding, Y. Kuang, A novel design of engineered multi-walled carbon nanotubes material and its improved performance in simultaneous detection of cd(II) and pb(II) by square wave anodic stripping voltammetry, Sens. Actuators B 236 (2016) 144–152, https://doi.org/10.1016/j.snb.2016.05.149.

[28] G. Liu, Y. Lin, Y. Tu, Z. Ren, Ultrasensitive voltammetric detection of trace heavy metal ions using carbon nanotube nanoelectrode array, Analyst 130 (7) (2005) 1098–1101, https://doi.org/10.1039/b419447k.

[29] D. Sahoo, A. Mandal, T. Mitra, K. Chakraborty, M. Bardhan, A.K. Dasgupta, Nanosensing of pesticides by zinc oxide quantum dot: an optical and electrochemical approach for the detection of pesticides in water, J. Agric. Food Chem. 66 (2) (2018) 414–423, https://doi.org/10.1021/acs.jafc.7b04188.

[30] L. Cui, J. Wu, H. Ju, Electrochemical sensing of heavy metal ions with inorganic, organic and bio-materials, Biosens. Bioelectron. 63 (2015) 276–286, https://doi.org/10.1016/j.bios.2014.07.052.

[31] E.K. Wujcik, H. Wei, X. Zhang, et al., Antibody nanosensors: a detailed review, RSC Adv. 4 (82) (2014) 43725–43745, https://doi.org/10.1039/c4ra07119k.

[32] T.O. Hara, B. Singh, Electrochemical biosensors for detection of pesticides and heavy metal toxicants in water: recent trends and Progress, ACS ES&T Water. 1 (3) (2021) 462–478, https://doi.org/10.1021/acsestwater.0c00125.

[33] W. Chen, Q. Zhu, Q. Tang, K. Zhao, A. Deng, J. Li, Ultrasensitive detection of diclofenac based on electrochemiluminescent immunosensor with multiple signal amplification strategy of palladium attached graphene oxide as bioprobes and ceria doped zinc oxide as substrates, Sens. Actuators B 268 (2018) 411–420, https://doi.org/10.1016/j.snb.2018.04.106.

[34] Z. Zhang, M. Yang, X. Wu, et al., A competitive immunosensor for ultrasensitive detection of sulphonamides from environmental waters using silver nanoparticles decorated single-walled carbon nanohorns as labels, Chemosphere 225 (2019) 282–287, https://doi.org/10.1016/j.chemosphere.2019.03.033.

[35] S. Oloketuyi, E. Mazzega, J. Zavašnik, et al., Electrochemical immunosensor functionalized with nanobodies for the detection of the toxic microalgae Alexandrium minutum using glassy carbon electrode modified with gold nanoparticles, Biosens. Bioelectron. 154 (2020) 112052, https://doi.org/10.1016/j.bios.2020.112052.

[36] F. Li, Z. Yu, X. Han, R.Y. Lai, Electrochemical aptamer-based sensors for food and water analysis: a review, Anal. Chim. Acta 1051 (2019) 1–23, https://doi.org/10.1016/j.aca.2018.10.058.

[37] W. Tang, Z. Wang, J. Yu, F. Zhang, P. He, Internal calibration potentiometric Aptasensors for simultaneous detection of Hg2+, Cd2+, and As3+ based on a screen-printed carbon electrodes Array, Anal. Chem. 90 (14) (2018) 8337–8344, https://doi.org/10.1021/acs.analchem.7b04150.

[38] J. Sun, Y. Gan, T. Liang, et al., Signal enhancement of electrochemical DNA biosensors for the detection of trace heavy metals, Curr. Opin. Electrochem. 17 (2019) 23–29, https://doi.org/10.1016/j.coelec.2019.04.007.

[39] M.R. Saidur, A.R.A. Aziz, W.J. Basirun, Recent advances in DNA-based electrochemical biosensors for heavy metal ion detection: a review, Biosens. Bioelectron. 90 (2017) 125–139, https://doi.org/10.1016/j.bios.2016.11.039.

[40] D. Soukarié, V. Ecochard, L. Salomé, DNA-based nanobiosensors for monitoring of water quality, Int. J. Hyg. Environ. Health 226 (2020) 113485, https://doi.org/10.1016/j.ijheh.2020.113485.

[41] P. Miao, Y. Tang, L. Wang, DNA modified Fe3O4@au magnetic nanoparticles as selective probes for simultaneous detection of heavy metal ions, ACS Appl. Mater. Interfaces 9 (4) (2017) 3940–3947, https://doi.org/10.1021/acsami.6b14247.

[42] O.S. Ahmad, T.S. Bedwell, C. Esen, A. Garcia-Cruz, S.A. Piletsky, Molecularly imprinted polymers in electrochemical and optical sensors, Trends Biotechnol. 37 (3) (2019) 294–309, https://doi.org/10.1016/j.tibtech.2018.08.009.

[43] X. Sun, C. Huang, L. Wang, et al., Recent Progress in graphene/polymer nanocomposites, Adv. Mater. 33 (6) (2021) 2001105, https://doi.org/10.1002/adma.202001105.

[44] S.K. Krishnan, E. Singh, P. Singh, M. Meyyappan, H.S. Nalwa, A review on graphene-based nanocomposites for electrochemical and fluorescent biosensors, RSC Adv. 9 (16) (2019) 8778–8781, https://doi.org/10.1039/c8ra09577a.

[45] Niu P, Fernández-Sánchez C, Gich M, Navarro-Hernández C, Fanjul-Bolado P, Roig A. Screen-Printed Electrodes Made of a Bismuth Nanoparticle Porous Carbon Nanocomposite Applied to the Determination of Heavy Metal Ions. Springer. doi:https://doi.org/10.1007/s00604-015-1684-4.

[46] S. Su, S. Chen, C. Fan, Recent advances in two-dimensional nanomaterials-based electrochemical sensors for environmental analysis, Green Energy Environ. 3 (2) (2018) 97–106, https://doi.org/10.1016/j.gee.2017.08.005.

[47] C. Wu, Q. Cheng, K. Wu, Electrochemical functionalization of N -Methyl-2-pyrrolidone-exfoliated graphene nanosheets as highly sensitive analytical platform for phenols, Anal. Chem. 87 (6) (2015) 3294–3299, https://doi.org/10.1021/ac504309j.

[48] H.Y. Chen, J. Wang, L. Meng, T. Yang, K. Jiao, Thin-layered MoS2/polyaniline nanocomposite for highly sensitive electrochemical detection of chloramphenicol, Chin. Chem. Lett. 27 (2) (2016) 231–234, https://doi.org/10.1016/j.cclet.2015.09.018.

[49] X. Fang, Z. Zeng, Q. Li, et al., Ultrasensitive detection of disinfection byproduct trichloroacetamide in drinking water with ag nanoprism@MoS2 heterostructure-based electrochemical sensor, Sens. Actuators B 332 (2021), https://doi.org/10.1016/j.snb.2021.129526, 129526.

[50] J.H. Hwang, M.A. Islam, H. Choi, et al., Improving electrochemical Pb2+ detection using a vertically aligned 2D MoS2 nanofilm, Anal. Chem. 91 (18) (2019) 11770–11777, https://doi.org/10.1021/acs.analchem.9b02382.

[51] P. Ramnani, N.M. Saucedo, A. Mulchandani, Carbon nanomaterial-based electrochemical biosensors for label-free sensing of environmental pollutants, Chemosphere 143 (2016) 85–98, https://doi.org/10.1016/j.chemosphere.2015.04.063.

[52] A.U. Alam, D. Clyne, H. Jin, N.X. Hu, M.J. Deen, Fully integrated, simple, and low-cost electrochemical sensor Array for *in situ* water quality monitoring, ACS Sensors. 5 (2) (2020) 412–422, https://doi.org/10.1021/acssensors.9b02095.

[53] M. Li, H. Gou, I. Al-Ogaidi, N. Wu, Nanostructured sensors for detection of heavy metals: a review, ACS Sustain. Chem. Eng. 1 (7) (2013) 713–723, https://doi.org/10.1021/sc400019a.

[54] L. Rassaei, F. Marken, M. Sillanpää, M. Amiri, C.M. Cirtiu, M. Sillanpää, Nanoparticles in electrochemical sensors for environmental monitoring, TrAC - Trends Anal. Chem. 30 (11) (2011) 1704–1715, https://doi.org/10.1016/j.trac.2011.05.009.

[55] M. Yang, X. Chen, T.J. Jiang, Z. Guo, J.H. Liu, X.J. Huang, Electrochemical detection of trace arsenic(III) by nanocomposite of nanorod-like α-MnO2 decorated with ~5 nm au nanoparticles: considering the change of arsenic speciation, Anal. Chem. 88 (19) (2016) 9720–9728, https://doi.org/10.1021/acs.analchem.6b02629.

[56] P.J. Vikesland, K.R. Wigginton, Nanomaterial enabled biosensors for pathogen monitoring - a review, Environ. Sci. Technol. 44 (10) (2010) 3656–3669, https://doi.org/10.1021/es903704z.

[57] G.N. Liu, R.D. Xu, R.Y. Zhao, et al., Hybrid copper iodide cluster-based pellet sensor for highly selective optical detection of o-nitrophenol and tetracycline hydrochloride in aqueous solution, ACS Sustain. Chem. Eng. 7 (23) (2019) 18863–18873, https://doi.org/10.1021/acssuschemeng.9b03963.

[58] S.A. Khan, J.A. DeGrasse, B.J. Yakes, T.R. Croley, Rapid and sensitive detection of cholera toxin using gold nanoparticle-based simple colorimetric and dynamic light scattering assay, Anal. Chim. Acta 892 (2015) 167–174, https://doi.org/10.1016/j.aca.2015.08.029.

[59] Y. Kim, R.C. Johnson, J.T. Hupp, Gold nanoparticle-based sensing of "spectroscopically silent" heavy metal ions, Nano Lett. 1 (4) (2001) 3–5, https://doi.org/10.1021/nl0100116.

[60] H. Yuan, W. Ji, S. Chu, et al., Mercaptopyridine-functionalized gold nanoparticles for fiber-optic surface plasmon resonance hg 2+ sensing, ACS Sensors. 4 (3) (2019) 704–710, https://doi.org/10.1021/acssensors.8b01558.

[61] L. Bao, P. Sheng, J. Li, S. Wu, Q. Cai, S. Yao, Surface enhanced Raman spectroscopic detection of polycyclic aromatic hydrocarbons (PAHs) using a gold nanoparticles-modified alginate gel network, Analyst 137 (17) (2012) 4010–4015, https://doi.org/10.1039/c2an35589b.

[62] J. Li, L. Yang, Y. Ruan, et al., Dual-mode optical Nanosensor based on gold nanoparticles and carbon dots for visible detection of as(III) in water, ACS Appl. Nano. Mater. 3 (8) (2020) 8224–8231, https://doi.org/10.1021/acsanm.0c01647.

[63] Y. Bhattacharjee, D. Chatterjee, A. Chakraborty, Mercaptobenzoheterocyclic compounds functionalized silver nanoparticle, an ultrasensitive colorimetric probe for hg(II) detection in water with picomolar precision: a correlation between sensitivity and binding affinity, Sens. Actuators B 255 (2018) 210–216, https://doi.org/10.1016/j.snb.2017.08.066.

[64] J. Zheng, S. Pang, T.P. Labuza, L. He, Semi-quantification of surface-enhanced Raman scattering using a handheld Raman spectrometer: a feasibility study, Analyst 138 (23) (2013) 7075–7078, https://doi.org/10.1039/c3an01450a.

[65] Z. Zhou, J. Lu, J. Wang, et al., Trace detection of polycyclic aromatic hydrocarbons in environmental waters by SERS, Spectrochim. Acta - Part A Mol. Biomol. Spectrosc. 234 (2020) 118250–118256, https://doi.org/10.1016/j.saa.2020.118250.

[66] R. Hu, R. Tang, J. Xu, F. Lu, Chemical nanosensors based on molecularly-imprinted polymers doped with silver nanoparticles for the rapid detection of caffeine in wastewater, Anal. Chim. Acta 1034 (2018) 176–183, https://doi.org/10.1016/j.aca.2018.06.012.

[67] W. Gu, X. Pei, Y. Cheng, et al., Black phosphorus quantum dots as the Ratiometric fluorescence probe for trace mercury ion detection based on inner filter effect, ACS Sensors. 2 (4) (2017) 576–582, https://doi.org/10.1021/acssensors.7b00102.

[68] Y. Zhang, J.Y. Xiao, Y. Zhu, et al., Fluorescence sensor based on biosynthetic CdSe/CdS quantum dots and liposome carrier signal amplification for mercury detection, Anal. Chem. 92 (5) (2020) 3990–3997, https://doi.org/10.1021/acs.analchem.9b05508.

[69] Y.F. Guan, B.C. Huang, C. Qian, H.Q. Yu, Quantification of humic substances in natural water using nitrogen-doped carbon dots, Environ. Sci. Technol. 51 (24) (2017) 14092–14099, https://doi.org/10.1021/acs.est.7b04430.

[70] A. Mehta, A. Mishra, S. Kainth, S. Basu, Carbon quantum dots/TiO2 nanocomposite for sensing of toxic metals and photodetoxification of dyes with kill waste by waste concept, Mater. Des. 155 (2018) 485–493, https://doi.org/10.1016/j.matdes.2018.06.015.

[71] B. Guven, N. Basaran-akgul, E. Temur, U. Tamer, SERS-based sandwich immunoassay using antibody coated magnetic nanoparticles for Escherichia coli enumeration, Analyst 136 (2011) 740–748, https://doi.org/10.1039/c0an00473a.

[72] S. Tang, Y. Li, H. Huang, et al., Efficient enrichment and self-assembly of hybrid nanoparticles into removable and magnetic SERS substrates for sensitive detection of environmental pollutants, ACS Appl. Mater. Interfaces 9 (8) (2017) 7472–7480, https://doi.org/10.1021/acsami.6b16141.

[73] J. Du, J. Xu, Z. Sun, C. Jing, Au nanoparticles grafted on Fe3O4 as effective SERS substrates for label-free detection of the 16 EPA priority polycyclic aromatic hydrocarbons, Anal. Chim. Acta 915 (2016) 81–89, https://doi.org/10.1016/j.aca.2016.02.009.

[74] M. Wu, Y. Zhuang, J. Liu, W. Chen, X. Li, R.J. Xie, Ratiometric fluorescence detection of 2,6-pyridine dicarboxylic acid with a dual-emitting lanthanide metal-organic framework (MOF), Opt. Mater. (Amst). 106 (2020) 110006, https://doi.org/10.1016/j.optmat.2020.110006.

[75] B. Wang, X.L. Lv, D. Feng, et al., Highly stable Zr(IV)-based metal-organic frameworks for the detection and removal of antibiotics and organic explosives in water, J. Am. Chem. Soc. 138 (19) (2016) 6204–6216, https://doi.org/10.1021/jacs.6b01663.

[76] Y. Zhou, Q. Yang, D. Zhang, N. Gan, Q. Li, J. Cuan, Detection and removal of antibiotic tetracycline in water with a highly stable luminescent MOF, Sens. Actuators B 262 (2018) 137–143, https://doi.org/10.1016/j.snb.2018.01.218.

[77] X.D. Zhu, K. Zhang, Y. Wang, et al., Fluorescent metal-organic framework (MOF) as a highly sensitive and quickly responsive chemical sensor for the detection of antibiotics in simulated wastewater, Inorg. Chem. 57 (3) (2018) 1060–1065, https://doi.org/10.1021/acs.inorgchem.7b02471.

[78] X.J. Zhang, F.Z. Su, D.M. Chen, et al., A water-stable Eu III -based MOF as a dual-emission luminescent sensor for discriminative detection of nitroaromatic pollutants, Dalton Trans. 48 (5) (2019) 1843–1849, https://doi.org/10.1039/c8dt04397c.

[79] N. Bhardwaj, S.K. Bhardwaj, J. Mehta, M.K. Nayak, A. Deep, Bacteriophage conjugated IRMOF-3 as a novel opto-sensor for: *S. arlettae*, New J. Chem. 40 (9) (2016) 8068–8073, https://doi.org/10.1039/c6nj00899b.

[80] L. Wen, X. Zheng, K. Lv, C. Wang, X. Xu, Two amino-decorated metal-organic frameworks for highly selective and quantitatively sensing of HgII and CrVI in aqueous solution, Inorg. Chem. 54 (15) (2015) 7133–7135, https://doi.org/10.1021/acs.inorgchem.5b00098.

[81] S. Mukherjee, M. Shah, K. Chaudhari, et al., Smartphone-based fluoride-specific sensor for rapid and affordable colorimetric detection and precise quantification at sub-ppm levels for field applications, ACS Omega. 5 (39) (2020) 25253–25263, https://doi.org/10.1021/acsomega.0c03465.

[82] V. Brahmkhatri, P. Pandit, P. Rananaware, A. D'Souza, M.D. Kurkuri, Recent progress in detection of chemical and biological toxins in water using plasmonic nanosensors, Trends Environ. Anal. Chem. 30 (2021) e00117, https://doi.org/10.1016/j.teac.2021.e00117.

[83] H. Wei, K. Rodriguez, S. Renneckar, W. Leng, P.J. Vikesland, Preparation and evaluation of nanocellulose–gold nanoparticle nanocomposites for SERS applications, Analyst 140 (2015) 5640–5649, https://doi.org/10.1039/c5an00606f.

[84] L.S. Lawson, W. Chan, T. Huser, A highly sensitive nanoscale pH-sensor using au nanoparticles linked by a multifunctional Raman- active reporter molecule, Nanoscale 6 (2014) 7971–7980, https://doi.org/10.1039/c3nr06277e.

[85] M. Chern, J.C. Kays, S. Bhuckory, A.M. Dennis, Sensing with photoluminescent semiconductor quantum dots, Methods Appl. Fluoresc. 7 (1) (2019) 1–36, https://doi.org/10.1088/2050-6120/aaf6f8.

[86] U. Resch-Genger, M. Grabolle, S. Cavaliere-Jaricot, R. Nitschke, T. Nann, Quantum dots versus organic dyes as fluorescent labels, Nat. Methods 5 (9) (2008) 763–775, https://doi.org/10.1038/nmeth.1248.

[87] D. Song, R. Yang, F. Long, A. Zhu, Applications of magnetic nanoparticles in surface-enhanced Raman scattering (SERS) detection of environmental pollutants, J. Environ. Sci. (China) 80 (2019) 14–34, https://doi.org/10.1016/j.jes.2018.07.004.

[88] D. Li, X. Cao, Q. Zhang, et al., Facile *in situ* synthesis of core–shell MOF@ag nanoparticle composites on screen-printed electrodes for ultrasensitive SERS detection of polycyclic aromatic hydrocarbons, J. Mater. Chem. A 7 (23) (2019) 14108–14117, https://doi.org/10.1039/C9TA03690C.

[89] H. Bagheri, A. Afkhami, M. Saber-Tehrani, H. Khoshsafar, Preparation and characterization of magnetic nanocomposite of Schiff base/silica/magnetite as a preconcentration phase for the trace determination of heavy metal ions in water, food and biological samples using atomic absorption spectrometry, Talanta 97 (2012) 87–95, https://doi.org/10.1016/j.talanta.2012.03.066.

[90] MIT Media Lab, Overview Argus: Water monitoring through nanosensors inside living plants, MIT Media Lab, 2021. Accessed April 1 https://www.media.mit.edu/projects/argus-sensors-inside-plants/overview/.

[91] H. Zhou, C.C. Mayorga-Martinez, S. Pané, L. Zhang, M. Pumera, Magnetically driven micro and nanorobots, Chem. Rev. 121 (2021) 4999–5041, https://doi.org/10.1021/acs.chemrev.0c01234.

[92] L. Kong, J. Guan, M. Pumera, Micro- and nanorobots based sensing and biosensing, Curr. Opin. Electrochem. 10 (2018) 174–182, https://doi.org/10.1016/j.coelec.2018.06.004.

[93] J. Orozco, V. García-Gradilla, M. D'Agostino, W. Gao, A. Cortés, J. Wang, Artificial enzyme-powered microfish for water-quality testing, ACS Nano 7 (1) (2013) 818–824, https://doi.org/10.1021/nn305372n.

[94] P.J. Vikesland, Nanosensors for water quality monitoring, Nat. Nanotechnol. 13 (8) (2018) 651–660, https://doi.org/10.1038/s41565-018-0209-9.

[95] M.M. Falinski, D.L. Plata, S.S. Chopra, T.L. Theis, L.M. Gilbertson, J.B. Zimmerman, A framework for sustainable nanomaterial selection and design based on performance, hazard, and economic considerations, Nat. Nanotechnol. 13 (8) (2018) 708–714, https://doi.org/10.1038/s41565-018-0120-4.

[96] GMI Research, Nanosensors Market Industry Report: Size, Opportunities & Forecast 2020–2027, GMI Research, 2020. Published 2020. Accessed March 31, 2021 https://www.gmiresearch.com/report/global-nanosensors-market/.

Chapter 4

Assessing the environmental impact of anthropogenic nanoparticles

Sherine O. Obare* and Chartanay D.J. Bonner

Joint School of Nanoscience and Nanoengineering, University of North Carolina Greensboro and North Carolina A&T, Greensboro, NC, United States

Corresponding author: e-mail address: soobare@uncg.edu

Introduction

The diversity in synthesis, production, and applications of silver nanoparticles (AgNPs) has led to a significant concern as an environmental contaminant [1,2]. AgNPs are known to display toxic behavior at varying concentrations toward microorganisms and aquatic species because of the diversity in adsorption, uptake, and stability [3]. As research has expanded to study the toxic action of NPs, limited studies have investigated the impact that multiple anthropogenic contaminants may have on the environment.

Anthropogenic contaminant lead (Pb^{+2}) transports easily through water, air, and soil sources, causing unknown levels of consumption by humans, animals, and other living organisms [4]. Some studies have reported the bioaccumulation of Pb^{+2} in plants, algae, and aquatic species promotes adverse effects [5,6]. Pb^{+2} concentration, oxidation state, and exposure time are directly related to toxicity levels. Pb^{+2} can alter the metabolic process of organisms at low concentrations and block uptake of essential micronutrients [5,7]. In aquatic species, Pb^{+2} can accumulate in major organs and tissues that lead to alternative functions.

As Pb^{+2} is known to accumulate in aquatic species, the presence of NPs can potentially impact their transport and fate [8]. The antimicrobial properties of NPs and heavy metals have been investigated when present in the environment [9]. As various anthropogenic contaminants are present in the environment, they can influence individual toxic responses within aquatic systems [10]. The coexposure of contaminants can lead to the alternative response because of how they interact. NPs have been investigated for the adsorption and uptake of Pb^{+2} to limit its toxic behavior in environmental systems [11,12]. However, limited studies have explored the impact these interactions can have when improperly released into the environment [13–16]. Therefore, it is pivotal to examine the effects of anthropogenic contaminants in the environment as they coexist.

In this study, we specifically investigate the interaction between Pb^{+2} and AgNPs. Considering the ionic character of Pb^{+2} when released in the environment and how it can impact the surface stability of NPs, we hypothesize that the interaction of Pb^{+2} can alter the toxic behavior of AgNPs toward bacterial strains. This study investigates the antimicrobial response of Pb^{+2} and AgNPs in the presence of gram-positive *Staphylococcus aureus* and gram-negative *Escherichia coli* at a 7.0 pH value. Natural organic matter, humic acid (HA), was used as a model to simulate an environmental setting.

Experimental details

Materials

Silver nitrate [$AgNO_3$], trisodium citrate dihydrate, sodium borohydride [$NaBH_4$], lead nitrate ($Pb(NO_3)_2$), and humic acid were purchased from the Sigma-Aldrich Company. For phosphate-buffered saline (PBS) buffer sodium chloride [NaCl], potassium chloride [KCl], sodium phosphate [Na_2HPO_4], and potassium phosphate [KH_2PO_4] were also purchased from the Sigma-Aldrich Company. All purchased chemicals were used as received, without further purification.

Separation Science and Technology, Volume 15. https://doi.org/10.1016/B978-0-323-90763-7.00007-X

Synthesis of silver nanoparticles (AgNPs)

AgNPs at a concentration of 2.5×10^{-4} M were prepared as follows: 0.01 M silver nitrate [$AgNO_3$], 0.01 M trisodium citrate dihydrate and 0.01 M $NaBH_4$ was prepared in Milli-Q water solvent; 0.05 mL of $AgNO_3$ stock solution and citrate capping agent was added to 18.5 mL of Milli-Q water; 0.05 mL of reducing agent $NaBH_4$ stock solution was added dropwise to the mixture to form AgNPs. The prepared solution was covered with aluminum foil, capped with a rubber stopper and parafilmed. The mixture was transferred to the refrigerator to sit undisturbed for 5 h to grow.

To monitor stability, AgNPs 0.01 M ($Pb(NO_3)_2$) was prepared with a Milli-Q water solvent. Five mL of prepared citrate-capped AgNPs was placed in a reaction vial covered with aluminum foil before adding heavy metal ions. ($Pb(NO_3)_2$) was added to AgNPs at various concentrations (70–200 μL) over 20 h.

Bacterial culture conditions

For growth of bacteria, bacteria strains and food broth were prepared. Auger broth and plate cultures of bacteria *S. aureus* and *E. coli* were prepared under a sterile environment. After the growth of bacteria plate cultures, within a sterile environment, *S. aureus* and *E. coli* were placed in a 50-mL reactor-centrifuged tube in the presence of auger broth. The solution was incubated at 37 °C and stirred for 16 h to allow the growth of bacteria in the presence of broth food.

Bacterial strains interaction with AgNPs and lead nitrate

For reaction medium for bacteria strains, AgNPs, and heavy metal ions, a phosphate buffered saline (PBS) buffer was prepared following standard conditions. The buffer was prepared by dissolving NaCl, KCl, Na_2HPO_4, and KH_2PO_4 salts in Milli-Q water. All reaction conditions were prepared in a sterile environment to limit contamination. Mixtures of AgNPs, ($Pb(NO_3)_2$) and bacteria strains (*S. aureus and E. coli*) were placed in a PBS buffer and tris-buffered saline (TBS) to promote bacterial growth. Each reaction condition was stirred and heated to improve interactions.

Bacterial strains interaction with Ag ions and lead nitrate

0.01 M $AgNO_3$ solution was prepared in Milli-Q water solvent. Mixtures of $AgNO_3$, ($Pb(NO_3)_2$) and bacterial strains interacted in a PBS buffer and TBS. Each reaction condition was stirred and heated to promote interactions.

Bacterial strains interaction with humic acid

A humic acid solution was prepared in a PBS buffer to maintain a stable pH and common solvent. A pH of 5.0 was adjusted with 1% hydrochloric acid and 1% sodium hydroxide. Bacteria were grown in PBS in the presence of humic acid, AgNPs, $AgNO_3$ and ($Pb(NO_3)_2$) with each bacterial strain. Stirring was maintained in all conditions, and the temperature remained at 37°C.

Characterization

The size of the AgNPs was measured using a JEOL transmission electron microscope (TEM), model JEM-1230. Samples were prepared by adding drops of AgNPs onto a copper grid and drying in a vacuum. UV–visible absorbance spectra of synthesized AgNPs and their interaction with heavy metal ions were recorded using a Varian Cary 50 Bio UV–visible spectrophotometer. Optical absorbance density of bacterial growth was recorded with a BioTek Epoch Microplate spectrophotometer.

Results and discussion

Monitoring surface plasmon resonance of AgNPs with increased Pb^{+2} to determine stability

The stability of AgNPs impacts their transport and function when released into the environment. AgNPs can undergo dissolution and aggregation in the presence of chemical species such as Pb^{+2}. To understand the behavior of AgNPs in the presence of Pb^{+2}, we investigated the stability and surface interaction of Pb^{2+} on AgNPs by UV–vis absorbance spectroscopy. To a solution of 5 mL of AgNPs, lead (II) nitrate was added at various concentrations (70–200 μL), ranging over a 20-h time frame. Upon adding Pb^{+2} to AgNPs, the solution changed colors from yellow to pale pink, indicating the AgNPs were undergoing surface modifications. Fig. 1 displays the UV–vis spectra of the AgNPs with increasing Pb^{+2} concentration. The initial interaction of Pb^{+2} with AgNPs surface led to the slight broadening of the SPR peak of AgNPs and caused the formation of a secondary peak around 555 nm at higher concentrations. After 20 h, a secondary peak became visible at each concentration, and higher concentrations resulted in quenching of the initial peak at 395 nm and a more defined secondary peak at 555 nm. The SPR data obtained gives insight into the impact of Pb^{+2} on the stability of AgNPs. It is evident that the surface's aggregation is present, but there is no clarity of secondary peak formation.

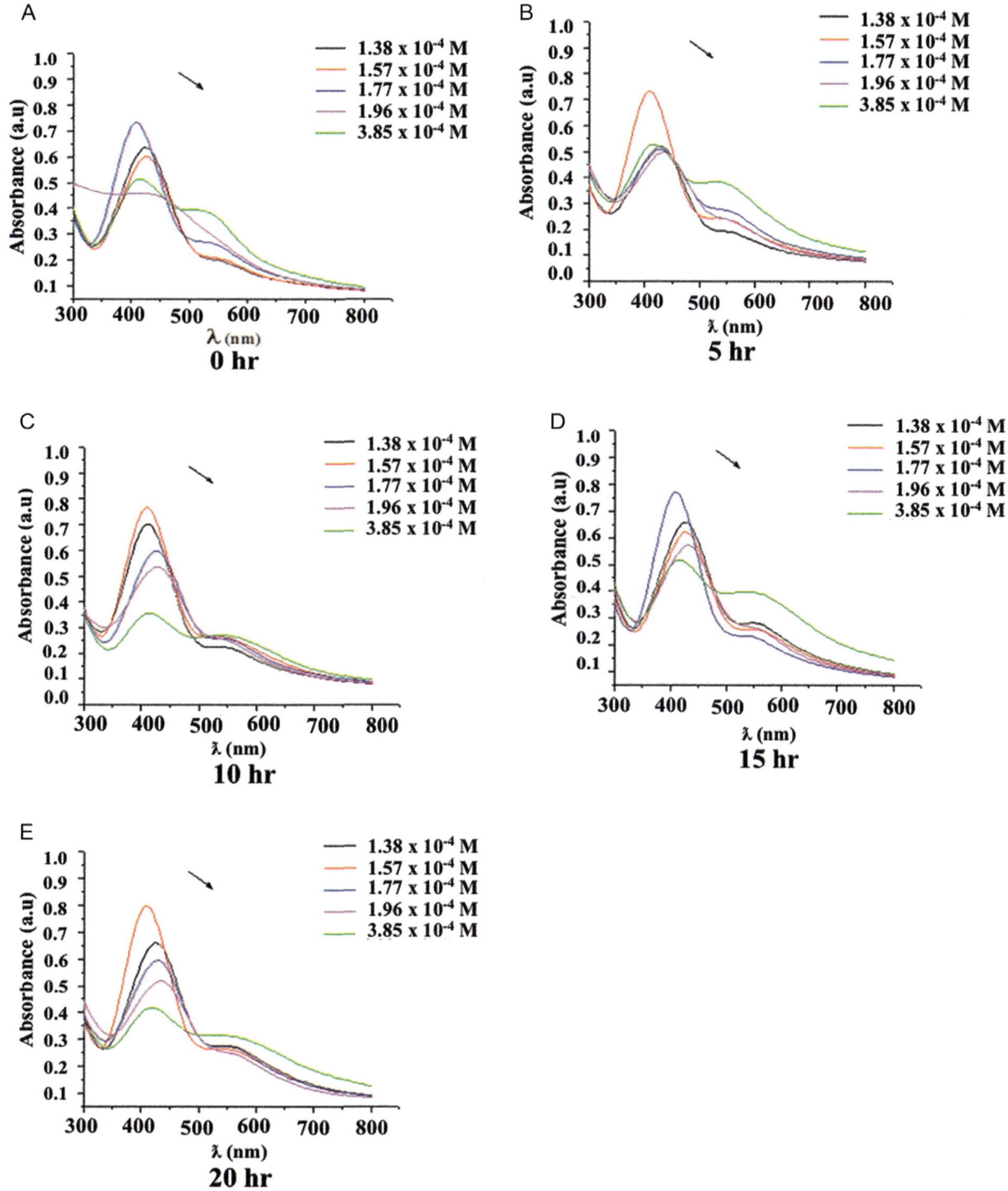

FIG. 1 UV–visible absorbance spectroscopy of the 4 nm AgNPs over 20 h of exposure to 1.38×10^{-4} M, 1.57×10^{-4} M, 1.77×10^{-4} M, 1.96×10^{-4} M, and 3.85×10^{-4} M Pb^{+2}. Exposure of Pb^{+2} occurred at (A) 0 h, (B) 5 h, (C) 10 h, (D) 15 h, and (E) 20 h.

Transmission electron microscopy (TEM) characterization was used to investigate AgNPs and heavy metal interaction, further displayed in Fig. 2. The surface functional groups of hydroxyl and amine present on citrate behave as a reducing metal agent [17–20]. Pb^{+2} alters the size and stability of AgNPs, displayed by aggregation. As time progressed, significant dissolution was seen.

Antimicrobial response of AgNPs and Pb^{+2} on bacterial strains

Pb^{+2} poses a threat to human and environmental health through accumulation and transport [21]. Microorganisms have learned to develop a defense against the toxicity of Pb^{+2} [22–27]. Through sorption, bacteria can resist the toxicity of Pb^{+2} at high concentrations over time. Compared to other heavy metals, Pb^{+2} has been studied to have little to no impact on the toxicity to bacterial cells [28]. Li et al. investigated the bioremediation of Pb^{+2} in soil by gram-negative bacteria *R. sphaeroides* [27]. *Rhodobacter* did not reduce total Pb^{+2} but change the speciation state to lead sulfate and lead sulfide. Levinson et al. identified that Pb^{+2} immobilizes within the intracellular process upon uptake, leading to the formation of nontoxic lead phosphate [25,26]. However, Pb^{+2} is present in highly contaminated sites and can interact with other anthropogenic contaminants such as NPs. These interactions can impact the bioremediation response of the bacteria by altering resistance [14,29]. The instability of NPs can promote the release of the heavy metal center, thus changing the physiochemical properties and behavior of NPs released into the environment [8,30]. Pb^{+2} and AgNPs promoted the instability of NPs. This project investigates the impact of Pb^{+2} on the antimicrobial properties of AgNPs. To investigate the toxic response, we measured the antimicrobial behavior of AgNPs and Pb^{+2} toward gram-positive and gram-negative strains *S. aureus* and *E. coli*, respectively, displayed in Figs. 3 and 5.

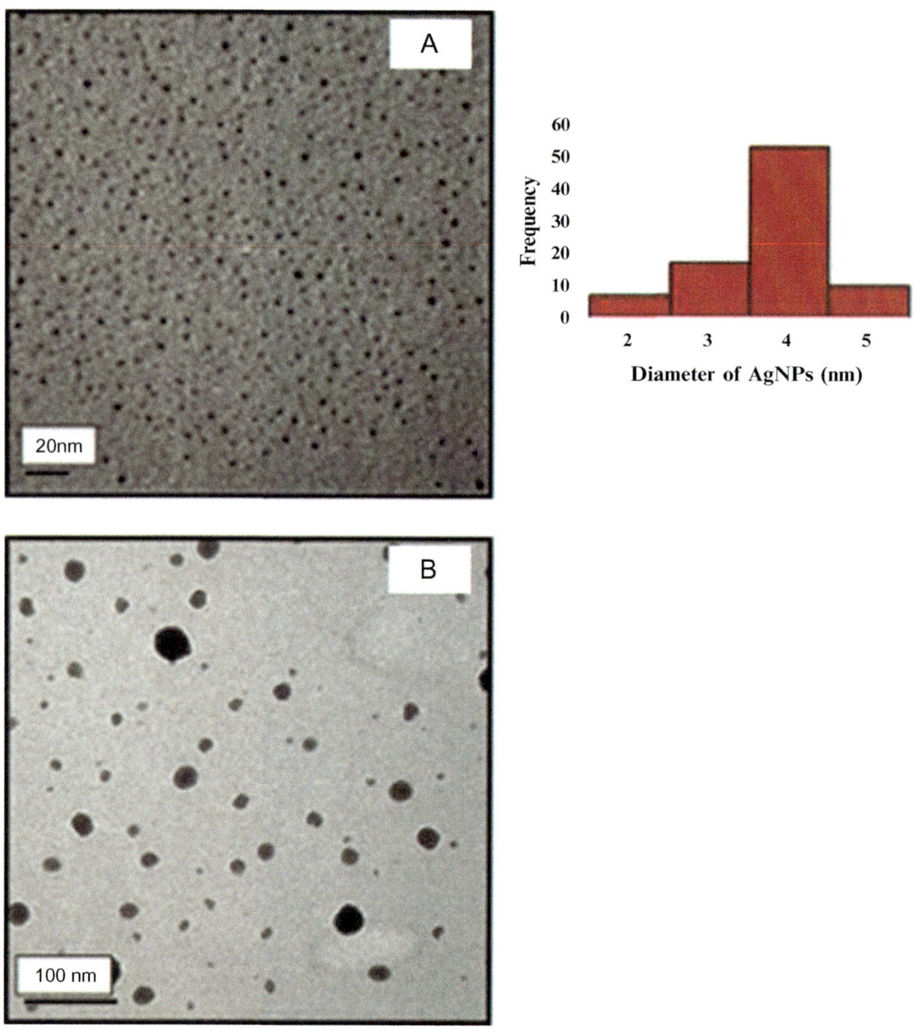

FIG. 2 (A) 4 nm citrate-AgNPs (scale bar = 50 nm) and (B) citrate-AgNPs and 3.85×10^{-4} M Pb^{2+} at 0 h of exposure (scale bar = 100 nm).

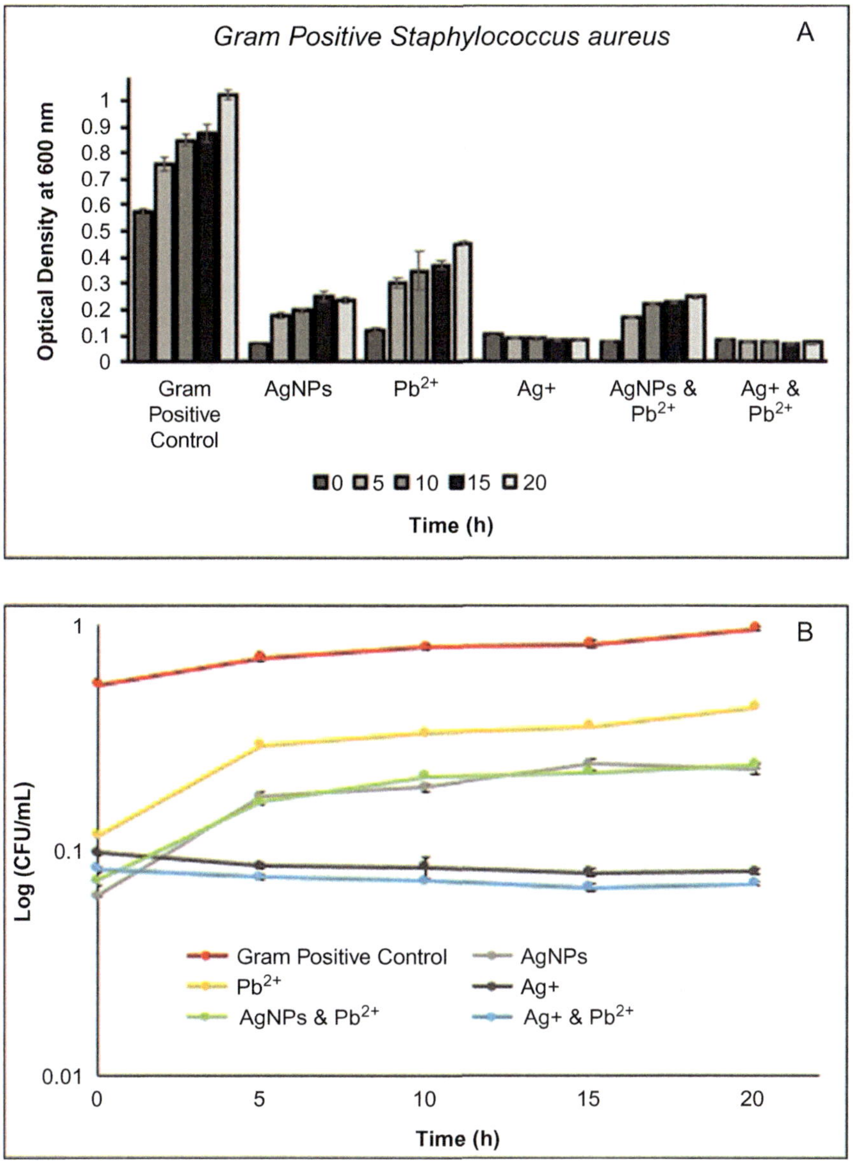

FIG. 3 Pb^{2+} exposure to (A) *S. aureus* values for time frame 0h, 5h, 15h, and 20h. Error bars support standard deviation. (B) *S. aureus* growth profile.

All reaction conditions caused a toxic response toward the growth of both bacterial strains. Pb^{+2} was the least toxic to bacteria growth. In the presence of *S. aureus*, there was an exponential growth of AgNPs in the presence and absence of Pb^{+2} upon initial 5h of exposure. AgNPs and Pb^{+2} displayed similar toxicity as AgNPs alone. Ag^{+} and Pb^{+2} slightly increased the toxicity of Ag^{+} alone over 20h. TEM characterization demonstrates the accumulation of AgNPs at the surface of *S. aureus* cells seen in Fig. 4. The AgNPs are aggregated from the initial size, which can impact the bacteria's function and morphology.

The presence of Pb^{+2} caused a slight reduction in AgNPs toxicity toward *E. coli* after 10h of exposure (4%) displayed in Fig. 5. The coexposure of AgNPs and Pb^{+2} was more toxic than Pb^{+2} alone as time progressed (>10%). At a longer time of exposure, the presence of AgNPs (20%) or Ag^{+} (30%) with Pb^{+2} was more toxic than Pb^{+2} alone. However, the toxicity response was similar to control groups of AgNPs and Ag^{+}.

Compared to gram-positive *S. aureus*, the presence of Pb^{+2} slightly decreased the toxic response of Ag^{+} to gram-negative *E. coli*. AgNPs and Pb^{+2} promoted alteration at *E. coli* bacterial strains' surface, as displayed in Fig. 6. The spherical structures are intact, yet there are changes visible at the surface. The presence of Pb^{+2} caused low yield and agglomerated AgNPs displayed in Fig. 6B. Alterations on AgNPs surface could be due to the promotion of dissolution of AgNPs by Pb^{+2} at a limited time of exposure [31–33].

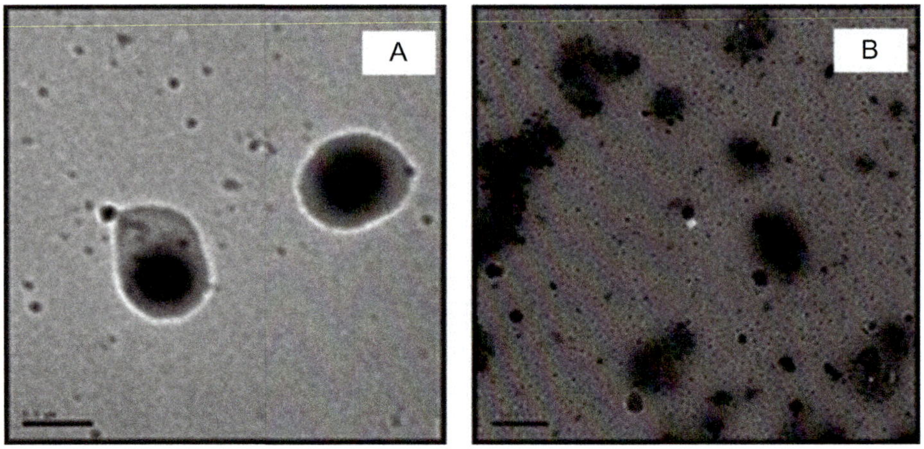

FIG. 4 (A) TEM characterization of *S. aureus* in the presence of Pb^{+2} and AgNPs. (B) TEM of AgNPs upon exposure of Pb^{+2}.

FIG. 5 Pb^{+2} exposure to (A) *E. coli* values for time frame 0h, 5h, 10h, 15h, and 20h. Error bars support standard deviation. (B) *E. coli* growth profile.

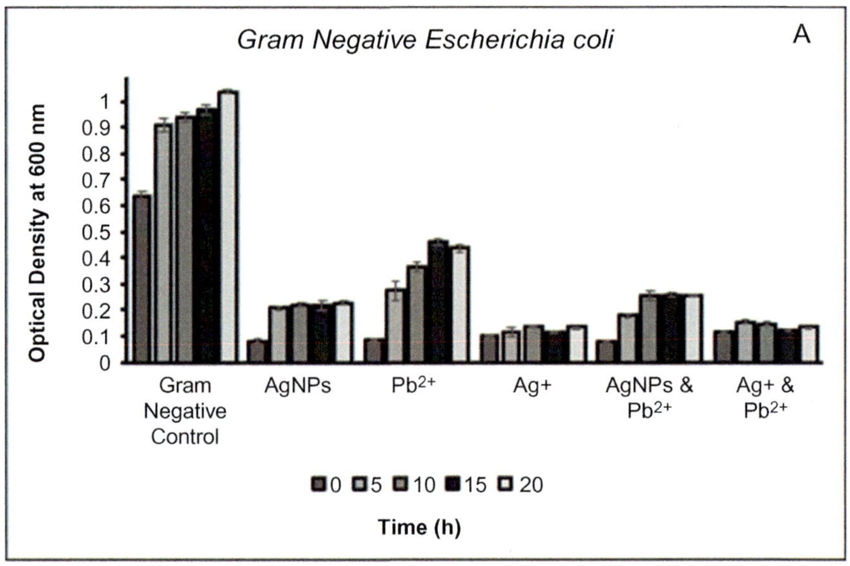

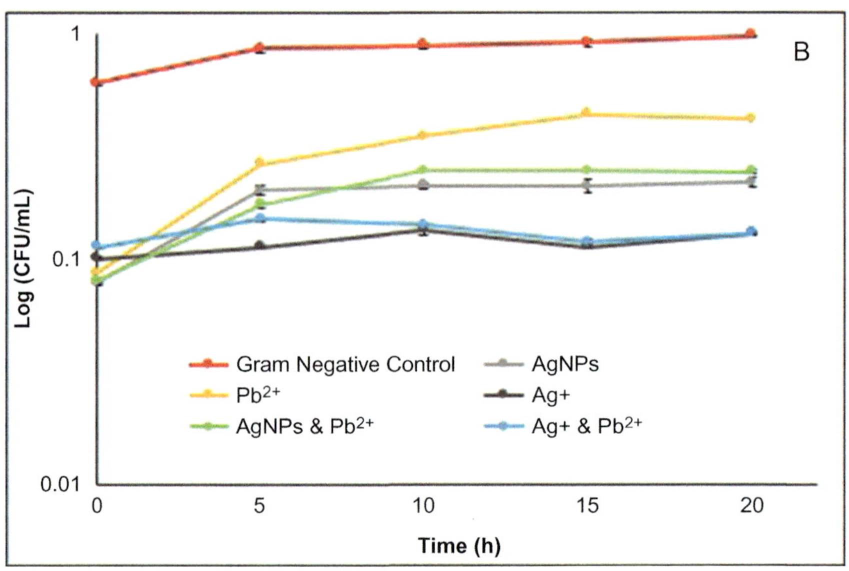

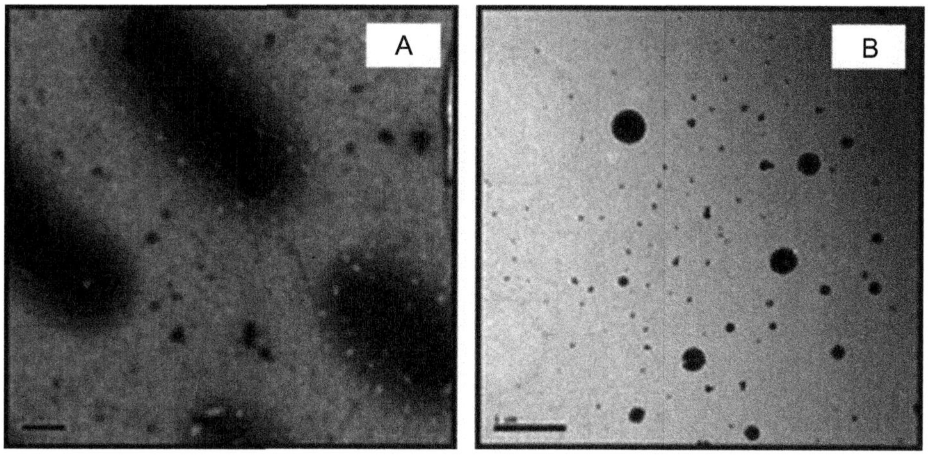

FIG. 6 (A) TEM characterization of *E. coli* in the presence of Pb^{2+} and AgNPs. (B) TEM of AgNPS upon exposure of Pb^{2+}.

Impact of environmental stressors: AgNPs and Pb^{+2}

Microorganisms have played an essential role in the bioremediation process in the environment [22–34]. The cationic character of Pb^{+2} allows sorption by a microorganism such as bacteria with negatively charged surfaces. Microorganisms within contaminated soils and industrial waste sites have developed a defense against heavy metals such as Pb^{+2}. Past studies have investigated expressed genes for the bioremediation of heavy metals. The resistance of Pb^{+2} is known to occur through the superfamily of cation-translocating P-type ATPases. For example, a Rensing et al. study revealed that some strains of *S. aureus* and *E. coli* expression of genes CadA and ZntA could resist Pb^{+2} at high concentrations, respectively [35]. The presence of bacterial strains in heavy metals determines their resistance and behavior for bioremediation [22,36,37]. Paredes-Paliz et al. determined that Pb^{+2} equally distributes inside and outside the cell of gram-negative strains as in the presence of gram-positive bacteria, Pb^{+2} showed lower accumulation [38]. Thus, it is evident that bacterial strains have pathways for the bioremediation of Pb^{+2}. However, the interaction of anthropogenic pollutants such as NPs and heavy metals can influence bacteria bioremediation properties. Table 1 displays the toxic response of individual and coexposure of AgNPs and Pb^{+2} to bacterial strains.

The coexposure of anthropogenic contaminants can alter the toxic response of individual contaminants when released into the environment [33]. As we explored the coexposure of AgNPs and Pb^{+2}, the presence of Pb^{+2} had little impact on the toxicity of AgNPs alone but was significantly more toxic than Pb^{+2} alone. The presence of Pb^{+2} positive charge can allow electrostatic interaction with the bacterial surface or AgNPs. Past studies have shown the difference in metal ion reduction potential, and NPs can dictate their interaction [39,40]. Pb^{+2} is known to immobilize on the surface of NPs sorption and reduction. However, in the presence of AgNPs, Pb^{+2} is known to interact through sorption with ligand surface functional groups [40]. The ligand coating of NPs can influence the binding affinity of Pb^{+2}. Uddin et al. developed Pb^{+2} containing Schiff bases whose antimicrobial response was dependent on the ligands used and the solvent's nature during interaction [41]. The similarity in toxicity to AgNPs alone could depend on the immobilization of Pb^{+2} by AgNPs and allow access of AgNPs to bacteria cells.

Adversely, NPs can promote damage of cellular barriers, thus leading to the increased bioavailability and toxicity of the heavy metal [29,42–44]. Miao et al. suggested that TiO_2 NPs increased the bioconcentration of Pb^{+2} which disrupted the thyroid endocrine and neural system in zebrafish [42]. Jia et al. studied the coexposure of groundwater pollutants ZnO NPs and Pb^{+2} on mice liver [16]. ZnO NPs enhanced the presence of Pb^{+2} in all major organs in overweight mice as there was no significant change in ZnO NPs uptake. He suggested toxicity was caused by NPs causing damage to organ barriers. Therefore, it can be assumed that the sorption of Pb^{+2} influenced the mobilization of Pb^{+2}. In our study, AgNPs had similar toxicity in the presence or absence of Pb^{+2} and could be because of damage to the bacterial surface and an increase in Pb^{+2} uptake.

The coexposure of heavy metal contaminants can present an alternative activity. Bishayi et al. investigated the synergetic impact of heavy metals As^{+3} and Pb^{+2} upon *S. aureus* [45]. The interaction of heavy metal caused a higher toxic response than each metal separately. They led to more prolonged exposure to *S. aureus* within blood and spleen, alteration in cell morphology, inhibition of cell adhesion, nitric oxide release, intracellular killing ability, and increase in DNA fragmentation. Past studies have identified AgNPs antimicrobial behavior is primarily due to the release of Ag^{+} during aggregation or dissolution [46,47]. Therefore, we investigated the interaction of Ag^{+} and Pb^{+2} to compare toxic response. Interestingly, the presence of Pb^{+2} increased the toxicity of Ag^{+} toward *S. aureus* strain as the toxicity of *E. coli* slightly

TABLE 1 Percentage by which environmental stressors AgNPs, Ag^{+}, and Pb^{+2} reduced the growth of bacterial strains (a) *Staphylococcus aureus* and (b) *Escherichia coli.*

Staphylococcus aureus					
	Percentage of Which Bacterial Growth Reduced				
1a	AgNPs	Pb^{2+}	Ag^{+}	AgNPs & Pb^{2+}	Ag^{+} & Pb^{2+}
t=0h	89±4%	79±2%	82±4%	87±4%	85±2%
t=5h	76±2%	60±3%	88±2%	78±2%	90±3%
t=10h	77±2%	59±2%	90±1%	74±2%	91±2%
t=15h	71±7%	58±2%	91±1%	74±2%	92±2%
t=20h	77±4%	56±2%	92±2%	76±3%	93±2%
Escherichia coli					
	Percentage of Which Bacterial Growth Reduced				
1b	AgNPs	Pb^{2+}	Ag^{+}	AgNPs & Pb^{2+}	Ag^{+} & Pb^{2+}
t=0h	87±2%	86±4%	84±3%	87±2%	82±2%
t=5h	77±1%	70±2%	87±1%	80±2%	83±3%
t=10h	77±1%	61±4%	85±2%	73±3%	84±2%
t=15h	78±2%	52±2%	88±2%	74±2%	87±3%
t=20h	78±2%	58±3%	87±1%	75±2%	87±2%

decreased. Our study displays a different behavioral response between AgNPs and Ag^{+} in the presence of Pb^{+2}. Based on HSAB theory, Ag^{+} is a soft acid with a high affinity to soft bases present within the cell wall of bacterial strains [9]. Pb^{+2}, on the other hand, is a neutral acid that has a lower binding affinity toward negatively charged bacteria. Both cations can promote toxicity due to selectivity to groups such as thiols that Ag^{+} did not bind with allowed the binding of $Pb^{+2.}$ The presence of Pb^{+2} with AgNPs or Ag^{+} resulted in a higher reduction of bacterial strain growth. The toxicity of heavy metals such as Pb^{+2} is dependent on oxidative stress. The binding affinity of Ag^{+} can be more selective to the surface.

Antimicrobial response of HA, AgNPs, and Pb^{+2} on bacterial strains

Natural organic matter (NOM) within contaminated sites can alter contaminants' adsorption, such as Pb^{+2} [48,49]. NOM has been studied for its impact on the toxicity of heavy metals when released into the environment. Worm et al. determined the addition of irradiation of Elliot humic acid (EHA) alters the availability of heavy metals toward green microalgae [50]. The presence of NOM can interfere with the uptake of Pb^{+2} by microorganisms. NOM is also known to inhibit the absorption of Pb^{+2} with remediation techniques such as the presence of NPs [51]. In the presence of NOM, the titanate nanotubes adsorption capacity of Pb^{+2} reduces because of a weak interaction with NOM [51]. Yang et al. analyzed the coexposure of Pb^{+2} and humic acid adsorption on multiwalled carbon nanotubes in aquatic systems [52]. Oxidative stress is a leading source of toxicity of heavy metals to cells. In the presence of PbO_2, NOM can promote reduction within a few hours [53]. Humic acid (HA) has hydrophilic surface groups (carboxyl, hydroxyl, phenol, and amines) prominently seen in the environment with a high affinity for heavy metals [53–55]. Here, we explore how HA can impact Pb^{+2} and AgNPs upon bacterial strains displayed in Figs. 7 and 8.

The toxicity response of HA and Pb^{+2} to *S. aureus* was dependent on the time of exposure as displayed in Fig. 7. The presence of HA reduced the toxicity of Pb^{+2} control groups at each time frame except at 0h and 20h. Ag^{+}, Pb^{+2}, and HA were significantly toxic to bacteria. Every other reaction condition displayed exponential toxicity significantly more than growth within 5h of exposure and then plateaued. When exposed to *E. coli*, HA significantly increased the toxicity above that of AgNPs and Pb^{+2}, with HA as displayed in Fig. 8.

All stressors were most toxic than Pb^{2+} alone, yet similar to AgNPs except at 10h. At 10h of exposure, Ag^{+}, Pb^{+2}, and HA were more toxic than Ag^{+}. Each reaction condition displayed exponential growth that plateaued in growth. HA and AgNPs and HA and Pb^{+2} exhibited a similar toxic response.

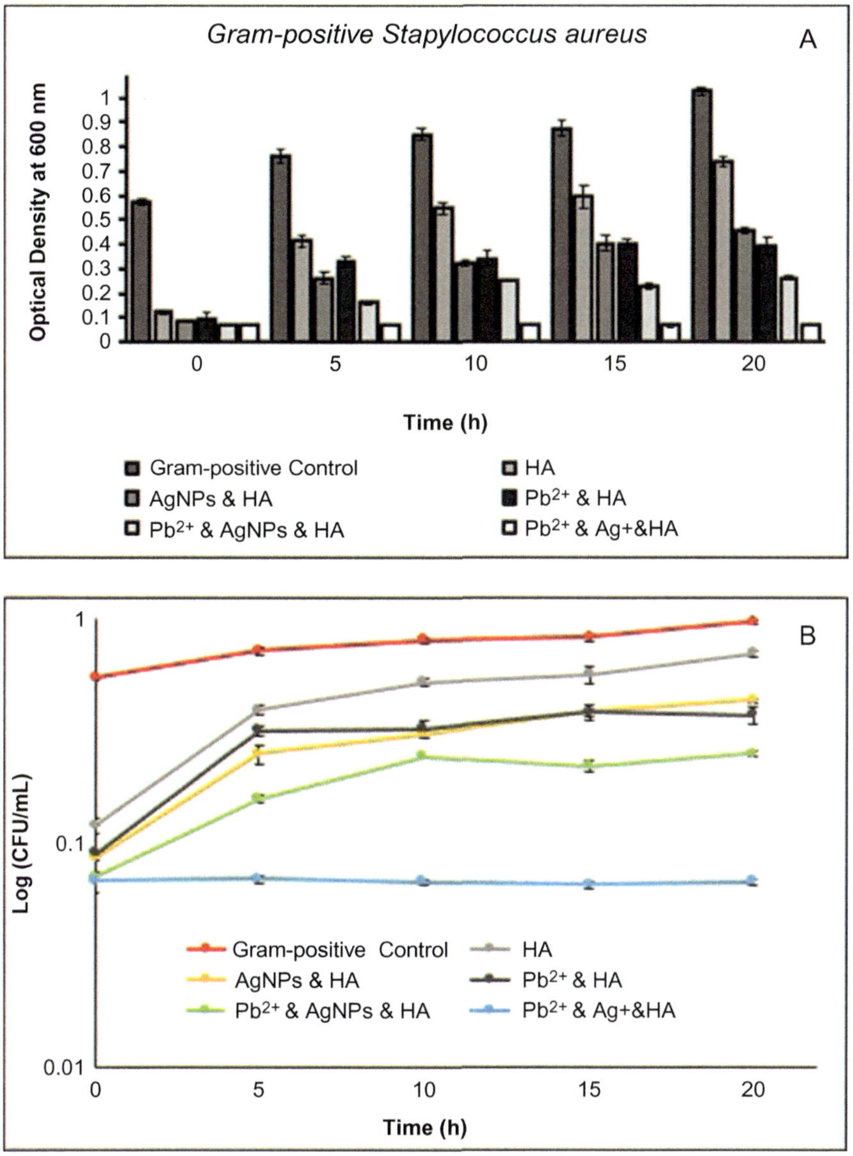

FIG. 7 Pb^{+2} exposure to (A) *S. aureus* values for time frame 0h, 5h, 10h, 15h, and 20h in the presence of environmental contaminants and HA. Error bars support standard deviation. (B) *S. aureus* growth profile.

Impact of environmental stressors: HA, AgNPs, and Pb^{+2}

Humic acid (HA) is an excellent depiction of NOM character present within the environment. HA is known to bind Pb^{2+} species through carboxylic acid and phenolic groups present within salicylate and phthalate [54]. The binding affinity of Pb^{2+} is dependent on concentration and pH levels because H^+ and Pb^{2+} compete for binding sites [54]. The interaction of NOM and AgNPs has been studied extensively for their environmental impact [56,57]. NOM promotes the stability of AgNPs reducing its antimicrobial behavior as Pb^{2+} can encourage instability and the release of Ag^+. As the interaction of all three environmental stressors is ubiquitous, we investigated their toxicity toward bacterial strain growth. Past studies have analyzed the impact of NOM on NPs removal of heavy metals. To our knowledge, this is the first study to investigate the effect of HA on the antimicrobial properties of AgNPs and Pb^{2+} Table 2 displays the toxic response of each reaction condition exposed to bacteria over time. HA had little impact on the growth rate of bacterial strains. However, in the presence of AgNPs, there was a significant reduction in the growth of both bacterial strains when compared to AgNPs control.

The presence of all three environmental stressors displayed higher toxicity to bacterial strains compared to individual contaminants. Alterations in toxic response could be a result of competition for multiple ligand binding groups. HA is known to alter adsorption and toxicity of different pollutants released within the environment. Multiple studies have suggested That the presence of HA reduces the toxicity of environmental stressors to microorganisms, yet conditions have also

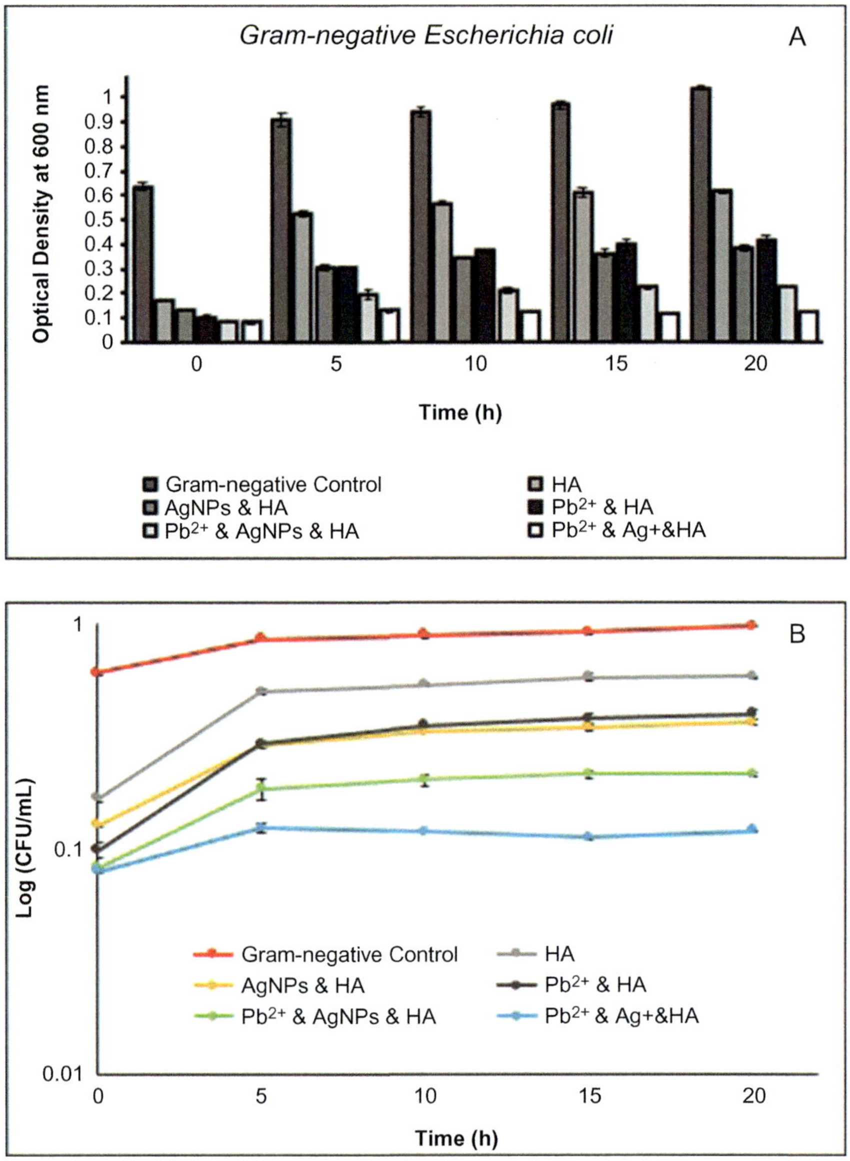

FIG. 8 Pb^{2+} exposure to (A) *E. coli* values for time frame 0h, 5h, 10h, 15h, and 20h in the presence of environmental contaminants and HA. Error bars support standard deviation. (B) *E. coli* growth profile.

been seen to promote toxicity. Zheng et al. investigated the impact of various ligand species on the toxicity and bioavailability of Pb^{2+} by mustard seeds [58]. HA was seen to promote toxicity and decreased the bioavailability. Kozlova et al. determined HA and bacteria together were seen to improve the diversity of soil-polluted regions [59]. However, in the case of dehydrogenase and phosphatase activities, the presence of HA and bacteria with Pb^{2+} did not have a positive impact on growth rate. Adversely, among aquatic species, the presence of NOM has been seen to increase bioavailability and toxicity to marine organisms. Sanchez-Marin et al. investigated Suwannee NOM and FA (fulvic acid) to compare the impact to HA on Pb^{2+} toxicity [60]. Each organic species displayed toxic response to urchin embryos. Therefore, it can be suggested the presence of multiple ligand species can compete for the sorption or complexation of Pb^{2+} and Ag character.

The presence of HA can influence the interaction of AgNPs and Pb^{2+} [53,54]. NOM character can promote complexation of the heavy metal ion to the surface of AgNPs. Past studies have determined pH, and ionic strength plays a prominent role in NPs adsorption of heavy metals in the presence of NOM [52,61–64]. Lin et al. studied the sorption of Pb^{2+} by multiwalled carbon nanotubes (MWCNTs) in the presence of HA [64]. HA promoted the electrostatic attraction between the outer spheres of MWCNTs and Pb^{2+}. HA provided oxygen-based functional groups that increased the negative surface charge. Ionic strength also increased the selective adsorption of titanate nanotubes to Pb^{2+} [65]. It is evident the interaction of multiple stressors can cause several interactions that alter the toxicity of AgNPs alone.

TABLE 2 Percentage of which environmental stressors HA, AgNPs, Ag^+, and Pb^{+2} reduced the growth of bacterial strain (a) *Staphylococcus aureus* and (b) *Escherichia coli.*

Staphylococcus aureus					
	Percentage of Which Bacterial Growth Reduced				
2a	HA	HA & AgNPs	HA & Pb^{2+}	AgNPs, Pb^{2+} & HA	Ag^+ Pb^{2+} & HA
$t=0$h	78±5%	84±4%	84±5%	87±2%	88±2%
$t=5$h	46±3%	66±2%	56±3%	79±4%	91±2%
$t=10$h	36±2%	62±3%	60±2%	70±3%	92±2%
$t=15$h	32±2%	54±2%	54±3%	74±2%	92±3%
$t=20$h	28±2%	56±2%	62±3%	75±2%	93±3%

Escherichia coli					
	Percentage of Which Bacterial Growth Reduced				
2b	HA	HA & AgNPs	HA & Pb^{2+}	AgNPs, Pb^{2+} & HA	Ag^+ Pb^{2+} & HA
$t=0$h	73±2%	79±4%	84±4%	87±3%	87±2%
$t=5$h	42±1%	67±4%	66±3%	79±2%	86±2%
$t=10$h	40±2%	63±3%	64±2%	73±4%	93±3%
$t=15$h	37±2%	63±2%	59±2%	77±2%	88±2%
$t=20$h	41±4%	63±2%	60±2%	78±2%	88±2%

Conclusions

As economic and industrial practices continue to expand, the release of anthropogenic contaminants will present a threat to environmental health. With the presence of multiple contaminants in the environment, they can interact with each other and environmental components based on physiochemical properties. Here, we explored leading contaminant Pb^{+2} interactions with AgNPs and HA. The presence of Pb^{2+} caused instability of the surface of AgNPs. To explore the antimicrobial impact of AgNPs and Pb^{2+} interaction, we measured their toxicity toward gram-positive and gram-negative bacteria, using optical density at 600nm. Our investigation revealed that the coexposure of AgNPs and Pb^{2+} played a role in the toxicity of each bacterial strains. There was a slight increase in toxicity when compared to Pb^{2+}, yet a decrease compared to AgNPs control groups. Alterations in toxicity could be based on the adsorption of Pb^{2+} on the surface of AgNPs that limited toxicity to bacterial strains.

Coexposure of heavy metal ions Ag^+ and Pb^{2+} were studied for toxic response to bacterial strains. The difference between Ag^+ and AgNPs could be based on the charge and the type of interaction with the surface of bacterial strains. The presence of NOM seemed to have the most significant impact on toxicity of cationic species and AgNPs toward bacterial strains. The toxic response of Pb^{2+} became significantly more toxic in the presence of other stressors. However, when HA is added to Ag character and Pb^{2+}, reduction in growth was similar to the presence of Ag and HA regardless of the existence of Pb^{2+}. The study revealed that measuring the interaction of environmental contaminants is significant in understanding true toxicity.

References

[1] J. Hou, M. Zhang, P. Wang, C. Wang, L. Miao, Y. Xu, G. You, B. Lv, Y. Yang, Z. Liu, Transport, retention, and long-term release behavior of polymer-coated silver nanoparticles in saturated quartz sand: the impact of natural organic matters and electrolyte, Environ. Pollut. 229 (2017) 49–59.

[2] T.M. Benn, P. Westerhoff, Nanoparticle silver released into water from commercially available sock fabrics, Environ. Sci. Tech. 42 (2008) 4133–4139.

[3] C. Marambio-Jones, E.M.V. Hoek, A review of the antibacterial effects of silver nanomaterials and potential implications for human health and the environment, J. Nanopart. Res. 12 (2010) 1531–1551.

[4] M. Sriuttha, B. Tengjaroenkul, S. Intamat, U. Phoonaploy, P. Thanomsangad, L. Neeratanaphan, Cadmium, chromium, and lead accumulation in aquatic plants and animals near a municipal landfill, Hum. Ecol. Risk Assess. Int. J. 23 (2017) 350–363.

[5] M. Patra, N. Bhowmik, B. Bandopadhyay, A. Sharma, Comparison of mercury, lead, and arsenic with respect to genotoxic effects on plant systems and the development of genetic tolerance, Environ. Exp. Bot. 52 (2004) 199–223.

[6] E.G. Spokas, B.W. Spur, H. Smith, F.W. Kemp, J.D. Bogden, Tissue Lead concentration during chronic exposure of *Pimephales promelas* (fathead minnow) to Lead nitrate in aquarium water, Environ. Sci. Tech. 40 (2006) 6852–6858.

[7] R. Raghuvanshi, 2-Ketogluconic acid and pyrroloquinoline quinone secreting probiotic *Escherichia coli* Nissle 1917 as a dietary strategy against heavy metal-induced damage in rats, J. Funct. Foods 37 (2017) 541–552.

[8] N. Akaighe, S.W. Depner, S. Banerjee, V.K. Sharma, M. Sohn, The effects of monovalent and divalent cations on the stability of silver nanoparticles formed from direct reduction of silver ions by Suwannee River humic acid/natural organic matter, Sci. Total Environ. 15 (2012) 277–289.

[9] J.A. Lemire, J.J. Harrison, R.J. Turner, Antimicrobial activity of metals: mechanisms, molecular targets and applications, Nat. Rev. Microbiol. 11 (2013) 371–384.

[10] R. Deng, D. Lin, L. Zhu, S. Majumdar, J.C. White, J.L. Gardea-Torresdey, B. Xing, Nanoparticle interactions with co-existing contaminants: joint toxicity, bioaccumulation and risk, Nanotoxicology 5 (2017) 591–612.

[11] C.J. Ollson, E. Smith, P. Herde, A.L. Juhasz, Influence of co-contaminant exposure on the absorption of arsenic, cadmium and lead, Chemosphere 168 (2017) 658–666.

[12] S. Kheirandish, M. Ghaedi, K. Dashtian, R. Jannesar, M. Montazerozohori, F. Pourebrahim, M.A. Zare, Simultaneous removal of Cd(II), Ni(II), Pb(II) and Cu(II) ions via their complexation with HBANSA based on a combined ultrasound-assisted and cloud point adsorption method using CSG-BiPO_4/FePO_4 as novel adsorbent: FAAS detection and optimization process, J. Colloid Interface Sci. 500 (2017) 241–252.

[13] P. Venkatachalam, M. Jayaraj, R. Manikandan, N. Geetha, E.R. Rene, N.C. Sharma, S.V. Sahi, Zinc oxide nanoparticles (ZnONPs) alleviate heavy metal-induced toxicity in *Leucaena leucocephala* seedlings: a physiochemical analysis, Plant Physiol. Biochem. 110 (2017) 59–69.

[14] E. Barbieri, J. Campos-Garcia, D.S. Martinez, J.R. da Silva, O.L. Alves, K.F. Rezende, Histopathological effects on gills of Nile tilapia (*Oreochromis niloticus*, Linnaeus, 1758) exposed to Pb and carbon nanotubes, Microsc. Microanal. 6 (2016) 1162–1169.

[15] D.S.T. Martinez, O.L. Alves, E. Barbieri, Carbon nanotubes enhanced the lead toxicity on the freshwater fish, J. Phys.; Conf. Ser. 429 (2013) 012043.

[16] J. Jia, F. Li, S. Zhai, H. Zhou, S. Liu, G. Jiang, B. Yan, Susceptibility of overweight mice to liver injury as a result of the ZnO nanoparticle-enhanced liver deposition of Pb^{2+}, Environ. Sci. Tech. 51 (2017) 1775–1784.

[17] M. Annadhasan, T. Muthukumarasamyvel, V.R.S. Babu, N. Rajendiran, Green synthesized silver and gold nanoparticles for colorimetric detection of Hg^{2+}, Pb^{2+}, and Mn^{2+} in aqueous medium, ACS Sust. Chem. Eng. 2 (2014) 887–896.

[18] F. Chai, C. Wang, T. Wang, L. Li, Z. Su, Colorimetric detection of Pb^{2+} using glutathione functionalized gold nanoparticles, ACS Appl. Mater. Interfaces 2 (2010) 1466–1470.

[19] Y.H. Chien, C.C. Huang, S.W. Wang, C.S. Yeh, Synthesis of nanoparticles: sunlight formation of gold nanodecahedra for ultrasensitive lead-ion detection, Green Chem. 13 (2011) 1162–1166.

[20] K. Yoosaf, B.I. Ipe, C.H. Suresh, K.G. Thomas, In situ synthesis of metal nanoparticles and selective naked-eye detection of lead ions from aqueous media, J. Phys. Chem. C 111 (2007) 12839–12847.

[21] G. Flora, D. Gupta, A. Tiwari, Toxicity of lead: a review with recent updates, Interdiscip. Toxicol. 5 (2012) 47–58.

[22] A. Jarostaweicka, Z. Piotrowska-Seget, Lead resistance in micro-organisms, Microbiology 160 (2014) 12–25.

[23] M.M. Naik, S.K. Dubey, Lead resistant bacteria: Lead resistance mechanisms, their applications in lead bioremediation and biomonitoring, Ecotoxicol. Environ. Saf. 98 (2013) 1–7.

[24] W. Wei, X. Liu, P. Sun, X. Wang, H. Zhu, M. Hong, Z.-W. Mao, J. Zhao, Simple whole-cell biodetection and bioremediation of heavy metals based on an engineered Lead-specific operon, Environ. Sci. Tech. 48 (2014) 3363–3371.

[25] H.S. Levinson, I. Mahler, P. Blackwelder, T. Hood, Lead resistance, and sensitivity in *Staphylococcus aureus*, FEMS Microbiol. Lett. 145 (1996) 421–425.

[26] H.S. Levinson, I. Mahler, Phosphatase activity and lead resistance in *Citrobacter freundii* and *Staphylococcus aureus*, FEMS Microbiol. Lett. 161 (1998) 135–138.

[27] X. Li, W. Peng, Y. Jia, L. Lu, W. Fan, Bioremediation of lead contaminated soil with *Rhodobacter sphaeroides*, Chemosphere 156 (2016) 228–235.

[28] C. Vijayadeep, P.S. Sastry, Effects of heavy metal uptake by *E. coli* and *Bacillus sps*, J. Bioremed. Biodegr. 5 (2014) 238.

[29] A.A. Shumakova, E.N. Trushina, O.K. Mustafina, S.K. Soto, I.V. Gmoshinsky, S.A. Khotimchenko, Lead toxicity in its joint administration with the aluminum oxide nanoparticles to rats, Vopr. Pitan. 84 (2015) 40–49.

[30] Y. Zhang, Y. Chen, P. Westerhoff, J. Crittenden, Impact of natural organic matter and divalent cations on the stability of aqueous nanoparticles, Water Res. 43 (2009) 4249–4257.

[31] V.V. Kumar, S. Anbarasan, L.R. Christena, N.S. Subramanian, S.P. Anthony, Bio-functionalized silver nanoparticles for selective colorimetric sensing of toxic metal ions and antimicrobial studies, Spectrochim Acta A: Mol. Biomol. Spectrosc. 129 (2014) 35–42.

[32] X. Li, J.J. Lenhart, H.W. Walker, Aggregation kinetics and dissolution of coated silver nanoparticles, Langmuir 28 (2012) 1095–1104.

[33] C. Levard, E.M. Hotze, G.V. Lowry, G.E. Brown, Transformation of silver nanoparticles: impact on stability and toxicity, Environ. Sci. Tech. 46 (2012) 6900–6914.

[34] C.-H. Kang, Y.-J. Kwon, J.-S. So, Bioremediation of heavy metals by using bacterial mixtures, Ecol. Eng. 89 (2016) 64–69.

[35] C. Rensing, Y. Sun, B. Mitra, B.P. Rosen, Pb(II)-translocating P-type ATPases, J. Biol. Chem. 273 (1998) 32614–32617.

[36] D.H. Nies, Microbial heavy-metal resistance, Appl. Microbiol. Biotechnol. 51 (1999) 730–750.

[37] A. Cabuk, T. Akar, S. Tunali, S. Gedikli, Biosorption of Pb(II) by industrial strain of *Saccharomyces cerevisiae* immobilized on the biomatrix of cone biomass of *Pinus nigra*: equilibrium and mechanism analysis, Chem. Eng. J. 131 (2007) 293–300.

[38] K.I. Paredes-Paliz, M.A. Caviedes, B. Doukkali, E. Mateos-Naranjo, I.D. Rodriguez-Llorente, E. Pajuelo, Screening beneficial rhizobacteria from *Spartina maritima* for phytoremediation of metal polluted salt marshes: comparison of gram-positive and gram-negative strains, Environ. Sci. Pollut. Res. 23 (2016) 19825–19837.

[39] X.-Q. Li, W.-X. Zhang, Sequestration of metal cations with zerovalent iron nanoparticles: a study with high resolution X-ray photoelectron spectroscopy (HR-XPS), J. Phys. Chem. C 111 (2007) 6939–6946.

[40] M.S. Bootharaju, T. Pradeep, Uptake of toxic metal ions from water by naked and monolayer protected silver nanoparticles: an X-ray photoelectron spectroscopic investigation, J. Phys. Chem. C 114 (2010) 8328–8336.

[41] M.N. Uddin, A. Salam, J. Sultana, Pb(II) complexes of Schiff bases derived from benzoylhydrazine as the antibacterial agents, Modern Chem. 3 (2015) 7–14.

[42] W. Miao, B. Zhu, X. Xiao, Y. Li, N.B. Dirbaba, Effects of titanium dioxide nanoparticles on lead bioconcentration and toxicity on thyroid endocrine system and neuronal development in zebrafish larvae, Aquat. Toxicol. 161 (2015) 117–126.

[43] C. Zhang, S. Zhai, L. Wu, Y. Bai, J. Jia, Y. Zhang, B. Zhang, B. Yan, Induction of size-dependent breakdown of blood-milk barrier in lactating mice by TiO_2 nanoparticles, PLoS One 10 (2015) e0122591.

[44] N.B. Hartmann, F. Von der Kammer, T. Hofmann, M. Baalousha, S. Ottofuelling, A. Baun, Algal testing of titanium dioxide nanoparticles-testing considerations, inhibitory effects and modification of cadmium bioavailability, Toxicology 269 (2010) 190–197.

[45] B. Bishayi, M. Sengupta, Synergism in immunotoxicological effects due to repeated combined, Int. Immunopharmacol. 6 (2006) 454–464.

[46] B. Reidy, A. Haase, A. Luch, K.A. Dawson, I. Lynch, Mechanisms of silver nanoparticle release, transformation and toxicity: a critical review of current knowledge and recommendations for future studies and applications, Mater. 6 (2013) 2295–2350.

[47] J. Dobias, R. Bernier-Latmani, Silver release from silver nanoparticles in natural waters, Environ. Sci. Tech. 47 (2013) 4140–4146.

[48] M. Jerzykiewicz, Formation of new radicals in humic acids upon interaction Pb(II) ions, Geoderma 122 (2004) 305–309.

[49] Y. Wang, F.M. Michel, Y. Choi, P.J. Eng, C. Levard, H. Siebner, B. Gu, J.R. Bargar, G.E. Brown, Pb, Cu, and Zn distributions at humic acid-coated metal-oxide surfaces, Geochim. Cosmochim. Acta 188 (2016) 407–423.

[50] I.A.M. Worms, D. Adenmatten, P. Mieville, J. Traber, V.I. Slaveykova, Photo-transformation of pedogenic humic acid and consequences for Cd(II), Cu(II) and Pb(II) speciation and bioavailability to green microalga, Chemosphere 138 (2015) 908–915.

[51] J. Ma, F. Li, T. Qian, H. Liu, W. Liu, D. Zhao, Natural organic matter resistant powder activated charcoal supported titanate nanotubes for adsorption of Pb(II), Chem. Eng. J. 315 (2017) 191–200.

[52] S.B. Yang, J. Hu, C.L. Chen, D.D. Shao, X.K. Wang, Mutual effect of Pb(II) and humic acid adsorption on multiwalled carbon nanotubes/polyacrylamide composites from aqueous solutions, Environ. Sci. Tech. 45 (2011) 3621–3627.

[53] L.D. Winning, B. Gorczyca, K. Brezinski, Effect of total organic carbon and aquatic humic substances on the occurrence of lead at the tap, Water Qual. Res. J. 52 (2017) 2–10.

[54] J. Xiong, L.K. Koopa, W.-F. Tan, L.-C. Fang, M.-X. Wang, W. Zhao, F. Liu, J. Zhang, L.-P. Weng, Lead binding to soil fulvic and humic acids: NICA-Donnan modeling and XAFS spectroscopy, Environ. Sci. Tech. 47 (2013) 11634–11642.

[55] R. Kautenburger, C. Hein, J.M. Sander, H.P. Beck, Influence of metal loading and humic acid functional groups on the complexation behavior of trivalent lanthanides analyzed by CE-ICP-MS, Anal. Chim. Acta 816 (2014) 50–59.

[56] N.F. Adegboyega, V.K. Sharma, K.M. Siskova, R. Vecerova, M. Kolar, R. Zboril, J.L. Gardea-Torresdey, Enhanced formation of silver nanoparticles in Ag^+-NOM –iron(II, III) systems and antibacterial activity studies, Environ. Sci. Tech. 48 (2014) 3228–3235.

[57] S.L. Chinnapongse, R.I. MacCuspie, V.A. Hackley, Persistence of singly dispersed silver nanoparticles in natural freshwaters, synthetic seawater, and simulated estuarine waters, Sci. Total Environ. 409 (2011) 2443–2450.

[58] S. Zheng, Y. Xu, G. Sun, D. Lin, Y. Sun, Effects of dissolved organic matter and pH on toxicity and bioavailability of lead, Nongye Huanjing Kexue Xuebao 25 (2006) 1413–1418.

[59] E. Kozlova, A.L. Stepanov, L.V. Lysak, The influence of bacterial-humus preparations on the biological activity of soils polluted with oil products and heavy metals, Eurasian Soil Sci. 48 (2015) 400–409.

[60] P. Sanchez-Marin, R. Beiras, Quantification of the increase in pb bioavailability to marine organisms caused by different types of DOM from terrestrial and river origin, Aquat. Toxicol. 110-111 (2012) 45–53.

[61] W.-W. Tang, G.-M. Zeng, J.-L. Gong, J. Liang, P. Xu, C. Zhang, B.-B. Huang, Impact of humic/fulvic acid on the removal of heavy metals from aqueous solutions using nanomaterials: a review, Sci. Total Environ. 468-469 (2014) 1014–1027.

[62] C.H. Wu, C.F. Lin, H.W. Ma, T.Q. His, Effects of fulvic acid on the sorption of Cu and Pb onto γAl_2O_3, Water Res. 37 (2003) 743–752.

[63] J. Li, S. Zhang, C. Chen, G. Zhao, X. Yang, J. Li, X. Wang, Removal of Cu(II) and fulvic acid by graphene oxide nanosheets decorated with Fe_3O_4 nanoparticles, ACS Appl. Mater. Interfaces 4 (2012) 4991–5000.

[64] D. Lin, X. Tian, T. Li, Z. Zhang, X. He, B. Xing, Surface-bound humic acid increased Pb^{2+} sorption on carbon nanotubes, Environ. Pollut. 167 (2012) 138–147.

[65] T. Wang, W. Liu, L. Xiong, N. Xu, J. Ni, Influence of pH, ionic strength, and humic acid on the competitive adsorption of Pb(II), Cd(II) and Cr(III) onto titanate nanotubes, Chem. Eng. J. 215-216 (2013) 366–374.

Chapter 5

Nanocellulose in membrane technology for water purification

Priyanka R. Sharma*, Sunil K. Sharma, and Benjamin S. Hsiao*

Department of Chemistry, Stony Brook University, Stony Brook, NY, United States

**Corresponding authors: e–mail address: priyanka.r.sharma@stonybrook.edu; benjamin.hsiao@stonybrook.edu*

Introduction

Nanostructured carbon (NC)-based membrane possesses specific properties such as high surface area, tunable porosity, immense functionality, flexibility, transparency, and excellent compatibility with the other organic and inorganic molecules, and has attracted their attention into the advance fields of digital world and energy storage such as sensors/security [1], CO_2 capture [2], high demand biomedical membranes [3], etc. NC-based membranes have been classified differently in the literature. For example, based on the structural qualities, NC-based membranes have been classified as self-standing, barrier-composite membranes and nanocomposite membranes; based on the type of NC used, these membranes have been presented as CNF-based membranes, CNC-based membrane, and BNC-based membranes. The self-standing membranes have found limited applications as compared to barrier membranes because of the certain constraints such as the lack of wet strength, hydrophilicity, low tortuosity, and inadequate qualities for a variety of applications; hence, various chemical treatments such as cross-linking, incorporation of organic and inorganic fillers, and surface modification have been considered to overcome these constraints. Exceptionally, the BNC (e.g., *Gluconacetobacter xylinus, Gluconacetobacter hansenii, Acetobacter xylinum*)-based membranes and *Cladophora*-based membranes have found to be excellent membranes without introducing any post-treatment, the membrane exhibited the properties of good dry/wet strength, excellent porosity for micro- and ultra-filtration, and high antifouling properties. Many studies have been done to strengthen the mechanical strength, antifouling competency, separation performance, without compromising the selectivity for their commercial implication at large-scale applications. The NC membrane has shown its good candidacy to tackle the individual or mixture of multiple impurities. Notably, real water contamination is a mixture of impurities with variation in their strength, chemical and physical environment, and concentrations. Consequently, the main contest of NC-based membrane to confront the wastewater problem is by exhibiting their diverse stability in competing environment, together with their equilibrium in between their high flux and excellent selectivity. Additionally, NC-based membrane has shown the excellent contention for the processing and buildup of most of the nanofibrous membranes such as (i) membrane driven by pressure (ΔP) which includes MF [4–8], UF [9–12], NF [13–18], and RO [19,20], (ii) membrane driven by concentration (ΔC) such as in FO [21], (iii) membrane driven by temperature (ΔT) like in membrane distillation MD [19,22–24], (iv) membrane driven by electric energy (ΔE) such as in electrolysis [25,26]. In the latter cases (FO, MD, and electrolysis) the separation by membranes is mostly based on size-exclusion principle, while in MF, UF, NF, and RO the separation performance is likely executed by physical and structural characteristics of membranes which involves porosity, thickness, pore size, and tortuosity. The following section has been compiled to explain the role of NC-based membranes in tackling various impurities like biological (viruses, bacteriophage, fouling proteins), organic impurities, dyes, metal ions, salt, and oil/water mixtures. We have tabulated (Table 1) all NC-based membranes, their method of preparation, and their role in tackling the multiple contaminants from water.

Removal of biological and organic impurities

Cladophora and bacterial cellulose are innately of a high aspect ratio, and possess the average length in μm and mean diameter ($\geq$30 nm), which have been promptly utilized to develop the MF and UF membranes for the remediation of

Separation Science and Technology, Volume 15. https://doi.org/10.1016/B978-0-323-90763-7.00011-1

TABLE 1 NC-based membranes, their method of preparation, and their role in tackling multiple contaminants from water.

NC type	Composition	Method of preparation	Contamination	Rejection	Membrane type	Refs.
Biological and organic impurities						
CNF	*Cladophora* CNF membrane	Vacuum filtration	Parvovirus	LRV > 4	MF	[27]
CNF	*Cladophora* CNF membrane	Cultivation	SIV virus	LRV ≥ 6.3	UF	[28]
CNF	*Cladophora* CNF membrane	Cultivation	xMuLV	LRV ≥ 5.25	MF	[29]
BNC	BNC membrane	Cultivation using banana peel	—	—	—	[30]
BNC	RGO incorporated *in situ* into BNC membrane	Cultivation	*E. coli*	100%	UF	[12]
BNC	Cultivation of BNC followed by cross-linking with poly(2-PMPC)	Cultivation and cross-linking	*S. aureus* *E. coli*	log CFU Reduction 4.3 1.8	—	[31]
CNF	TEMPO CNF on ePAN	Coating	BSA	>98%	UF	[32]
CNF	TEMPO CNF and PVA coated on PES membrane	Coating	BSA *E. coli*	75% —	—	[9]
CNF	TEMPO CNF coated on PES membrane modified with PAHCl	Coating	BSA *E. coli*	49% —	—	[33]
CNF	CNF coated on PES membrane	Coating	HA BSA	15% 2.73%	UF	[34]
CNF	TEMPO-CNF as a barrier layer on TFNC membrane comprised of ePAN and PET	Coating	BSA	~97%	UF	[35]
CNF	Cationic CNF membrane solvent exchanged by ethanol	Casting	HA	55 mg/L	UF	[36]
CNF	TEMPO CNF grafted with PVA and cast on ePAN/nonwoven PET support layer	Casting	Bacteriophage (MS2)	LRV $= 4$	UF	[7]
CNF	CNF	Phase inversion method	BSA	91–95%	UF	[37]
CNF	CNF membrane cross-linked with PAE	Papermaking	BSA	79.7%	UF	[11]
CNC	CNC coated on PES membrane modified with PAHCl	Coating	BSA *E. coli*	41% —	—	[33]
CNC	CNC coated on PES membrane	Coating	HA BSA	Fouling ratio 23 5.44	UF	[34]

CNC	CNC infused into cellulose diacetate membrane	Phase inversion method	BSA	89%–92%	UF	[38]
CNC	TEMPO CNC infused in ePAN	Casting	*E. coli*	LRV = 6	MF	[8]
CNC	CNC and PVDF membrane	Phase inversion	BSA	88.2%	UF	[39]
CNC	CNC and PVDF membrane	Phase inversion	BSA	92.5%	UF	[40]
NC	Micro/NC grafted with zwitterionic PCysMA	Vacuum filtration	BSA *S. aureus*	84% 86.92%	—	[41]
NC	NC as a barrier layer on a tissue paper support layer	Vacuum filtration	BSA	97.14%	UF	[42]
Dye impurities						
BNC	BNC membrane and Ag NP	Vacuum filtration	Rhodamine 6G Methyl orange 4-Nitro phenol	99% ≥80% 90–100%	—	[43]
BNC	Pd NP/GO/BNC membrane	GO blending with BNC network and *in situ* growth of Pd NP	Methylene orange	99.3%	UF	[44]
BNC	BNC-MoS_2 aerogel membrane	Cultivation of BNC membrane followed by growth of MoS_2 by hydrothermal process	Methylene blue	96%	—	[45]
BNC	BNC embedded with mesoporous PDA and decorated with Pd NP	Cultivation of BNC membrane followed by *in-situ* incorporation of mPDA followed by *in-situ* Pd growth.	Methyl orange 4-Nitro phenol Methylene blue	93–99% 94–99% 95–99%	UF	[46]
BNC	BNC membrane cross-linked with poly(2-PMPC)	Cultivation of BNC membrane followed by crosslinking	Methylene blue Methyl orange	101 mg/g 266 mg/g	—	[31]
BNC	PDA and BNC membrane	Cultivation of BNC membrane and *in-situ* deposition of PDA	Rhodamine 6G	100%	UF	[47]
BNC	TiO_2/PDA on BNC membrane	Coating of TiO_2/PDA	Methyl orange Methylene blue Rhodamine B	95.1% 100% 99.5%	—	[48]
CNF	TEMPO CNF and PVA coated on PES membrane	Casting	Victoria blue Methylene orange	97% 6%	UF	[9]
CNF	Fe_3O_4 infused in CNF membrane	Vacuum filtration	Rhodamine B	94.9%	—	[49]
CNF	CuO *in situ* grown on amine modified CNF	Vacuum filtration	4-Nitro phenol	—	—	[50]
CNF	CNF and GO membrane	Vacuum filtration	Victoria blue B Methyl violet 2B Rhodamine 6G	98.8% 97.6% 92.3%	UF	[51,52]

Continued

TABLE 1 NC-based membranes, their method of preparation, and their role in tackling multiple contaminants from water.—cont'd

NC type	Composition	Method of preparation	Contamination	Rejection	Membrane type	Refs.
CNF	Tissue paper coated with ODA/GA modified TEMPO CNF	Coating	Methyl orange	~99%	—	[53]
CNF	CNF aerogel membrane cross-linked with PEI, decorated with Ag NP	Freeze-drying	Methylene blue Congo red	97–99% 96.4%	—	[54]
CNF	ZIF-8 *in situ* synthesized at CNF membrane	Grafting	Janus green B Methylene blue Orange G Methyl orange Bromothymol blue	98.9% 93.8% 8.1% 5.8% 1.3%	—	[55]
CNF	Carboxymethyl cellulose and CNF	Solvent casting	Basic red	100%	—	[56]
CNC	CNC incorporated with chitosan, cross-linked with GA	Vacuum filtration	Victoria blue 2B Methyl violet 2B Rhodamine 6G	98% 84% 70%	UF	[57]
CNC	Amine functionalized CNC and PES	Casting	Direct red 16	99%	—	[58]
CNC	APTES-modified CNC and PES	Electrospinning	Colored compound from processing industry	94%	—	[59]
CNC	CNC and chitosan and PHA	Electrospinning	Congo red	30.9%	—	[60]
NC	Zeolite (H-ZSM-5 type) integrated NC film	Casting	Reactive black dye	30 mg/g 45%	—	[61]
NC	Nano C nitride (C_3N_2–6) on NC membrane	Vacuum filtration	Organic pollutant	—	—	[62]
Metal ions impurities						
BNC	BNC	papermaking	Au(II) NP (10 to 20 nm)	100%	UF	[63]
BNC	BNC-MoS_2 aerogel membrane	Cultivation of BNC membrane followed by growth of MoS_2 by hydrothermal process	Cr(VI)	88%	—	[45]
BNC	BNC and PDA membrane	Cultivation of BNC membrane and *in-situ* deposition of PDA	Pb(II)	~35%	—	[47]
BNC	Mercerized BNC modified with EDTA	Freeze-drying	Sr^{2+}	44.86 mg/g	—	[64]

CNF	CNF membrane cross-linked with PAE	Papermaking	Lime nanoparticles	99%	UF	[11]
CNF	CNF and PCL membrane	Electrospinning	Fe Cr	38–75% 42–99%	UF	[65]
CNF	TEMPO-CNF	Vacuum filtration	Ca (II)	34%	NF	[66]
CNF	TEMPO CNF dipped, coated on cellulose microfibers membrane, followed by solvent exchange	Coating and vacuum filtration	Ag(I) Cu(II) Fe(II)/ Fe(III)	100% 19% 30%	UF	[67]
CNF	TEMPO CNF grafted with PVA	Casting of CNF/PVA on ePAN/ nonwoven PET support layer	Cr(VI) Pb(II)	100 mg/g 260 mg/g	UF	[7]
CNF	TEMPO CNF	Casting of CNF on ePAN/ nonwoven PET support layer followed by interfacial polymerization	$MgSO_4$ (2000 ppm)	>99%	NF	[17]
CNF	Meldrum acid-modified CNF	Impregnated into PVDF	Fe_2O_3 NP	>99%	—	[5]
CNF	Thiol-modified CNF	Infusion of modified CNF into ePAN/nonwoven PET support layer	Cr(VI) Pb(II)	87.5 mg/g 137.7 mg/g	—	[68]
CNF	TEMPO CNF on ePAN membrane	Coating	Silica NP	>99.8%	NF	[18]
CNF	TEMPO CNF coated with titanate-bismuth oxide	Vacuum filtration	Cs(I), Sr(II). Pb(II). and anions	100%	MF	[69]
CNC	CNC as support layer and CNC/ gelatin as barrier layer	Vacuum filtration	Cu(II) Ag(I) Fe(III)/ Fe(II)	36% 13% 86%	MF	[6]
CNC	CNC as support layer and CNC from BNC as barrier layer	Vacuum filtration	Cu(II) Ag(I) Fe(III)/ Fe(II)	33% 14% 74%	MF	[6]
CNC	CNC as support layer and phosphorylated CNC as barrier layer	Vacuum filtration	Cu(II) Ag(I) Fe(III)/ Fe(II)	100% 100% 100%	MF	[6]
CNC	CNC modified with BTCA deposited on PVA-co-PE nanofibrous membranes	Coating	Cu(II) Pb(II) Mn(II) Cr(VI)	95.8% 52.1% 9.94% 3.91%	—	[70]
CNC	Amine functionalized CNC and PES	Casting	Cu(II)	90%	—	[58]

Continued

TABLE 1 NC-based membranes, their method of preparation, and their role in tackling multiple contaminants from water.—cont'd

NC type	Composition	Method of preparation	Contamination	Rejection	Membrane type	Refs.
CA NF	CA NF (electrospun) modified with citric acid	Electrospinning	Cr(VI)	~23%	UF	[71]
NC	NC fibers prepared in NMMO supported on alumina (200 nm thick)	Solvent casting	Ferritin Au NP (10 nm)	93.8% 82.6%	UF	[72]
NC	NC fibers prepared in NMMO supported on 200-nm pore filter, performed interfacial polymerization using PEI and TMC	Solvent casting followed by interfacial polymerization	$MgCl_2$ $MgSO_4$ NaCl Na_2SO_4	89.7% 65.3% 43.6% 39.1%	NF	[73]
Salt impurities						
BNC	Silane-modified BNC aerogel membrane	Supercritical drying	NaCl	99.9%	MD	[74]
BNC	FTCS fluoro-silanized PDA containing BNC aerogel membrane	Multiple steps	NaCl	≥99.9%	PMD	[22,75]
BNC	BNC/microcellulose/silica membrane	Flat-sheet preparation	NaCl	4.89%	MD	[76]
CNC	CNC infused in cellulose triacetate membrane	Casting	NaCl	99.9%	Desalination	[23,77]
CNC	CNC/polyamide as a barrier layer in TFN membrane	Interfacial polymerization of CNC/polyamide on PES support layer	Monovalent ions Divalent ions	16.19% >98%	Desalination	[13]
CNC	CNC/polyamide as a barrier layer in TFN membrane	Interfacial polymerization of CNC/polyamide on PES support layer	NaCl (brackish water)	97.8%	RO	[20]
CNC	CNC as a barrier layer in TFN membrane	Coating of CNC followed by interfacial polymerization of diamine and acyl chloride	Na_2SO_4	>97%	NF	[14]
CNC	CNC on polysulfone support layer	Casting	NaCl Water flux	96%	RO	[19]
CNC	CMC on polysulfone support layer	Casting	NaCl	94.4%	RO	[19]
CNC	CNC and PVDF-HFP	Electrospinning	NaCl	99.9%	MD	[24]
NC	$PVDF/TiO_2/NC$ Post-treatment with ODTS	Casting	NaCl	>97%	MD	[78]

Separation of oil/water and solvent/water mixtures						
BNC	Water infused BNC membrane	Vacuum filtration	*n*-decane/water	>99.9%	—	[79]
BNC	BNC membrane	Cultivation of BNC membrane on matrix	*n*-decane	≥99.9%	UF	[80]
BNC	Silane-modified BNC aerogel membrane	Conventional drying	*n*-hexane hexadecane kerosene Petroleum ether	99.6% 98.3% 99.4% 99.8%	—	[81]
CNF	TEMPO CNF cross-linked with Al^{3+} before casting into membrane	Vacuum filtration	THF *n*-hexane	— —	NF	[16]
CNF	TEMPO CNF membrane	Casting of CNF on ePAN/nonwoven PET support layer	Oil/water 1350 ppm	>99.6%	UF	[82]
CNF	PVA and TEMPO CNF	Casting of CNF on ePAN/nonwoven PET support layer	Oil/water 1350 ppm	>99.5%	UF	[83]
CNF	TEMPO CNF	Casting of CNF on ePAN/nonwoven PET support layer	Oil/water 1350 ppm	99.6%	UF	[84]
CNF	CA NF membrane	Electrospinning	Mineral oil Kerosene Hexane Petroleum ether	99.9% 99.75% 99.8% 99.5%	—	[85]
CNF	CNF and NRL membrane	Mixing and compression molding	Benzene Toluene *p*-xylene	—	—	[65]
CNF	Tissue paper coated with ODA/GA modified TEMPO CNF	Coating	Toluene Chloroform Hexane Heptane Dicholoroethane Kerosene Diesel Silicon Oil Soybean Oil	90–99%	—	[53]
CNF	CNF membrane coated with GO	Electrospinning	Oil	>99%	—	[86]
CNF	TEMPO CNF and CNF infused in cellulose membrane	Infusion	Oil	98.5–99.98%	—	[87]
CNF	TEMPO CNF cross-linked with PEG as barrier layer	Coating of CNF/PEG on ePAN/PET support layer	PEG4600	>90%	UF	[59]
CNF	TEMPO CNF coated with titanate-bismuth oxide	Vacuum filtration	Oil	>99.8%	MF	[69]

Continued

TABLE 1 NC-based membranes, their method of preparation, and their role in tackling multiple contaminants from water.—cont'd

NC type	Composition	Method of preparation	Contamination	Rejection	Membrane type	Refs.
CNF	Tissue paper coated with ODA/ GA-modified TEMPO CNF	Coating	Toluene Chloroform Hexane Heptane Dicholoroethane Kerosene Diesel Silicon Oil Soybean Oil	90–99%	—	[53]
CNC	TiO_2, NP, and CNC	Vacuum filtration	Hexadecane Soybean oil Pump oil	99.60% 99.48% 99.54%	MF	[88]
CNC	Tunicate CNC membrane	Vacuum filtration	Toluene Isooctane Chloroform *n*-hexane	≥99.9% ≥99.9% 93.3% ≥99.9%	—	[89]
NC	Tubular porous alumina membrane with three layers: Codoped Zn-based MOF/PDA/NC gel/hydrotalcite	Multiple steps	Alkane/arene mixture	—	—	[90]
NC	Blend of NC and chitin nanofibers	Casting of CNF on ePAN/ nonwoven PET support layer	Oil/water 1350 ppm	99.5%	UF	[84]
CNF	TEMPO CNF on ePAN membrane	Coating	Oil	>99.5%	UF	[18]
CNF	SBS matrix and CNF	Solvent casting evaporation method	Ethanol recovery	$PSI = 89.4 \pm 12.0\ g^{-2}\ h^{-1}$	Pervaporation membrane	[91]
Miscellaneous separation performance						
NC	PSF/NC/Ag NP/ Pt NP as a supporting layer for TFC polyamide membrane	NIPS process	Microorganism Wastewater Urea	67–80%	FO	[21]
CNF	TEMPO CNF and deacetylated chitin NF membrane coated with HFTD-modified nano SiO_2	Vacuum filtration and screen printing	—	—	Superhydrophobic (CA = 150°)	[92]
CNF	Silica grafted on TEMPO CNF	Grafting and vacuum filtration	—	—	Superhydrophobic (CA = 153°)	[93]
CNF	Silica coated on TEMPO CNF	Coating and vacuum filtration	—	—	Superhydrophobic (CA = 165°)	[93]
CNF	Pulp CNF aerogel modified with FTCS	Vacuum oven	—	—	Superhydrophobic (CA = 158°)	[94]

viruses, bacteriophages, and some metal ions [27,28,66]. For example, a CNF membrane from *Cladophora* has been found effective in removing the parvovirus [27], SIV [28] and xMuLV [29] with LRV of 4, 6.3, 5.25, respectively. Similarly, the cultivation of BNC membrane with fiber diameter in the range of 30 to 50 nm, using banana peel, was found to be a good MF membrane [30]. Furthermore, the three-dimensional network was developed using BNC cross-linked with PMPC where the MPC used is polym-erizable, environment-friendly, and possesses the zwitterionic moiety with a trimethylammonium cation and a phosphate anion which are responsible for the microbial growth. For, BNC cross-linked with PMPC demonstrated the log CFU reduction of 4.3 for *Stphylococcus aureus* and 1.8 for *Escherichia coli* [31]. The antifouling properties in BNC membranes were developed by incorporating the RGO into BNC *in situ* during its growth. This membrane has shown excellent aqueous stability under an environmentally relevant pH range, high pressure, and vigorous stirring conditions, and is effective in the removal of 100% *E. coli* because of its superior photothermal property [12]. NC obtained natively and from chemical modification (TEMPO-oxidation, acid hydrolysis, cationic modification, etc.) infused and applied as a barrier layer on various supporting membranes such as ePAN, PES, and ePAN/PET to induce the bactericidal and antifouling properties in NC-based membranes. The infusion of CNF (which has a cross section of 2 to 10 nm in diameter and up to a few μm in length) into a microfibrous scaffold is an effective way to fine-tune the pore size without significant changes in porosity of the membrane, whereby the loading of CNF is considered to alter the mean pore size and pressure drop. On the other side, the NC barrier layer in composite membrane scaffold also provides the adsorption functionality, mainly beneficial to enhance the antifouling property and in removal of charged impurities. For example, TEMPO CNC infused in ePAN resulted in effective removal of *E. coli* with LRV of 6 as the infusion of CNC has decreased the mean pore size of ePAN from 0.38 to 0.22 μm [8]. Similarly, the composite nanofibrous membrane produced by infusing modified CNF into ePAN/PET supporting layer has shown effective reduction in mean flow pore size from 0.66 to 0.32 μm, which has allowed the retention of a bacteriophage (MS2) with LRV of 4 [7].

The organic contaminants in water enhance the fouling in membranes, responsible for drastic decrease in flux and increasing the operation cost of membranes, are consider an intrinsic problem in water treatment. The hydrophilicity of native NC membranes and its enhancement by various chemical modification on NC is proven to be an effective method to promote the antifouling properties. There very rare examples that dealt with the application of NC-based membranes to treat the real wastewater. However, to show the candidacy of NC-based membranes to tackle the various organic contaminants in water, the model foulants such as BSA and HA have been demonstrated. For example, TEMPO-CNF which has a carboxylate functional group when coated as a barrier layer on the ePAN/PET support layer has shown the 97% removal of BSA [35], while CNC when infused into a cellulose acetate membrane has demonstrated 89.92% removal of BSA with flux of 173.8 LMH [38]. CNC and PVDF membranes, when prepared by phase inversion method, showed 92.5% removal of BSA with flux of 230.8 LMH at 1.1 kPa [40].

Removal of dye impurities

Dye impurities in water are organic in nature and have cationic, anionic, or neutral functionality. NC-based membranes work on the principal of electrostatic interaction and adsorption to perform dye removal from water. For the most part, BNC membrane networks are prepared by cultivating bacteria in a suitable nutrient medium. A BNC membrane does not possess any surface charge; hence, to activate the surface to effectively trap or degrade the dye impurities, organic and inorganic additives have been employed. For example, BNC membrane is incorporated with PDA. The addition of PDA enhanced the porosity of membrane, and its catecholamine group prompted the membrane surface activation for the adsorption of methylene blue and methylene orange, with the maximum adsorption capacity of 101 mg/g and 266 mg/g, respectively [31]. In another study, BNC and PDA the membrane network showed 100% removal of rhodamine 6G [47]. Gholami [46] et al. have further upgraded the BNC/PDA membrane network to photocatalytic membrane by incorporating the Pd NP. The incorporation of photocatalytic groups such as Pd NP [46], TiO_2 NP [48], Ag NP [43,45], and MoS_2 into BNC membrane network has enhanced their separation performance, as the above composite membrane simultaneously adsorb and cause the photocatalytic reduction of impurities on light illumination. For example, BNC/PDA membrane has shown the removal of rhodamine 6G dye; however, the integration of Pd NP in BNC/PDA has presented the reduction of 4-nitro phenol along with adsorption of methyl orange and methylene blue with removal capacity between 93 and 99%. Similarly, TiO_2, when added to BNC/PDA network has shown the fast photocatalytic degradation of methylene blue, methylene orange and rhodamine B within 60 min. With efficiencies between 95 and 100% [48]. Interestingly, the blending of GO into a BNC network and *in situ* growth of Pd NP has developed the capillary channel in its developed lamellar structure that has allowed the fast and effective removal of methylene orange with an efficiency of 99.3% [44].

Most popular, the TEMPO CNF and carboxymethyl cellulose–based membrane possesses the negative surface charge in the form of carboxylate group; hence, it has shown excellent capacity in between 95 and 100% to remove the cationic dyes such as victoria blue, methylene orange, basic red via electrostatic interaction [9,56]. The infusion of Fe_3O_4 NP in CNF membrane has induced the superparamagnetic behavior with the capability of catalyzing the sulfate radical–based advanced oxidation process. The membranes show the successful activation of PMS to degrade rhodamine B with an efficiency of 94.9% in 300 min [49]. Similarly, the incorporation of an other inorganic metal NP such as Ag NP [54], CuO NP [50] has induced the photocatalytic capability in an NC-based membrane that has improved their separation performance. For example, CNF membrane cross-linked with PEI followed by decoration with Ag NP, showed excellent continuous catalytic discoloration of an aqueous cationic and anionic dye with a discoloration efficiency of 98% even after reuse 10 times with water permeance as high as 5×10 [4] LMH [54]. Furthermore, the fabrication of an ultrathin GO membrane on CNF membrane fillers has exhibited significantly high water flux of $18{,}123 \pm 574$ LMH/bar with the capability to remove the anionic and cationic dyes with the high removal capacity of 92 to 97%. The creation of nanochannels in the CNF/GO membrane nanocomposite membrane due to the anisotropic layered arrangement of GO has enhanced its performance through a combination of electrostatic and hydrophobic interactions, and size exclusion [51]. Kokol et al. have demonstrated the coating of zeolite (H-ZSM-5 type) as an adsorbent on a CNF membrane network to tackle the black anionic reactive dye from textile effluent. This membrane was capable of removing 45% of the dye with the flux rate of 11 kL/m [2]hMPa [61]. In a similar study, ZIF-8 was synthesized *in situ* at a CNF membrane network whereby ZIF-8 acted as a spacer and was successfully wrapped around CNF because of their strong interaction. The effective anchoring of ZIF-8 in a CNF network has allowed it to maintain the 85 to 92 LMH; for 24 h and to highly selectively remove the cationic dyes, the removal efficiency for Janus green B was 98% [55]. Some research has also presented electrospinning to prepare the CNF-based nanocomposite membrane with control pore size and porosity to remove the dye impurities [59,60]. The tissue paper, when coated with TEMPO CNF modified with ODA/GA, exhibited 99% removal of methyl orange [53]. This type of study with model Au particles is used to determine the mean pore size of a constructed membrane [63].

Removal of metal ion impurities

NC-based MF and UF membranes have been found to be efficient in the separation of metal ion impurities from water through the combination of electrostatic interaction, adsorption, and size exclusion. The BNC membrane can be inherently modified as a UF that has shown 100% retention of Au (II) NP having a particle size of 10 to 20 nm. This type of study with model Au particles is utilized to determine the mean pore size of constructed membrane [63]. This membrane is then upgraded by depositing the PDA, and the presence of PDA has enhanced the porosity of membrane, permitting the removal of Pb(II) with a removal efficiency of 35% [47]. However, the modification of the BNC network with EDTA, followed by freeze-drying has shown the removal of Sr^{2+} with a capacity of 44.86 mg/g [64]. A UF BNC aerogel membrane is upgraded to photocatalytic membrane by introducing the MoS_2 particle. The introduction of MoS_2 developed the porous morphology and photocatalytic activation of a composite membrane that has allowed a fast and efficient removal of Cr(VI) with a capacity of 88% within 120 min [45]. TEMPO CNF, when coated on ePAN, has exhibited the removal of >99.8% of silica NP; however, when TEMPO CNF membrane was coated with titanate-bismuth oxide, the membrane showed 100% removal of Cs(I), Sr(II), Pb(II), and anions, indicating its capability to handle complex metal ion impurities [69]. The modification of CNF membrane functionality is mostly created to augment its selective interactions and physiochemical properties. Karim et al. [95] introduced the TEMPO modification on CNF membrane, which caused a several times increase in adsorption capacity for Ag(I), Cu(II), and Fe(II)/Fe(III), when compared with an unmodified CNF membrane. CNC modified with BTCA coated on PVA-co-PE membrane was also evaluated to find its adsorption efficiency for Cu(II), Pb(II), Mn(II) and Cr(VI) ions. The higher negative charge on the membrane surface has led to the effective removal of positively charged metal ions [70]. Furthermore, Karim et al. [6] demonstrated the layered cellulose-based nanocomposite membrane using cellulose microfiber as a support layer and CNC in a gelatin matrix as the barrier layer. Three forms of CNC have been studied: (i) CNC, (ii) CNC from BNC, and (iii) phosphorylated CNC. The phosphorylated CNC has shown excellent adsorption capacity of 100% for Cu(II), Ag(I), and Fe(III)/Fe(II) with excellent permeability (900–4000 LMH) at <1.5 bars. The modification of NC to amino functionality is another way to modify NC-based membranes. The amino group at low pH appeared as a charged group suitable for attracting the negatively charged metal ion impurities. The amine functionalized CNC and PES cast into a membrane to remove the Cu(II) impurities with removal efficiency of 90%. Similarly, thiol modified CNF, when infused into an ePAN/PET support layer, exhibited higher adsorption capacities for both Pb(II) and Cr(VI), with a maximum removal capacity of 87.5 and 137.7 mg/g, respectively. [68] Zhang et al. [72] showed the removal of ferritin and Au NP by a self-standing NC membrane supported on alumina by ultrafiltration. This type of membrane was then upgraded by Soyekwo et al. to nanofiltration by interfacial polymerization using PEI and TMC. This process created the interconnected nanochannels responsible for the transport of water molecules. The membrane, with a thickness of 77 nm with a mean pore size of ~0.45 nm, has exhibited water flux of 32.7 LMH/bar that is an order of magnitude higher than previously reported similar NF membranes [73].

Removal of salt impurities

Mostly, the solid salt (e.g., NaCl, Na_2SO_4, $MgSO_4$) is performed by a high-pressure membrane filtration process such as NF or RO, or by a thermal driven separation process (e.g., MD). The development of an NC-based membrane to remove the salt impurities involves the preparation of a nanocomposite membrane for NF, RO, and MD. For example, Wang et al. have presented TFC nanofiltration membranes comprised of polyamide layer on CNC interlayer supported on a PES scaffold. The presence of a CNC interlayer facilitated the interfacial polymerization at a low monomer concentration (50 mg/100 mL). The prepared membrane showed permeation of 204 LMH under 0.6 MPa, with a Na_2SO_4 rejection of 97.6% [13,14]. In another study, CNC was incorporated into the polyamide active layer synthesized by *in-situ* interfacial polymerization of MPD and TMC. The prepared CNC/TFN membrane, comprised of 0.1% CNC loading, showed higher water flux of 63 ± 10 LMH at 20 bar, with salt rejection of 97.8% [20]. Leitch et al. fabricated silane-modified BNC aerogel membranes in a direct contact MD system. The unsupported modified BNC aerogels exhibited higher porosity and delivered significantly higher intrinsic membrane permeability with salt rejection of 99.9% and flux of 22.92 Kg m^{-2} h^{-1} [74]. Wu et al. demonstrated a solar-driven PMD system that comprised a photothermal membrane composed of PDA particles and BNC, which is optically and photothermally active and possesses high porosity (~93%) and reduced conductive heat transfer, thus increasing the thermal efficiency of the membrane. The FTCS modification on a PDA/BNC membrane provides high hydrophobicity resulting in a salt rejection of >99.9% [22]. Kadhom et al. demonstrated the addition of CMCs and CNC to the polysulfone support layer in a TFN membrane and studied their desalination performance. The best results indicated that loading of 2% CNC in the support layer enhanced the water flux by > 40%, and salt rejection was maintained at ~95% [19]. Furthermore, a PVDF membrane was upgraded, using TiO_2, NP, and NC for MD; prior to the addition of TiO_2 NP and NC, the PVDF modified by ODTS using phase inversion technique. The dispersion of TiO_2 NP in the membrane allowed the formation of a large pore size and high porosity, which led to higher water permeate flux with salt rejection of >97% [78].

Separation of oil/water and solvent/water mixtures

Membrane filtration techniques have been found to be an efficient way to segregate the oil/water and solvent/water mixtures. The membrane for this purpose mostly operates on the principle of size exclusion. However, the integration of some specific physiochemical properties such as hydrophobicity by incorporating inorganic fillers has augmented the separation performance. BNC cross-linked 3D network, because of its strong chemical and physical interactions of water molecules, permits efficient liquid separation with super wettability. In another case, an NW BNC membrane was cultured on custom-designed 3D-printed holder ring made of PLA that had overcome the problem of stretching the thin layer BNC. NW BNC was successful in separating of stable oil-in-water micro- and nano-emulsion with flux of 5000 LMH and 3500 LMH at 20 psi, respectively. The removal efficiency observed for microemulsion and nanoemulsion was 99.5% and 98% respectively [80]. Furthermore, BNC modified with an SiO_2 3D porous structure has shown a removal efficiency for *n*-hexane, hexadecane, kerosene, and petroleum ether in the range of 98.3 to 99.8% with significantly high flux of 34,000 ± 50 LMH/MPa [81]. Ma et al. [82] have studied the use of CNF with a diameter of 5 to 10 nm as barrier layers in TFNC membranes. The addition of CNF as a barrier layer into ePAN/PET support layer showed a 10-fold higher permeation flux, as compared to two commercial UF membranes (PAN10 and PAN400) with a greater-than 99.5% rejection ratio for ultrafiltration of oil/water. In another study, Ma et al. demonstrated the PVA–cross-linked CNF as a barrier layer on ePAN/PET support layer demonstrated the presence of directed water channels that allowed the oil/water emulsion separation with the permeation flux five times higher than the commercial PAN_{10} UF membranes at low pressure of 30 psi with the rejection ratio of ~99.7%. Furthermore, TEMPO CNF was fabricated as a top barrier layer on ePAN/PET support layer while maintaining the surface at 55 nm, has shown the permeate flux of about eightfold higher than that of commercial PAN_{10} for ultrafiltration of oil/water [84].

The addition of inorganic fillers such as TiO_2 NP, GO, titanate-bismuth oxide, and/or silane has demonstrated the preparation of NC-based membranes for oil/water removal. Roy et al. have exhibited the use of tissue paper in an effective separation membrane by coating with an ODA/GA/TEMPO CNF mixture. The modified tissue paper showed that the water contact angle of 161° (±3°) was capable of separating various solvents including toluene, chloroform, hexane heptane, kerosene, and diesel oil with an efficiency between 90 and 99% [53]. Furthermore, the layer-by-layer structure of CNF/titanate nanofibers/bismuth oxide membrane was fabricated to tackle the oil/water-contaminated water. The upper layer of titanate nanofibers showed a pore size of >100 nm that allowed oil/water separation with an efficiency of >99.8%, with flux of 25 LMH by gravity. [69] Kamtsikakis et al. have demonstrated the preparation of dense nanocomposite membranes comprised of an SBS matrix and CNF, by a solvent casting evaporation method. The loading of 15 wt% CNF has shown a threefold increase in mass flux accompanied by 40% reduction in separation factor, which elevated PSI to 89.4 g^{-2} h^{-1} for 10 wt% ethanol/water mixture [91].

Miscellaneous separation performances

NC-based composite membranes containing Ag, Pt, and NP as additives were incorporated in a TFC membrane in an FO filtration system. The fabrication of an NC-based upper layer was done by the NIPS technique, and the membrane exhibited the diverse pores with a finger-like morphology. Interestingly, higher water flux of 11.5 to 11.8 MPH/bar and three- to eightfold reduction in bacteria were observed when used with wastewater. This type of membrane exhibited electro-chemical and antimicrobial properties [21]. Several other studies indicated the incorporation of inorganic fillers such as silica and nano silica [92,93], and chemical modification with FTCS [94], etc. for fabricating the superhydrophobic membranes, suitable for an MD process. CNF hydrogel is also found effective in capturing microplastics (100 nm and 1.1 μm) [96,97].

Conclusions and future perspective

The use of NC-based membranes in water purification has shown an excellent contention to remove multicontaminations in water and shows the potential to be incorporated at large-scale water treatment systems. NC-based membranes have shown promising features over the most prominently used synthetic polymers such as (i) the preparation of NC membrane not involving the use of toxic chemicals, (ii) method of preparation is low-cost, simple and easily applicable, (iii) excellent compatibility with other organic and inorganic fillers that allow the NC membrane to reflect synergetic properties, (iii) high surface functionality (e.g., three hydroxyls groups in one anhydroglucose unit) allows their facile surface modification to generate more active sites, (iv) easy tunable porosity and pore size, (v) high dry strength, (vi) the wet strength of NC-based membranes can effectively improve by various cross-linking agents along with further modification and incorporation of fillers, and (vii) innate antifouling property due to the presence of hydrophilic nature of NC.

Additionally, the current developments for NC-based membranes showed better performance in terms of harmonizing the separation performance and the permeance competency by transforming the self-standing membrane to nanocomposite membranes by incorporation of the compatible inorganic/organic fillers and by suitable chemical modifications. However, the unprecedented membrane-processing challenges and unpredictable rise in final cost are still unknown. Therefore, the life cycle analysis (LCA) and techno-economic analysis (TEA) for the developing and existing NC-based processes should be accomplished in the coming years. The addition of inorganic fillers (e.g., TiO_2 NP, Ag NP, Au NP, etc.) onto NC or NC-based membranes has shown excellent dispersity without losing innate performance and regulated the secondary contaminations challenges that are associated in utilizing the metal NP in water purification. Therefore, the strategies for using hybrid membranes comprised of NC metal NP should be further promoted to explore advanced membranes.

Acknowledgment

The financial support for this work was provided by a grant from the Polymer Program of the Division of Materials Science in the National Science Foundation (DMR-1808690).

Abbreviations

APTES	(3-aminopropyl) triethoxysilane
BNC	bacterial nanocellulose
BSA	bovine serum albumin
BTCA	1,2,3,4-butanetetracarboxylic acid
CA NF	cellulose acetate nanofibers
CFU	colony forming units
CMCs	cellulose microcrystals
CNC	cellulose nanocrystals

CNF	cellulose nanofibers
COD	chemical oxygen demand
EDTA	ethylenediaminetetraacetic acid
ePAN	electrospun polyacrylonitrile
FO	forward osmosis
FTCS	tridecafluoro-1,1,2,2-tetrahydrooctyl-trichlorosilane
GA	glutaraldehyde
GO	graphene oxide
HA	humic acid
HFTD	heptadecafluoro-1,1,2,2-tetrahydrodecyl dimethylchlorosilane
LMH	$Lm^{-2} h^{-1}$
LRV	log reduction value
MD	membrane distillation
MF	microfiltration
μm	micrometer
MOF	metal organic framework
MoS_2	molybdenum disulfide
MPC	2-methacryloyloxyethyl phosphorylcholine
NF	nanofiltration
NF	nanofiber
NIPS	non–solvent-induced phase separation
nm	nanometer
NMMO	*N*-methylmorpholine n-oxide
NP	nanoparticles
NW	nonwoven
ODA	octadecyl amine
ODTS	octadecyltrichlorosilane
PAE	polyamide-amine-epichlorohydrin
PAHCl	polyallylamine hydrochloride
PAN	polyacrylonitrile
PCL	polycaprolactone
PCysMA	poly(cysteine methacrylate)
PDA	polydiacetylenes
PEI	poly(ethylene imine)
PES	polyethersulfone
PET	poly(ethylene terephthalate)
PHA	poly(hydroxyalkanoate)
PLA	poly lactic acid
PMD	photothermal membrane distillation
PMPC	poly(2-methacryloyloxyethyl phosphorylcholine)
PMS	peroxymonosulfate
PSF	polysulfone
PSI	pervaporation separation index
PVA	polyvinylamine
PVDF	polyvinylidene fluoride
PVDF-HFP	polyvinylidenefluoride-co-hexafluoropropylene
(PVA-co-PE)	poly(vinyl alcohol-co-ethylene)
RGO	reduced graphene oxide
RO	reverse osmosis
SBS	polystyrene-block-polybutadiene-block-polystyrene
SiO_2	silicon dioxide
SIV	simian immunodeficiency virus
TEMPO	(2,2,6,6-tetramethylpiperidin-1-yl)oxyl
TFC	thin film composite
TFNC	thin film nanocomposite
UF	ultrafiltration
xMuLV	model retro virus

References

[1] W.S. Lee, S. Jeon, S.J. Oh, Wearable sensors based on colloidal nanocrystals, Nano Converg. 6 (1) (2019) 10.

[2] L. Ansaloni, J. Salas-Gay, S. Ligi, M. Giacinti Baschetti, Nanocellulose-based membranes for CO_2 capture, J. Membr. Sci. 522 (2016).

[3] A. Razaq, G. Nyström, M. Strømme, A. Mihranyan, L. Nyholm, High-capacity conductive nanocellulose paper sheets for electrochemically controlled extraction of DNA oligomers, PLoS One 6 (12) (2011) e29243.

[4] M.-X. Hu, H.-M. Niu, X.-L. Chen, H.-B. Zhan, Natural cellulose microfiltration membranes for oil/water nanoemulsions separation, Colloids Surf. A Physicochem. Eng. Asp. 564 (2019) 142–151.

[5] D.A. Gopakumar, D. Pasquini, M.A. Henrique, L.C. de Morais, Y. Grohens, S. Thomas, Meldrum's acid modified cellulose nanofiber-based polyvinylidene fluoride microfiltration membrane for dye water treatment and nanoparticle removal, ACS Sustain. Chem. Eng. 5 (2) (2017) 2026–2033.

[6] Z. Karim, A.P. Mathew, V. Kokol, J. Wei, M. Grahn, High-flux affinity membranes based on cellulose nanocomposites for removal of heavy metal ions from industrial effluents, RSC Adv. 6 (25) (2016) 20644–20653.

[7] R. Wang, S. Guan, A. Sato, X. Wang, Z. Wang, R. Yang, B. Hsiao, B. Chu, Nanofibrous microfiltration membranes capable of removing bacteria, viruses and heavy metal ions, J. Membr. Sci. 446 (2013) 376–382.

[8] H. Ma, C. Burger, B.S. Hsiao, B. Chu, Nanofibrous microfiltration membrane based on cellulose Nanowhiskers, Biomacromolecules 13 (1) (2012) 180–186.

[9] A. Aguilar-Sanchez, B. Jalvo, A. Mautner, V. Rissanen, K.S. Kontturi, H.N. Abdelhamid, T. Tammelin, A.P. Mathew, Charged ultrafiltration membranes based on TEMPO-oxidized cellulose nanofibrils/poly(vinyl alcohol) antifouling coating, RSC Adv. 11 (12) (2021) 6859–6868.

[10] Y. Liang, H. Ma, A.A. Taha, B.S. Hsiao, High-flux anti-fouling nanofibrous composite ultrafiltration membranes containing negatively charged water channels, J. Membr. Sci. 612 (2020) 118382.

[11] M.L. Hassan, S.M. Fadel, R.E. Abouzeid, W.S. Abou Elseoud, E.A. Hassan, L. Berglund, K. Oksman, Water purification ultrafiltration membranes using nanofibers from unbleached and bleached rice straw, Sci. Rep. 10 (1) (2020) 11278.

[12] Q. Jiang, D. Ghim, S. Cao, S. Tadepalli, K.-K. Liu, H. Kwon, J. Luan, Y. Min, Y.-S. Jun, S. Singamaneni, Photothermally active reduced graphene oxide/bacterial nanocellulose composites as biofouling-resistant ultrafiltration membranes, Environ. Sci. Technol. 53 (1) (2019) 412–421.

[13] L. Bai, Y. Liu, N. Bossa, A. Ding, N. Ren, G. Li, H. Liang, M.R. Wiesner, Incorporation of cellulose nanocrystals (CNCs) into the polyamide layer of thin-film composite (TFC) nanofiltration membranes for enhanced separation performance and antifouling properties, Environ. Sci. Technol. 52 (19) (2018) 11178–11187.

[14] J.-J. Wang, H.-C. Yang, M.-B. Wu, X. Zhang, Z.-K. Xu, Nanofiltration membranes with cellulose nanocrystals as an interlayer for unprecedented performance, J. Mater. Chem. A 5 (31) (2017) 16289–16295.

[15] A. Quellmalz, A. Mihranyan, Citric acid cross-linked nanocellulose-based paper for size-exclusion nanofiltration, ACS Biomater Sci. Eng. 1 (4) (2015) 271–276.

[16] A. Mautner, K.Y. Lee, P. Lahtinen, M. Hakalahti, T. Tammelin, K. Li, A. Bismarck, Nanopapers for organic solvent nanofiltration, Chem. Commun. 50 (43) (2014) 5778–5781.

[17] X. Wang, D. Fang, B.S. Hsiao, B. Chu, Nanofiltration membranes based on thin-film nanofibrous composites, J. Membr. Sci. 469 (2014) 188–197.

[18] X. Cao, M. Huang, B. Ding, J. Yu, G. Sun, Robust polyacrylonitrile nanofibrous membrane reinforced with jute cellulose nanowhiskers for water purification, Desalination 316 (2013) 120–126.

[19] M. Kadhom, N. Albayati, S. Salih, M. Al-Furaiji, M. Bayati, B. Deng, Role of cellulose Micro and Nano crystals in thin film and support layer of nanocomposite membranes for brackish water desalination, Membranes 9 (8) (2019).

[20] F. Asempour, D. Emadzadeh, T. Matsuura, B. Kruczek, Synthesis and characterization of novel cellulose nanocrystals-based thin film nanocomposite membranes for reverse osmosis applications, Desalination 439 (2018) 179–187.

[21] P. Cruz-Tato, E.O. Ortiz-Quiles, K. Vega-Figueroa, L. Santiago-Martoral, M. Flynn, L.M. Díaz-Vázquez, E. Nicolau, Metalized nanocellulose composites as a feasible material for membrane supports: design and applications for water treatment, Environ. Sci. Technol. 51 (8) (2017) 4585–4595.

[22] X. Wu, S. Cao, D. Ghim, Q. Jiang, S. Singamaneni, Y.-S. Jun, A thermally engineered polydopamine and bacterial nanocellulose bilayer membrane for photothermal membrane distillation with bactericidal capability, Nano Energy 79 (2021) 105353.

[23] I. Prihatiningtyas, A. Volodin, B. Van der Bruggen, 110th anniversary: cellulose nanocrystals as organic nanofillers for cellulose triacetate membranes used for desalination by pervaporation, Ind. Eng. Chem. Res. 58 (31) (2019) 14340–14349.

[24] B.S. Lalia, E. Guillen, H.A. Arafat, R. Hashaikeh, Nanocrystalline cellulose reinforced PVDF-HFP membranes for membrane distillation application, Desalination 332 (1) (2014) 134–141.

[25] N. Ferraz, D.O. Carlsson, J. Hong, R. Larsson, B. Fellström, L. Nyholm, M. Strømme, A. Mihranyan, Haemocompatibility and ion exchange capability of nanocellulose polypyrrole membranes intended for blood purification, J. Royal Soc., Interface 9 (73) (2012) 1943.

[26] N. Ferraz, A. Leschinskaya, F. Toomadj, B. Fellström, M. Strømme, A. Mihranyan, Membrane characterization and solute diffusion in porous composite nanocellulose membranes for hemodialysis, Cellul. 20 (6) (2013) 2959.

[27] S. Gustafsson, A. Mihranyan, Strategies for tailoring the pore-size distribution of virus' retention filter papers, ACS Appl. Mater. Interfaces 8 (22) (2016) 13759–13767.

[28] G. Metreveli, L. Wågberg, E. Emmoth, S. Belák, M. Strømme, A. Mihranyan, A size-exclusion nanocellulose filter paper for virus removal, Adv. Healthc. Mater. 3 (10) (2014) (1546-50, 1524).

[29] M. Asper, T. Hanrieder, A. Quellmalz, A. Mihranyan, Removal of xenotropic murine leukemia virus by nanocellulose based filter paper, Biologicals 43 (6) (2015) 452–456.

[30] E.K. Sijabat, A. Nuruddin, P. Aditiawati, B.S. Purwasasmita, Synthesis and characterization of bacterial nanocellulose from Banana Peel for water filtration membrane application, J. Phys. Conf. Ser. 1230 (2019) 012085.

[31] C. Vilela, C. Moreirinha, A. Almeida, A.J.D. Silvestre, C.S.R. Freire, Zwitterionic nanocellulose-based membranes for organic dye removal, Materials (Basel) 12 (9) (2019) 1404.

[32] M. Yang, P. Hadi, X. Yin, J. Yu, X. Huang, H. Ma, H. Walker, B.S. Hsiao, Antifouling nanocellulose membranes: how subtle adjustment of surface charge lead to self-cleaning property, J. Membr. Sci. 618 (2021) 118739.

[33] A. Aguilar Sanchez, B. Jalvo, A. Mautner, S. Nameer, T. Pöhler, T. Tammelin, A.P. Mathew, Waterborne nanocellulose coatings for improving the antifouling and antibacterial properties of polyethersulfone membranes, J. Membr. Sci. 620 (2020) 118842.

[34] L. Bai, Y. Liu, A. Ding, N. Ren, G. Li, H. Liang, Surface coating of UF membranes to improve antifouling properties: a comparison study between cellulose nanocrystals (CNCs) and cellulose nanofibrils (CNFs), Chemosphere 217 (2019) 76–84.

[35] P. Hadi, M. Yang, H. Ma, X. Huang, H. Walker, B. Hsiao, Biofouling-resistant nanocellulose layer in hierarchical polymeric membranes: synthesis, characterization and performance, J. Membr. Sci. 579 (2019).

[36] H. Sehaqui, B. Michen, E. Marty, L. Schaufelberger, T. Zimmermann, Functional cellulose nanofiber filters with enhanced flux for the removal of humic acid by adsorption, ACS Sustain. Chem. Eng. 4 (9) (2016) 4582–4590.

[37] P. Qu, H. Tang, Y. Gao, L. Zhang, S. Wang, Polyethersulfone composite membrane blended with cellulose fibrils, Bioresources 5 (2010).

[38] J. Lv, G. Zhang, H. Zhang, F. Yang, Exploration of permeability and antifouling performance on modified cellulose acetate ultrafiltration membrane with cellulose nanocrystals, Carbohydr. Polym. 174 (2017) 190–199.

[39] J. Lv, G. Zhang, H. Zhang, C. Zhao, F. Yang, Improvement of antifouling performances for modified PVDF ultrafiltration membrane with hydrophilic cellulose nanocrystal, Appl. Surf. Sci. 440 (2018) 1091–1100.

[40] H. Bai, X. Wang, Y. Zhou, L. Zhang, Preparation and characterization of poly(vinylidene fluoride) composite membranes blended with nanocrystalline cellulose, Prog. Nat. Sci. 22 (3) (2012) 250–257.

[41] L. Valencia, S. Kumar, B. Jalvo, A. Mautner, G. Salazar-Alvarez, A.P. Mathew, Fully bio-based zwitterionic membranes with superior antifouling and antibacterial properties prepared via surface-initiated free-radical polymerization of poly(cysteine methacrylate), J. Mater. Chem. A 6 (34) (2018) 16361–16370.

[42] Z. Wang, W. Zhang, J. Yu, L. Zhang, L. Liu, X. Zhou, C. Huang, Y. Fan, Preparation of nanocellulose/filter paper (NC/FP) composite membranes for high-performance filtration, Cellul. 26 (2) (2019) 1183–1194.

[43] Y. Yang, Z. Chen, X. Wu, X. Zhang, G. Yuan, Nanoporous cellulose membrane doped with silver for continuous catalytic decolorization of organic dyes, Cellul. 25 (4) (2018) 2547–2558.

[44] T. Xu, Q. Jiang, D. Ghim, K.-K. Liu, H. Sun, H.G. Derami, Z. Wang, S. Tadepalli, Y.-S. Jun, Q. Zhang, S. Singamaneni, Catalytically active bacterial nanocellulose-based ultrafiltration membrane, Small 14 (15) (2018) 1704006.

[45] E.P. Ferreira-Neto, S. Ullah, T.C.A. da Silva, R.R. Domeneguetti, A.P. Perissinotto, F.S. de Vicente, U.P. Rodrigues-Filho, S.J.L. Ribeiro, Bacterial nanocellulose/MoS2 hybrid aerogels as bifunctional adsorbent/photocatalyst membranes for in-flow water decontamination, ACS Appl. Mater. Interfaces 12 (37) (2020) 41627–41643.

[46] H. Gholami Derami, P. Gupta, R. Gupta, P. Rathi, J.J. Morrissey, S. Singamaneni, Palladium nanoparticle-decorated mesoporous polydopamine/bacterial nanocellulose as a catalytically active universal dye removal ultrafiltration membrane, ACS Appl. Nano Mater. 3 (6) (2020) 5437–5448.

[47] H. Gholami Derami, Q. Jiang, D. Ghim, S. Cao, Y.J. Chandar, J.J. Morrissey, Y.-S. Jun, S. Singamaneni, A robust and scalable polydopamine/bacterial nanocellulose hybrid membrane for efficient wastewater treatment, ACS Appl. Nano Mater. 2 (2) (2019) 1092–1101.

[48] L. Yang, C. Chen, Y. Hu, F. Wei, J. Cui, Y. Zhao, X. Xu, X. Chen, D. Sun, Three-dimensional bacterial cellulose/polydopamine/TiO2 nanocomposite membrane with enhanced adsorption and photocatalytic degradation for dyes under ultraviolet-visible irradiation, J. Colloid Interface Sci. 562 (2020) 21–28.

[49] N. Amiralian, M. Mustapic, M.S.A. Hossain, C. Wang, M. Konarova, J. Tang, J. Na, A. Khan, A. Rowan, Magnetic nanocellulose: a potential material for removal of dye from water, J. Hazard. Mater. 394 (2020) 122571.

[50] Z.Y. Lin, Guihua, J. Chen, M. He, Ji, Dexian Modified CNF Membrane Capable of Catalytically Degrading 4-Nitrophenol, Preparation Method Therefor and Application Thereof, 2021. WIPO Patent CN2020076726.

[51] P. Liu, C. Zhu, A.P. Mathew, Mechanically robust high flux graphene oxide - nanocellulose membranes for dye removal from water, J. Hazard. Mater. 371 (2019) 484–493.

[52] L. Valencia, S. Monti, S. Kumar, C. Zhu, P. Liu, S. Yu, A.P. Mathew, Nanocellulose/graphene oxide layered membranes: elucidating their behaviour during filtration of water and metal ions in real time, Nanoscale 11 (46) (2019) 22413–22422.

[53] S. Roy, L. Zhai, L. Van Hai, J.W. Kim, J.H. Park, H.C. Kim, J. Kim, One-step nanocellulose coating converts tissue paper into an efficient separation membrane, Cellul. 25 (9) (2018) 4871–4886.

[54] W. Zhang, X. Wang, Y. Zhang, B. van Bochove, E. Mäkilä, J. Seppälä, W. Xu, S. Willför, C. Xu, Robust shape-retaining nanocellulose-based aerogels decorated with silver nanoparticles for fast continuous catalytic discoloration of organic dyes, Sep. Purif. Technol. 242 (2020) 116523.

[55] Y. Song, J.Y. Seo, H. Kim, K.-Y. Beak, Structural control of cellulose nanofibrous composite membrane with metal organic framework (ZIF-8) for highly selective removal of cationic dye, Carbohydr. Polym. 222 (2019) 115018.

[56] S. Gorgieva, R. Vogrinčič, V. Kokol, The effect of membrane structure prepared from carboxymethyl cellulose and cellulose nanofibrils for cationic dye removal, J. Polym. Environ. 27 (2) (2019) 318–332.

[57] Z. Karim, A.P. Mathew, M. Grahn, J. Mouzon, K. Oksman, Nanoporous membranes with cellulose nanocrystals as functional entity in chitosan: removal of dyes from water, Carbohydr. Polym. 112 (2014) 668–676.

[58] F. Rafieian, M. Jonoobi, Q. Yu, A novel nanocomposite membrane containing modified cellulose nanocrystals for copper ion removal and dye adsorption from water, Cellul. 26 (5) (2019) 3359–3373.

[59] M. Jonoobi, A. Ashori, V. Siracusa, Characterization and properties of polyethersulfone/ modified cellulose nanocrystals nanocomposite membranes, Polym. Test. 76 (2019).

[60] C.Y. Soon, N.A. Rahman, Y.B. Tee, R.A. Talib, C.H. Tan, K. Abdan, E.W.C. Chan, Electrospun biocomposite: nanocellulose and chitosan entrapped within a poly(hydroxyalkanoate) matrix for Congo red removal, J. Mater. Res. Technol. 8 (6) (2019) 5091–5102.

[61] V. Kokol, V. Vivod, S. Arnuš, U. Černigoj, B. Galičič, K.O. Vazner, B. Neral, A. Mihelič, Zeolite integrated nanocellulose films for removal of loose anionic reactive dye by adsorption vs. filtration mode during textile laundering, Fibers Polym. 19 (7) (2018) 1556–1566.

[62] Y.Z. Wu, Lilong, G. Fan, H. Hu, Yang, Mingde Nanocellulose Carbon Nitride Composite Membrane and Preparation Method and Application Thereof in Photocatalytic Degradation of Organic Pollutants, 2020. CN 111167493 A 20200519.

[63] A. Mautner, A. Bismarck, Bacterial nanocellulose papers with high porosity for optimized permeance and rejection of nm-sized pollutants, Carbohydr. Polym. 251 (2021) 117130.

[64] R. Cheng, M. Kang, S. Zhuang, L. Shi, X. Zheng, J. Wang, Adsorption of Sr(II) from water by mercerized bacterial cellulose membrane modified with EDTA, J. Hazard. Mater. 364 (2019) 645–653.

[65] H. Palacios Hinestroza, H. Urena-Saborio, F. Zurita, A.A. Guerrero de León, G. Sundaram, B. Sulbarán-Rangel, Nanocellulose and polycaprolactone Nanospun composite membranes and their potential for the removal of pollutants from water, Molecules (Basel, Switzerland) 25 (3) (2020) 1–13.

[66] A. Mautner, K.-Y. Lee, T. Tammelin, A.P. Mathew, A. Nedoma, K. Li, A. Bismarck, Cellulose nanopapers as tight aqueous ultra-filtration membranes, React. Funct. Polym. 86 (2015) 209–214.

[67] Z. Karim, S. Claudpierre, M. Grahn, K. Oksman, A.P. Mathew, Nanocellulose based functional membranes for water cleaning: tailoring of mechanical properties, porosity and metal ion capture, J. Membr. Sci. 514 (2016) 418–428.

[68] R. Yang, K.B. Aubrecht, H. Ma, R. Wang, R.B. Grubbs, B.S. Hsiao, B. Chu, Thiol-modified cellulose nanofibrous composite membranes for chromium (VI) and lead (II) adsorption, Polymer 55 (5) (2014) 1167–1176.

[69] Y. Xiong, C. Wang, H. Wang, C. Jin, Q. Sun, X. Xu, Nano-cellulose hydrogel coated flexible titanate-bismuth oxide membrane for trinity synergistic treatment of super-intricate anion/cation/oily-water, Chem. Eng. J. 337 (2018) 143–151.

[70] Q. Zhu, Y. Wang, M. Li, K. Liu, C. Hu, K. Yan, G. Sun, D. Wang, Activable carboxylic acid functionalized crystalline nanocellulose/PVA- co -PE composite nanofibrous membrane with enhanced adsorption for heavy metal ions, Sep. Purif. Technol. 186 (2017).

[71] D. Zhang, W. Xu, J. Cai, S.Y. Cheng, W.P. Ding, Citric acid-incorporated cellulose nanofibrous mats as food materials-based biosorbent for removal of hexavalent chromium from aqueous solutions, Int. J. Biol. Macromol. 149 (2020) 459–466.

[72] Q.G. Zhang, C. Deng, F. Soyekwo, Q.L. Liu, A.M. Zhu, Sub-10 nm wide cellulose nanofibers for ultrathin Nanoporous membranes with high organic permeation, Adv. Funct. Mater. 26 (5) (2016) 792–800.

[73] F. Soyekwo, Q. Zhang, R. Gao, Y. Qu, C. Lin, X. Huang, A. Zhu, Q. Liu, Cellulose nanofiber intermediary to fabricate highly-permeable ultrathin nanofiltration membranes for fast water purification, J. Membr. Sci. 524 (2017) 174–185.

[74] M.E. Leitch, C. Li, O. Ikkala, M.S. Mauter, G.V. Lowry, Bacterial nanocellulose aerogel membranes: novel high-porosity materials for membrane distillation, Environ. Sci. Technol. Lett. 3 (3) (2016) 85–91.

[75] Y.-S.S. Jun, Srikanth, X. Wu, Q. Jiang, Novel materials and methods for photothermal membrane distillation, 2021. United States Patent 20210023505A1.

[76] E.K. Sijabat, A. Nuruddin, P. Aditiawati, B. Sunendar Purwasasmita, Flat sheet membrane composite for desalinisation applications based on bacterial nanocellulose (BNC) from banana peel waste, cellulose, and silica, Mater. Res. Express 7 (10) (2020) 105004.

[77] I. Prihatiningtyas, Y. Li, Y. Hartanto, A. Vananroye, N. Coenen, B. Van der Bruggen, Effect of solvent on the morphology and performance of cellulose triacetate membrane/cellulose nanocrystal nanocomposite pervaporation desalination membranes, Chem. Eng. J. 388 (2020) 124216.

[78] H.-F. Tan, W.-L. Tan, N. Hamzah, M.H.K. Ng, B.S. Ooi, C.P. Leo, Membrane distillation crystallization using PVDF membrane incorporated with TiO2 nanoparticles and nanocellulose, Water Supply 20 (5) (2020) 1629–1642.

[79] Z. Ashrafi, Z. Hu, L. Lucia, W. Krause, Bacterial Superoleophobic fibrous matrices: a naturally occurring liquid-infused system for oil–water separation, Langmuir 37 (8) (2021) 2552–2562.

[80] Z. Ashrafi, L. Lucia, W. Krause, Underwater Superoleophobic matrix-formatted liquid-infused porous biomembranes for extremely efficient Deconstitution of Nanoemulsions, ACS Appl. Mater. Interfaces 12 (45) (2020) 50996–51006.

[81] Y. Hou, C. Duan, G. Zhu, H. Luo, S. Liang, Y. Jin, N. Zhao, J. Xu, Functional bacterial cellulose membranes with 3D porous architectures: conventional drying, tunable wettability and water/oil separation, J. Membr. Sci. 591 (2019) 117312.

[82] H. Ma, C. Burger, B.S. Hsiao, B. Chu, Ultrafine polysaccharide nanofibrous membranes for water purification, Biomacromolecules 12 (4) (2011) 970–976.

[83] H. Ma, K. Yoon, L. Rong, M. Shokralla, A. Kopot, X. Wang, D. Fang, B.S. Hsiao, B. Chu, Thin-film nanofibrous composite ultrafiltration membranes based on polyvinyl alcohol barrier layer containing directional water channels, Ind. Eng. Chem. Res. 49 (23) (2010) 11978–11984.

[84] H. Ma, C. Burger, B.S. Hsiao, B. Chu, Fabrication and characterization of cellulose nanofiber based thin-film nanofibrous composite membranes, J. Membr. Sci. 454 (2014) 272–282.

[85] S.K. Hong, S. Bae, H. Jeon, M. Kim, S.J. Cho, G. Lim, An underwater superoleophobic nanofibrous cellulosic membrane for oil/water separation with high separation flux and high chemical stability, Nanoscale 10 (6) (2018) 3037–3045.

[86] C. Ao, W. Yuan, J. Zhao, X. He, X. Zhang, Q. Li, T. Xia, W. Zhang, C. Lu, Superhydrophilic graphene oxide@ electrospun cellulose nanofiber hybrid membrane for high-efficiency oil/water separation, Carbohydr. Polym. 175 (2017) 216–222.

[87] A. Halim, Y. Xu, K.-H. Lin, M. Kobayashi, M. Kajiyama, T. Enomae, Fabrication of cellulose nanofiber-deposited cellulose sponge as an oil-water separation membrane, Sep. Purif. Technol. 224 (2019) 322–331.

[88] H. Zhan, N. Peng, X. Lei, Y. Huang, D. Li, R. Tao, C. Chang, UV-induced self-cleanable TiO2/nanocellulose membrane for selective separation of oil/water emulsion, Carbohydr. Polym. 201 (2018) 464–470.

[89] Q. Cheng, D. Ye, C. Chang, L. Zhang, Facile fabrication of superhydrophilic membranes consisted of fibrous tunicate cellulose nanocrystals for highly efficient oil/water separation, J. Membr. Sci. 525 (2017) 1–8.

[90] X.T. Jiang, Yong, J. Tan, High-Performance Tubular Hybrid Membrane, Preparation Method Thereof and Application Thereof to Separate Alkane/arene Mixture, 2021. CN 112403285 A 20210226.

[91] A. Kamtsikakis, S. McBride, J.O. Zoppe, C. Weder, Cellulose nanofiber nanocomposite pervaporation membranes for ethanol recovery, ACS Appl. Nano Mater. 4 (1) (2021) 568–579.

[92] J. Xu, X. Deng, Y. Dong, Z. Zhou, Y. Zhang, J. Yu, J. Cai, Y. Zhang, High-strength, transparent and superhydrophobic nanocellulose/nanochitin membranes fabricated via crosslinking of nanofibers and coating F-SiO2 suspensions, Carbohydr. Polym. 247 (2020) 116694.

[93] Q. Chen, J. Xiong, G. Chen, T. Tan, Preparation and characterization of highly transparent hydrophobic nanocellulose film using corn husks as main material, Int. J. Biol. Macromol. 158 (2020) 781–789.

[94] H. Jin, M. Kettunen, A. Laiho, H. Pynnönen, J. Paltakari, A. Marmur, O. Ikkala, R.H.A. Ras, Superhydrophobic and Superoleophobic nanocellulose aerogel membranes as bioinspired cargo carriers on water and oil, Langmuir 27 (5) (2011) 1930–1934.

[95] Z. Karim, M. Hakalahti, T. Tammelin, A.P. Mathew, In situ TEMPO surface functionalization of nanocellulose membranes for enhanced adsorption of metal ions from aqueous medium, RSC Adv. 7 (9) (2017) 5232–5241.

[96] I. Leppänen, T. Lappalainen, T. Lohtander, C. Jonkergouw, S. Arola, T. Tammelin, Capturing the Colloidal Microplastics with Plant-Based Nanocellulose Networks, 2021, https://doi.org/10.21203/rs.3.rs-286541/v1.

[97] P.M. Tahir, L.H. Zaini, M. Jonoobi, H.A. Khalil, Preparation of nanocellulose from kenaf (Hibiscus cannabinus L.) via chemical and chemo-mechanical processes, in: Handbook of Polymer Nanocomposites. Processing, Performance and Application, Springer, Berlin, Heidelberg, 2015, pp. 119–144.

Chapter 6

Plant-derived carboxycellulose: Highly efficient bionanomaterials for removal of toxic lead from contaminated water

Sunil K. Sharma[a,*], Priyanka R. Sharma[a], Ken I. Johnson[a], Yogita Madan[b], Songtao Li[a], George Cai[a], Isha Brahmbhatt[a], William Borges[a], and Benjamin S. Hsiao[a,*]

[a]*Department of Chemistry, Stony Brook University, Stony Brook, NY, United States,* [b]*Amity School of Applied Sciences, Amity University, Jaipur, Rajasthan, India*

[*]*Corresponding authors: e–mail address: sunil.k.sharma@stonybrook.edu, benjamin.hsiao@stonybrook.edu*

Introduction

Lead contamination in the environment is a grave concern worldwide, which causes large number of deaths annually [1]. This is because of toxic effects of lead on human body, as it affects the central nervous system by crossing the blood brain barrier as an ancillary for calcium. The main side effects of lead exposure among the human beings are disorders in neurodevelopment, behavioral problems, and neurotoxicity, and lead exposure can also result in increased risk of hypertension, heart attacks or strokes, kidney dysfunction, and hormonal inequities [2–7]. Lead is mainly used in the infrastructure of water hauling around the world; it is highly posing an increasing threat to many areas, because lead dissolution would increase over time due to the infrastructure corrosion, resulting in an increase in lead concentration in drinking water. The recent use of chloramine to enhance disinfection in water treatment plants has further increases the risk of high lead concentrations in drinking water due to the reaction between chlorine and lead piping [8].

Numerous methods to tackle the lead problem have been developed. Among these methods, activated charcoal with decent lead adsorption capacity that can separated from various plant-based materials, such as sugar cane waste, shells of nuts; and some fruit components is very interesting for us [9–15]. Because that solutions is based on biomass which is most abundant and biodegradable polymer on earth [16]. It can offer low cost and sustainable way to tackle this rising challenge for the society around the world. Recently, nanomaterials extracted from various wooden sources (nanocellulose) using various physicochemical process have been recognized as an effective medium to remediate toxic metal ion impurities from drinking water [17]. For example, carbon aerogel (a nanosized material with high surface area) obtained from cellulose were found to be highly efficient to remove heavy metal ions impurities for example Cr (VI) and Pb(II) [18]. In another class of cellulose derived nanomaterials that also reveals remarkable toxic heavy metal ion removal capability is carboxycellulose nanofiber (CNF) [19–26].

CNF is a functionalized or oxidized form of cellulose nanofibers, in that fiber surface having ample carboxylate groups (COO^-) that can act as active adsorption surface to interact with various positively charged toxic metals ions for example Cadmium, Chromium, lead [27–31].

There various chemical methods reported in the literature to isolate CNF from various types of biomass, For example TEMPO mediated oxidation, carboxymethylation, phosphorylation, acetylation, and silylation [32–36]. Mostly, these methods are highly effective in conversion of pure cellulose in to CNF; here pure cellulose means cellulose without any lignin and hemicellulose component. Thus, these methods require various chemical pretreatment practices to remove cellulose from raw biomass. Due to these drawbacks of above-mentioned procedures, here we have used a simple method for direct extraction of CNF from raw biomass using a mixture of nitric acid (HNO_3) and sodium nitrite ($NaNO_2$), which is called "nitro-oxidation" method. As per our hypothesis, the presence of nitric acid initiates the fibrillation process of raw

Separation Science and Technology, Volume 15. https://doi.org/10.1016/B978-0-323-90763-7.00004-4

biomass by eliminating noncellulosic components, whereby the reaction of HNO_3 and $NaNO_2$ creates HNO_2 and releases nitrosonium ions (NO^+). The produced nitrosonium ion can reacts with the primary hydroxyl group of cellulose at the C6 position and produce aldehyde groups (transitional) and then the carboxylate groups. This method significantly reduces the need for multichemicals, and offers significantly lowering the consumption of water and electricity [37].

In this study, CNF extracted from raw jute fiber using the nitro-oxidation method was examined against the removal of lead(II) ions from contaminated water. It was observed that the maximum removal efficiency (~2200 mg/g) of lead(II) ions by this CNF was around two times higher than that of the most of the lead (II) absorbent reported till now [38]. In this study, we found that the reason for such a high removal efficiency of CNF was not only due to adsorption, which was dominated till $\leq$1000 ppm concentration of lead (II) ions, but also due to the mineralization of lead crystals at high lead(II) ion concentration (>1000 ppm) on the CNF surface.

Experimental materials and methods

Untreated jute fibers samples were provided by Toptran Bangladesh Ltd. (Bangladesh). Nitric acid (ACS reagent, 70 wt%), sodium nitrite (ACS reagent $\geq$97 wt%), lead(II) acetate (tetrahydrate), sodium hydroxide, hydrochloric acid (36% assay) were purchased from Fisher Scientific. All above chemicals were used without further purification. Sodium nitrite, nitric acid and jute fibers were used for the preparation of CNF. Heavy metals were used for sample preparation in electron microscopic measurement and creation of solutions for remediation test.

Preparation and characterization of CNF

CNF were prepared from raw jute fibers by using the nitro-oxidation method (by mixture of nitric acid/sodium nitrite) as reported in previous work [39]. In brief, the reaction conditions included the use of 22.2 mmol of nitric acid and 14 mmol of sodium nitrite, for 12 h and at 50°C temperature. The separated nanofibers possessed 1.15 mmol/g carboxylate content, measured by conductometric titration method. The surface negative charge for CNF measured using Zetaprobe analyzer was −70 mV.

The spectroscopic and surface characterization of nanofibers before and after lead(II) absorption were characterized by using Fourier transform infrared spectroscopy (FTIR, Perkin Elmer instrument ATR mode), transmission electron microscopy (TEM, FEI Tecnai G2 Spirit BioTWIN instrument), scanning electron microscopy (SEM, Zeiss LEO 1550 SFEG-SEM) with energy dispersive X-ray spectroscopy (EDS) capability, ultraviolet (UV)-visible spectroscopy (CARY-300 UV–visible spectroscopy). The structural characterization of CNF was performed using X-ray diffraction techniques (WAXD) and adsorption experiments.

Wide-angle X-ray diffraction (WAXD)

WAXD measurements were carried out using a Benchtop Rigaku MiniFlex 600 instrument with Cu Kα radiation (λ = 0.154 nm), which was generated at 40 kV and 40 mA using a Ni filter. Data collection was carried out using a flat holder in Bragg–Brentano geometry (5–50°; 10°/min). The crystallinity index (CI) of CNF was estimated by the equation:

$$\mathrm{CI} = \frac{(I_{total} - I_{am})}{I_{total}} \quad (1)$$

where I_{total} is the integrated intensity of the diffraction spectrum, and I_{am} is the integrated intensity of the amorphous background.

Static adsorption study

The stock solutions of lead acetate were prepared by using various lead(II) ion concentrations in the range from 50 to 5000 ppm for this study. The adsorption study was carried out as follows. Two milliliter of the lead solution was mixed with 2 mL of a 0.23 wt% CNF suspension. The clear water above the settled gel/precipitant was removed from the test tubes and diluted by a factor of 100. The resulting solution was placed in UV–visible spectroscopy for basic observation of lead(II) ions and later analyzed using ICP-MS with required dilution in the ppb range. Based on the adsorption efficiency at the different concentrations of lead(II) ions, and at various pH, isotherm was generated. The maximum adsorption efficiency of lead by CNF was determined using the Langmuir model.

Results and discussion

Characterization of CNF extracted from untreated jute using nitro-oxidation approach

Fig. 1 demonstrates the TEM image of nanofibers extracted from raw jute fiber using the nitro-oxidation approach. Based on these images, the estimated average fiber length (L) was found to be around 320 nm, and the average fiber width (D) was around 5 nm, using the ImageJ software.

Fig. 2 shows the WAXD profile of CNF extracted from raw jute biomass. The diffraction pattern showed two important peaks: 2 θ at 22.6°, related to the (200) reflection plane, and 2 θ at 34.5°, indicating the (004) plane. The two weak peaks around 2 θ of 15° could attributed to the ($1\bar{1}10$) and (110) planes. The crystallinity index (CI) of the dried nanofiber (CNF) sample using Eq. (1) was 35%, which was significantly lower than that of the raw jute fibers (CI ~62%). We assume that the presence of nitric acid initiates the fibrillation process of raw biomass by eliminating noncellulosic part ultimately removing some noncrystalline regions from the cellulose and introducing carboxylate groups on the surface of fibers.

The presence of negatively charged carboxylate group was further verified by the FTIR analysis, as depicted in Fig. 3(i)A. In this spectrum, the distinguishing peaks at 3410 cm^{-1} was due to presence of the hydroxyl (—OH) stretching, 2883 cm^{-1}

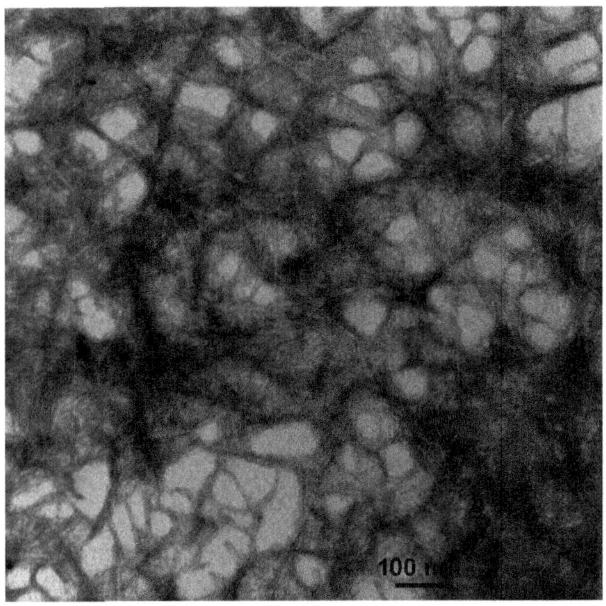

FIG. 1 (i) TEM of CNF obtained from untreated jute using the nitro-oxidation method.

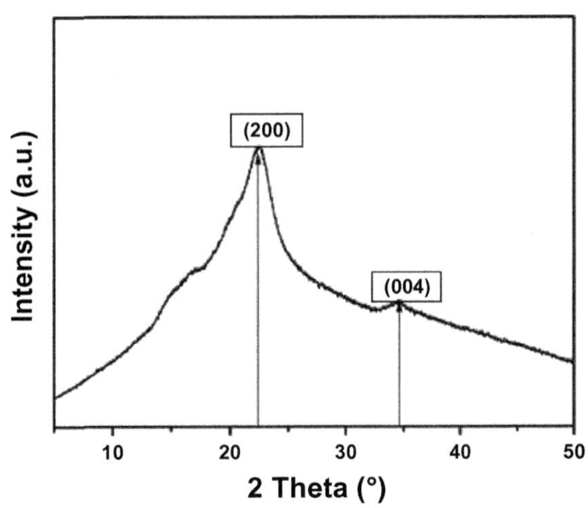

FIG. 2 WAXD profile of CNF obtained from untreated jute using the nitro-oxidation method.

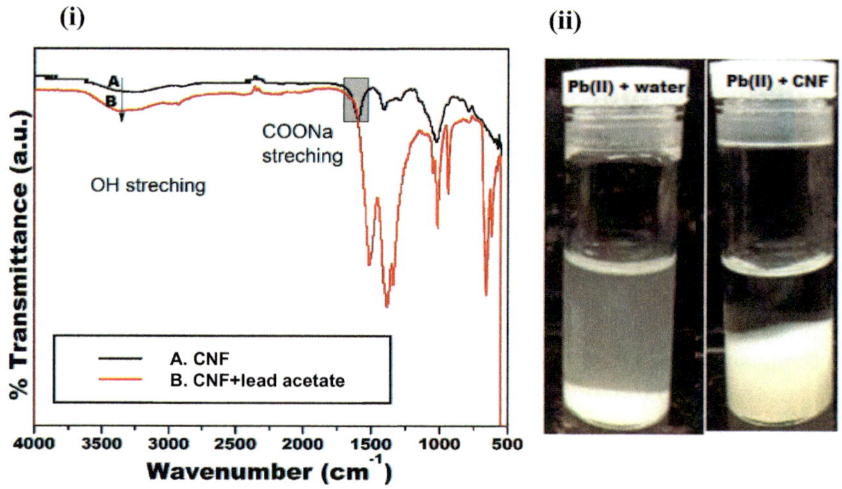

FIG. 3 FTIR spectra obtained from the dried CNF specimen prepared from (A) 0.23 wt% CNF suspension, and (B) suspended floc in the mixture of 0.23 wt% CNF and 750 ppm of lead(II) ions; (ii) Photograph showing the mixture containing 5 g distilled water and 1 g (109,000 ppm) lead acetate (*left*), and the mixture containing 5 g of 0.23 wt% CNF and 1 g (109,000 ppm) lead acetate (*right*). (The mixing time was 5 min).

was due to the C—H symmetrical stretching, 1232 cm^{-1} was due to the C—OH bending at C6 position, 1204 cm^{-1} was due to the C—O—C symmetric stretching. Most importantly the peak (highlighted) at 1563 cm^{-1} can be assigned to asymmetric stretching of CO_2 (carboxylate group) [40].

Mechanism for interactions among CNF surface and lead(II) ions

Strong interactions between negatively charged nanofiber surface and positively charged lead(II) metal ions were quite evident by visual examination of mixing 0.23 wt% CNF suspension with various concentration of lead solutions. It was observed that at lead concentrations less than 1000 ppm, the CNF suspension formed a uniform gel but when concentration is more than 1000 ppm, the CNF suspension started to aggregate formed precipitate at the bottom of the vial. The mechanism of gelation and precipitation can be explained as below.

The CNF suspension acts as a polyelectrolyte due to presence of negative surface charge (−70 mV) due to presence of surface carboxylated groups (1.15 mmol/g). The precipitation of positively charged metal ions with polyelectrolyte has been studied previously by Hirasawa et al. [41]. They have investigated the phenomenon to get nanosized lead crystals from lead sulfate in the presence of polyethylenimine (PEI) polyelectrolyte. In this study, the component of PEI acted as a nucleating agent in the aqueous environment. In the early stage of lead sulfate addition to the PEI polyelectrolyte, no noticeable change in lead concentration was observed. However, with the continuing addition of lead sulfate, the lead concentration exhibited a sudden decrease. At this point, the solution changed from clear to white, due to formation of large number of nanocrystals. In the present study, a similar physiochemical change was also found after addition of lead(II) ions to the CNF suspension.

Fig. 3(ii) demonstrates the photographs of two mixtures, containing 1 g (109,000 ppm) lead acetate in 5 g of DI water, and 5 g of 0.23 wt% CNF suspension and 1 g (109,000 ppm) lead acetate, respectively. Because of the high lead acetate concentration, the mixture without CNF was slightly milky but remained consistent. In the existence of CNF, white gel-like precipitant was formed at the bottom of the vial. The above physiochemical observations indicates that CNF obtained by the nitro-oxidation method work as good substrate for toxic lead(II) ion removal, from water, whereby the mechanism of adsorption included the interaction between negatively charged CNF surface (carboxylate groups on the surface) and lead(II) ions (as indicated in FTIR spectrum, Fig. 3(i)B), followed by the precipitation of lead crystals, probably in the form of $Pb(OH)_2$.

The FT-IR spectrum of the floc obtained by mixing the CNF with lead solution is depicted in Fig. 3(i)B, which exhibited very different characteristics peaks in FT-IR compare to pure CNF, especially in the region of around 1500 cm^{-1}. The peak at 1563 cm^{-1} results from asymmetric stretching of CO_2 group in pure CNF was shifted to 1500 cm^{-1} (lower value); which is due to interaction of negatively charged carboxylate groups on CNF surface and metallic lead(II) ions. As we understand that the real lead removal capacity of CNF in suspension was much higher than the expected value calculated from presence of carboxylate groups on CNF, we concluded that the overall lead removal of by CNF involved not only adsorption but also formation of lead crystals which is discussed later.

The mechanism of toxic lead(II) ion removal by CNF surface in suspension was revealed by the electron microscopy methods (which involved SEM/EDS and TEM) as depicted in Figs. 4 and 5, respectively. Fig. 4 shows SEM images of the floc formed from the mixture having 750 ppm of lead (II) ions, and (ii) that from the mixture having 1200 ppm lead(II) ions. In the sample with lead concentration below 1000 ppm, the image exhibited the typical aggregation of CNF in a very uniform fashion without any sign of crystallization. However, in the sample with high lead concentration ($\geq$ 1000 ppm), the flower shape aggregate structure (highlighted in red circle) was observed, indicating the formation of lead crystals after precipitation. The corresponding EDS profiles from SEM images (Fig. 4(i) and (ii)) are shown in inset left of each image). It was found that the CNF floc formed with less than 1000 ppm lead(II) ion exhibited less concentration of lead (Pb), comparatively to the CNF floc formed with more than 1000 ppm lead(II) ions. However, the prominent present of the lead element in Fig. 4(ii) (*left* inset) indicated that the adsorbed lead ions on the CNF behaved as nuclei, which initiated lead crystallization resulting in precipitation. The results obtained here are almost similar to the formation of nano size lead crystals by adding lead precursor to the polyethyleneimine polyelectrolyte [41]. So we have concluded that at low lead(II) ion concentration, the removal mechanism was mainly dominated by the electrostatic interactions between lead(II) ions and CNF, while at high lead(II) ion concentration (above 1000 ppm), it was dominated by the crystallization of lead possibly in the form of lead hydroxide [42].

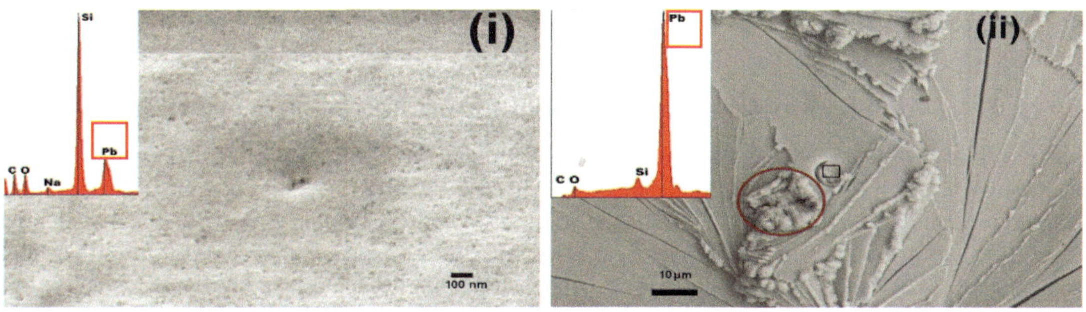

FIG. 4 SEM images of (i) the floc formed from the mixture with 750 ppm (<1000 ppm) of lead(II) ions (inset *left*: EDS spectra), and (ii) the floc formed from the mixture with 1200 ppm (>1000 ppm) lead(II) ions (*red circle* indicates the formation of lead crystals) (inset *left*: EDS spectra).

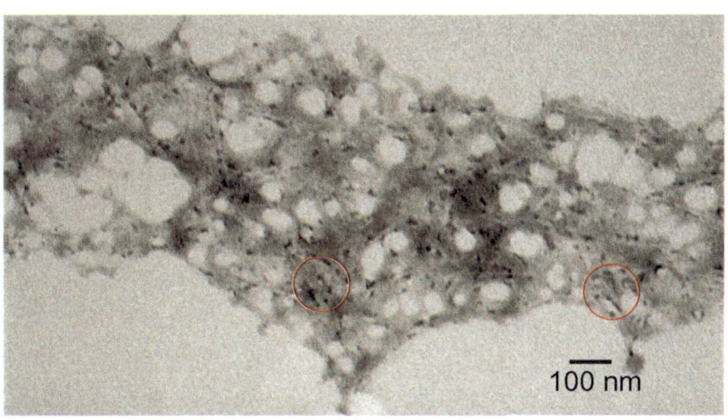

FIG. 5 TEM image of the floc sample containing CNF and lead(II) ions (the lead concentration was 300 ppm) aggregates, taken at 263,000X. The black spots on the edge of the fiber indicate the formation of lead nanocrystals.

The creation of lead nanocrystals on the surface of CNF could be confirmed by TEM analysis. Fig. 5 demonstrates the TEM image of the floc containing CNF and lead(II) ion nanocrystals. This TEM image shows that CNF acted as an effective nucleating agent, for the formation of lead nanocrystals (*red circles* in Fig. 5) having an average dimension of 4 nm. These lead nanocrystals were dispersed on the surface of the nanofiber. The formation nanocrystals by metal ions (ferrite, Ag, Au, ZnO, TiO_2) on CNF surface has been also reported previously [43–45].

Static adsorption study

The lead ion removal efficiency of nanofibers was revealed by the static adsorption study in a wide range of lead(II) ion concentration: 50–5000 ppm. In current study, the Langmuir isotherm model was used to find the effect of CNF on the remediation of lead(II) ions from contaminated water.

Fig. 6(i) demonstrates the experimentally measured removal capacity (Qe) of CNF against the original lead(II) ion concentration (Ce) from 50 to 5000 ppm. Based on the Langmuir model the value of Ce/Qe was plotted against Ce, and data could be fitted by a straight line with very good R [2] value of 0.978 (Fig. 6(ii). This indicated the lead ion adsorption results of CNF in the chosen lead(II) ions concentration region obeyed the Langmuir model. Based on this approach, the inverse of the slope against Ce/Qe versus Ce is equal to the adsorbent's maximum adsorption capacity (Qm). Thus, the maximum lead adsorption capacity of CNF was 2270 mg/g.

The adsorption capacity (Qe) is the product of adsorption efficiency and ideal adsorption ability of CNF (i.e., mg of Pb(II) per gram of CNF with carboxylate content of 1.15 mmol/g), where the adsorption efficiency of CNF could be calculated as the difference between the original lead concentration and the final lead concentration divided by the original lead concentration. The corresponding values are demonstrated in Table 1. It was observed that the adsorption results for CNF at very dilute lead concentrations (50 to 125 ppm) could not be fitted by the Langmuir model (Fig. 6(ii)), when the adsorption efficiency was less than 40% (Table 1). Maybe this was due to the kinetics impact because we have used very short time for adsorption study. Also at very high lead ion concentrations (>2500 ppm), the adsorption efficiency of CNF was also observed to be comparatively low as shown in Table 1). The possible reason for that can be the growth of lead nanocrystals on CNF surface, which differed from the adsorption mechanism. There were similar reports in literature for adsorption of Pb^{2+} ions on chitin/cellulose beads [46].

Comparison of CNF with other materials as adsorbent for lead(II) ion removal

After above studies it is clear that CNF can be a good substrate for absorptive removal of lead(II) ions from contaminated water. The evaluation of the maximum adsorption capacity (Qm) of CNF with different types of adsorbent reported in the literature is shown in Table 2. It was fascinating to see that CNF used in this work has two times higher efficiency than the maximum adsorption capacity of the most effective organic adsorbent ever reported crab shell [38]. Furthermore, CNF are about six times more effective than that of inorganic adsorbent, for example, nanostructured magnesium oxide and magnesium silicates spheres [47,48]. The recent development of several surface-modified (e.g., thiol-, ethylenediamine-, and diethylenediamine-modified) cellulose nanofibers only exhibited the lead removal capability of 50 to 131 mg/g, which is 55 to 17 times lower than the maximum adsorption capacity of CNF developed in this study [49–51]. The maximum adsorption capability of CNF against lead(II) ions is substantially higher than those of activated charcoals, which can also be derived from a variety of biomass sources and modified lignin [52,53]. Most likely this is because lead mineralization (in nanocrystal form) induced by the lead ion adsorption onto the charged CNF surface has significantly increased the remediation ability of lead(II) ions in water.

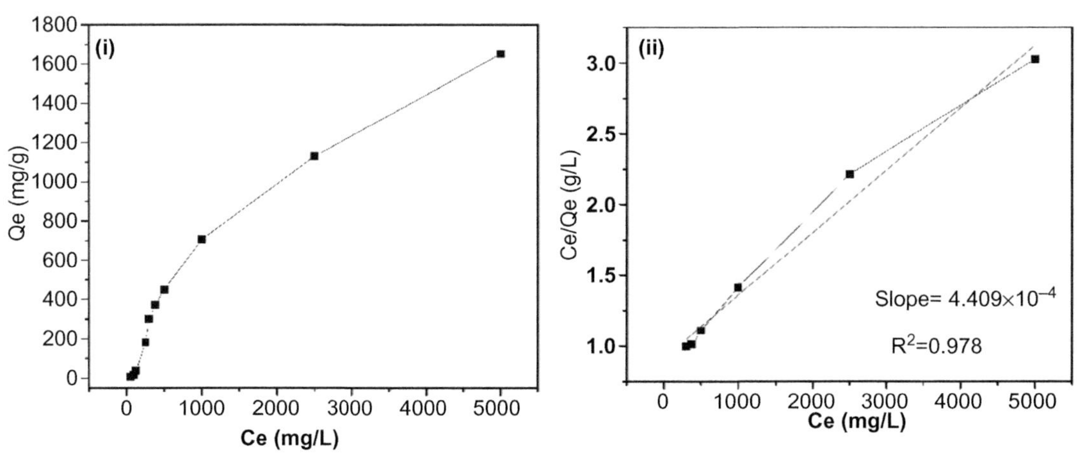

FIG. 6 (i) Lead(II) ions removal at concentration regime (50 to 5000 ppm), (ii) the adsorption data fitted by the Langmuir isotherm model, and (iii) effect of pH at the lead concentration of 20,000 ppm, Qe refers to the adsorption capacity measured in milligrams of lead(II) ions per gram of CNF; Ce represents the concentration of lead(II) ions (mg) per liter of solution.

TABLE 1 Calculation of ideal removal capacity of lead(II) ions by CNF in the concentration range of 50 to 5000 ppm.

Pb (II) ion (mg)	Original Pb(II) ion concentration, before ICP-MS analysis (ppb)	Final Pb(II) concentration, after ICP-MS analysis (ppb)	Total CNF used (g)[a]	Percent adsorption efficiency (%)[b]	Ideal adsorption (mg/g)[c]	Experimental adsorption (mg/g), Qe	Ce/Qe
20	5000	3096	0.0046	38	4348	1652	3.027
10	2500	1197	0.0046	52	2173	1130	2.212
5	1000	349	0.0046	65	1087	707	1.414
2.5	500	85	0.0046	83	543	451	1.109
1.875	375	34	0.0046	91	408	371	1.011
1.5	300	40	0.0046	92	326	300	1
1	250	85	0.0046	84	217	182	1.374
0.5	125	61	0.0046	32	109	35	3.57
0.4	100	31	0.0046	39	43	17	5.89
0.2	50	31	0.0046	38	21.7	8	6.25

[a] *Amount of CNF used in 2 mL of 0.23 wt% suspension.*
[b] *Adsorption efficiency = (original lead ion concentration − final lead ion concentration)/original lead ion concentration.*
[c] *Ideal adsorption capability of CNF was calculated based on carboxylate (—COO) content of 1.15 mmol/g and the total amount of CNF used during the remediation.*

TABLE 2 Comparison of the maximum adsorption capacity (Qm) value obtained from CNF with those of different types of adsorbent reported in the literature.

Adsorbent	Maximum adsorption efficiency (mg/g)	References
CNF produced by the nitro-oxidation method	**2270**	**This study**
Flower-like magnesium oxide	1980	[54]
CMC–g–PAA	990	[55]
Hybrid sorbent HMO–001	395	[47]
Magnesium silicate hollow nanospheres	300	[48]
Thiol-modified CNF	131	[49]
Ethylenediamine-modified cellulose	50.0	[50]
Diethylenetriamine bacterial cellulose	31.4	[51]
Modified lignin	95.8	[53]
Crab shell	1300	[38]
Sago waste	46.6	[52]

CMC, carboxymethyl cellulose; g, grafted; PAA, polyacrylic acid; APT, attapulgite hydrogel; HMO, hydrous magnesium oxide.

Conclusions

CNF extracted from raw jute using the one step nitro-oxidation method were found to be highly effective material for remediation of lead(II) ions from contaminated water. The maximum lead removal capacity of CNF, calculated using the Langmuir model, was found to be higher than the materials reported in the literature so far. The adsorption mechanism at low ($\leq$1000 ppm) lead(II) ion concentration is mainly due to electrostatic interactions between lead(II) ions and CNF, while that at high ($\geq$1000 ppm) lead(II) ion concentration is dominated by the growth of lead nanocrystals on CNF surface. Now, it is clear that CNF can be used either in the form of coagulants, adsorbents, or flocculants in both dispersion and membrane formats to removal toxic metal ions form drinking water.

Acknowledgments

The authors would like to thank the SusChEM Program of the National Science Foundation (DMR-1409507) for financial support.

References

[1] J.A. Carter, R.J. Erhardt, B.T. Jones, G.L. Donati, Survey of Lead in drinking water from schools and child care centers operating as public water suppliers in North Carolina, USA: implications for future legislation, Environ. Sci. Technol. 54 (22) (2020) 14152–14160.

[2] H.E. Gruber, H.C. Gonick, F. Khalil-Manesh, T.V. Sanchez, S. Motsinger, M. Meyer, C.F. Sharp, Osteopenia induced by long-term, low- and high-level exposure of the adult rat to lead, Miner. Electrolyte Metab. 23 (2) (1997) 65–73.

[3] R. Kim, A. Rotnitzky, D. Sparrow, S.T. Weiss, C. Wager, H. Hu, A longitudinal study of low-level Lead exposure and impairment of renal function: the normative aging study, JAMA 275 (15) (1996) 1177–1181.

[4] H.L. Needleman, C.A. Gatsonis, Low-level Lead exposure and the IQ of children: a Meta-analysis of modern studies, JAMA 263 (5) (1990) 673–678.

[5] J. Schwartz, Lead, blood pressure, and cardiovascular disease in men and women, Environ. Health Perspect. 91 (1991) 71–75.

[6] J. Schwartz, Societal benefits of reducing Lead exposure, Environ. Res. 66 (1) (1994) 105–124.

[7] A.J. McMichael, H.M. Johnson, Long-term mortality profile of heavily-exposed lead smelter workers, J. Occup. Med. 24 (5) (1982) 375–378.

[8] A. Kuch, I. Wagner, A mass transfer model to describe lead concentrations in drinking water, Water Res. 17 (10) (1983) 1303–1307.

[9] J. Liu, J. Si, Q. Zhang, J. Zheng, C. Han, G. Shao, Preparation of negatively charged hybrid adsorbents and their applications for Pb2+ removal, Ind. Eng. Chem. Res. 50 (14) (2011) 8645–8657.

[10] R.R. Bansode, J.N. Losso, W.E. Marshall, R.M. Rao, R.J. Portier, Adsorption of metal ions by pecan shell-based granular activated carbons, Bioresour. Technol. 89 (2) (2003) 115–119.

[11] W. Kermit, Y. Hong, W.S. Chung, E.M. Wayne, Select metal adsorption by activated carbon made from peanut shells, Bioresour. Technol. 97 (18) (2006) 2266–2270.

[12] I. Gulnaziya, A. Mohamed Kheireddine, S. Nik Meriam Nik, Removal of lead from aqueous solutions on palm shell activated carbon, Bioresour. Technol. 97 (18) (2006) 2350–2355.

[13] A. Mohamed, E.M. Wayne, A.H. Abdo, M.R. Ramu, G. Ipek, The use of nutshell carbons in drinking water filters for removal of trace metals, Water Res. 38 (4) (2004) 1062–1068.

[14] D.J. Malik, V. Strelko, M. Streat, A.M. Puziy, Characterisation of novel modified active carbons and marine algal biomass for the selective adsorption of lead, Water Res. 36 (6) (2002) 1527–1538.

[15] R. Mallampati, L. Xuanjun, A. Adin, S. Valiyaveettil, Fruit peels as efficient renewable adsorbents for removal of dissolved heavy metals and dyes from water, ACS Sustain. Chem. Eng. 3 (6) (2015) 1117–1124.

[16] D. Klemm, B. Heublein, H.-P. Fink, A. Bohn, Cellulose: fascinating biopolymer and sustainable raw material, Angew. Chem. Int. Ed. 44 (22) (2005) 3358–3393.

[17] A. Kardam, K.R. Raj, S. Srivastava, M.M. Srivastava, Nanocellulose fibers for biosorption of cadmium, nickel, and lead ions from aqueous solution, Clean Technol. Environ. Policy 16 (2) (2014) 385–393.

[18] S.-M. Alatalo, F. Pileidis, E. Mäkilä, M. Sevilla, E. Repo, J. Salonen, M. Sillanpää, M.-M. Titirici, Versatile cellulose-based carbon aerogel for the removal of both cationic and anionic metal contaminants from water, ACS Appl. Mater. Interfaces 7 (46) (2015) 25875–25883.

[19] C. Zhan, P.R. Sharma, H. He, S.K. Sharma, A. McCauley-Pearl, R. Wang, B.S. Hsiao, Rice husk based nanocellulose scaffolds for highly efficient removal of heavy metal ions from contaminated water, Environ. Sci. Water Res. Technol. 6 (11) (2020) 3080–3090.

[20] S.K. Sharma, P.R. Sharma, H. Chen, K. Johnson, C. Zhan, R. Wang, B. Hsiao, Cellulose-supported Nanosized zinc oxide: highly efficient bionanomaterial for removal of arsenic from water, in: Contaminants in our Water: Identification and Remediation Methods, American Chemical Society, Vol. 1352, 2020, pp. 253–267.

[21] P.R. Sharma, S.K. Sharma, W. Borges, H. Chen, B.S. Hsiao, Remediation of UO_2^{2+} from water by nitro-oxidized Carboxycellulose nanofibers: performance and mechanism, in: Contaminants in our Water: Identification and Remediation Methods, American Chemical Society, Vol. 1352, 2020, pp. 269–283.

[22] P.R. Sharma, S.K. Sharma, T. Lindström, B.S. Hsiao, Water purification: Nanocellulose-enabled membranes for water purification: perspectives (Adv. Sustainable Syst. 5/2020), Adv. Sustain. Syst. 4 (5) (2020) 2070009.

[23] H. Chen, S.K. Sharma, P.R. Sharma, H. Yeh, K. Johnson, B.S. Hsiao, Arsenic(III) removal by nanostructured Dialdehyde cellulose–cysteine microscale and nanoscale fibers, ACS Omega 4 (26) (2019) 22008–22020.

[24] P.R. Sharma, S.K. Sharma, R. Antoine, B.S. Hsiao, Efficient removal of arsenic using zinc oxide nanocrystal-decorated regenerated microfibrillated cellulose scaffolds, ACS Sustain. Chem. Eng. 7 (6) (2019) 6140–6151.

[25] C. Zhan, Y. Li, P.R. Sharma, H. He, S.K. Sharma, R. Wang, B.S. Hsiao, A study of TiO2 nanocrystal growth and environmental remediation capability of TiO2/CNC nanocomposites, RSC Adv. 9 (69) (2019) 40565–40576.

[26] B.S. Hsiao, B. Chu, P.R. Sharma, Production of Carboxylated Nanocelluloses. US 10,894,838 B2, 01/19/2021, 2021.

[27] N. Mahfoudhi, S. Boufi, Nanocellulose as a novel nanostructured adsorbent for environmental remediation: a review, Cellul. 24 (3) (2017) 1171–1197.

[28] N. Sankararamakrishnan, R. Singh, I. Srivastava, Performance of novel MgS doped cellulose nanofibres for cd(II) removal from industrial effluent – mechanism and optimization, Sci. Rep. 9 (1) (2019) 12639.

[29] V.E. Pakade, N.T. Tavengwa, L.M. Madikizela, Recent advances in hexavalent chromium removal from aqueous solutions by adsorptive methods, RSC Adv. 9 (45) (2019) 26142–26164.

[30] S. Nitesh, K. Balasubramanian, An effective technique for removal and recovery of uranium(vi) from aqueous solution using cellulose–camphor soot nanofibers, RSC Adv. 4 (53) (2014) 27691–27701.

[31] P.R. Sharma, A. Chattopadhyay, S.K. Sharma, L. Geng, N. Amiralian, D. Martin, B.S. Hsiao, Nanocellulose from spinifex as an effective adsorbent to remove cadmium(II) from water, ACS Sustain. Chem. Eng. 6 (3) (2018) 3279–3290.

[32] A. Isogai, T. Saito, H. Fukuzumi, TEMPO-oxidized cellulose nanofibers, Nanoscale 3 (1) (2011) 71–85.

[33] A. Naderi, T. Lindström, T. Pettersson, The state of carboxymethylated nanofibrils after homogenization-aided dilution from concentrated suspensions: a rheological perspective, Cellul. 21 (4) (2014) 2357–2368.

[34] V. Kokol, M. Božič, R. Vogrinčič, A.P. Mathew, Characterisation and properties of homo- and heterogenously phosphorylated nanocellulose, Carbohydr. Polym. 125 (2015) 301–313.

[35] E. Abraham, D. Kam, Y. Nevo, R. Slattegard, A. Rivkin, S. Lapidot, O. Shoseyov, Highly modified cellulose nanocrystals and formation of epoxy-Nanocrystalline cellulose (CNC) nanocomposites, ACS Appl. Mater. Interfaces 8 (41) (2016) 28086–28095.

[36] Z. Zhang, G. Sèbe, D. Rentsch, T. Zimmermann, P. Tingaut, Ultralightweight and flexible Silylated Nanocellulose sponges for the selective removal of oil from water, Chem. Mater. 26 (8) (2014) 2659–2668.

[37] P.R. Sharma, R. Joshi, S.K. Sharma, B.S. Hsiao, A simple approach to prepare Carboxycellulose nanofibers from untreated biomass, Biomacromolecules 18 (8) (2017) 2333–2342.

[38] M.-Y. Lee, J.M. Park, J.-W. Yang, Micro precipitation of lead on the surface of crab shell particles, Process Biochem. 32 (8) (1997) 671–677.

[39] P.R. Sharma, B. Zheng, S.K. Sharma, C. Zhan, R. Wang, S.R. Bhatia, B.S. Hsiao, High aspect ratio Carboxycellulose nanofibers prepared by nitro-oxidation method and their Nanopaper properties, ACS Appl. Nano Mater. 1 (8) (2018) 3969–3980.

[40] P.R. Sharma, A.J. Varma, Thermal stability of cellulose and their nanoparticles: effect of incremental increases in carboxyl and aldehyde groups, Carbohydr. Polym. 114 (2014) 339–343.

[41] I. Hirasawa, T. Mikami, A. Katayama, T. Sakuma, Strategy to obtain nm size crystals through precipitation in the presence of polyelectrolyte, Chem. Eng. Technol. 29 (2) (2006) 212–214.

[42] W.-J. Huang, J.-S. Lo, Synthesis and efficiency of a new chemical fixation agent for stabilizing MSWI fly ash, J. Hazard. Mater. 112 (1) (2004) 79–86.

[43] S. Galland, R.L. Andersson, M. Salajková, V. Ström, R.T. Olsson, L.A. Berglund, Cellulose nanofibers decorated with magnetic nanoparticles – synthesis, structure and use in magnetized high toughness membranes for a prototype loudspeaker, J. Mater. Chem. C 1 (47) (2013) 7963–7972.

[44] A.M. Adel, Incorporation of nano-metal particles with paper matrices, Interdiscip. J. Chem. 1 (2) (2016) 36–46.

[45] S. Boufi, A.M. Ferraria, A.M.B. do Rego, N. Battaglini, F. Herbst, M.R. Vilar, Surface functionalisation of cellulose with noble metals nanoparticles through a selective nucleation, Carbohydr. Polym. 86 (4) (2011) 1586–1594.

[46] D. Zhou, L. Zhang, S. Guo, Mechanisms of lead biosorption on cellulose/chitin beads, Water Res. 39 (16) (2005) 3755–3762.

[47] Q. Su, B. Pan, B. Pan, Q. Zhang, W. Zhang, L. Lv, X. Wang, J. Wu, Q. Zhang, Fabrication of polymer-supported nanosized hydrous manganese dioxide (HMO) for enhanced lead removal from waters, Sci. Total Environ. 407 (21) (2009) 5471–5477.

[48] Y. Wang, G. Wang, H. Wang, C. Liang, W. Cai, L. Zhang, Chemical-template synthesis of Micro/nanoscale magnesium silicate hollow spheres for waste-water treatment, Chem. – A Eur. J. 16 (11) (2010) 3497–3503.

[49] R. Yang, K.B. Aubrecht, H. Ma, R. Wang, R.B. Grubbs, B.S. Hsiao, B. Chu, Thiol-modified cellulose nanofibrous composite membranes for chromium (VI) and lead (II) adsorption, Polymer 55 (5) (2014) 1167–1176.

[50] S.M. Musyoka, J.C. Ngila, B.B. Mamba, Remediation studies of trace metals in natural and treated water using surface modified biopolymer nanofibers, Phys. Chem. Earth, Parts A/B/C 66 (2013) 45–50.

[51] W. Shen, S. Chen, S. Shi, X. Li, X. Zhang, W. Hu, H. Wang, Adsorption of cu(II) and Pb(II) onto diethylenetriamine-bacterial cellulose, Carbohydr. Polym. 75 (1) (2009) 110–114.

[52] S.Y. Quek, D.J. Wase, C.F. Forster, The use of sago waste for the sorption of lead and copper, Water SA 24 (3) (1998) 251–256.

[53] A. Demirbas, Adsorption of lead and cadmium ions in aqueous solutions onto modified lignin from alkali glycerol delignication, J. Hazard. Mater. 109 (1) (2004) 221–226.

[54] C.-Y. Cao, J. Qu, F. Wei, H. Liu, W.-G. Song, Superb adsorption capacity and mechanism of flowerlike magnesium oxide nanostructures for Lead and cadmium ions, ACS Appl. Mater. Interfaces 4 (8) (2012) 4283–4287.

[55] Y. Liu, W. Wang, A. Wang, Adsorption of lead ions from aqueous solution by using carboxymethyl cellulose-g-poly (acrylic acid)/attapulgite hydrogel composites, Desalination 259 (1) (2010) 258–264.

Chapter 7

Remediation of heavy metals with nanomaterials

Sriparna Dutta[a], Ranjana Dixit[b], and R.K. Sharma[a,*]
[a]*Green Chemistry Network Centre, Department of Chemistry, University of Delhi, Delhi, India,* [b]*Ramjas College, Department of Chemistry, University of Delhi, Delhi, India*
[*]*Corresponding author: e-mail address: rksharmagreenchem@hotmail.com*

Introduction

"Heavy metal is immortal, but we are not"—the underlying significance of this quote by Rob Halford is being felt deeply today, and the realization is growing stronger with every passing day [1]. When we review the past, we realize that the world has made a massive progress, but all the success accomplished so far has come at an exorbitant cost, to the extent of risking the lives of many. Heavy metals have been with us since the prehistoric times, meeting our wants and needs. However, our excessive reliance on their capability has exposed us to their potential toxicity [2]. Series of metal positioning case studies, particularly instances of Itai-Itai [3] and Minamata incidences [4], clearly reveal the fact that it was our sheer negligence that had driven us toward the critical situations. What negligence implies here is "not assessing the hazardous impacts of our activities." For instance, in case of mercury contamination, the Chisso Company, located at the Minamata City, south-west region of Japan's Kyushu Island, kept on dumping the wastes comprising of mercury (Hg) which was utilized as a catalyst for the manufacture of polyvinyl chloride, without thinking about the repercussions associated with that action. This negligence ultimately led to the concentration of mercury in the food chain, resulting in the death of several people. Similarly, arsenic-based compounds which were and still are being extensively employed in the manufacture of wood preservatives, insecticides, LEDs, and laser diodes have been identified as highly toxic, acquiring the first position in the ATSDR/EPA priority list of hazardous substances [5]. Long-term exposure to these substances is also known to cause deadly diseases such as arsenicosis, cancer, hyperkeratosis, and pigmentation changes [6]. A few years back, a shocking tale of India's cancer train drew the attention of all. Quoting the lines from the article *"Everyday, the dimly lit and shabby Bhatinda railway station in the Indian state of Punjab comes to life sharp at 9 pm."* [7] The excessive use of heavy metals and pesticides as a consequence of the 1970s' Industrial Green Revolution was ascribed as the prime reason for this grievous situation. The complexity of the problem grew manifold in face of expansive population growth catalyzed by improved agricultural practices, better infrastructure and enhanced industrial production. So, what we thoughtfully realize from these incidences is that economic progress should come simultaneously with environmental benefits. We cannot keep jeopardizing the existence of all the living creatures by virtue of our actions. Simultaneously, it is also true that the damage once done also needs to be altered.

Among the various environmental matrices infected with heavy metals, their presence in water has particularly proven dangerous. Thus, some of the international organizations such as World Health Organization (WHO) [8] and Environmental Protection Agency (EPA) [9] have outlined specific standards and limits of the concentration of various metals, which if exceeded may lead to serious health impacts. In this chapter, we will comprehensively throw light on the problems emerging out of heavy metal contamination due to their toxic, persistent, recalcitrant and bioaccumulative nature. Also, we will acquire knowledge pertaining to the strategies, especially nano based devised to combat the issues.

An analysis over the last few decades shows that a number of diverse methodologies have been explored and subsequently applied for wastewater treatment, especially for the heavy metal remediation. The conventional technologies including chemical precipitation, ion-exchange, membrane separation process and filtration, electrochemical techniques and adsorption have definitely given a direction to the researchers for coming up with solutions to remediate heavy metals [10]. However, factors such as lack of sensitivity, selectivity, poor economic and environmental credentials have

Separation Science and Technology, Volume 15. https://doi.org/10.1016/B978-0-323-90763-7.00009-3

necessitated the development of competent technologies that would be greener and sustainable. Thankfully, with the advent of nanotechnology, it has been possible to amalgamate the requirements of high efficiency, durability, economic viability and environmental benefits in one single platform [11]. Nano-based treatment technologies contribute an inertia toward the paradigm shift called for meeting the global water challenges by allowing the engineering of materials at the nano dimensions with the desired functionalities [12,13]. The size effect becomes apparent owing to the additional prospects offered as a result of increasing surface area to volume ratio and tunable pore size, causing a drastic enhancement in the activity. Additional advantages arising out of the possibility to integrate the nanomaterials with existing treatment processes facilitate the real time applicability for complete remediation of heavy metal contaminated water.

The chapter gives due consideration to the nano-based solutions devised so far for treating wastewater poisoned with heavy metals (especially the most toxic ones such as Cd, Hg, Pb, and As) from differentially charged sources. We present the compelling needs for providing clean and drinkable water through nanotechnology. We have also assessed the pros and cons of using these materials and where the researchers need to focus upon or rethink in future for the implementable solutions at large scale for the common people.

Heavy metals and their toxic impacts

.Isn't it a dismaying situation that lately, as per survey conducted by Central Water Commission, "42 rivers in India have been found to be containing high concentration of at least two neurotoxic heavy metals?" [14] Also according to an article published in the "Down to Earth" journal, river water in the Gangetic belt linked to gall bladder cancer was found to be contaminated with heavy metals [15]. Before we understand the fate of the contaminants and the factors behind their toxicity, it would be pertinent to elucidate what are essentially heavy metals. Unlike the other class of contaminants, heavy metals have been tagged as "indestructible poisons" as they strongly interfere with the body's usual metabolic functions. Although there are diverse definitions for the term "heavy metals," yet broadly speaking these are classified as either metals or metalloids possessing high density and known to cause toxic effects to the flora and fauna, even at low concentrations [16]. Examples of extremely toxic heavy metals include lead, arsenic, mercury and cadmium while other heavy metals of concern are chromium, copper, manganese, zinc, and nickel. These can virtually end up reaching all the environmental matrices once dispersed into the biosphere. Primarily, the sources of heavy metal dispersion have been displayed in Fig. 1.

The irony of the situation is that untreated effluents are deliberately disposed into the ground which in turn contaminates the underground aquifers. Ultimately, they find a way to pave through the entire food chain through contaminated soil and water [17]. Therefore, not only the marine [18], but also the terrestrial life is also disrupted as a consequence of the direct inhalation or ingestion of the heavy metals. Table 1 shows the highly toxic heavy metals, their sources of contamination, toxicity caused and prescribed limits (guideline value of concentration of heavy metals in mg/L) by WHO. The table includes data pertaining to the elements posing threat to human health.

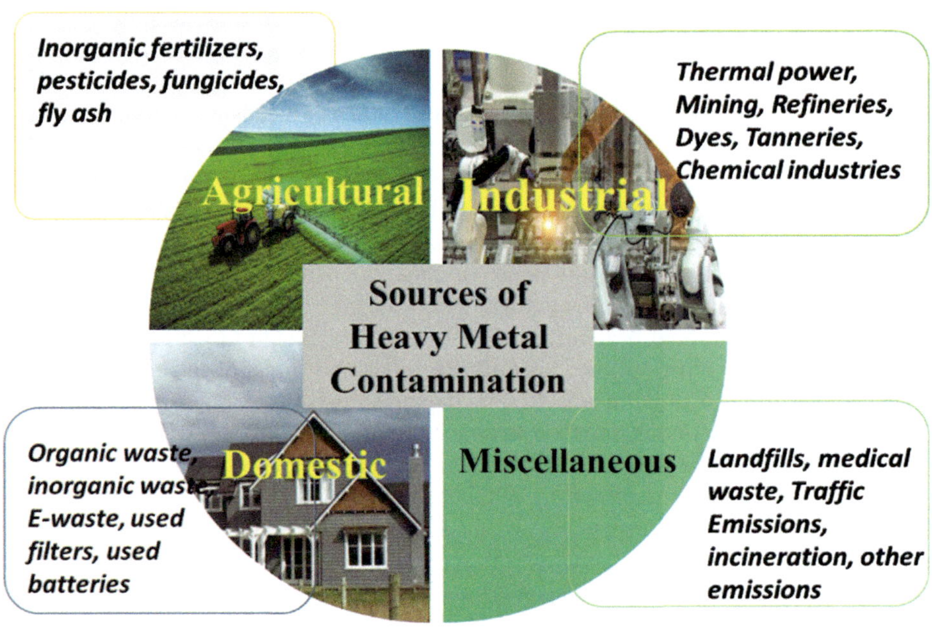

FIG. 1 Sources of heavy metal contamination.

TABLE 1 Heavy metals, their uses, source of contamination, toxicity caused, and guideline value [19]

Sl. No	Heavy metal	Uses	Source of contamination	Toxicity	Guideline value in terms of mg/L (WHO)
1.	Arsenic	Wood preservation, insecticides, laser diodes, LEDs, in lead alloys for ammunition	Geogenic/natural processes, smelting operations, thermal power plants fuel	Cancer in liver, lung, kidneys and skin lesions, neurotoxins, kidney damage	0.01
2.	Cadmium	Batteries, alloys, coatings (electroplating), solar cells, plastic stabilizers, and pigments	Smelting and refining of nonferrous metals, fossil fuel combustion and municipal waste incineration Geogenic sources such as volcanic eruptions	Damage caused to skeletal, reproductive, cardiovascular, central and peripheral nervous, and respiratory systems	0.003
3.	Lead	Batteries, pipes, pigments, insecticides, hair dyes and as an antiknocking additive for petrol	Automobile emissions, smoking, mining and plumbing, batteries, paints, e-waste, smelting operations, coal-based thermal power plants, ceramics, bangle industry	Anemia, seizures, unconsciousness and death	0.01
4.	Mercury	Electrical switches, thermometers, batteries, fluorescent lights	Chlor-alkali plants, mining and refining of mercury, thermal power plants, fluorescent lamps, hospital waste (damaged thermometers, barometers, sphygmomanometers), electrical appliances	Neurotoxins, kidney damage	0.006

While it is quite apparent from the data presented in the table that elements such as Pb, Cd, As, and Hg are toxic even at a very low/trace level, it would be informative to note the mechanism of toxicity associated with each of these metals.

Mechanism of toxicity

(i) *Lead (Pb):*

Lead poisoning primarily occurs either *via* ingestion through food or water contaminated by Pb. It may be toxic in either its inorganic or organic form; however, the organic form is highly dangerous as it is highly soluble in lipids which may result in its hasty consequences. This is one of those elements which affect almost every organ of the body. It exhibits toxicity by either inhibiting the activity of the –SH containing enzymes or by interfering with the key enzymes involved in the heme synthesis which leads to anemia [20].

(ii) *Mercury (Hg):*

Mercury shows its presence in the environment in both organic as well as inorganic forms. It is inherently different from rest of the elements in the periodic table, being the only element which is liquid at room temperature. Inhaled elemental mercury may end up getting either retained in the lungs, or distributed throughout the plasma and erythrocytes or may even cross the blood brain barrier. It is striking to note that the elemental mercury changes its toxicological form rather very fast as it gets oxidized into the mercuric form. In fact, the industrial release of inorganic mercury into the marine ecosystems has shown evidences of uptake by the marine organisms. After ingestion, mercury shows high propensity to get converted into the highly toxic form-"methyl mercury," that has a low rate of breakdown and thus gets accumulated in the entire food chain, reaching potentially toxic levels at the species at the top of the food chain [21]. In fact, organic mercury may also cross the placental barrier and from epidemiological studies it has been found that excessive exposure to this element may cause serious neurological disorders. The mechanism behind the toxicity of Hg(II) can be attributed to its high affinity for the deprotonated form of the thiol ligand. On binding, there might be conformational changes of the active site of the enzyme, blocking its activity [20].

(iii) *Cadmium (Cd):*

Cadmium principally affects our kidneys and bones [22]. As evident from the Itai-Itai disease [23] caused by Cd poisoning that resulted in osteomalacia (softening of the bones) and osteoporosis (loss of bone mass and weakness) reported during 1912, Cd can be quite toxic. Similar to Hg (II), Cd (II) also being a soft cation has a tendency to strongly bind to soft –SH groups of cysteine residues of enzymes due to stronger soft-soft interactions (e.g., Carbonic anhydrase, dipeptidase, carboxypeptidase, etc.) impacting their active confirmation as a result of the strong binding [20].

(iv) *Arsenic (As):*

One of the greatest threats to the general population comes from arsenic contaminated groundwater. Today, it is recognized that more than 140 million people across at least 50 countries have been consuming water contaminated with As at levels above the WHO guideline value [24]. The toxicity of As can also be judged from the fact that it has acquired the first position in the ATSDR/EPA priority list of hazardous substances. In fact, inorganic arsenic has been confirmed as a carcinogen and regarded as the most significant chemical contaminant detected globally in water. As per studies, rice plants can intake approximately 10-folds more As as compared to the other cereal crops. In India, West Bengal is one of those regions which is suffering intensely from the Arsenic contamination and people who are severely affected by As show symptoms of arsenicosis and other skin lesions such as pigmentation, melanosis, etc. [25] From the environmental as well as human health standpoints, As(III) and As(V) have been considered the most toxic forms. As (III) shows toxic effects by binding strongly with the SH containing enzymes [26]. Besides, arsenate can replace phosphate in many of the reactions [27].

Traditional strategies devised for the remediation of heavy metal–contaminated water

In order to combat issues pertaining to the heavy metal contamination, increasing efforts have been directed by the scientific community toward devising technologies that can eliminate/remove the toxic metals from diverse contaminated water channels as well as other environmental strata. Traditionally, methods such as chemical coprecipitation, ion-exchange, membrane separation, filtration, electrochemical techniques and adsorption have been extensively utilized for this purpose [28]. Some of these technologies have proven to be competent and demonstrated promising prospects for the industrial scale up and for real-time applications. Table 2 depicts each of these techniques by summarizing the treatment method and also deliberates upon the boons and limitations associated.

Review of these conventional methodologies divulges the need for developing integrated treatment methods that can offer high flexibility in system size, show high economic competence with simultaneous benefits of sustainability.

Nanotechnology: The competent tool for improved heavy metal remediation

The emergence of nanotechnology has provided glorious opportunities to the researchers for rendering contaminant free water to the humanity [41,42]. By adapting to the "nano based treatment technologies (integration of nano with the existing methodologies)," we have succumbed to live and breathe well. Quite appreciably, utilizing the boons of nanoscience and nanotechnology, we have been able to develop a promising solution for the removal rather challenging poisonous toxicant "arsenic" from drinking water [43]. This technology is able to deliver clean and affordable water to about 1 million people each day and has been approved by the Government of India. It has astonishingly provided a ray of hope for another 80 million people of the country who are facing the hazards of arsenic contamination.

Overall conceptualized on the fundamentals of "size matters," going down to the nanoscale can indeed be magical as it enables the scientists to access the extraordinary physical, chemical, mechanical, and optical properties of materials that naturally occur at that scale (1–100 nm in diameter) [44]. Nanomaterials have obvious advantages over their bulk counterparts owing to enhanced surface area to volume ratio, excellent thermal stability, presence of important functionalities that render them profitable for diverse applications [45,46]. Having diverse classifications, nanomaterials have been commonly defined on the basis of their morphologies and dimensions. For instance, nanoparticles are referred as zero dimensional, nanowires as one dimensional, nanosheets as two dimensional and nanocubes as three dimensional (Fig. 2) [47]. For the ease of comprehension, we have provided a succinct overview of the nanomaterials on the basis of their structural configuration in this chapter.

Chiefly, two approaches (bottom-up) and top-down are utilized for the synthesis of the nanomaterials (Fig. 3) [48,49]. The bottom up approach involves the self-assembly of smaller building blocks such as atoms and molecules to generate

TABLE 2 Different heavy metal treatment technologies employed for removal of heavy metals

Sl. No	Method of removal	Description	Examples illustrating promising applicability	Boons	Limitations	Reference
1.	Adsorption	Conceptualized on the principle of mass transfer in which the dissolved solids from a liquid phase tend to get deposited the surface of solid matrices owing to attractive chemical forces or *via* physical adsorption Most common adsorbent: Activated Carbon Other examples include adsorbents derived from agricultural waste, industrial by-product, natural material or modified biopolymers	Using cellulose fiber extracted from pineapple leaves and modifying it further with EDTA carboxymethyl (CM), Seubsai et al. managed to remove Pb (II) and Cd (II) from wastewater [29]	Economical (low-operation cost) Simple design Greener and sustainable Less requirement of control systems Most of the metals can be removed using this method	Performance depends on the chosen adsorbent Low regenerative capacity at times Low efficiency	[30]
2.	Chemical precipitation	Principally relies on the precipitation of heavy metals in the form of their complexes (insoluble precipitates of hydroxides, sulfides, carbonates, phosphate, etc.) *via* addition of a suitable precipitant	Zhang et al successfully treated waste water contaminated with heavy metals such as Fe^{3+} and Cr^{3+} with magnesium hydroxyl carbonate prepared from low grade magnesite [31]	Relatively Simple Economical Ease of operation, Automatic pH control	Requirement of enormous amounts of chemicals for formation of metal precipitates, Environmental impacts associated with the posttreatment due to excessive sludge formation, Tendency for the metal precipitates to undergo aggregation	[32,33]
3.	Electroanalytical	Fundamentally makes use of electricity for treatment of heavy metals present in water. Examples include electrodeposition, electrocoagulation, electro-floatation, and electrooxidation. Essentially, it leads to the precipitation of the heavy metals in weak acidic or neutralized catholyte as hydroxides when electricity is passed through an aqueous metal-bearing solution containing a cathode plate and an insoluble anode	Cui et al. developed an electrochemical method for removal of both low as well high concentration heavy metal pollution using either direct or alternating current electrodeposition with graphene-oxide-modified carbon felt electrode (CF-GO) [34]	Increased energy efficiency as it is performed at low temperature Unwanted side reactions can be prevented by adjusting the voltage	Sacrificial anodes require periodic replacement High cost of electricity Requires minimum solution conductivity that limits its use with effluents containing low dissolved solids	[35]

Continued

TABLE 2 Different heavy metal treatment technologies employed for removal of heavy metals—cont'd

Sl. No	Method of removal	Description	Examples illustrating promising applicability	Boons	Limitations	Reference
4.	Ion-Exchange	Based on the exchange of similarly charged cation attached to a solid support (ion-exchanger that are water insoluble substances)	Wang and coworkers developed a potential-responsive hybrid ion-exchange film system comprising of α-ZrP nanosheets a conducting polyaniline intercalated chain for heavy metal removal [36]	Cost-effective Convenient	Effective in treating water that has relatively low concentration of heavy metals (This is because if high concentration is taken, the matrix gets fouled by organics and other types of substances present in wastewater) Nonselective Shows high sensitivity toward pH of solution, thus unfit for large scale commercial applicability	[37]
5.	Membrane Filtration Technology	The technique utilizes permeable membrane for the separation of the inorganic contaminants (heavy metals) on the basis of pore size and molecular weight of the separating species	Cao et al. utilized the ultrafiltration technique by making use of extracellular polymeric substances which could remove 94.8% Pb (II) [38]	High removal efficiency (greater than 90% removal efficiency for metal concentration ranging from 10 to 112 mg/L)	Not economically viable owing to the possibility of membrane fouling	[39,40]

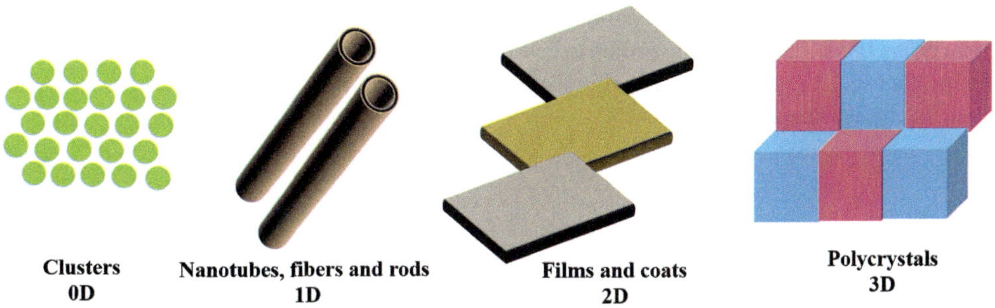

FIG. 2 Types of nanomaterials based on their dimensions.

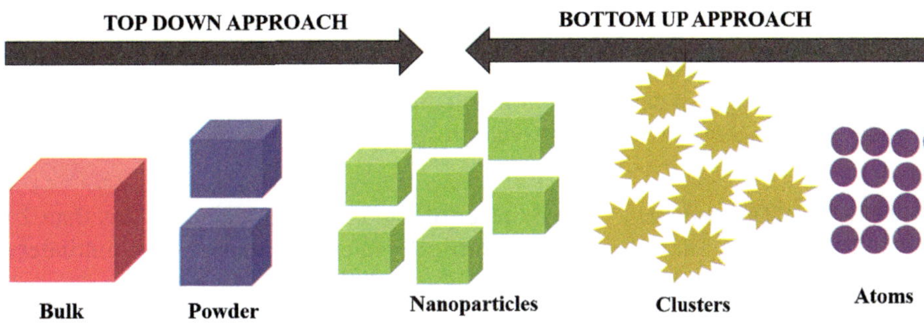

FIG. 3 Depiction of top down and bottom up approaches.

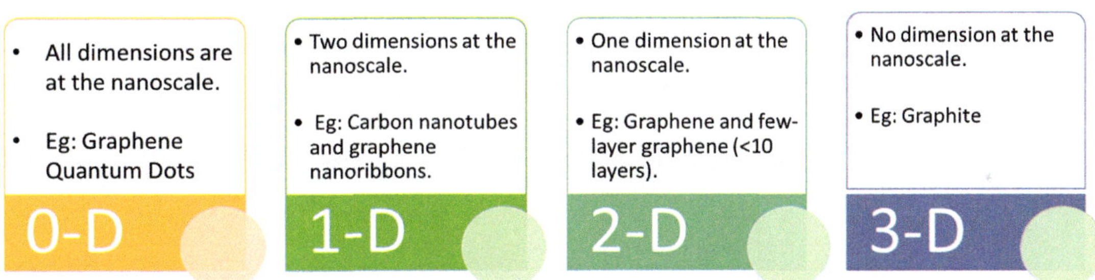

FIG. 4 Different types of carbon-based materials based on their dimensions.

larger nanostructures. Sol-gel processing, chemical vapor deposition (CVD), laser pyrolysis, and plasma or flame spraying synthesis are some of the common examples of the bottom-up approach. The top-down strategy on the other hand, involves the breakdown of larger system into correspondingly smaller structures. Generally different types of lithographic or X-cutting, etching, grinding, ball milling, etc. represent good examples of this synthetic methodology [50].

Different types of nanomaterials have been designed and explored for the removal of heavy metals which are being discussed in detail on the basis of their structural configuration in the ensuing subsections.

Carbon-based nanomaterials

Carbon has been utilized since antiquity in fabricating water filters and purifiers. Tracing back to 3750 BC, we have strong evidences of Egyptians and Sumerians employing carbon in the form of wood charcoal for water purification. Since long carbon was known to exist in two forms graphite and diamond; however, progressively many other allotropes were discovered as a result of the tendency of carbon to form covalent bonds with other carbon atoms. The element–"carbon" continues to proclaim its glorious victory over the years, showing its outstanding applicability in the wastewater treatment. The efficacy of carbonaceous materials has gained further momentum with the synergistic integration of nanoscience. The scientific exploration in this regard has led to the development of a diverse array of carbon nanomaterials, the most popular ones include carbon nanotubes (CNTs), fullerenes, graphene, graphene oxide and activated carbon which have been included under different dimensions of nanomaterials depicted in Fig. 4 [51].

Among a large number of complex carbon nanostructures, the chapter focuses on carbon nanotubes and graphene nanosheets particularly, as these have been documented as the most frequently employed nanomaterials for the removal of heavy metals. Properties such as high specific surface area, tunable surface properties and high chemical stability endure them highly efficient in the wastewater treatment.

Carbon nanotubes

Ever since the pioneering discovery by Iijima in 1991 at the NEC Laboratory in Tsukuba, Japan, CNTs have witnessed an exponential growth in their use in almost every field of science [52]. Embarking remarkable revolutions particularly in the field of water purification, CNTs stand out as the material of choice for the fabrication of nanoadsorbents [53], nanofilters [54] and nanomembranes [55] owing to tunable physical, chemical, electrical and structural properties, large surface area (100–300 $m^2 g^{-1}$), abundant pore structures and high pore volume that enhances its surface accessibility, ease of functionalization, high aspect ratio, etc. (Fig. 5).

It is their interesting cylindrical morphology comprising of rolled-up sheets of single-layer carbon atoms (graphene sheets) which demarcates them from the rest of the carbon-based materials [56]. CNTs in general are highly durable which can be primarily attributed to their sp^2 bonds that form strong form of molecular interaction. Broadly, these are categorized into two types:

(i) Single Walled CNTs (SWCNTs): consist of a single layer of nanotube with diameter less than 1 nm
(ii) Multi-walled CNTs (MWCNTs): consist of several concentrically interlinked nanotubes (multilateral graphene sheets) having diameter more than 100 nm.

Even though both SWCNTs and MWCNTs are used for adsorption of heavy metals, yet SWCNTs possess higher adsorption capacity, while MWCNTs are easy to synthesize. In case of the SWCNTs, four different sites of adsorption have been identified (Fig. 6) which are external surface, grooves (formed at the periphery of bundle due to contact between different nanotubes), internal hollow sites and the interstitial channel formed between tubes in bundles.

Adsorption affinity of the CNTs for a particular metal ion depends on the combination of several factors some of which include ionic radius and electronegativity of the metal.

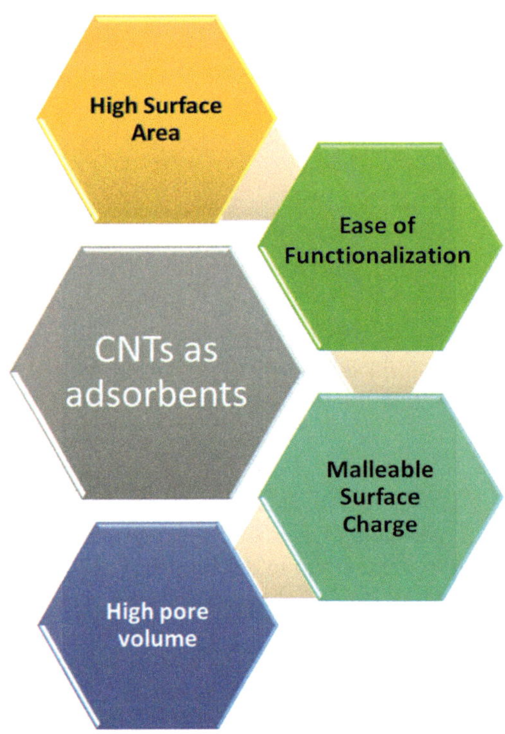

FIG. 5 Advantages of CNTs as adsorbents.

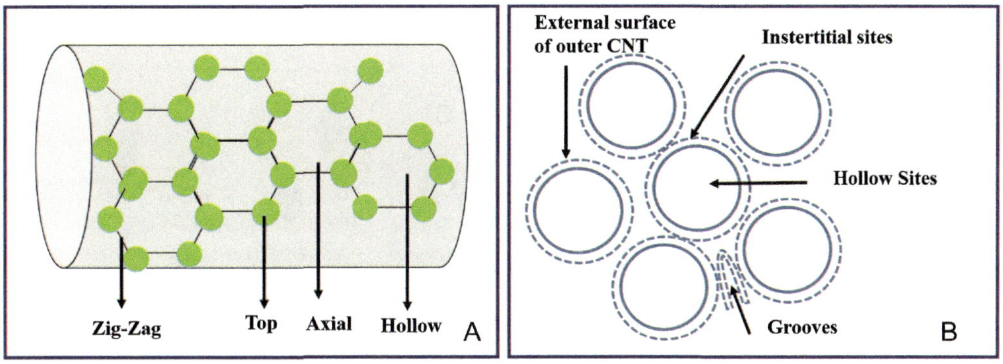

FIG. 6 Diagrammatic illustration of adsorption sites on (A) isolated carbon nanotubes (B) carbon nanotube bundles.

One of the major challenges encountered with the CNTs is the presence of impurities including amorphous carbon, small fullerenes, graphitic wrapped up sheets, metal catalysts, etc. [57,58] Thus, they are purified by using either physical or chemical methods. Chemical purification results in the introduction of new functional moieties onto the surface of the CNTs and can be achieved by using acids such as nitric acid, oxidizing agents (hydrogen peroxide or potassium permanganate) or *via* the use of mixture of oxidizers ($HCl–H_2O_2$ or $H_2SO_4–HNO_3$). Using oxidized CNTs, Li and coworkers could successfully adsorb Pb^{2+} and Cd^{2+}, exhibiting adsorption capacities of 63.29 and 11.01 mg/g, respectively [59]. Not only acids, but surface modification using other chemical linkers/functionalizing agents of the pristine CNTs renders it more selective for adsorption of heavy metals. There are promising examples that support this fact.

The functional group can be tethered onto the side wall or tip of the nanomaterial surface *via* either covalent or noncovalent modes of interaction/binding. Although a wide variety of functionalizing agents have been incorporated till date for functionalizing the CNTs [60], yet the anchoring of polymers have particularly been beneficial in augmenting the adsorption capacity of the native nanotubes owing to higher surface area, greater porosity and excellent crystallinity [61]. Recently, Bankole and coworkers utilized two different nanosorbents (i) pure CNTs (PCNTs) and (ii) polyhydroxybutyrate (PHB) functionalized CNTs using catalytic vapor deposition method for the removal of selected heavy metals from industrial electroplating wastewater [62]. For preparing the functionalized CNTs, a $Fe\text{-}Co/CaCO_3$ catalyst that was taken in a ceramic boat was placed in a horizontal tubular quartz reactor within a furnace. The furnace was subjected to heating at a rate of 10 °C/min while the argon (carrier gas) was allowed to flow over the catalyst. The flow rates of Ar gas were adjusted, and ethyne was introduced following which reaction was allowed to proceed for 45 min. Finally, reaction contents were cooled at room temperature and the obtained CNTs were removed from the quartz reactor. Thereafter, a mixture of nitric acid and sulfuric acid was utilized for the purification of the CNTs and the next step was introduction of functional group which was done by first dispersing CNTs in dimethyl formamide (DMF) and then adding thionyl chloride and sieving with tetrahydrofuran (THF) (Fig. 7). A host of characterization techniques were then used for structural and morphological analysis of the developed adsorbents. HRTEM image projected in Fig. 8 revealed that the inner outer diameter of MWCNTs increased after functionalization with the PHBs. This could be attributed to the increase in the layer on the outer wall of nanotube as a result of polymer placement. Both the nanoadsorbents worked efficiently with an optimum contact time of 10 min and could effectively remove 99% of Pb with only 20 mg of adsorbent used. Through experimental studies it could be established that the sorption isotherm suited the temkin model, while the kinetic data fitted to the pseudo-second order.

Apart from this, several other functionalizing agents have been exploited by the researchers engaged in the fabrication of nanomaterials which have been projected in Table 3.

Xu et al. have thoroughly reviewed the functionalized carbon nanotubes for adsorbing heavy metals from wastewater, covering the preparation, application, and mechanism of modified CNTs [60]. Recently, Gosain and coworkers have investigated the adsorptive capability of multiwalled carbon nanotube functionalized with molybdenum sulfide (MOS_2)/thiol nanocomposite in removal of Cd(II) and Pb(II) from industrial mine water [70]. The nanosorbent preparation strategy involved acidic treatment followed by sulfurization of MWCNTs and finally a hydrothermal reaction involving sodium molybdate and diethyldithiocarbamate as the precursors (Fig. 9). Using this nanocomposite, very high adsorption capacities of 90.0 and 66.6 mg/g could be attained for Pb (II) and Cd(II), respectively. The adsorption process was predominantly

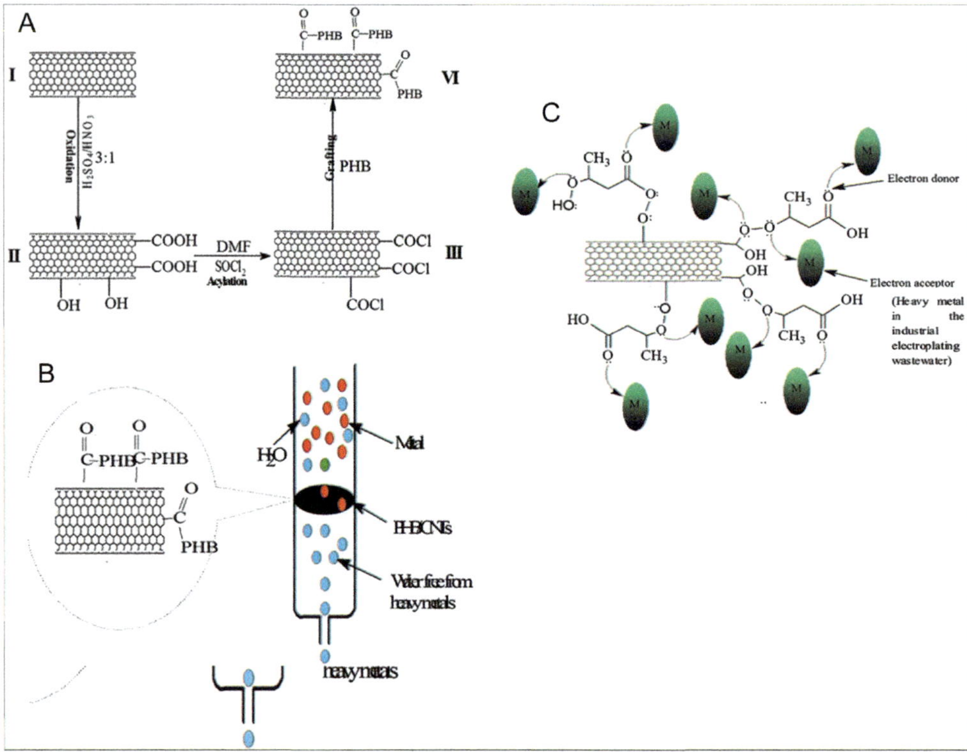

FIG. 7 (A) Schematic pathway/mechanism of purified and polymer functionalized carbon nanotubes production, (B) Removal of heavy metals from the electroplating wastewater using polymer functionalized carbon nanotubes (PHB-CNTs) *via* batch adsorption process and (C) Chemical mechanism pathway for the adsorption of heavy metals from the electroplating wastewater using polymer functionalized carbon nanotubes (PHB-CNTs) *via* batch adsorption process. *Adapted from M.T. Bankole, A.S. AbdulkareemI, I.A. Mohammed, S.S. Ochigbo, J.O. Tijani, O.K. Abubakre, W.D. Roos, Selected heavy metals removal from electroplating wastewater by purified and polyhydroxylbutyrate functionalized carbon nanotubes adsorbents, Sci. Rep. 9 (1) (2019) 1–19.*

facilitated by the metal sulfur complex formation and electrostatic interactions. The spent sorbent could be utilized further for photocatalytic and other electro-based applications.

Also CNTs are being modified further and used in preparation of nanocomposites *via* either the incorporation of magnetic moieties or addition of dopants or integrating other type of nanomaterial.

Graphene-based nanomaterials

Graphene, the basic building block of all the graphitic nanomaterials represents conceptually a novel class of two-dimensional nanomaterials that has shown phenomenal capability in removal of heavy metals from diverse wastewater streams [71]. A single graphene nanosheet is structurally composed of a single layer of carbon atoms that are beautifully stacked in a honeycomb like structure. There is a covalent linkage between each carbon atom. Strikingly, graphene was first discovered *via* the mechanical exfoliation of 3D graphite crystals and peeling away a single layer of graphene with a scotch tape. If several graphene sheets are stacked one after another, multilayer graphene is formed, and after a certain period of time, the material becomes graphitic (a three-dimensional structural compound). The parent graphene can also be chemically modified to generate graphene oxide which contains hydroxyl and epoxide functional group on its basal plane and carboxyl groups on its edges.

GO-based nanomaterials

Despite showing multiple benefits, commercialization of graphene has remained challenging owing to its high cost. Henceforth, graphene oxide has materialized as the preferred cheaper alternative [72]. Apart from being economic, GO displays excellent water-solubility and has an enormous surface area that results in its superlative adsorption behavior. Since GO possesses a number of oxygen-containing functional groups on its surface such as epoxide, carbonyl, carboxylic acid and hydroxyl, it offers interesting prospects of covalent modification with various chelating ligands that show great propensity to coordinate with the metal ions in order to develop superior water treatment agents [73,74]. This interesting property of

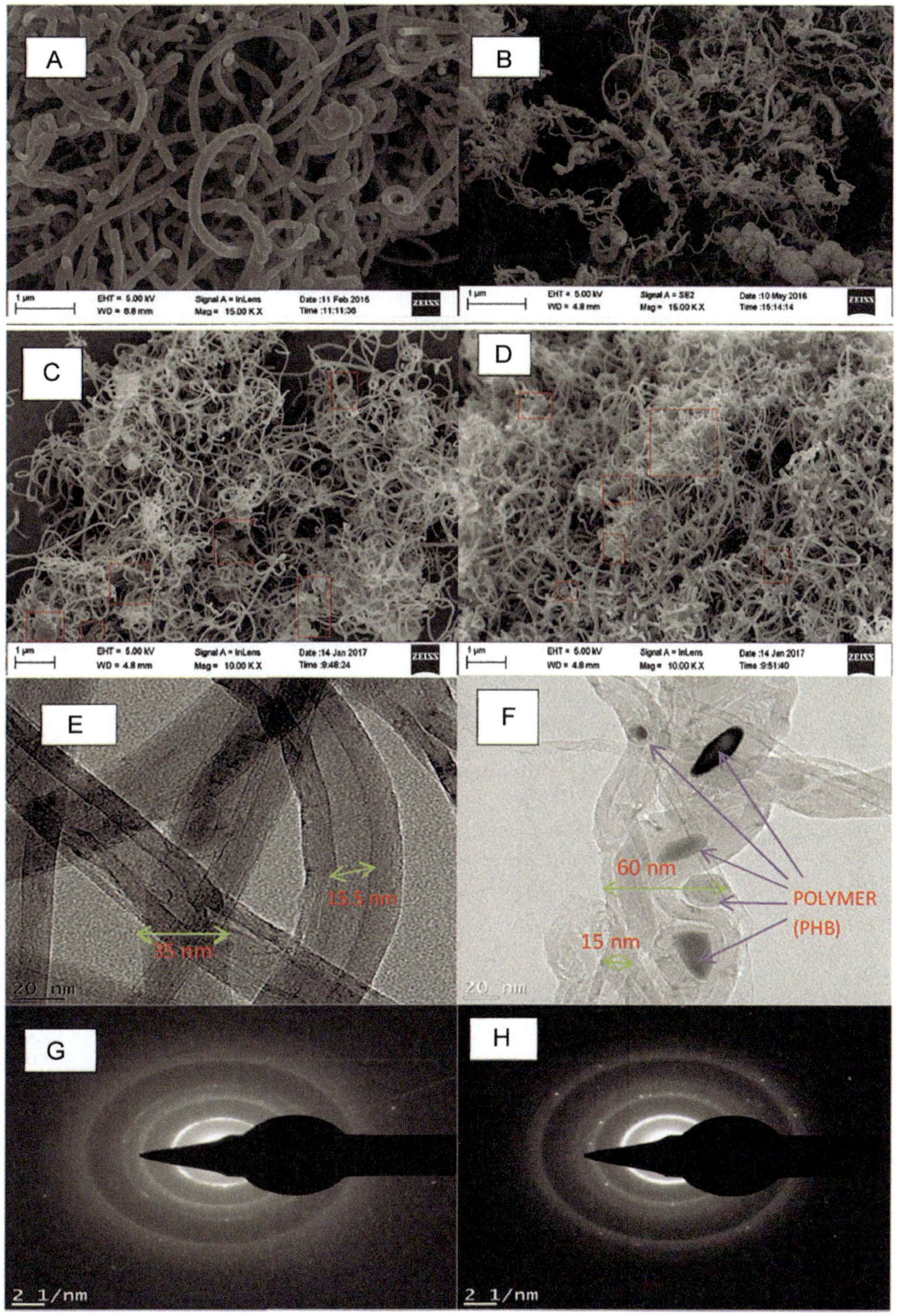

FIG. 8 High resolution scanning electron microscope images of the (A) P-CNTs and (B) PHB-CNTs (C) P-CNTs utilized for electroplating wastewater treatment and (D) PHB-CNTs utilized for electroplating wastewater treatment. HRTEM images of the (E) P-CNTs and (F) PHB-CNTs. SAED patterns of (G) P-CNTs and (H) PHB-CNTs. *Adapted from M.T. Bankole, A.S. Abdulkareeml, I.A. Mohammed, S.S. Ochigbo, J.O. Tijani, O.K. Abubakre, W.D. Roos, Selected heavy metals removal from electroplating wastewater by purified and polyhydroxylbutyrate functionalized carbon nanotubes adsorbents, Sci. Rep. 9 (1) (2019) 1–19.*

TABLE 3 Types of functionalized CNTs in removal of heavy metal ions

Sl. No	Type of functionalization	Functionalizing reagent	Removal of heavy metal	Reference
1.	Acid Functionalization (COOH, —C=O, and —OH functional groups)	Concentrated HNO_3 or mixtures of H_2SO4 with HNO_3, H_2O_2 or $KMnO_4$	Cd^{2+}	[63]
2.	Amine Functionalization	Polydopamine (PDA)	Cu^{2+}	[64]
3.	Dendrimer Functionalization	Triazole dendrimers (TD) Poly-amidoamine dendrimer	Hg^{2+} Pb^{2+} and Cu^{2+}	[65] [66]
4.	Fluorination	Elemental Fluorine	Zn^{2+} and Hg^{2+}	[67]
5.	Sulfonate functionalization	Conc. H_2SO_4	Cu^{2+}	[68]
6.	Thiol Functionalization	3-Mercaptopropyltriethoxysilane (MPTS)	Hg^{2+} and Pb^{2+}	[69]

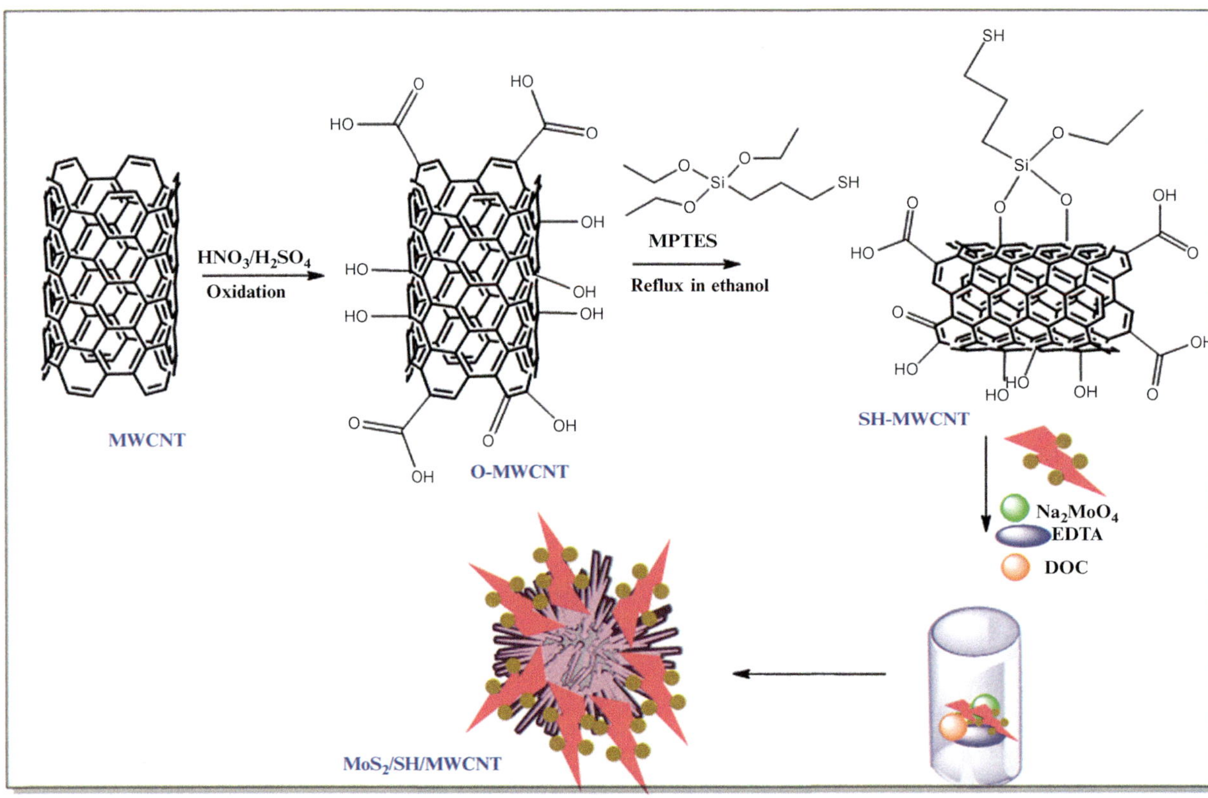

FIG. 9 Schematic presentation of the synthetic route of the MoS_2/SH-MWCNT nanocomposite.

GO was utilized by Awad and team of researchers for preparing a chelating adsorbent that comprised of 2-imino-4-thiobiuret covalently grafted onto partially reduced GO [75]. A modified Hummer's approach was incorporated for the synthesis of GO which was followed by the activation of the carboxylic groups of GO by $SOCl_2$ (Fig. 10). In the next step, in order to increase the number of carboxylic groups, Cl-CH_2COOH was added in presence of tetra butyl ammonium bromide as a catalyst which led to the introduction of the amidoxime (AO) functional group within the GO surface.

This nanomaterial displayed exceptional performance in the extraction of heavy metals from waste and polluted water (particularly it showed 100% removal capacity for Hg^{2+} at concentration of 100 ppm and contact time of 90 min). In fact, the reason behind the high selectivity toward mercury could be attributed to the presence of AO group that demonstrated high affinity toward this heavy metal. A set of physicochemical characterization tools employed such as FTIR, Raman, XRD and XPS authenticated the successful attachment of the amidinothiourea groups within the GO nanosheets which occurred primarily through the nucleophilic substitution of the acyl chloride (CH_3COCl) groups present in the modified GO. On the basis of sorption isotherms, a monolayer adsorption behavior was suggested that fitted well with the Langmuir model and kinetic studies revealed a pseudo second order kinetic model, indicating a chemisorption mechanism.

Recently, the melamine thiourea chelating ligand was incorporated on the surface of partially reduced graphene oxide (MT-PRGO) by Bakry et al. [76] This material could capture Hg(II) ions from water, showing removal efficiency of 100% up to concentration of 100 ppm. The combination of the heavier donor atom sulfur with amine and nitrogen functional groups of triazine motif attached to the partially reduced GO nanosheets worked effectively in the capture of Hg(II) ions from water.

Considering the current surge to devise new nanosystems for efficient removal of heavy metals from water, Sánchez et al. reported a self-propelled graphene oxide-based microbot technology which was named as the "GOx-microbots" for the capture, removal of and recovery of Pb (II) (Fig. 11) [77]. The microbot comprised of several layers of GO, Ni and Pt, each of these functionalities had a vital role to play. While the outer layer captured lead on its surface, the inner layer consisting of Pt worked as the self-propulsion system and the middle Ni layer allowed the magnetic control of the microbots. The material demonstrated excellent performance in adsorption of Pb which was primarily attributed to the strong complexation between Pb (II) ions and the oxygen moieties on GO.

The applicability of GO-based materials is not just limited to adsorption, but a slight modification can also render it as a superlative material for nanofiltration as proven by Alam et al. They developed a carboxylated graphene oxide-incorporated polyphenylsulfone nanofiltration membrane and investigated the capability of this nanofilter in the removal of arsenic, chromium, cadmium, lead, and zinc (Fig. 12) [78]. A phase inversion technique was utilized for the fabrication of PPSU and PPSU/carboxylated-GO nanocomposite membranes. For the nanofilter preparation, carboxylated GO was allowed to undergo dispersion under ultrasonic bath using N-methylpyrrolidone as the solvent and thereafter PPSU was added to the mixture and subjected to stirring for 24 h. The SEM analysis revealed that the membrane possessed a sponge-like porous support layer having open pores. Due to large availability of carboxylic groups, the membrane had a high molecular weight cut-off and hydraulic permeability. For the investigated cations, maximum removal rate of 80% could be achieved using this nanofilter.

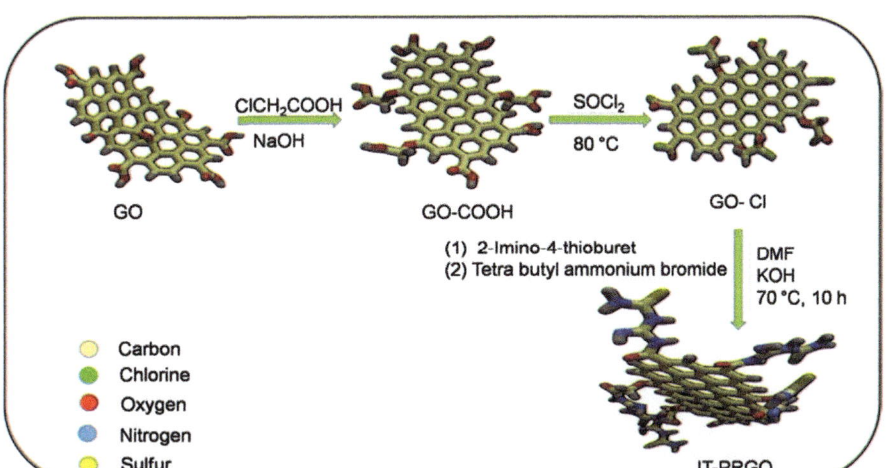

FIG. 10 Fabrication of 2-imino-4-thiobiuret partially reduced graphene oxide (IT-PRGO) sorbent. *Adapted from F.S. Awad, K.M. AbouZeid, W.M.A. El-Maaty, A.M. El-Wakil, M.S. El-Shall, Efficient removal of heavy metals from polluted water with high selectivity for mercury (II) by 2-imino-4-thiobiuret–partially reduced graphene oxide (IT-PRGO), ACS Appl. Mater. Interfaces 9 (39) (2017) 34230–34242.*

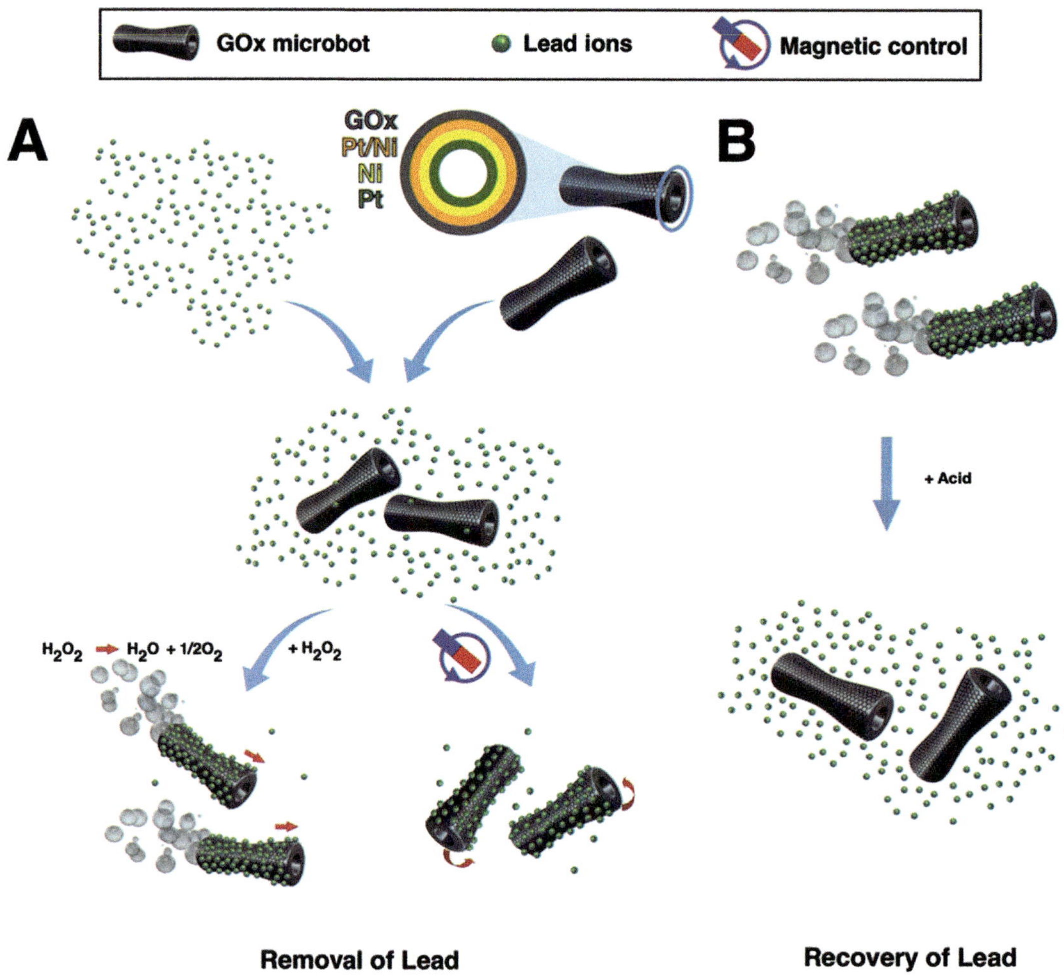

FIG. 11 Fabrication of GO_x-microbots for removal and recovery of lead. *Adapted from D. Vilela, J. Parmar, Y. Zeng, Y. Zhao, S. Sánchez, Graphene-based microbots for toxic heavy metal removal and recovery from water, Nano Lett. 16 (4) (2016) 2860–2866.*

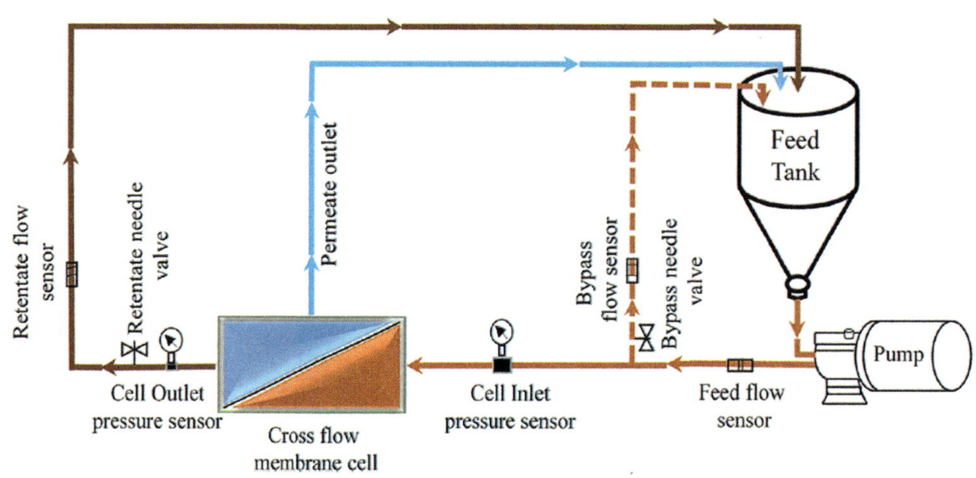

FIG. 12 Schematic illustration of the removal process *via* a cross-flow module. *Adapted from A. K. Shukla, J. Alam, M. Alhoshan, L.A. Dass, F.A.A. Ali, U. Mishra, U. Ansari. Removal of heavy metal ions using a carboxylated graphene oxide-incorporated polyphenylsulfone nanofiltration membrane, Environ. Sci.: Water Res. Technol. 4 (3) (2018) 438–448.*

GO derivatives-based nanomaterials

In the span of past few years, the derivatives of GO have also drawn the synthetic pursuit of the researchers for the large scale water purification [79]. For further modification, GO is often integrated with other nanomaterials such as ferrite [80], MOF [81], polymers [82], aluminosilicates [83], etc., and as a result of efficient coupling, improved results have been obtained. For example, Xu et al. synthesized graphene oxide supported organomontmorillonite composite (GO-OM) using a facile ultrasonic-assisted method that could simultaneously remove the most noxious contaminants such as Pb (II), Cd (II) and As (V) (Fig. 13) [84]. It is well established that montmorillonite is a clay mineral that comprises of octahedral sheets of alumina placed within two tetrahedral silicate layers. The further modification of interlayers can be done using various surfactants. The use of only montmorillonite has been associated with low removal efficiency and hence the material was amalgamated with GO to improve its adsorption capacity. The nanocomposite was prepared by firstly preparing GO using a modified Hummer's method and then purifying the clay mineral followed by its thermal treatment. Thereafter, in the next step, both these materials were taken in desired amount and then reacted under ultrasonication. The nanocomposite exhibited appreciable sorption capacities of 227.06 for Pb(II), 70.10 for Cd (II) and 80.20 mg/g for As (V). The Freundlich and pseudo-second-order kinetic models could successfully explain the simultaneous removal of these heavy metals from the contaminated matrices.

Silica-based nanomaterials

Nanosilica (nanoform of silicon dioxide possessing size <100 nm) has also emerged as a prevalent material in the field of water purification owing to its interesting physiochemical properties including high surface area, nanometer size, rigid framework, with excellent chemical, thermal and mechanical stability [85]. Structurally, silica is composed of a three-dimensional polytetrahedral network structure comprising of SiO_4 units and a bulk structure that terminates through either siloxane groups (µSi–O–Siµ) or silanol groups (µSi–OH) [86]. Based on porosity, silica NPs may be classified as: (i) Conventional nonporous (solid silica spheres ranging from 20 to 1000 nm) [87] and (ii) Mesoporous (pores ranging in diameter from 2 to 50 nm; e.g., MCM-48, MCM-41, HMS and SBA-15) [88,89]. On the basis of morphology, nanosilica may be classified as:-

(i) Silica Nanospheres [90]
(ii) Silica Nanorods [91]
(iii) Hollow Mesoporous Silica NPs [92]
(iv) Core-Shell Silica [93]

Three different silanol groups may be present on the surface of silica which include (i) vicinal (hydrogen-bonded silanols), (ii) germinal (two hydroxyl groups attached to the same silicon atom) and (iii) isolated (no hydrogen bonds possible) [94]. The presence of large concentration of surface hydroxyl groups increases the prospects of functionalizing both the exterior

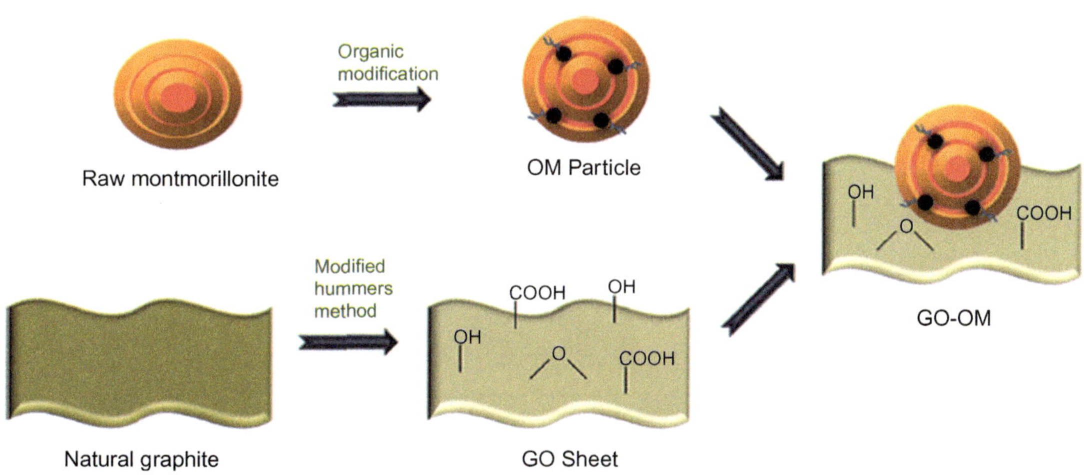

FIG. 13 Schematic illustration of the synthesis of GO-OM composite. *Adapted from J. Wei, M. F. Aly Aboud, I., Shakir, Z. Tong, Y. Xu, Graphene oxide-supported organo-montmorillonite composites for the removal of Pb (II), Cd (II), and As (V) contaminants from water, ACS Appl. Nano Mater. 3 (1) (2019) 806–813.*

as well as interior surfaces of nanosilica with diverse functional groups like $-NH_2$, –SH, etc., which upsurges its applicability in large scale water treatment. Cocondensation (one-pot synthesis) and grafting (postsynthetic strategy) are the two common routes adopted for functionalization of nanosilica.

Silica NPs-based nanomaterials

Using the postsynthetic strategy, recently, Garcia et al. functionalized the surface of silica NPs with pyridylpyrazole-β-ketoenol receptor and investigated the adsorption efficiency of this material in the removal of toxic metals (Pb, Zn, Cu and Cd) from water [95]. The chelating nanoadsorbent was fabricated in a stepwise manner beginning from the synthesis of the ligand "(Z)-1-(1,5-dimethyl-1H-pyrazol-3-yl)-3-hydroxy-3-(pyridine-2-yl)prop-2-en-1-one" *via* the reaction between the precursors—["pyrazole-3-carboxylate," "1-(pyridine-2-yl) ethanone," and Na] under ambient conditions in presence of anhydrous toluene as the solvent. In the next step, 3-aminopropyltriethoxy silane (APTES) was grafted onto the surface of silica using toluene as the solvent (Fig. 14). Finally, the ligand was allowed to react with the NH_2 functionalized silica NPs using a condensation technique. The results of the adsorption experiment revealed that the material was highly selective for Pb (II), displaying adsorption capacity of 110 mg/g rapidly within 20 min. The functionalized silica-based adsorbent also demonstrated good recyclability for at least five consecutive runs.

Constant modifications are on-way and different linkers are being exploited for the purpose of tuning the selectivity of the materials. Table 4 sheds light on the various linking agents utilized for functionalizing the surface of SiNPs and engineering them for the extraction of a particular metal ion in contaminated matrices comprising of several contaminants.

Mesoporous silica–based nanomaterials

Apart from the conventional silica NPs, mesoporous silica–based materials have emanated as excellent adsorbents as they have abundant surface active adsorption sites, allowing the attainment of high adsorption capacity and well-engineered pore size and geometry that can promote faster kinetics [100]. Reports have revealed that amine functionalized mesoporous silica NPs work exceptionally well in the removal of heavy metals [101,102]. Aguado and his competent team of researchers developed amine functionalized porous SBA-15–based mesoporous adsorbents which performed well in the extraction of Cu(II) [103]. For this, first pure SBA-15 was prepared using Pluronic 123 triblock-copolymer and then post-amine functionalization was carried out by making use of both grafting strategy as well as the condensation methodology.

Lately, a new type of mesoporous silica material known as "fibrous silica nanosphere (KCC-1)" has also been applied as an adsorbent for adsorption of Pb (II) [104]. The fibrous morphology of this nanomaterial enhanced the accessibility of active sites enabled 83.06% removal of Pb. The N_2 physisorption analysis confirmed the presence of Pb(II) on the surface and pore of KCC-1. The prospects of reusing the material for five cycles also affirmed the high potential of this material for lead adsorption from diverse contaminated streams.

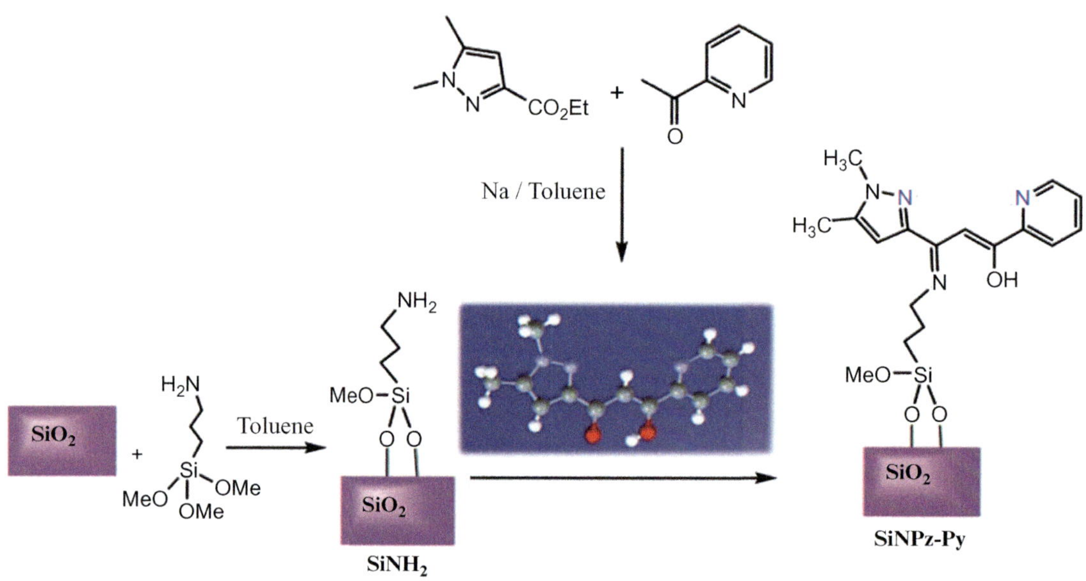

FIG. 14 Preparation of modified chelating material (SiNPz-Py). *Adapted from S. Tighadouini, S. Radi, M. Ferbinteanu, Y. Garcia, Highly selective removal of Pb (II) by a pyridylpyrazole-β-ketoenol receptor covalently bonded onto the silica surface, ACS Omega 4 (2) (2019) 3954–3964.*

TABLE 4 Functionalizing agents used for surface modification of silica NPs and subsequent application in removal of heavy metals

Sl. No	Type of functionalization	Functionalizing reagent	Removal of heavy metal	Reference
1.	Amine Functionalization	(3-Aminopropyl)triethoxysilane (APTES)	Pb (II) and Hg (II)	[96]
2.	Bromine and Azide using click chemistry	3-Bromopropyl propiolate and azide-modified silica nanoparticles	Pb (II)	[97]
3.	Mercapto functionalization	(3-Mercaptopropyl)triethoxysilane (MPTES)	Pb (II) and Hg (II)	[18]
4.	Norepinephrine functionalization using biomimetic approach	Norepinephrine	Pb (II)	[98]
5.	Plasma Polymer	Liquid monomer thiophene	Cu (II) and Zn (II)	[99]

TABLE 5 Examples of functionalizing agents used for surface modification of mesoporous silica NPs and subsequent application in removal of heavy metal

Sl. No	Functionalization	Functionalizing reagent	Removal of heavy metal	Reference
1.	Amine (NH_2)	Chitosan	Pb (II)	[105]
2.	Amine (NH_2)	APTES	Zn (II) and Cu(II) Cr(VI)	[106] [107]
3.	Amine (NH_2)	N-propyl-2-pyridylimine and ethylenediaminepropyl-2-pyridylimine	Pb (II) and Cu (II)	[108]
4.	Amine and carboxyl	2-Cyanoethyltriethoxysilane (CTES) and 3-[2-(2-Aminoethylamino) ethylamino]propyl-trimethoxysilane	Ni (II)	[109]
5.	Amine and mercapto	Aminopropyltrimethoxysilane (APTMS) and mercaptopropyltrimethoxysilane (MPTMS)	Pb (II) and Cu (II)	[110]
6.	Carboxyl	Iminodiacetic acid	Pb (II) and Cu(II)	[111]
7.	Dithiocarbamate	Dithiocarbamate	Hg (II)	[112]
8.	Ionic liquid	1-[3-(trimethoxysilyl)propyl]-3- or −4-undecanoylpyridinium chloride, and 3- or 4-[1-(hydroxylimine)undecyl]-1-[3-(trimethoxysilyl)propyl]pyridinium chloride	Cd (II) and Pb (II)	[113]
9.	Mercapto	2-Mercaptothiazoline	Hg (II)	[114]
10.	Phosphonic	Phosphoric Acid	U (VI)s	[115]

To further tune the applicability of the silica-based nanomaterials for better removal of heavy metals, the surface of the mesoporous silica NPs has been functionalized using diverse linkers; examples of some have been projected in Table 5. Literature documents the use of thiol functionalities on mesoporous silica-based materials that showed great propensity to remove Hg (II).

Dendritic mesoporous silica NPs

Very recently, dendritic mesoporous silica nanoparticles functionalized with thiol moieties were utilized as tailor-made nanoadsorbent (TDMSNs) for trapping Hg (II) (Fig. 15) [116]. The nanosorbent exhibited significantly high adsorption

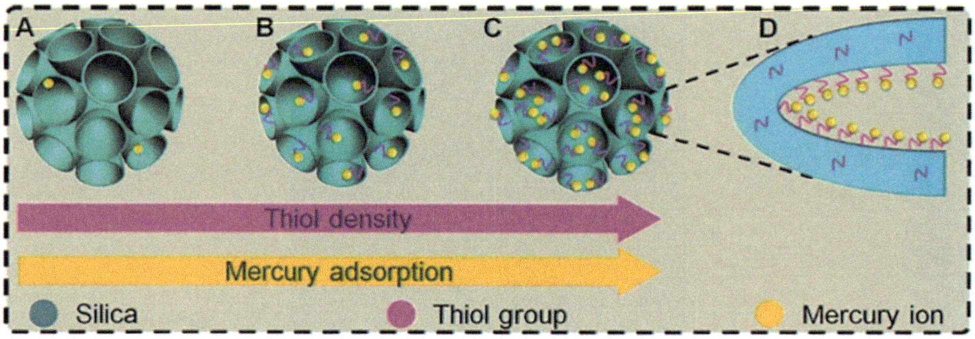

FIG. 15 Schematic illustration of the adsorption performance of (A) pure DMSNs, (B) TDMSNs with a low content of thiol moieties, and (C) TDMSNs with a high content of accessible thiol moieties. (D) The schematic illustration of well exposed adsorption sites; the large yet short-length mesopores make all adsorption sites accessible for mercury ions. *Adapted from M. Kalantari, Z. Gu, Y., Cao, C. Lei, J. Zhang, Thiolated silica nanoadsorbents enable ultrahigh and fast decontamination of mercury (ii): understanding the contribution of thiol moieties' density and accessibility on adsorption performance, Environ. Sci. Nano 7 (3) (2020) 851–860.*

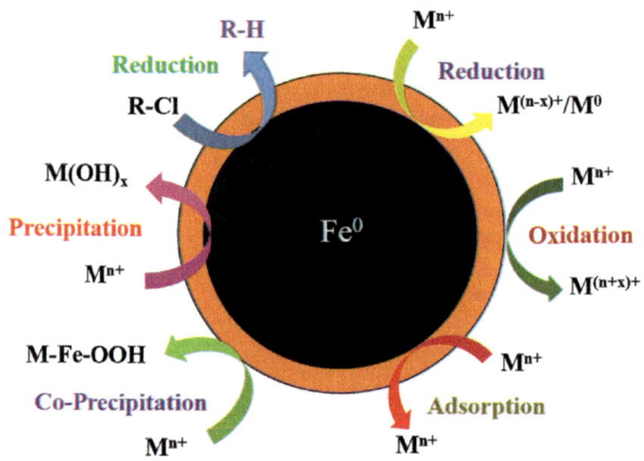

FIG. 16 Core–shell structure of nZVI depicting various mechanisms for the removal of metals.

capacity of 1502.4 mg g^{-1}, excellent reusability and faster kinetics which could be accredited to the presence of well-exposed mesopores of TDMSNs and highly accessible thiol groups that enabled the more efficient binding of Hg (II), allowing the concentration of Hg (II) to drop from 10 ppm to 2 ppb.

Iron-based nanomaterials

The cutting edge discovery of iron-based nanomaterials for water remediation [117], predominantly as potent sorbents has opened new horizons for researchers across the globe because of their ready availability, cost-effectiveness, large specific surface area and most importantly their excellent magnetic response facilitating ease of separation. Zero-Valent Iron [118], maghemite (γ-Fe_2O_3) [119], hematite (α-Fe_2O_3) [120] and magnetite (Fe_3O_4) [121] are examples of most frequently utilized iron-based materials used for adsorption of heavy metals.

Nanoscale zero valent iron (nZVI)

Nanoscale zero valent iron (nZVI) is a core-shell composite made up of Fe(O) and ferric oxide coating (Fig. 16) that has gained a rapid surge in interest as material for removal of Hg (II), Cr (VI), Cu (II), Ni (II), Cd (II), etc. [122,123] In fact, the use of nZVI for remediation of heavy metals can be traced to almost two decades back. While Fe (0) has a phenomenal reducing capability, the ferric oxide shell offers abundant accessible sites for reactivity and electrostatic interaction with the various heavy metals.

As compared to the conventional iron powders, the nanoscale zero valent iron is preferred owing to controllable size, larger specific surface area and high reactivity. Depending on the standard electrode potential of the heavy metals, mechanism for removal can differ. However, in general, for a heavy metal that has a higher E° value as compared to Fe (II), the removal mechanism can be summarized as below:

$$\text{Reduction: } Fe^0 + M^{2+} \longrightarrow Fe^{2+} + M^0$$
$$\text{Sorption: } FeOOH + M^{2+} \longrightarrow FeOOM^+ + H^+$$

Liquid phase reducing method was traditionally utilized to generate pristine zero valent iron [124]; however with rapid improvement in synthetic technologies, new techniques have been devised. Some examples of the top-down approach involved in the synthesis of nZVI include pulsed laser ablation, noble gas sputtering and milling processes, while bottom up usually start from either dissolved iron salts, nanosized iron oxides, or iron-containing molecules [125]. Boparai and coworkers had employed the "bottom-up" method strategy for preparing pristine nZVI by adding $NaBH_4$ to $FeCl_3$ solution dropwise that was applied for the successful removal of Cd^{2+} ions from aqueous solutions [126]. Using a similar approach, Ramos and coworkers prepared nZVI for removal of arsenic ions from groundwater. Wheizhou and team concluded through their studies that undried nZVI works well in removal of Pb(II), as it has more active sites due to its weaker oxidation (Fig. 17) [127].

As a step further toward achieving greater adsorption, Zhu and coworkers coupled potassium carbonate activated porous biochar with nano zero valent iron/nano-α-hydroxy-iron oxide (nHIO) and found a three times enhancement in Cd removal capacity as compared to the pristine biochar. Complexation-reduction was thought to be involved in the coupling mechanism of Cd on nZVI@KBC [128].

In view of potential toxicity associated with the chemical reductants, recently greener methods involving plant extracts as reductants have been developed [129]. Lynch et al. have revealed a leaf extract "Mentha piperita" mediated greener method for synthesizing nZVI under microwave irradiation conditions that could be synthesized within 3 min and successfully remove 79.33% lead along with other inorganic ions from the water samples that were procured from the Worcester and Birmingham canal located at United Kingdom [130]. The fabricated nZVI-NPs were found to possess a spherical morphology with size lying between 5 and 10 nm.

Despite showing multiple benefits, pristine nZVI can pose serious complexity in removal of contaminants because of its tendency to undergo agglomeration (this happens due to can weak van der Waals forces, high surface energy and intrinsic magnetic interactions) [131]. Fortunately, this problem has been resolved by either coating the surface of the NPs with strong stabilizers/protecting agents (metal, metal oxides, polymers, etc.) or incorporation of suitable functionalizing agents *via* covalent or noncovalent interactions [132]. Surface modification can help in decreasing the electrostatic forces of attraction between the pristine NPs, thereby preventing aggregation and facilitating the mobility of nZVI for environmental heavy metal remediation.

nZVI has shown its efficiency in not only the fabrication of superior nanosorbents, but also in the generation of low-cost nanofilters. For instance, Danila et al. utilized nZVI as a nanofilter for the removal of Cd(II), Cu(II), Ni(II), and Pb(II) ions from their aqueous solutions with adsorption capacities of 79.33–102.00, 111.11–142.85, 107.30–137.96 and 110.97–142.68 mg/g, respectively, for these heavy metals [133].

Working on similar front, Cabrera and his research team synthesized nZVI using ferric chloride salt and sodium borohydride as the reducing agent under ambient conditions and exposed this nanomaterial to 6 ppm of Cd^{2+} [134]. The results

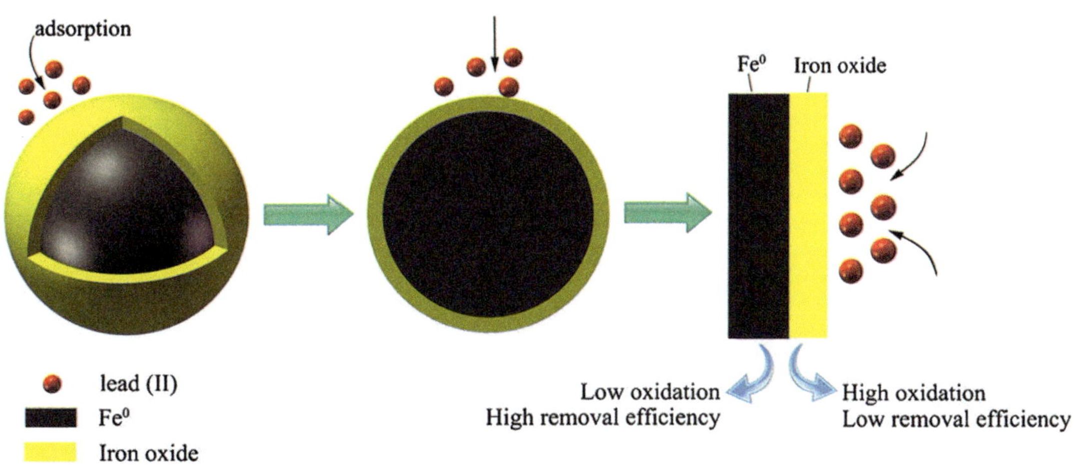

FIG. 17 Removal of Pb (II) using nZVI. *Adapted from Z. Dongsheng, G. Wenqiang, C. Guozhang, L. Shuai, J. Weizhou, L. Youzhi, Removal of heavy metal lead (II) using nanoscale zero-valent iron with different preservation methods, Adv. Powder Technol. 30 (3) (2019) 581–589.*

indicated that 90% of cadmium was removed from water within 5 h which integrated to the surface of the nZVI. Interestingly, the HR-TEM images divulged that after interaction with the cadmium ions, there was a nanofiber formation which comprised of a mixture containing Fe(0), iron oxyhydroxides and iron oxides. Probably the interactions between Cd^{2+} and Fe^{3+} promoted the formation of unintended $CdFe_2O_4$ which displays bright prospects of reusing the remediation products for applications related to light harvesting. The various physicochemical characterization techniques also revealed the formation of Fe(0), α-Fe_2O_3 and FeOOH. The conceptual model of cadmium adsorption process on nZVI nanostructures with cadmium iron oxides on the surface has been projected in Fig. 18. The experimental results signified that the removal of Cd^{2+} was enhanced due to the formation of a FeOOH shell and other iron oxides in nZVI.

Zou et al. have penned a comprehensive review on environmental remediation and application of nanoscale zero-valent Iron and its composites for the removal of heavy metal ions [123]. Table 6 provides a few examples of nZVI-based adsorbents that have shown good performance in the removal of heavy metal ions.

Maghemite NPs

Maghemite NPs (γ-Fe_2O_3) represent a promising class of iron oxide NPs used extensively for the selective removal of heavy metals [144,145]. Way back in 2009, Tutijarvi et al. synthesized maghemite NPs and utilized them successfully for the removal of As (V) ions [146]. Attempts have been directed toward the synthesis of stable iron oxide nanostructures

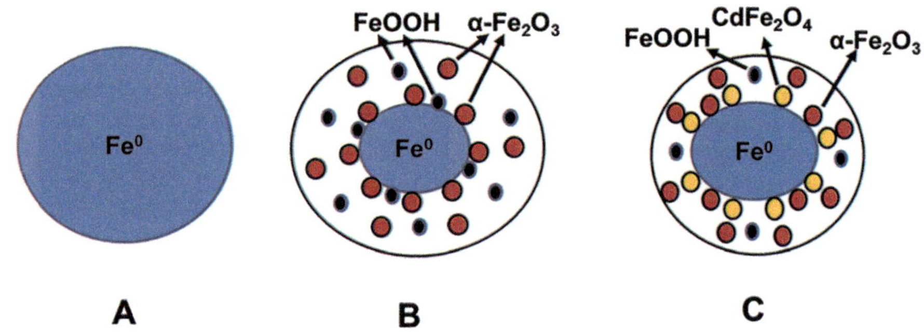

FIG. 18 Conceptual model of cadmium adsorption process on nZVI nanostructures with cadmium iron oxides on the surface. *Adapted from K.T.S. Hidalgo, R. Guzmán-Blas, E.O. Ortiz-Quiles, E.R. Fachini, J. Corchado-García, E. Larios, et al., Highly organized nanofiber formation from zero valent iron nanoparticles after cadmium water remediation, RSC Adv. 5 (4) (2015) 2777–2784.*

TABLE 6 nZVI-based adsorbents in removal of heavy metals (Prime parameters comparative analysis)

Sl. No	Adsorbent	Precursors	Target metals	Adsorption capacity (mg/g)	Reference
1.	Activated carbon supported nZVI	$FeCl_3 \cdot 6H_2O$ and $NaBH_4$	As(III)	18.2	[135]
2.	Amino Functionalized nZVI	$FeCl_3 \cdot 6H_2O$ and CTAB	Pb (II)	111	[136]
3.	Au doped nZVI	$FeCl_3$ and $NaBH_4$	Cd (II)	188	[137]
4.	Bare nZVI	$FeCl_3 \cdot 6H_2O$	As (V)	38.2	[138]
5.	Bare nZVI	Nano iron	Cr (VI)	47.2	[139]
6.	Bare nZVI	$FeCl_3$ and $NaBH_4$	Cd(II)	769.2	[36]
7.	G-NZVI	$FeCl_3 \cdot 6H_2O$ and $NaBH_4$	Pb (II)	585.2	[140]
8.	nZVI-Fe_3O_4	$FeCl_3 \cdot 6H_2O$ and CTAB $FeSO_4 \cdot 7H_2O$	Cr (VI)	29.83	[141]
9.	Magnetic Fe_3O_4/Graphene supported nZVI	$FeCl_3 \cdot 6H_2O$ and $FeSO_4 \cdot 7H_20$	Cr (VI)	66.2	[142]
10.	PVP-modified nZVI	$FeSO_4 \cdot 7H_20 + NaBH_4$	Sb(III)/Sb (V)	6.99/1.65	[143]

possessing well controllable morphology for enhancing their capability to remove heavy metals from wastewater. Among the diverse morphologies accomplished so far, 3-D hierarchically nanostructured Mn doped α-Fe_2O_3 was prepared magnificently by Fu et al. using a hydrothermal approach wherein calcined carbon spheres comprising of Fe^{2+} and Mn^{3+} ions were utilized [147]. The nanomaterial exhibited flower-like nanostructures having diameter ranging between 0.5 and 1 mm that were formed from irregular nanorods (Fig. 19) and worked as a phenomenal adsorbent as it could successfully remove Pb(II), As(III) and Cr(VI) from aqueous solutions. The authors highlighted the significance of Mn substitution that led to an enhancement in the magnetic properties of a-Fe_2O_3.

Madrakian and team synthesized a new mercaptoethylamino monomer-modified maghemite nanoparticles (Fig. 20) which exhibited excellent adsorption capacities of 260.55, 237.60, 118.51, and 91.55 mg/g in the removal of Ag (I), Hg (II), Pb (II), and Cd (II), respectively [148].

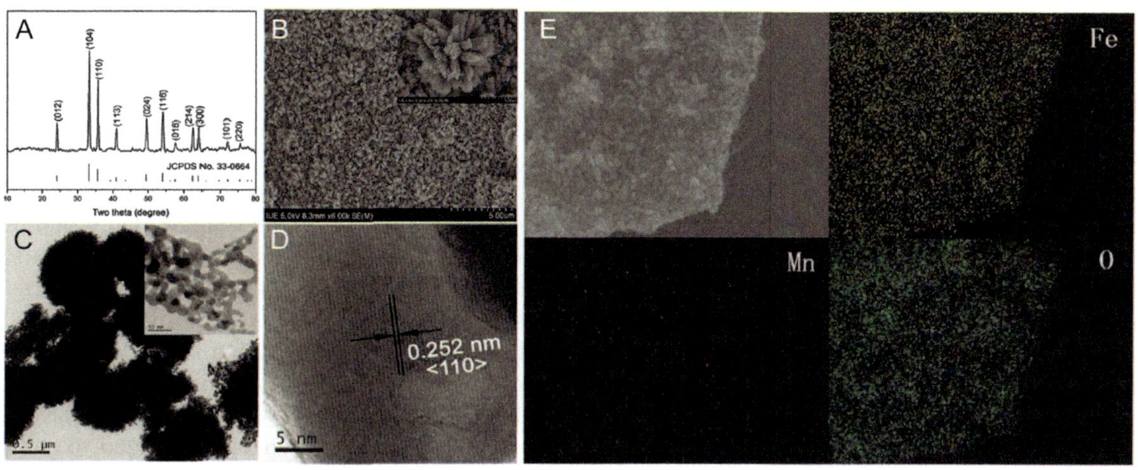

FIG. 19 (A) XRD pattern (B) SEM images (C and D) TEM images of the synthesized Mn doped a-Fe_2O_3 nanostructures and (E) Low-magnification SEM image of the Mn doped α-Fe_2O_3 nanostructures and SEM elemental distribution mapping for Fe, Mn, and O elements. *Adapted from H.J. Cui, J.K. Cai, J.W. Shi, B. Yuan, C.L. Ai, M.L. Fu, Fabrication of 3D porous Mn doped α-Fe2O3 nanostructures for the removal of heavy metals from wastewater, RSC Adv. 4 (20) (2014) 10176–10179.*

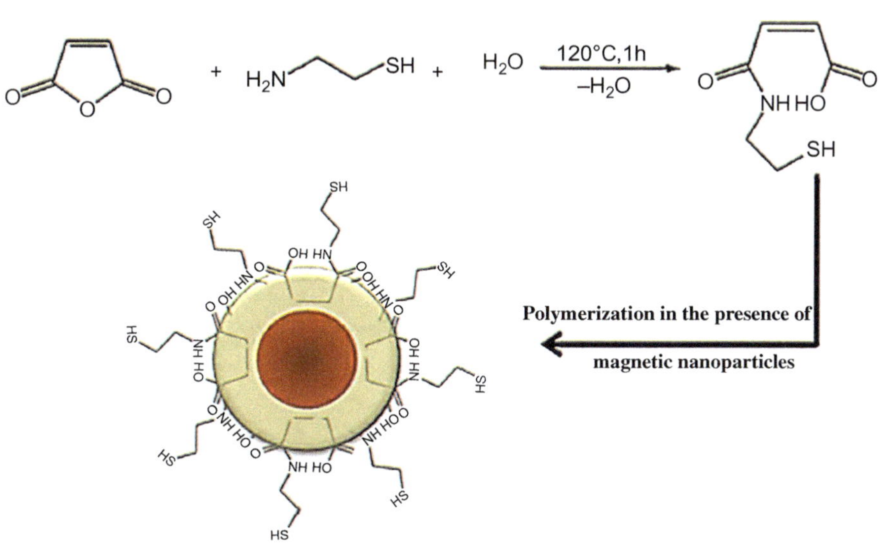

FIG. 20 Preparation of mercaptoethylamino monomer-modified maghemite nanosorbent. *Adapted from T. Madrakian, A. Afkhami, B. Zadpour, M. Ahmadi, New synthetic mercaptoethylamino homopolymer-modified maghemite nanoparticles for effective removal of some heavy metal ions from aqueous solution, J. Ind. Eng. Chem. 21 (2015) 1160–1166.*

Magnetite NPs

Magnetite NPs (Fe_3O_4 NPs) have spread a radical revolution in the field of wastewater treatment owing to the myriad of advantages they offer including a very high surface area to volume ratio, ease of availability, cost-effectiveness, ready dispersibility in solutions, fast adsorption kinetics and their magical superparamagnetic properties that aid in their facile separation from a reaction mixture using a simple magnet (they possess a very high saturation magnetization value) [149]. Additionally, they are not toxic, can be reused for several cycles which renders them economical and also can be synthesized very readily using simple approaches. Among the various classes of iron oxide NPs, these are undeniably preferred for the fabrication of the nanosorbents having preferable M_s values and also comparatively are more durable. Some of the most common synthetic routes employed for the synthesis of Fe_3O_4 NPs include coprecipitation, hydrothermal/solvothermal, microemulsion and thermal decomposition [150]. However, efforts have been directed toward developing novel synthetic routes for fabrication of magnetite NPs (MNPs)and some of them include the template assisted methodology, ultrasonic assisted technique, utility of large scale continuous flow reactors and microwave reactors that have resulted in higher yield, better quality, controllable morphology and are faster as compared to the conventional techniques. In our recent review article, we have elaborately explained these techniques and also elucidated the utility of the MNPs in removal of a diverse variety of heavy metal as well as other contaminants [149]. The prime concerns arising due to the agglomeration of the MNPs have also been dealt with and suitable coating agents (e.g., metals, metal oxides, polymers, surfactants) have been utilized to impart them durability and also prevent them from getting aggregated or oxidized (the strategies are similar to the ones described earlier in case of nZVI) [151]. In addition, remarkable endeavors have also been focused toward the surface engineering of the native Fe_3O_4 NPs for tuning their properties for superior uptake capacity. Broadly outlining, two approaches have been incorporated for the modification of the MNPs, embracing the noncovalent binding of various molecules (surfactants, polymers, etc.) and covalent attachment of different grafting agents using suitable linkers such as silane-based moieties (APTES, MPTES, APTMS), phosphates, sulfonates, cyclodextrins, dendrimers, diols, etc. [152] Selective adsorption of heavy metals have been facilitated *via* the further covalent binding of chelating ligands with the functionalized support material or *via* the use of a molecular imprinting methodology. Since over the span of past two decades, a multitude of reports have been published on the MNPs-based adsorbents which clearly emphasize the role and utility of these NPs, we have projected a few examples (the most effective ones) to enable the readers to enlighten the readers to understand how to fabricate these materials and subsequently apply them in the water purification process.

A surfactant free greener approach was reported for obtaining 3,4-dihydroxyphenethylcarbamodithioate (DHPCT)–coated Fe_3O_4 nanoparticles using watermelon extract (Fig. 21) that demonstrated a very high Hg(II) removal of 98% with maximum adsorption capacity of 52.1 mg/g [153]. The adsorption process was facilitated by the excellent binding capacity of the ligand with Hg (II) which was primarily based on soft–soft interactions that resulted in selective mercury removal.

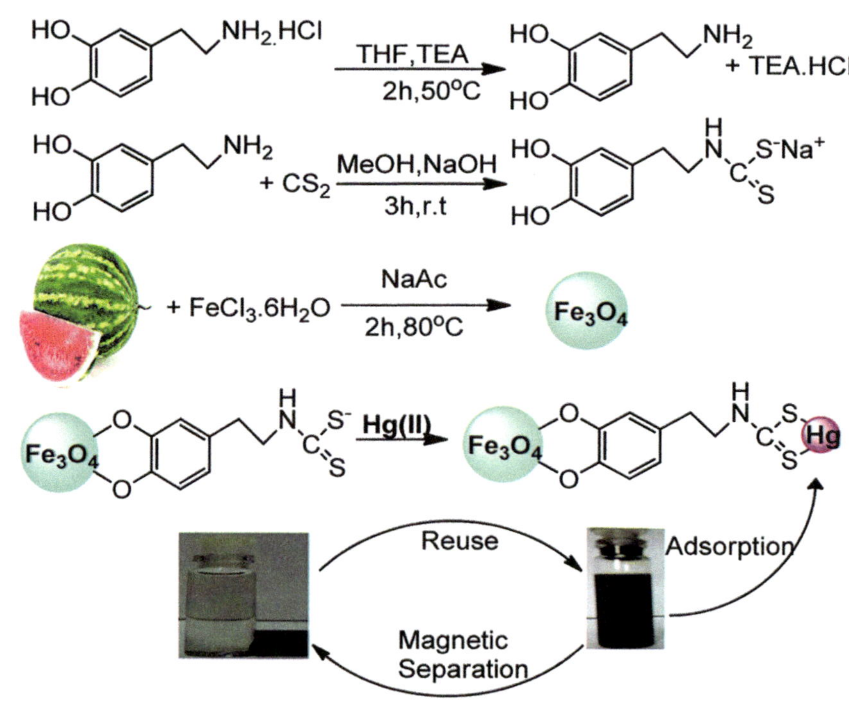

FIG. 21 Procedure for the synthesis of DHPCT and functionalized DHPCT@Fe_3O_4 MNPs for Hg(II) adsorption. *Adapted from S. Venkateswarlu, M. Yoon, Surfactant-free green synthesis of Fe 3 O 4 nanoparticles capped with 3, 4-dihydroxyphenethylcarbamodithioate: stable recyclable magnetic nanoparticles for the rapid and efficient removal of Hg (II) ions from water, Dalton Trans. 44 (42) (2015) 18427–18437.*

By exploiting the superparamagnetic properties of the resultant material, the adsorbent could be recovered readily by magnetic attraction and reused for several times without substantial decline in its activity.

As a step further, Sharma et al. covalently immobilized a chelating ligand (acetanilide) onto the surface of amine functionalized silica coated magnetite NPs. The developed nanosorbent was used for the selective and cyclic removal of Pb^{2+} ions from different charged wastewaters [154]. The sorbent was fabricated in a stepwise manner beginning from the synthesis of the pristine Fe_3O_4 NPs using a coprecipitation method, followed by its coating with a silica shell using the sol-gel method and thereafter functionalized with APTES which resulted in the incorporation of the amine groups (Fig. 22). Finally, a Schiff's base condensation method was adopted by grafting the chelating ligand onto the surface of the functionalized MNPs. The magnetic nanomaterial could remove Pb^{2+} effectively from rainwater, river water, and mycorrhizal-treated ash samples, showing quick removal efficacy and enhanced adsorption rate. The maximum enrichment capacity was found to be 392.2 mg/g at 318 K and it was remarkable to note that the adsorbent could be reused for 10 consecutive runs.

Working on similar lines, many other Fe_3O_4 supported adsorbents have been fabricated and successfully employed for the removal of majority of the troublesome heavy metals which have been summarized in Table 7.

Subsequently, mesoporous (organo) silica decorated with MNPs [specifically, magnetic mobile crystalline material-41 (MCM-41) functionalized by chlorosulfonic acid (MMCM-41-SO_3H)] (Fig. 23) prove to be a potent adsorbent for removal of As (V) [164].

In the year 2018, a new type of magnetic chelating resin bearing iminodiacetate group derived from acrylamide – N, N′-methylenebis(acrylamide) copolymer (Fig. 24) was reported for the removal of Cu(II), Pb(II), Zn(II), and Cd(II) from aqueous solutions [165]. The adsorption followed a pseudo-second-order kinetics and was primarily based on the Langmuir monolayer chemical adsorption model.

On similar grounds, a novel core-shell ferromagnetic nanorod based on amine polymer composite (Fe_3O_4@DAPF) was fabricated by carrying out the surface modification of the ferromagnetic Fe_3O_4 nanorods using a 2,3-diaminophenol and formaldehyde (DAPF)-based polymer (Fig. 25) [166]. The developed nanomaterial could remove Pb displaying a magnificent adsorption capacity of 83.3 mg/g which is comparable to the largest value obtained so far.

Recently, Ding et al. developed an eco-fabricated magnetic filter possessing mesh like architecture and an adsorbent made up of poly(acrylic acid) (PAA)-coated MNPs (Fig. 26) [167]. While the coated MNPs exhibited good adsorption capacity and faster kinetics for Pb, Ni, Co, and Cu, the nanofilter could capture >99% of the nanoparticles under a low magnetic field due to van der Waals force of interaction between particle/surfactant and adsorbent surface. Both these materials could be recovered for the purpose of reuse simply *via* ultrasonic washing.

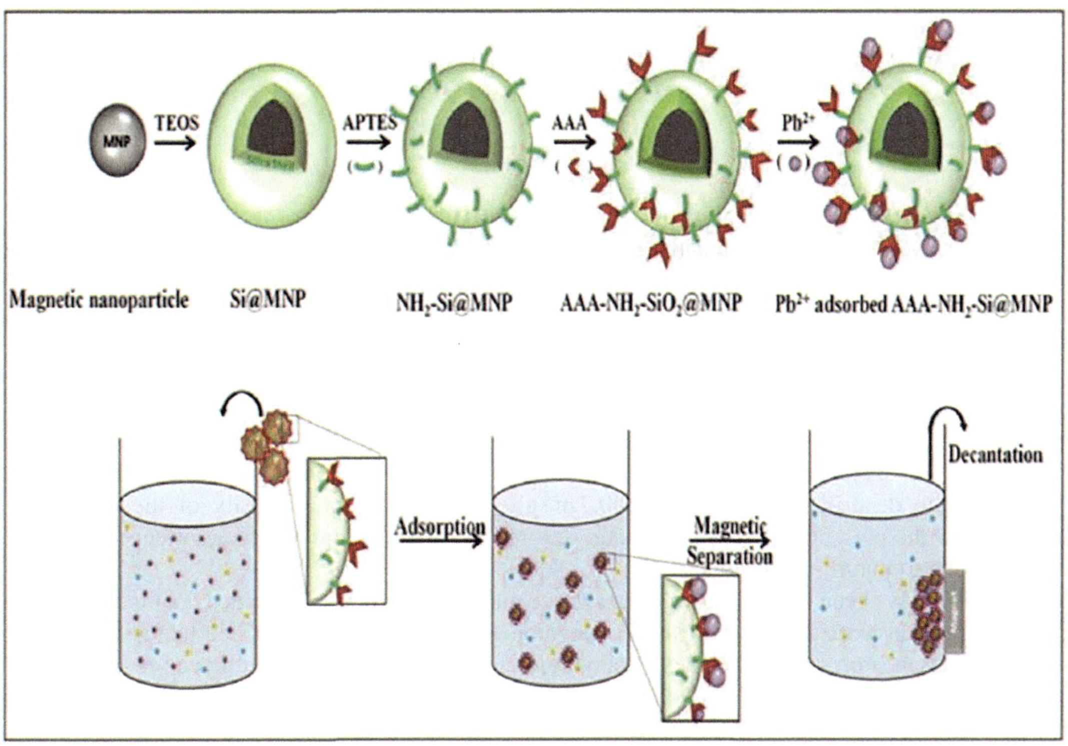

FIG. 22 Schematic illustration of the synthetic route of AAA-NH_2-Si@MNPs followed by Pb (II) removal sequence. *Adapted from R.K. Sharma, A. Puri, Y. Monga, A. Adholeya, Acetoacetanilide-functionalized Fe3O4 nanoparticles for selective and cyclic removal of Pb 2+ ions from different charged wastewaters. J. Mater. Chem. A 2 (32) (2014) 12888–12898.*

TABLE 7 Fe_3O_4 supported adsorbent in removal of diverse heavy metals

Sl. No	Adsorbent	Heavy metal removed	Adsorption capacity	% Removal	Reusability	Reference
1.	DAPTS@NH_2@SiO_2@Fe_3O_4	Pd (II)	64.8 mg/g	Up to 99%	9 runs	[155]
2.	Ethylenediamine modified magnetic cross-linking chitosan NPs (EMCN)	Pd(II) and Pt(IV)	Excellent adsorption loading capacities for Pt(IV) and Pd(II) at pH 2 (i.e., 171 and 138 mg/g, respectively)	>90%	5 runs	[156]
3.	Fe_3O_4@SiO_2@PEI-NTDA	Pb(II)	285.3 mg/g	–	6 runs	[157]
4.	Fe_3O_4@BFR	Cd (II)	92.6 mg/g	88.5%	3 runs	[158]
5.	Fe_3O_4-Functionalized Boron Nitride Nanosheets	As (III)	30.3 mg/g	–	5 runs	[159]
6.	Fe_3O_4@Al_2O_3@Zn-Fe LDH	As	67.57 mg/g	97.3%	–	[160]
7.	Hollow Polyaniline Microsphere/Fe_3O_4 Nanocomposite	As(III) and As(V)	28.27 and 83.08 mg/g, respectively, for As(III) and As(V)	98%–99% removal of As(III) and As(V)	3 runs	[161]
8.	PEG-4000 coated APTES functionalized Fe_3O_4 NPs	Pb (II)	37 mg/g	81.39 ± 2.5% and 62.27% industrial wastewater	–	[162]
9.	PHN nano-adsorbent comprising of pomegranate-like hybrid nanosphere (PHN) with a polydopamine (PDA) peel and Fe_3O_4 nanocrystals as seeds	Cr (VI)	454.55 mg/g	92.5%	5 cycles	[163]

Hematite NPs

Hematite NPs (α-Fe_2O_3 NPs) have also attracted attention from the academic and industrial community as they represent one of the most durable forms of iron oxide [168]. An extensive literature survey reveals that (α-Fe_2O_3 NPs) have shown their worth in the elimination of heavy metals such as lead, copper and strontium from the aqueous solutions. Shipley et al examined the efficacy of the commercially prepared α-Fe_2O_3 NPs in the removal of lead and found that this material could eliminate 100% Pb (II) ions using only 500 mg of nanosorbent [169]. Moving a step further, Ponpandian et al. reported the fabrication of hydroxyapatite (HAp) NPs dispersed on dendritic α-Fe_2O_3 nanostructures for adsorption of lead ions (97.5%) at pH 4 with a contact time of 120 min from industrial wastewater (Fig. 27) [170]. The schematic illustration provides an overview of the hydrothermal methodology undertaken for the preparation of the nanocomposites. The high surface area of the adsorbent due to the dendritic hematite NPs (80.7 m^2/g), good adsorption capacity of the final nanocomposites (754.14 mg/g) owing to the presence of HAp and ready desorption owing to the magnetic effect rendered this protocol a viable attractive to the literature precedents.

Subsequently, efforts have been directed toward the surface modification of the hematite NPs for enhancing the removal efficiency. For example, Hashemzadeh and coworkers developed oleic acid functionalized hematite NPs using a hydrothermal approach wherein iron (III) chloride hexahydrate and oleic acid were utilized as precursors (Fig. 28).

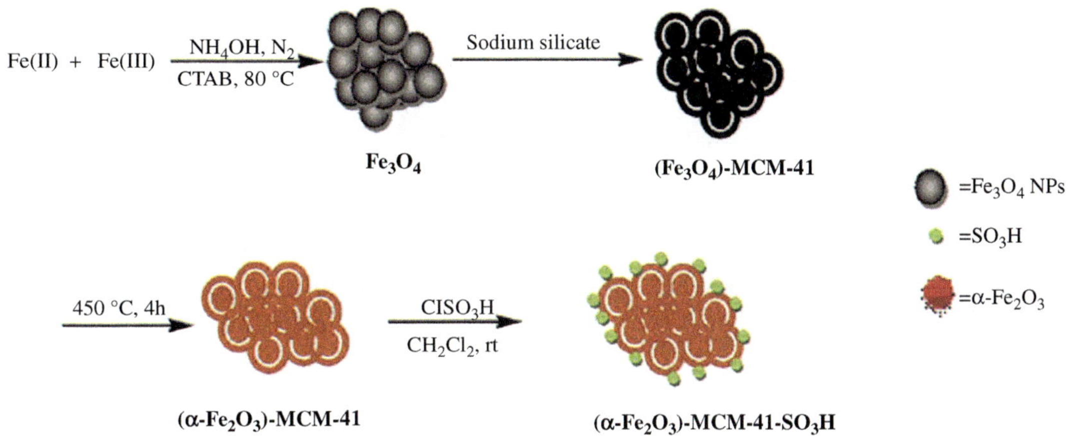

FIG. 23 Synthesis of MMCM-41-SO_3H sorbent and its application in the removal of As (V). *Adapted from M. Hasanzadeh, F. Farajbakhsh, N. Shadjou, A. Jouyban, Mesoporous (organo) silica decorated with magnetic nanoparticles as a reusable nanoadsorbent for arsenic removal from water samples, Environ. Technol. 36 (1) (2015) 36–44.*

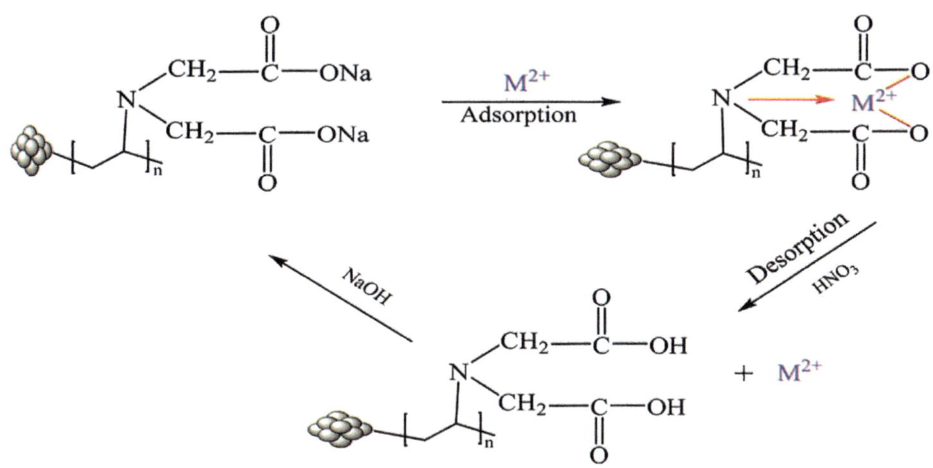

FIG. 24 Synthesis of iminodiacetate chelating resin-functionalized Fe_3O_4 nanosorbent in removal of Cu(II), Pb(II), Zn(II), and Cd(II). *Adapted from S.M. El-Bahy, New iminodiacetate chelating resin-functionalized Fe3O4 nanoparticles: synthesis, characterization, and application for the removal of some noxious metal ions from wastewater, J. Chem. Eng. Data 63 (6) (2018) 2299–2313.*

The developed surface modified nanomaterial worked as a promising adsorbent for the removal of lead ions, showing a maximum uptake capacity of 111 mg/g, contact time (4 h) at room temperature. A linear pseudo-second order model was suggested on the basis of the studies and the adsorption process was facilitated through the sharing or exchange of electrons between Pb (II) ions and the synthesized nano α-Fe_2O_3–based sorbent.

Other metal oxide–based nanomaterials

Apart from various forms iron oxide and silica discussed above, other metal oxides such as CeO_2, Al_2O_3, TiO_2, etc., have also shown interesting prospects in wastewater remediation. "Ceria," the name used to describe CeO_2 and its composites, has proven useful in the elimination of extremely toxic arsenic, lead and cadmium ions due to large surface area, meticulous and controllable morphology, tunable porosity and abundance of functional groups. Seal et al. synthesized ceria–graphene

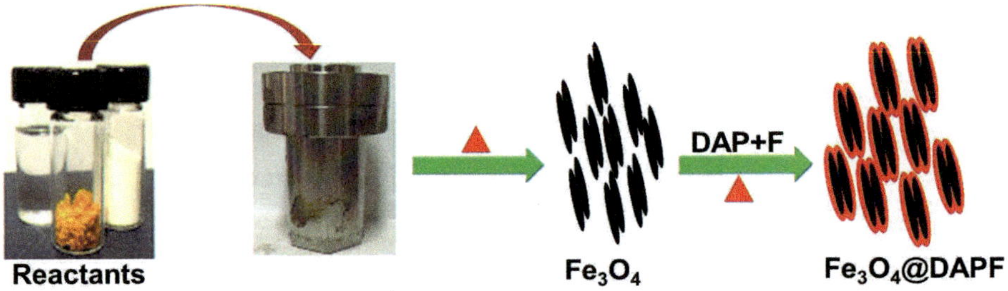

FIG. 25 Schematic illustration of the synthesis of Fe_3O_4@DAPF nanosorbent. *Adapted from S. Venkateswarlu, M. Yoon, Core–shell ferromagnetic nanorod based on amine polymer composite (Fe3O4@ DAPF) for fast removal of Pb (II) from aqueous solutions. ACS Appl. Mater. Interfaces, 7 (45) (2015) 25362–25372.*

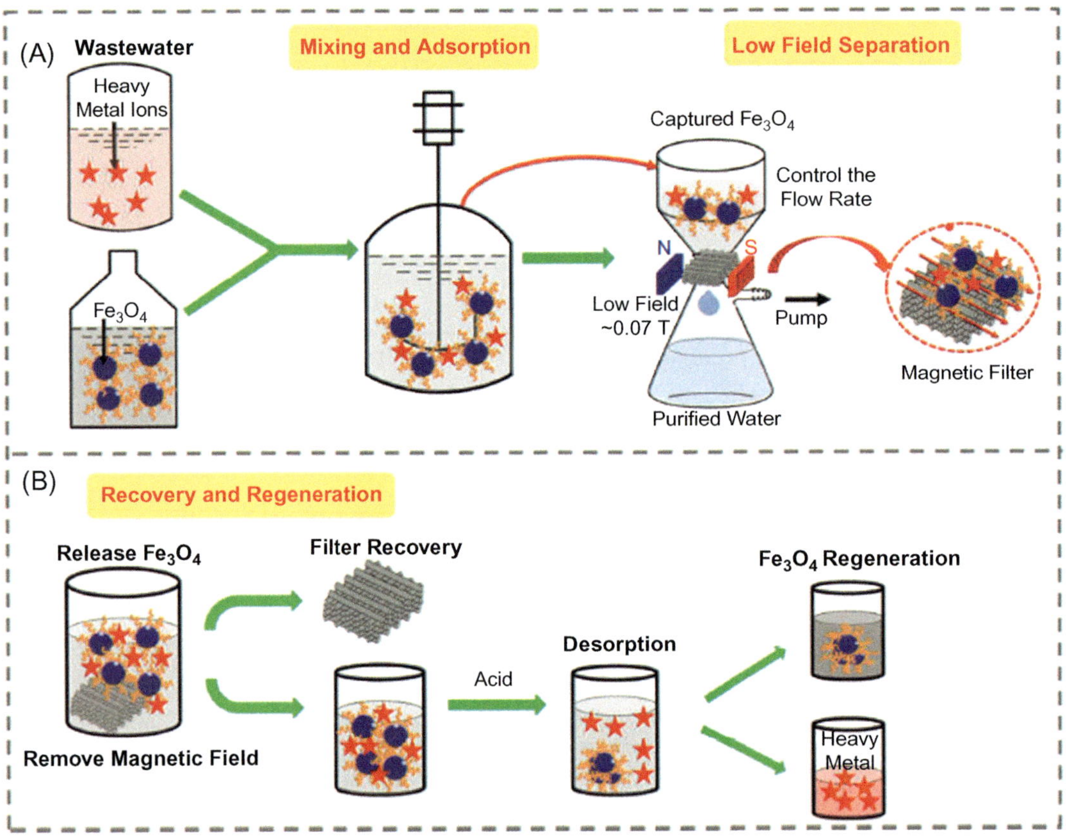

FIG. 26 Schematic illustration of heavy metal removal combined with adsorption and low-field and dynamic separation technique. (A) Decontamination of Pb(II) from wastewater using Fe_3O_4 nanoparticles and low-field magnetic filtration. (B) Recovery and regeneration of the magnetic filter and Fe_3O_4 nanoparticles (NPs). *Adapted from X. Wei, P.J. Sugumaran, E. Peng, X.L. Liu, J. Ding, Low-field dynamic magnetic separation by self-fabricated magnetic meshes for efficient heavy metal removal, ACS Appl. Mater. Interfaces 9 (42) (2017) 36772–36782.*

oxide (ceria–GO) nanocomposites using one-pot hydrothermal approach by dispersing GO in deionized water under ultrasonication and then adding cerium (III) nitrate hexahydrate to the GO solution and subjecting the reaction contents to stirring under basic conditions (Fig. 29A) [171]. The highly crystalline oxides of cerium was inferred from the comparative X-ray diffraction data (Fig. 29B), while the flake like morphology was revealed from the SEM and the TEM micrographs of the composite (Fig. 29C and D). The nanocomposite demonstrated notable capability in the removal of over 99.99% of arsenic species (both As(III) and As (V)) rapidly within 0.1 mg/L of the initial concentration. Strikingly, by exhibiting the adsorption capacities of 185 mg/g for As (III) and 212 mg/g for As (V), respectively, this nanomaterial proclaimed a clear winner in the race of carbon-based adsorbents. XPS analysis confirmed the presence of both Ce (III) as well as Ce (IV) in the adsorbent before arsenic adsorption. The concentration of Ce (III) however increased after arsenic adsorption which was primarily considered as the active site.

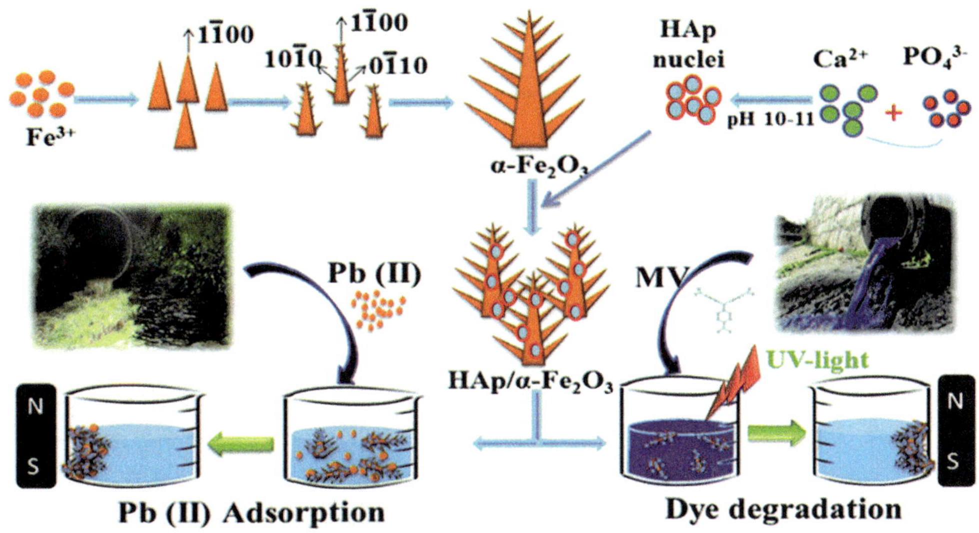

FIG. 27 Schematic illustration for the formation mechanism of HAp dispersed on dendritic α-Fe_2O_3 hierarchical architecture by controlled growth along six crystallographically equivalent directions and Pb(II) adsorption. *Adapted from G. Bharath, N. Ponpandian, Hydroxyapatite nanoparticles on dendritic α-Fe_2O_3 hierarchical architectures for a heterogeneous photocatalyst and adsorption of Pb (II) ions from industrial wastewater, RSC Adv. 5 (103) (2015) 84685–84693.*

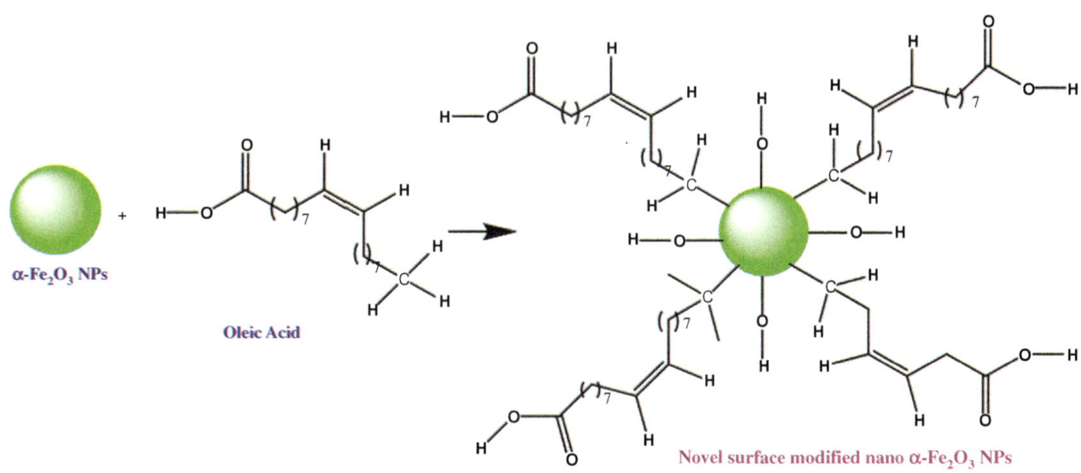

FIG. 28 Surface modification of synthesized α-Fe_2O_3 NPs using oleic acid.

Al_2O_3 or alumina has also emerged as a good support material for the design and development of suitable adsorbents. Poursani et al. reported the synthesis of nano-γ-Al_2O_3 using a novel sol-gel strategy which was employed for the continuous removal of heavy metals (Cr (VI), Pb (II), Ni (II) and Cd (II) ions) [172]. The adsorbents, preparative method involved the dispersion of $AlCl_3{\cdot}6H_2O$ in ethanol followed by addition of 1,2 epoxybutane under ultrasonic irradiation. Finally, deionized water was added followed by calcination of the gel-solution produced in a furnace at 700 °C for 5 h. The synthesized sorbent showed good efficiency in adsorbing chromium and lead particularly.

Hydrated form of Zirconia (HZO) has also emerged as an adsorbent of choice owing its excellent adsorption capacity of heavy metals facilitated by the presence of hydroxyl-rich surface. In addition to this, it also possesses high chemical as well as thermal stability that renders it effective in controlling heavy metal pollution. Yang et al. designed and synthesized nano-HZO based on polymer hydrogel (HZO-P(TAA/HEA)) using an irradiation polymerization methodology (Fig. 30) which was characterized suitably using a number of physicochemical approaches and subsequently utilized as an adsorbent for the removal of Pb (II), Cu (II), Cd (II) and Ni (II) [173]. The adsorption kinetics followed a pseudo second order model while the adsorption isotherms adhered to the Langmuir equation. On the basis of the results obtained from FTIR and XPS, it could be concluded that the adsorption process was principally facilitated by the complexation between zirconia and the heavy metals.

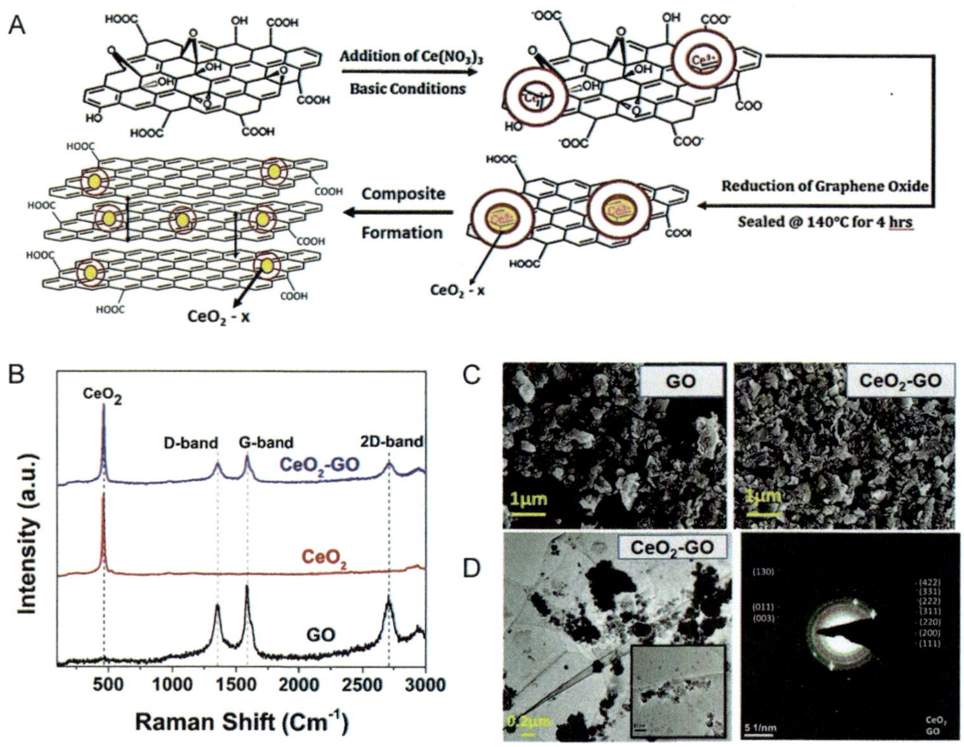

FIG. 29 (A) One-pot synthesis of ceria-GO nanocomposite, (B) Comparative X-ray diffraction (C) SEM images of GO (*left*) and Ceria-GO composite (*right*) revealing a flake like morphology (D) TEM images of Ceria-GO composite and SAED pattern. *Adapted from T.S. Sakthivel, S. Das, C.J. Pratt, S. Seal, One-pot synthesis of a ceria–graphene oxide composite for the efficient removal of arsenic species, Nanoscale 9 (10) (2017) 3367–3374.*

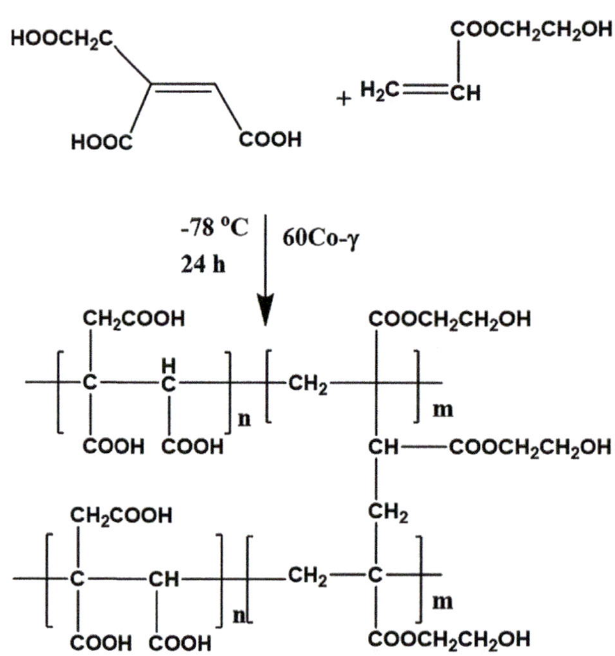

FIG. 30 Synthesis of HZO-P(TAA/HEA) hydrogel.

Considering the quality research work accomplished so far on metal oxide supported adsorbents, Zou and his competent research fellows compiled all of these in the form of a review article published in Nanoscale journal of RSC Publications [174]. In this review, they have described the adsorption properties of various metal oxides, emphasizing the major factors that play a key role in effectuating the adsorption performance (e.g., particle/pore size, morphology, surface modification, etc.).

Metal organic framework–based nanocomposites

The past decades have witnessed a rapid upsurge in interest on the use of metal organic frameworks (MOFs)—"a fascinating class of hybrid crystalline nanoporous substances" as promising adsorbents for the removal of inorganic pollutants owing to their ultrahigh internal surface areas, multifunctionalization capability that aid in postsynthetic modification, abundant exposed metal sites, remarkable thermal stability, optimizable pore volumes and pore size distributions [175]. MOFs form periodic three-dimensional network structures through strong co-ordination between secondary building units (SBUs) which include inorganic metal ions and organic linkers. Among the several attributes, particularly the high porosity and large surface area offer prospects of greater adsorption site accessibility and diffusion of contaminants through the framework. Considering these advantages, a multitude of MOF-based nanosorbents have been designed and explored for the elimination of the absolutely hazardous metals, which have been given due coverage in review articles published over the years [176,177]. For the rapid and selective removal of Pb^{2+} and Hg^{2+}, Queen et al. designed a water stable MOF/polymer composite made up of Fe-BTC (also known as MIL-100), featuring triangular Fe_3O clusters interlinked by BTC^{3-} (1,3,5-benzenetricaboxylate) ligands that had polydopamine (PDA) embedded within the channels (Fig. 31) [178]. For obtaining this composite, desolvated form of MOF was mixed with dopamine that underwent spontaneous polymerization to form PDA within its pores *via* the Fe^{3+} open metal sites. The BET analysis divulged the surface area of the final material to be 1134 m^2/g showing no signs of loss of crystallinity. HAADF-STEM images provided evidence that nitrogen is indeed located throughout the MOF crystal (Fig. 32). Wondrously, as a consequence of the presence of abundant

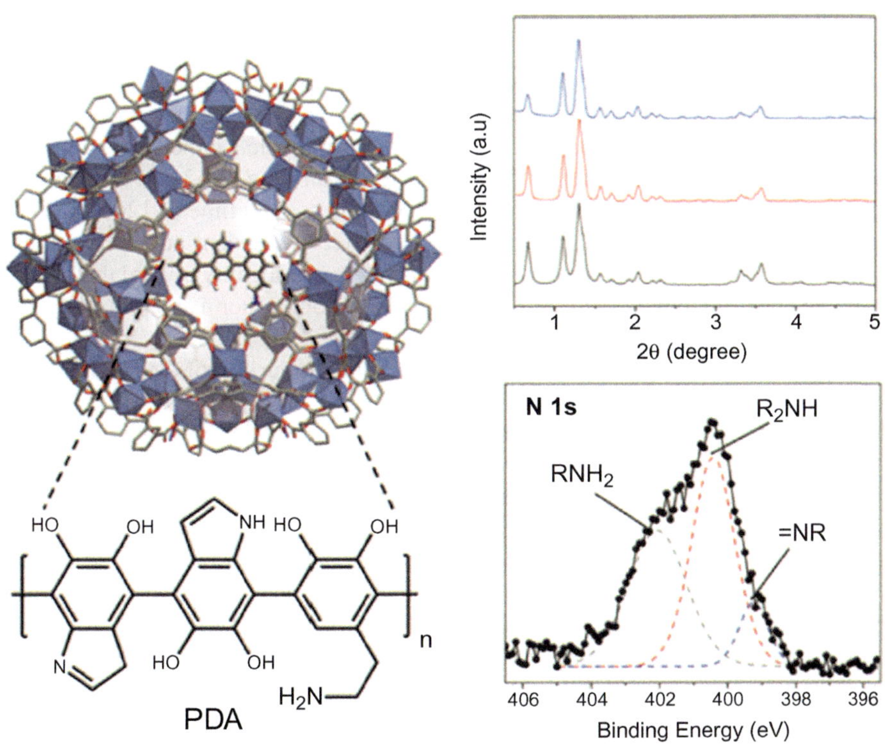

FIG. 31 Structure and characterization of Fe-BTC/PDA (*purple sphere* represents the void space inside of the cage prior to the dopamine addition). (*Top*) The simulated XRD pattern of Fe-BTC (*black*) compared to the synchrotron XRD data (λ = 0.50084 Å) of Fe-BTC (*red*) and Fe-BTC/PDA-19 (*blue*). X-ray photoelectron spectroscopy data obtained from the N 1s spectrum of Fe-BTC/PDA-19. *Adapted from D.T. Sun, L. Peng, W.S. Reeder, S.M. Moosavi, D. Tiana, D.K. Britt, et al., Rapid, selective heavy metal removal from water by a metal–organic framework/polydopamine composite, ACS Cent. Sci. 4 (3) (2018) 349–356.*

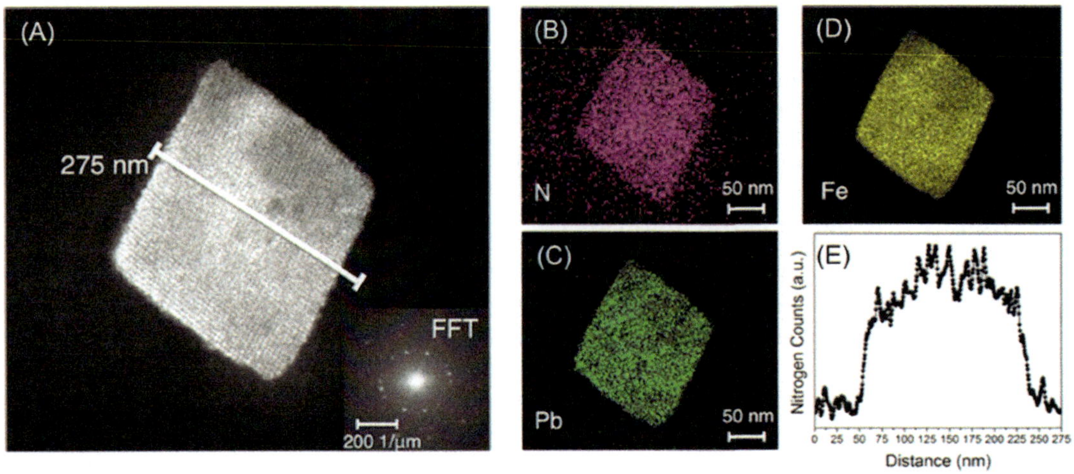

FIG. 32 HAADF-STEM images. (A) HAADF-STEM image of a sliced single crystalline of Fe-BTC/PDA-19 and (B–D) the corresponding EDX elemental mapping. (E) STEM-EDX line profile of nitrogen across the region indicated in (A). *Adapted from D.T. Sun, L. Peng, W.S. Reeder, S.M. Moosavi, D. Tiana, D.K. Britt, et al., Rapid, selective heavy metal removal from water by a metal–organic framework/polydopamine composite, ACS Cent. Sci. 4 (3) (2018) 349–356.*

amine and catechol functionalities in PDA, capable of scavenging metals, the composite displayed its efficacy by removing more than 99.8% of Hg (II) and Pb (II) from 1 ppm solution, rendering drinkable levels within seconds. The MOF shell prevented the dispersion of the hydrophilic PDA into water.

Recently, a novel carboxylic acid-functionalized layered double hydroxide/MOF nanocomposites has been prepared by Marjani and coworkers by growing UiO-66-(Zr)-$(COOH)_2$ MOF nanocrystals homogeneously over the entire surface of the functionalized LDH nanosheets ($Ni_{50}Co_{50}$-LDH-COOH) using water as the solvent, which is an environmentally benign alternative to the rather toxic DMF and DMSO commonly employed for MOF synthesis [179]. To achieve the desired nanosorbent, initially, the surface functionalization of the LDH nanosheets were done using pyromellitic acid molecules for allowing stronger binding between secondary building units in MOF and carboxylate functional groups on the LDH (Fig. 33). This step was accompanied by the growth of the UiO-66(Zr)-$(COOH)_2$ nanocrystals. The synthesized nanomaterial served as a potential sorbent for the eradication of toxic Cd(II) and Pb(II) from water, exhibiting 415.3 and 301.4 mg/g adsorption capacities for these metals, respectively, with only 0.02 g of adsorbent dosage and contact time of 120 min. The reason behind the superlative adsorption characteristics could be attributed to the presence of pendant adsorption cites comprising of carboxylic functional groups that decorated the pores of MOF and displayed strong affinity toward the heavy metals. The adsorption and kinetics model showed best agreement with the Langmuir adsorption isotherm and pseudo-first-order model.

It is also worth highlighting the possible modes of interactions between heavy metals and the surface of the MOF-based adsorbent (Fig. 34).

In another study reported by Hmadeh et al, defect engineered water-stable Zr-metal-organic framework (MOF) (UiO-66) was synthesized and subsequently employed for the elimination of arsenate ions from water [180]. The synthetic strategy involved the introduction of a monocarboxylic acid modulator that resulted in the induced defects and interestingly, as the concentration of the modulator was changed, the concentration of the defects could be tuned and controlled accordingly (Fig. 35). The SEM micrographs showed octahedral shaped crystals of the samples at high modulator concentrations, typically above 12 equivalent with average particle size lying around 112 nm. The arsenate uptake capacity could be drastically enhanced by more than 100% introducing the defects as evident from the comparison between the adsorption data of the defect-free UiO-66. This increase could be accredited to the presence of free Lewis acid sites generated in the MOF due to the missing linker defects. In addition to exhibiting high removal efficiency, the nanomaterial also exhibited high selectivity for the target ions and excellent reusability up to five consecutive regeneration cycles.

Table 8 provides an overview of the MOFs designed for the removal of different heavy metals.

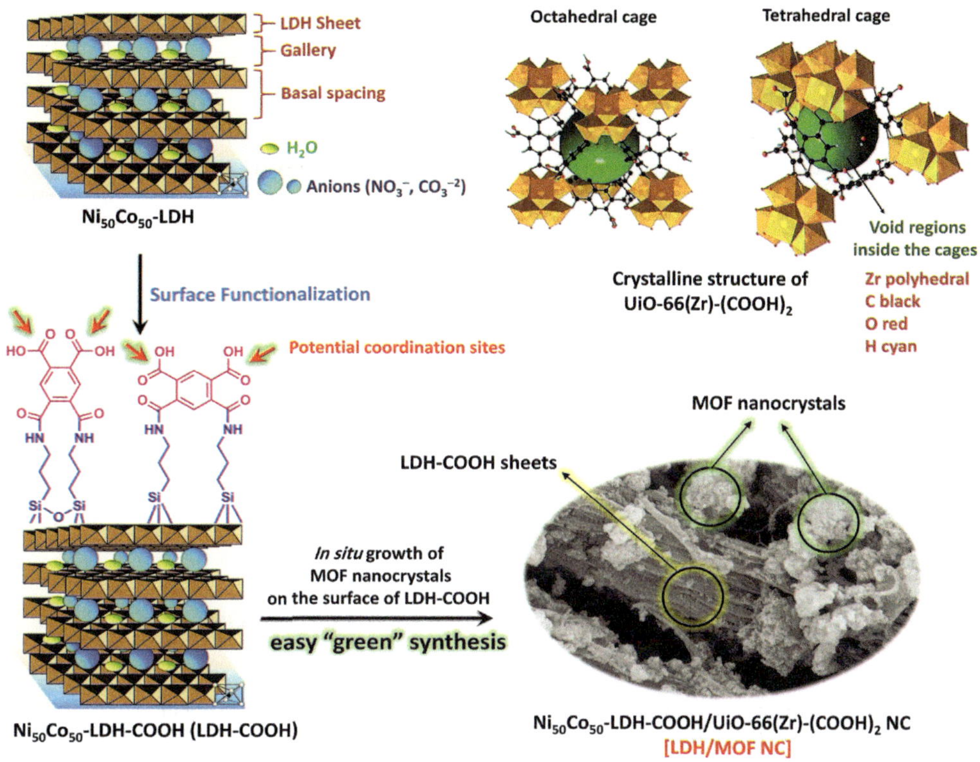

FIG. 33 The overall strategy for the synthesis of $Ni_{50}Co_{50}$-LDH-COOH/UiO-66(Zr)-$(COOH)_2$NC [LDH/MOF NC]. *Adapted from R. Soltani, R. Pelalak, M. Pishnamazi, A. Marjani, A.B. Albadarin, S.M. Sarkar, S. Shirazian, A novel and facile green synthesis method to prepare LDH/MOF nanocomposite for removal of Cd (II) and Pb (II), Sci. Rep. 11 (1) (2021) 1–15.*

Nanobiocomposites

With the current emphasis on rendering a waste-free planet, attempts have been directed toward utilization of diverse waste materials for deriving value added products. "Circular economy" is the buzzword that has enabled the researchers to explore new techniques for converting waste to resource. Consequently, agro based by products including rice husk, coconut shell, maize cob, rice straw, jackfruit, hazelnut shell, maize husk (blessed with abundant and distinct functional constituents including lignin, cellulose and carbohydrates) have been put to extensive use as biosorbents for the removal of heavy metals [191]. Additionally, inexpensive biological molecules like chitin, chitosan and lignin have also witnessed popularity as promising adsorbents in eradicating heavy metals and purifying wastewater [192]. The natural abundance and prospects of bio-renewability of polysaccharides such as cellulose has also rendered them as biosorbents of choice. To further enhance the utility of the bio-based adsorbents, these are being integrated with other nano-based components/materials to generate nanobiocomposites that have greater adsorption efficiency, lesser toxicity, enhanced reusability and facile recoverability [193,194]. In view of these incredible features along with their superlative biodegradable nature, nanobiocomposites have played a predominant role in accelerating the effective large scale wastewater purification. Chitosan-based polymer bionanocomposites, clay hydrogel nanocomposites, gum polysaccharide–based bionanocomposites, cellulosic nanocomposites are some attractive examples of this category of composite materials employed for the purpose of noxious contaminant removal [195].

Hitzky and research team developed low cost and eco-friendly functional bionanocomposites by assembling/intercalating cationic starch (CST) biopolymers modified with quaternary ammonium groups on double layer silicates comprising of montmorillonite and natural bentonite that exhibited magnificent adsorption properties in the removal of chromate oxyanions [196]. The cationic starch was prepared by refluxing a mixture containing potato starch and glycidyltrimethylammonium chloride (2,3-epoxypropyl)trimethyl ammonium chloride (GTAC) under alkaline conditions (Fig. 36). This was followed by the synthesis of the nanobiocomposite by reacting the clay mineral with the CST in distilled water. The purpose

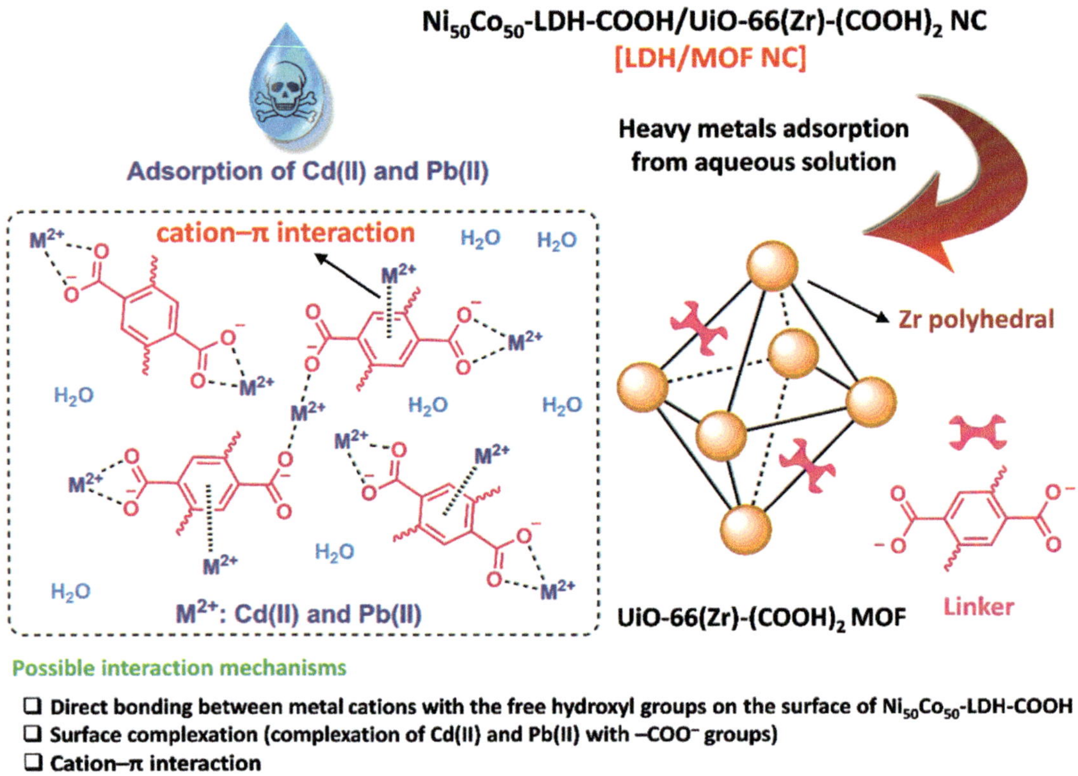

FIG. 34 Possible interactions between heavy metals, Cd(II) and Pb(II), and the surface of the adsorbent. *Adapted from R. Soltani, R. Pelalak, M. Pishnamazi, A. Marjani, A.B. Albadarin, S.M. Sarkar, S. Shirazian, A novel and facile green synthesis method to prepare LDH/MOF nanocomposite for removal of Cd (II) and Pb (II), Sci. Rep. 11 (1) (2021) 1–15.*

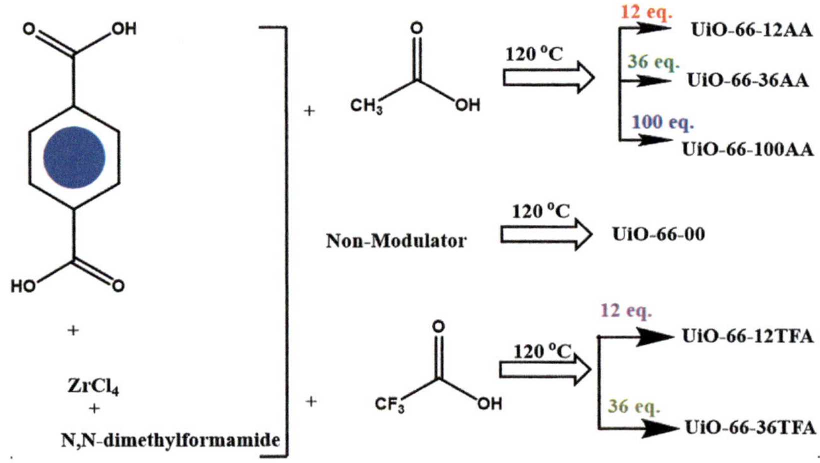

FIG. 35 Schematic presentation of the synthesis conditions of the Six MOF.

for choosing starch as the biopolymer was apparently evident in consideration of its abundance along with ease of modification *via* covalent grafting of functional groups that lead to improved adsorption capacity of the biosorbent. While, on the other hand, the silicate layers played a key role in the adsorption process by increasing the stability in water of the cationic starch, allowing its ready recovery from the aqueous solution postadsorption. As compared to some of the reported biopolymer-based environmentally benign adsorbents, the CST/clay bionanocomposite had a higher chromate adsorption capacity (20.6mg/g) and easier recovery owing to the presence of the layered solid.

TABLE 8 Examples of metal organic framework–based sorbents developed during 2015–2020 for removal of heavy metals (important comparative parameters have been included)

Sl. No	MOF	Description	Target metal/s	Adsorption capacity	% Removal	Reusability	Reference
1.	BS-HMT	MOF-808 functionalized with EDTA through solvent-assistant linker exchange method	22 heavy metals including Hg(II), Cd(II), Pb(II), Zn(II), Co(II), Ni(II) La (III)	205, 592, and 313 mg/g for La^{3+}, Hg^{2+}, and Pb^{2+}, respectively	>99%	–	[181]
2.	CeO_2@UiO-66-(SH)2 MOF bead composite	Spherical microbeads generated in one step by continuous-flow spray-drying	As(III and V), Cd(II), Cr(III and VI), Cu(II), Pb(II), and Hg(II)	Pb(II) (128 mg/g), Cr(VI) (122 mg/g), Cd(II) (93 mg/g), and As(V) (17 mg/g)	Up to 99%	3 runs	[182]
3.	Cr-MOF/AC	An efficient adsorbent-based on an activated carbon and metal–organic framework	Pb (II) and Co (II)	127 and 138 mg/g	–	5 cycles	[183]
4.	Fe_3O_4@SiO_2@UiO-66-Urea	Amino-decorated Zr-based magnetic Metal-Organic Frameworks composites (Zr-MFCs)	Pb (II)	102 mg/g	80%	6 cycles	[184]
5.	MIL-53(Al)-GO nanocomposites	An aluminum-based MOF MIL-53(Al) ($AlC_8H_5O_5$) and GO at varying ratios were taken in different ratios for synthesis of MOF-GO nanocomposites	As(III)	65 mg/g	94.8%	–	[185]
6.	MCNC@Zn-BTC	Magnetic Cellulose Nanocrystal/Metal–Organic Framework Composite–based on Zn(II) and benzene-1,3,5-tricarboxylic acid (Zn-BTC), using a simple mechanical agitation method	Pb (II)	558.66 mg/g	>80	5 cycles	[186]
7.	MCCR-350	Magnetic calcium carbonate adsorbent with nanostructure	Pb(II) and Cd(II)	1179 and 821 mg/g, respectively, for Pb(II) and Cd (II), respectively	–	–	[187]
8.	PtNP@UiO-66-NH	Platinum Nanoparticle Encapsulated Metal–Organic Frameworks	Hg (II)	206.25 mg/g	99%	4 cycles	[188]
9.	Sulfonic acid functionalized MOF, $Cu_3(BTC)^{2-}SO3H$	SO_3H functional groups onto the pore surface of MOF using sequential postsynthetic modification and oxidation	Cd (II)	88.7 mg/g	95.1%	6 cycles	[189]
10.	ZIF-67 modified bacterial cellulose/ chitosan composite aerogel	Combination of physical mixture, in situ synthesis, and lyophilization used for synthesizing zeolitic imidazolate framework-67 (ZIF-67) modified bacterial cellulose/ chitosan (BC/CH) composite aerogel	Cr (VI) and Cu (II)	200.6 and 152.1 mg/g for Cu^{2+} and Cr^{6+}, respectively	72% and 81% for removal of Cu^{2+} and Cr^{6+}, respectively	5 cycles	[190]

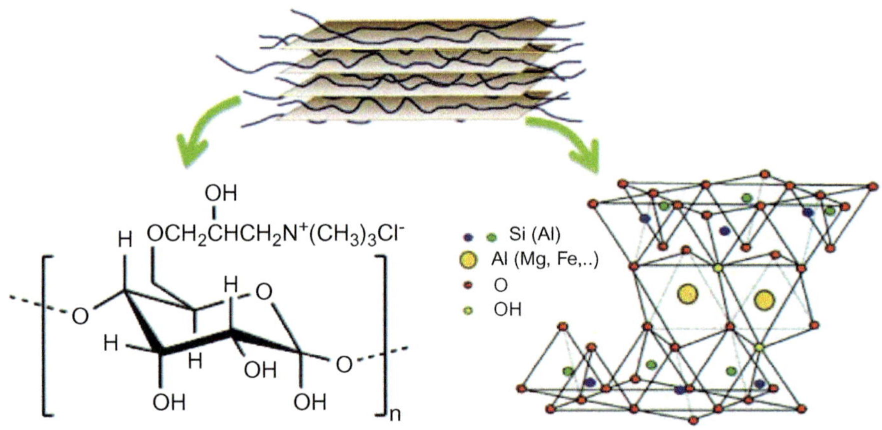

FIG. 36 Bionanocomposite based on the intercalation of cationic starch functionalized with quaternary ammonium groups in the layered silicate montmorillonite. *Adapted from Y. Koriche, M. Darder, P. Aranda, S. Semsari, E. Ruiz-Hitzky, Bionanocomposites based on layered silicates and cationic starch as eco-friendly adsorbents for hexavalent chromium removal, Dalton Trans. 43 (27) (2014) 10512–10520.*

For the effective removal of Pb(II), Cu(II) and Cr(VI), Maleki et al. designed a magnetic guanidinylated chitosan (MGC) nanobiocomposite for accessing the benefits of magnetic separability along high adsorption performance of chitosan-based materials (Fig. 37) [197]. The surface mapping analysis revealed that the synthesized MNPs had a roughly spherical morphology with size lying around 30 to 34 nm, while the magnetic studies revealed the saturation magnetization value (Ms) of 43.66 emu/g which was good enough for an efficient separation. Under optimized conditions, the magnetic biosorbent could adsorb 100%, 98.64%, and 33.76% of Cu (II), Pb (II), and Cr (VI), respectively. As a consequence of introduction of NH_2-CN molecules to chitosan biopolymer, it exhibited higher removal efficacy as in comparison to the unmodified chitosan biopolymer.

Very recently, a bionanocomposite of eggshell decorated with Ag-Fe NPs was synthesized and utilized for the removal of lead, arsenic, and mercury from water [198]. Chemically, egg-shell is made up of 94% calcium carbonate, 1% magnesium carbonate, 1% calcium phosphate, and 4% organic matter, which can be subjected to biomineralization to particularly furnish calcium carbonate that display a phenomenal ability to affect the chemistry of aquatic systems. Thus, eggshells were employed as adsorbent for removal of heavy metals from aqueous contaminated streams; however, it was integrated with NPs of zero valent silver (Ag^0) and iron oxide (Fe^0) for enhancement in its adsorption efficiency. The adsorbent preparative method involved the immersion of requisite amount of dried eggshell into a solution containing silver nitrate and iron (II) sulfate and subsequent exposure to microwave irradiation. The fabricated bionanocomposite exhibited high adsorption capacities of 17.7 27.8 for and 15.88 mg/g for As (v), Pb (II) and Hg (II), respectively, within an equilibrium time ranging between 4 and 8 h.

Sustainable development goals and future direction in line with research needs

With the design of sustainable development goals (SDGs) by the various member countries of UN in 2015, the world got unified by these common goals to work for a better and sustainable future [199,200]. Among the seventeen pertinent goals, SDG-6 acquires a special significance as it aims at the improvement of water quality and provide a safer access to drinkable water to the entire population [201]. The accomplishment of this goal is essential for unlocking economic growth, prosperity and survival of every individual globally [202,203]. This chapter highlights the actions and key steps being taken for accomplishment of this goal. Metal contamination undoubtedly remains a major challenge and despite efforts being devoted for the remediation, there still remains a number of collective issues to tackle especially in terms of large scale implementation of the various technologies being devised. A more integrated solution toward managing and allocating water resources is the need of the hour which can be attained if every person comes forward to help in this endeavor [204]. It is not the duty of any one individual, but rather everyone needs to contribute toward saving our water resources from getting polluted. The research directions should be lined up with the needs and focus on resolution of the impediments. For instance, although nanotechnology has provided a promising tool for fabricating nanostructures that can prevent water contamination, yet there are inherent challenges associated with their potential toxicity, tendency to undergo aggregation, long-term performance issues, operational cost, biocompatibility, etc. Apart from the clean water shortage, another critical issue related to the depletion of metal reserves also needs to be tackled. Consequently, more dedicated efforts are required and international collaborations between the researchers from multidisciplinary field having rich expertise can foster fruitful results, helping in overcoming the various blockages and ultimately rendering safer water [205].

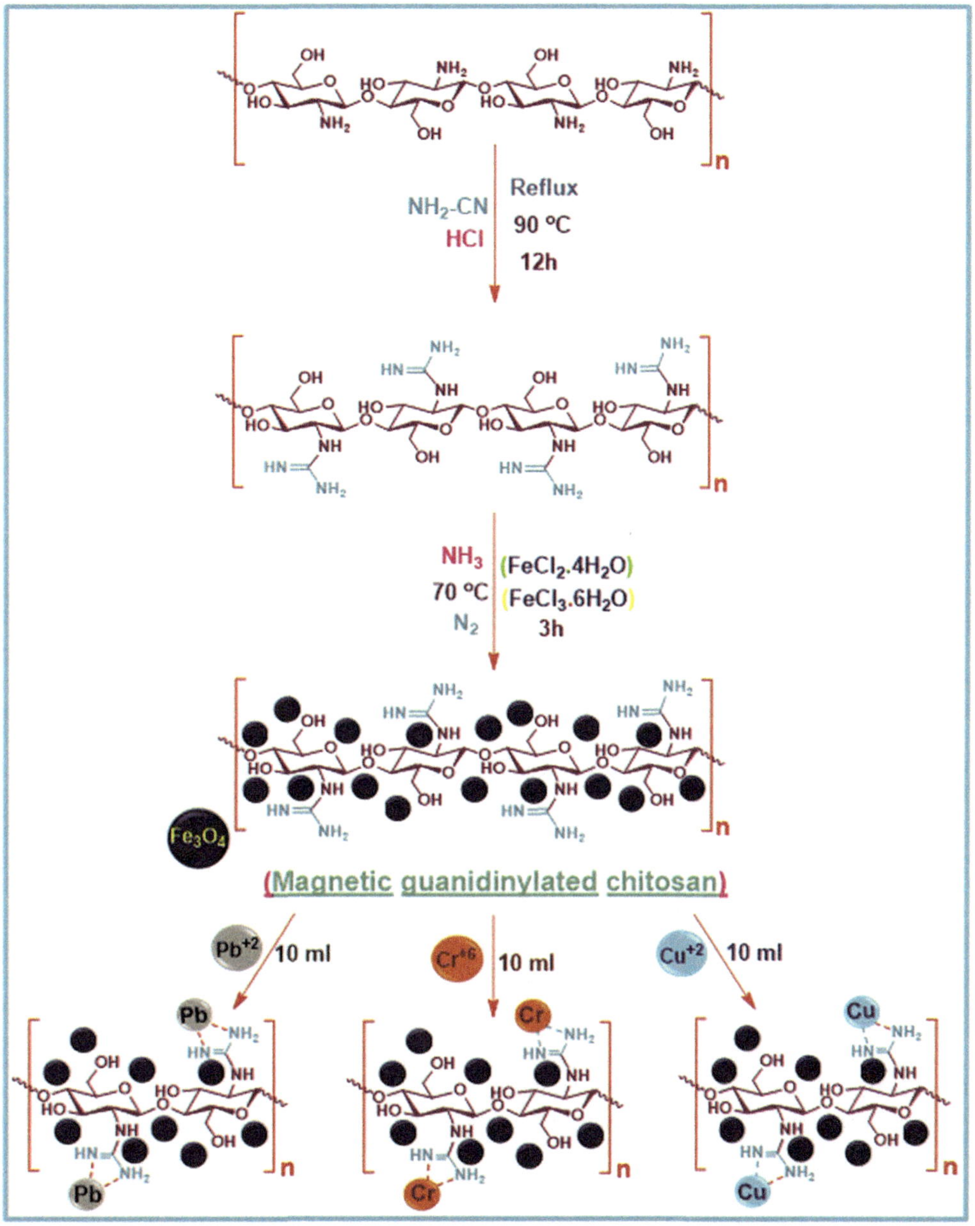

FIG. 37 Synthesis of MGC nanobiocomposite and its applicability in adsorption of metal ions. *Adapted from R. Eivazzadeh-Keihan, F. Radinekiyan, S. Asgharnasl, A. Maleki, H. Bahreinizad, A natural and eco-friendly magnetic nanobiocomposite based on activated chitosan for heavy metals adsorption and the in-vitro hyperthermia of cancer therapy, J. Mater. Res. Technol. 9 (6) (2020) 12244–12259.*

Conclusions

Emergence of nanomaterials have comprehensively helped in not only extending the vision of sustainable and affordable clean water, but also provided strong means of tackling the current global challenges by offering new, cost-effective, environmentally acceptable technologies. In comparison to the chemically and energy intensive traditional water treatment techniques, nanomaterials allow the effective tuning of their physicochemical properties by surface engineering that further enhances their selectivity toward a particular contaminant. Additionally, nanotechnology permits the utilization of molecular and bottom-up design approaches for the creation of complex nano-systems that exhibit superior performance. In this chapter, we have covered the design and potential applications of diverse nano structured materials including carbon-based (graphene oxide, carbon nanotubes), silica-based (silica NPs, mesoporous silica, dendritic mesoporous silica), iron-based (zero-valent iron NPs, magnetite, maghemite, hematite), other metal oxides, MOF-based nanomaterials and nanobiocomposites including the most recent ones that have shown marvelous efficacy in the removal of undesirable heavy metal contaminants from wastewater. There is a fruitful discussion on the sustainable development goals and steps being taken toward the fulfillment with the aid of nanotechnology. Even though the last decade has witnessed a tremendous expansion in interest pertaining to the fabrication of nanomaterials, yet the nano-based technologies have a long way to completely conquer the market because of challenges related to higher operational cost and field-level issues. Much research will be required to overcome these issues, and also evaluating the sustainability of the nanoengineered materials at all levels is highly desirable.

Acknowledgment

The authors express their due thanks to DST, India, for providing funding for our project, and one of the authors, Dr. Sriparna Dutta, would like to gratefully acknowledge CSIR for providing her research associateship (RA).

References

[1] https://www.brainyquote.com/authors/rob-halford-quotes.
[2] V.K. Gaur, P. Sharma, P. Gaur, S. Varjani, H.H. Ngo, W. Guo, et al., Sustainable mitigation of heavy metals from effluents: toxicity and fate with recent technological advancements, Bioengineered 12 (1) (2021) 7297–7313.
[3] H. Baba, K. Tsuneyama, M. Yazaki, K. Nagata, T. Minamisaka, T. Tsuda, et al., The liver in itai-itai disease (chronic cadmium poisoning): pathological features and metallothionein expression, Mod. Pathol. 26 (9) (2013) 1228–1234.
[4] D'itra, F. M., Mercury contamination-what we have learned since Minamata, in: Fourth Symposium on our Environment, Springer, Dordrecht, 1991, pp. 165–182.
[5] M.F. Hughes, B.D. Beck, Y. Chen, A.S. Lewis, D.J. Thomas, Arsenic exposure and toxicology: a historical perspective, Toxicol. Sci. 123 (2) (2011) 305–332.
[6] Y.S. Hong, K.H. Song, J.Y. Chung, Health effects of chronic arsenic exposure, J. Prev. Med. Public Health 47 (5) (2014) 245.
[7] https://www.ofai.org/the-shocking-tale-of-indias-cancer-train.
[8] https://www.who.int/water_sanitation_health/publications/2011/9789241548151_ch12.pdf.
[9] A. Fairbrother, R. Wenstel, K. Sappington, W. Wood, Framework for metals risk assessment, Ecotoxicol. Environ. Saf. 68 (2) (2007) 145–227.
[10] N. Pugazhenthiran, S. Anandan, M. Ashokkumar, Removal of heavy metal from wastewater, in: Handbook of Ultrasonics and Sonochemistry, Springer, 2016, p. 813.
[11] N.B. McGuinness, M. Garvey, A. Whelan, H. John, C. Zhao, G. Zhang, et al., Nanotechnology solutions for global water challenges, in: Water Challenges and Solutions on a Global Scale, American Chemical Society, 2015, pp. 375–411.
[12] L. Liu, X.B. Luo, L. Ding, S.L. Luo, Application of nanotechnology in the removal of heavy metal from water, in: Nanomaterials for the Removal of Pollutants and Resource Reutilization, Elsevier, 2019, pp. 83–147.
[13] M.S. Mauter, I. Zucker, F. Perreault, J.R. Werber, J.H. Kim, M. Elimelech, The role of nanotechnology in tackling global water challenges, Nat. Sustain. 1 (4) (2018) 166–175.
[14] https://www.downtoearth.org.in/news/water/huge-amounts-of-toxic-heavy-metals-swim-in-indian-rivers-60545#:~:text=India's%2042%20rivers %20have%20at,of%20lead%20in%2069%20rivers.
[15] https://www.downtoearth.org.in/news/intrinsic-link-11514.
[16] P.B. Tchounwou, C.G. Yedjou, A.K. Patlolla, D.J. Sutton, Heavy metal toxicity and the environment, Mol. Clin. Environ. Toxicol. 101 (2012) 133–164.
[17] A. Khan, S. Khan, M.A. Khan, Z. Qamar, M. Waqas, The uptake and bioaccumulation of heavy metals by food plants, their effects on plants nutrients, and associated health risk: a review, Environ. Sci. Pollut. Res. 22 (18) (2015) 13772–13799.
[18] G. Pandey, S. Madhuri, Heavy metals causing toxicity in animals and fishes, Res. J. Anim. Vet. Fish. Sci. 2 (2) (2014) 17–23.
[19] S.J. Cobbina, A.B. Duwiejuah, R. Quansah, S. Obiri, N. Bakobie, Comparative assessment of heavy metals in drinking water sources in two small-scale mining communities in northern Ghana, Int. J. Environ. Res. Public Health 12 (9) (2015) 10620–10634.
[20] R.K. Sharma, Bioinorganic Chemistry E-Book Module, National Science Digital Library, 2005. http://niscair.res.in/ispui/handle/123456789/230.
[21] N.J. Langford, R.E. Ferner, Toxicity of mercury, J. Hum. Hypertens. 13 (10) (1999) 651–656.
[22] D. Đukić-Ćosić, K. Baralić, D. Javorac, A.B. Djordjevic, Z. Bulat, An overview of molecular mechanisms in cadmium toxicity, Curr. Opin. Toxicol. 19 (2020) 56–62.
[23] G.F. Nordberg, Health hazards of environmental cadmium pollution, Ambio 3 (1974) 55–66.
[24] https://www.who.int/news-room/fact-sheets/detail/arsenic.
[25] D.G. Mazumder, U.B. Dasgupta, Chronic arsenic toxicity: studies in West Bengal, India, Kaohsiung J. Med. Sci. 27 (9) (2011) 360–370.
[26] S. Shen, X.F. Li, W.R. Cullen, M. Weinfeld, X.C. Le, Arsenic binding to proteins, Chem. Rev. 113 (10) (2013) 7769–7792.
[27] D.S. Tawfik, R.E. Viola, Arsenate replacing phosphate: alternative life chemistries and ion promiscuity, Biochemistry 50 (7) (2011) 1128–1134.

[28] A. Renu, M., & Singh, K., Methodologies for removal of heavy metal ions from wastewater: an overview, Interdiscip. Environ. Rev. 18 (2) (2017) 124–142.

[29] A. Daochalermwong, N. Chanka, K. Songsrirote, P. Dittanet, C. Niamnuy, A. Seubsai, Removal of heavy metal ions using modified celluloses prepared from pineapple leaf fiber, ACS Omega 5 (10) (2020) 5285–5296.

[30] R. Arora, Adsorption of heavy metals—a review, Mater. Today Proc. 18 (2019) 4745–4750.

[31] Y. Zhang, X. Duan, Chemical precipitation of heavy metals from wastewater by using the synthetical magnesium hydroxy carbonate, Water Sci. Technol. 81 (6) (2020) 1130–1136.

[32] M.M. BrbootI, B.A. AbiD, N.M. Al-ShuwaikI, Removal of heavy metals using chemicals precipitation, Eng. Technol. J. 29 (3) (2011) 595–612.

[33] M.A. Barakat, New trends in removing heavy metals from industrial wastewater, Arabian J. Chem. 4 (4) (2011) 361–377.

[34] C. Liu, T. Wu, P.C. Hsu, J. Xie, J. Zhao, K. Liu, et al., Direct/alternating current electrochemical method for removing and recovering heavy metal from water using graphene oxide electrode, ACS Nano 13 (6) (2019) 6431–6437.

[35] E. Bazrafshan, L. Mohammadi, A. Ansari-Moghaddam, A.H. Mahvi, Heavy metals removal from aqueous environments by electrocoagulation process—a systematic review, J. Environ. Health Sci. Eng. 13 (1) (2015) 1–16.

[36] Z. Wang, G. Monge, F.J. Jimenez-Espejo, A. García-Alix, F. Martínez-Ruiz, N. Mattielli, C. Finlayson, et al., Earliest evidence of pollution by heavy metals in archaeological sites, Sci. Rep. 5 (1) (2015) 1–9.

[37] S.K. Gunatilake, Methods of removing heavy metals from industrial wastewater, Methods 1 (1) (2015) 14.

[38] D.Q. Cao, X. Wang, Q.H. Wang, X.M. Fang, J.Y. Jin, X.D. Hao, et al., Removal of heavy metal ions by ultrafiltration with recovery of extracellular polymer substances from excess sludge, J. Membr. Sci. 606 (2020), 118103.

[39] N.N. Maximous, G.F. Nakhla, W.K. Wan, Removal of heavy metals from wastewater by adsorption and membrane processes: a comparative study, Int. J. Environ. Ecol. Eng. 4 (4) (2010) 125–130.

[40] K.C. Khulbe, T. Matsuura, Removal of heavy metals and pollutants by membrane adsorption techniques, Appl. Water Sci. 8 (1) (2018) 1–30.

[41] A. Nagar, T. Pradeep, Clean water through nanotechnology: needs, gaps, and fulfillment, ACS Nano 14 (6) (2020) 6420–6435.

[42] X. Qu, P.J. Alvarez, Q. Li, Applications of nanotechnology in water and wastewater treatment, Water Res. 47 (12) (2013) 3931–3946.

[43] T. Pradeep, A. Baidya, B.B. Rath, A.A. Kumar, IIT, M., Cellulose Nanocrystals Templated Iron Oxyhydroxide Based Adsorbent for Arsenic Removal From Water and a Device Thereof, 2016. Ind. Pat. Appl.

[44] C. de Mello Donegá, The nanoscience paradigm:"size matters!", in: Nanoparticles, Springer, Berlin, Heidelberg, 2014, pp. 1–12.

[45] A. Alagarasi, Chapter-Introduction to Nanomaterials, Indian Institute of Technology Madras, 2013, pp. 1–24.

[46] Z.H. Khan, A. Kumar, S. Husain, M. Husain, Introduction to nanomaterials, in: Advances in Nanomaterials, Springer, New Delhi, 2016, pp. 1–23.

[47] C. Buzea, I. Pacheco, Nanomaterials and their classification, in: EMR/ESR/EPR Spectroscopy for Characterization of Nanomaterials, Springer, New Delhi, 2017, pp. 3–45.

[48] P. Iqbal, J.A. Preece, P.M. Mendes, Nanotechnology: the "top-down" and "bottom-up" approaches, in: Supramolecular Chemistry: From Molecules to Nanomaterials, Wiley, 2012.

[49] D. Vollath, Nanomaterials an introduction to synthesis, properties and application, Environ. Eng. Manage. J. 7 (6) (2008) 865–870.

[50] R.K. Sharma, Y. Monga, S. Sharma, M. Yadav, S. Dutta, R. Gaur, et al., Silica-Based Organic-Inorganic Hybrid Nanomaterials: Synthesis, Functionalization and Applications in the Field of Catalysis, vol. 4, World Scientific, 2019.

[51] R. Baby, B. Saifullah, M.Z. Hussein, Carbon nanomaterials for the treatment of heavy metal-contaminated water and environmental remediation, Nanoscale Res. Lett. 14 (1) (2019) 1–17.

[52] R. Vidu, M. Rahman, M. Mahmoudi, M. Enachescu, T.D. Poteca, I. Opris, Nanostructures: a platform for brain repair and augmentation, Front. Syst. Neurosci. 8 (2014) 91.

[53] X. Ren, C. Chen, M. Nagatsu, X. Wang, Carbon nanotubes as adsorbents in environmental pollution management: a review, Chem. Eng. J. 170 (2–3) (2011) 395–410.

[54] M.C. Nayak, A.M. Isloor, B. Lakshmi, H.M. Marwani, I. Khan, Polyphenylsulfone/multiwalled carbon nanotubes mixed ultrafiltration membranes: fabrication, characterization and removal of heavy metals $Pb2+$, $Hg2+$, and $Cd2+$ from aqueous solutions, Arab. J. Chem. 13 (3) (2020) 4661–4672.

[55] J. Sun, L. Wu, Y. Li, Removal of lead ions from polyether sulfone/Pb (II)-imprinted multi-walled carbon nanotubes mixed matrix membrane, J. Taiwan Inst. Chem. Eng. 78 (2017) 219–229.

[56] S. Mallakpour, E. Khadem, Carbon nanotubes for heavy metals removal, in: Composite Nanoadsorbents, Elsevier, 2019, pp. 181–210.

[57] A. Aqel, K.M. Abou El-Nour, R.A. Ammar, A. Al-Warthan, Carbon nanotubes, science and technology part (I) structure, synthesis and characterisation, Arab. J. Chem. 5 (1) (2012) 1–23.

[58] M. Bassyouni, A.E. Mansi, A. Elgabry, B.A. Ibrahim, O.A. Kassem, R. Alhebeshy, Utilization of carbon nanotubes in removal of heavy metals from wastewater: a review of the CNTs' potential and current challenges, Appl. Phys. A 126 (1) (2020) 1–33.

[59] Y.H. Li, J. Ding, Z. Luan, Z. Di, Y. Zhu, C. Xu, et al., Competitive adsorption of $Pb2+$, $Cu2+$ and $Cd2+$ ions from aqueous solutions by multiwalled carbon nanotubes, Carbon 41 (14) (2003) 2787–2792.

[60] J. Xu, Z. Cao, Y. Zhang, Z. Yuan, Z. Lou, X. Xu, X. Wang, A review of functionalized carbon nanotubes and graphene for heavy metal adsorption from water: preparation, application, and mechanism, Chemosphere 195 (2018) 351–364.

[61] G. Zhao, X. Huang, Z. Tang, Q. Huang, F. Niu, X. Wang, Polymer-based nanocomposites for heavy metal ions removal from aqueous solution: a review, Polym. Chem. 9 (26) (2018) 3562–3582.

[62] M.T. Bankole, A.S. Abdulkareem, I.A. Mohammed, S.S. Ochigbo, J.O. Tijani, O.K. Abubakre, W.D. Roos, Selected heavy metals removal from electroplating wastewater by purified and polyhydroxylbutyrate functionalized carbon nanotubes adsorbents, Sci. Rep. 9 (1) (2019) 1–19.

[63] Y.H. Li, S. Wang, Z. Luan, J. Ding, C. Xu, D. Wu, Adsorption of cadmium (II) from aqueous solution by surface oxidized carbon nanotubes, Carbon 41 (5) (2003) 1057–1062.

[64] X. Zhang, Q. Huang, M. Liu, J. Tian, G. Zeng, Z. Li, et al., Preparation of amine functionalized carbon nanotubes *via* a bioinspired strategy and their application in Cu2+ removal, Appl. Surf. Sci. 343 (2015) 19–27.

[65] D. Iannazzo, A. Pistone, I. Ziccarelli, C. Espro, S. Galvagno, S.V. Giofré, et al., Removal of heavy metal ions from wastewaters using dendrimer-functionalized multi-walled carbon nanotubes, Environ. Sci. Pollut. Res. 24 (17) (2017) 14735–14747.

[66] B. Hayati, A. Maleki, F. Najafi, H. Daraei, F. Gharibi, G. McKay, Super high removal capacities of heavy metals (Pb2+ and Cu2+) using CNT dendrimer, J. Hazard. Mater. 336 (2017) 146–157.

[67] A. Panahi, A. Shomali, M.H. Sabour, E. Ghafar-Zadeh, Molecular dynamics simulation of electric field driven water and heavy metals transport through fluorinated carbon nanotubes, J. Mol. Liq. 278 (2019) 658–671.

[68] Y. Ge, Z. Li, D. Xiao, P. Xiong, N. Ye, Sulfonated multi-walled carbon nanotubes for the removal of copper (II) from aqueous solutions, J. Ind. Eng. Chem. 20 (4) (2014) 1765–1771.

[69] C. Zhang, J. Sui, J. Li, Y. Tang, W. Cai, Efficient removal of heavy metal ions by thiol-functionalized superparamagnetic carbon nanotubes, Chem. Eng. J. 210 (2012) 45–52.

[70] R. Gusain, N. Kumar, E. Fosso-Kankeu, S.S. Ray, Efficient removal of Pb (II) and Cd (II) from industrial mine water by a hierarchical MoS2/SH-MWCNT nanocomposite, ACS Omega 4 (9) (2019) 13922–13935.

[71] L. Xu, J. Wang, The application of graphene-based materials for the removal of heavy metals and radionuclides from water and wastewater, Crit. Rev. Environ. Sci. Technol. 47 (12) (2017) 1042–1105.

[72] X. Liu, R. Ma, X. Wang, Y. Ma, Y. Yang, L. Zhuang, et al., Graphene oxide-based materials for efficient removal of heavy metal ions from aqueous solution: a review, Environ. Pollut. 252 (2019) 62–73.

[73] K. Aghdam, H.A. Panahi, E. Alaei, A.H. Hasani, E. Moniri, Preparation of functionalized graphene oxide and its application as a nanoadsorbent for Hg 2+ removal from aqueous solution, Environ. Monit. Assess. 188 (4) (2016) 223.

[74] C.J. Madadrang, H.Y. Kim, G. Gao, N. Wang, J. Zhu, H. Feng, et al., Adsorption behavior of EDTA-graphene oxide for Pb (II) removal, ACS Appl. Mater. Interfaces 4 (3) (2012) 1186–1193.

[75] F.S. Awad, K.M. AbouZeid, W.M.A. El-Maaty, A.M. El-Wakil, M.S. El-Shall, Efficient removal of heavy metals from polluted water with high selectivity for mercury (II) by 2-imino-4-thiobiuret–partially reduced graphene oxide (IT-PRGO), ACS Appl. Mater. Interfaces 9 (39) (2017) 34230–34242.

[76] A.M. Bakry, F.S. Awad, J.A. Bobb, A.A. Ibrahim, M.S. El-Shall, Melamine-based functionalized graphene oxide and zirconium phosphate for high performance removal of mercury and lead ions from water, RSC Adv. 10 (62) (2020) 37883–37897.

[77] D. Vilela, J. Parmar, Y. Zeng, Y. Zhao, S. Sánchez, Graphene-based microbots for toxic heavy metal removal and recovery from water, Nano Lett. 16 (4) (2016) 2860–2866.

[78] A.K. Shukla, J. Alam, M. Alhoshan, L.A. Dass, F.A.A. Ali, U. Mishra, M.A. Ansari, Removal of heavy metal ions using a carboxylated graphene oxide-incorporated polyphenylsulfone nanofiltration membrane, Environ. Sci.: Water Res. Technol. 4 (3) (2018) 438–448.

[79] İ. Duru, D. Ege, A.R. Kamali, Graphene oxides for removal of heavy and precious metals from wastewater, J. Mater. Sci. 51 (13) (2016) 6097–6116.

[80] M.U. Farooq, M.I. Jalees, Application of magnetic graphene oxide for water purification: heavy metals removal and disinfection, J. Water Process Eng. 33 (2020), 101044.

[81] Z. Rao, K. Feng, B. Tang, P. Wu, Surface decoration of amino-functionalized metal–organic framework/graphene oxide composite onto polydopamine-coated membrane substrate for highly efficient heavy metal removal, ACS Appl. Mater. Interfaces 9 (3) (2017) 2594–2605.

[82] Y.L.F. Musico, C.M. Santos, M.L.P. Dalida, D.F. Rodrigues, Improved removal of lead (II) from water using a polymer-based graphene oxide nanocomposite, J. Mater. Chem. A 1 (11) (2013) 3789–3796.

[83] M.A. Farghali, M.M. Abo-Aly, T.A. Salaheldin, Modified mesoporous zeolite-A/reduced graphene oxide nanocomposite for dual removal of methylene blue and Pb2 + ions from wastewater, Inorg. Chem. Commun. 126 (2021) 108487.

[84] J. Wei, M.F. Aly Aboud, I. Shakir, Z. Tong, Y. Xu, Graphene oxide-supported organo-montmorillonite composites for the removal of Pb (II), Cd (II), and As (V) contaminants from water, ACS Appl. Nano Mater. 3 (1) (2019) 806–813.

[85] M.E. Mahmoud, N.A. Fekry, M.M. El-Latif, Nanocomposites of nanosilica-immobilized-nanopolyaniline and crosslinked nanopolyaniline for removal of heavy metals, Chem. Eng. J. 304 (2016) 679–691.

[86] S. Dutta, R. Gaur, S. Sharma, R.K. Sharma, Silica nanoparticles: a smart and promising support material for the design of heterogeneous catalytic systems, in: Silica-Based Organic-Inorganic Hybrid Nanomaterials: Synthesis, Functionalization and Applications in the Field of Catalysis, vol. 4, World Scientific Publishing, 2019, p. 33.

[87] L. Tang, J. Cheng, Nonporous silica nanoparticles for nanomedicine application, Nano Today 8 (3) (2013) 290–312.

[88] R.K. Kankala, Y.H. Han, J. Na, C.H. Lee, Z. Sun, S.B. Wang, et al., Nanoarchitectured structure and surface biofunctionality of mesoporous silica nanoparticles, Adv. Mater. 32 (23) (2020) 1907035.

[89] L. Han, Y. Zhou, T. He, G. Song, F. Wu, F. Jiang, J. Hu, One-pot morphology-controlled synthesis of various shaped mesoporous silica nanoparticles, J. Mater. Sci. 48 (17) (2013) 5718–5726.

[90] J.L. Gole, J.D. Stout, W.L. Rauch, Z.L. Wang, Direct synthesis of silicon nanowires, silica nanospheres, and wire-like nanosphere agglomerates, Appl. Phys. Lett. 76 (17) (2000) 2346–2348.

[91] A. Chen, M. Komura, K. Kamata, T. Iyoda, Highly ordered arrays of mesoporous silica nanorods with tunable aspect ratios from block copolymer thin films, Adv. Mater. 20 (4) (2008) 763–767.

[92] F. Chen, H. Hong, S. Shi, S. Goel, H.F. Valdovinos, R. Hernandez, et al., Engineering of hollow mesoporous silica nanoparticles for remarkably enhanced tumor active targeting efficacy, Sci. Rep. 4 (1) (2014) 1–10.

[93] M. Marini, B. Pourabbas, F. Pilati, P. Fabbri, Functionally modified core-shell silica nanoparticles by one-pot synthesis, Colloids Surf. A Physicochem. Eng. Asp. 317 (1–3) (2008) 473–481.

[94] R.K. Sharma, S. Sharma, S. Dutta, R. Zboril, M.B. Gawande, Silica-nanosphere-based organic–inorganic hybrid nanomaterials: synthesis, functionalization and applications in catalysis, Green Chem. 17 (6) (2015) 3207–3230.

[95] S. Tighadouini, S. Radi, M. Ferbinteanu, Y. Garcia, Highly selective removal of Pb (II) by a pyridylpyrazole-β-ketoenol receptor covalently bonded onto the silica surface, ACS Omega 4 (2) (2019) 3954–3964.

[96] X.F. Kong, B. Yang, H. Xiong, Y. Zhou, S.G. Xue, B.Q. Xu, S.X. Wang, Selective removal of heavy metal ions from aqueous solutions with surface functionalized silica nanoparticles by different functional groups, J. Cent. South Univ. 21 (9) (2014) 3575–3579.

[97] W. Li, Y. Xu, Y. Zhou, W. Ma, S. Wang, Y. Dai, Silica nanoparticles functionalized via click chemistry and ATRP for enrichment of Pb (II) ion, Nanoscale Res. Lett. 7 (1) (2012) 1–7.

[98] J. Gao, X. Guo, W. Tao, D. Chen, J. Lu, Y. Chen, Norepinephrine-functionalised nanoflower-like organic silica as a new adsorbent for effective Pb (II) removal from aqueous solutions, Sci. Rep. 9 (1) (2019) 1–10.

[99] B. Akhavan, K. Jarvis, P. Majewski, Plasma polymer-functionalized silica particles for heavy metals removal, ACS Appl. Mater. Interfaces 7 (7) (2015) 4265–4274.

[100] E. Da'na, Adsorption of heavy metals on functionalized-mesoporous silica: a review, Microporous Mesoporous Mater. 247 (2017) 145–157.

[101] A. Heidari, H. Younesi, Z. Mehraban, Removal of Ni (II), Cd (II), and Pb (II) from a ternary aqueous solution by amino functionalized mesoporous and nano mesoporous silica, Chem. Eng. J. *153* (1–3) (2009) 70–79.

[102] T. Yokoi, Y. Kubota, T. Tatsumi, Amino-functionalized mesoporous silica as base catalyst and adsorbent, Appl. Catal. Gen. 421 (2012) 14–37.

[103] J. Aguado, J.M. Arsuaga, A. Arencibia, M. Lindo, V. Gascón, Aqueous heavy metals removal by adsorption on amine-functionalized mesoporous silica, J. Hazard. Mater. 163 (1) (2009) 213–221.

[104] R. Hasan, H.D. Setiabudi, Removal of Pb (II) from aqueous solution using KCC-1: optimization by response surface methodology (RSM), J. King Saud Univ. Sci. 31 (4) (2019) 1182–1188.

[105] F. Pourebrahim, M. Ghaedi, K. Dashtian, F. Heidari, S. Kheirandish, Simultaneous removing of Pb2+ ions and alizarin red S dye after their complexation by ultrasonic waves coupled adsorption process: spectrophotometry detection and optimization study, Ultrason. Sonochem. 35 (2017) 51–60.

[106] K. Nakanishi, M. Tomita, K. Kato, Synthesis of amino-functionalized mesoporous silica sheets and their application for metal ion capture, J. Asian Ceramic Soc. 3 (1) (2015) 70–76.

[107] E.H. Jang, S.P. Pack, I. Kim, S. Chung, A systematic study of hexavalent chromium adsorption and removal from aqueous environments using chemically functionalized amorphous and mesoporous silica nanoparticles, Sci. Rep. 10 (1) (2020) 1–20.

[108] Y. Zhang, X. Cao, J. Sun, G. Wu, J. Wang, D. Zhang, Synthesis of pyridyl Schiff base functionalized SBA-15 mesoporous silica for the removal of Cu (II) and Pb (II) from aqueous solution, J. Sol-Gel Sci. Technol. 94 (3) (2020) 658–670.

[109] L. Zhang, F. Song, Y. Wu, L. Cheng, J. Qian, S. Wang, et al., A novel amino and carboxyl functionalized mesoporous silica as an efficient adsorbent for Nickel (II), J. Chem. Eng. Data 64 (1) (2018) 176–188.

[110] J.Y. Lee, C.H. Chen, S. Cheng, H.Y. Li, Adsorption of Pb (II) and Cu (II) metal ions on functionalized large-pore mesoporous silica, Int. J. Environ. Sci. Technol. 13 (1) (2016) 65–76.

[111] A.M. Alswieleh, H.Y. Albahar, A.M. Alfawaz, A.S. Alsilme, A.M. Beagan, A.M. Alsalme, et al., Evaluation of the adsorption efficiency of glycine-, iminodiacetic acid-, and amino propyl-functionalized silica nanoparticles for the removal of potentially toxic elements from contaminated water solution, J. Nanomater. 2021 (2021) 6664252, https://doi.org/10.1155/2021/6664252.

[112] K. Venkatesan, T. Srinivasan, P. Vasudeva Rao, Removal of complexed mercury by dithiocarbamate grafted on mesoporous silica, J. Radioanal. Nucl. Chem. 256 (2) (2003) 213–218.

[113] K. Wieszczycka, K. Filipowiak, I. Wojciechowska, T. Buchwald, K. Siwińska-Ciesielczyk, B. Strzemiecka, et al., Novel highly efficient ionic liquid-functionalized silica for toxic metals removal, Sep. Purif. Technol. 265 (2021) 118483.

[114] D. Pérez-Quintanilla, I. Del Hierro, M. Fajardo, I. Sierra, 2-Mercaptothiazoline modified mesoporous silica for mercury removal from aqueous media, J. Hazard. Mater. 134 (1–3) (2006) 245–256.

[115] H. Sarafraz, G. Alahyarizadeh, A. Minuchehr, H. Modaberi, A. Naserbegi, Economic and Efficient phosphonic functional groups mesoporous silica for uranium selective adsorption from aqueous solutions, Sci. Rep. 9 (1) (2019) 1–10.

[116] M. Kalantari, Z. Gu, Y. Cao, C. Lei, J. Zhang, Thiolated silica nanoadsorbents enable ultrahigh and fast decontamination of mercury (ii): understanding the contribution of thiol moieties' density and accessibility on adsorption performance, Environ. Sci. Nano 7 (3) (2020) 851–860.

[117] P.N. Dave, L.V. Chopda, Application of iron oxide nanomaterials for the removal of heavy metals, J. Nanotechnol. 2014 (2014) 398569, https://doi.org/10.1155/2014/398569.

[118] X.Q. Li, D.W. Elliott, W.X. Zhang, Zero-valent iron nanoparticles for abatement of environmental pollutants: materials and engineering aspects, Crit. Rev. Solid State Mater. Sci. 31 (4) (2006) 111–122.

[119] J. Hu, G. Chen, I.M. Lo, Selective removal of heavy metals from industrial wastewater using maghemite nanoparticle: performance and mechanisms, J. Environ. Eng. 132 (7) (2006) 79–82.

[120] M.A. Ahmed, S.M. Ali, S.I. El-Dek, A. Galal, Magnetite–hematite nanoparticles prepared by green methods for heavy metal ions removal from water, Mater. Sci. Eng. B *178* (10) (2013) 744–751.

[121] L. Giraldo, A. Erto, J.C. Moreno-Piraján, Magnetite nanoparticles for removal of heavy metals from aqueous solutions: synthesis and characterization, Adsorption 19 (2) (2013) 465–474.

[122] R.A. Crane, T.B. Scott, Nanoscale zero-valent iron: future prospects for an emerging water treatment technology, J. Hazard. Mater. 211 (2012) 112–125.

[123] Y. Zou, X. Wang, A. Khan, P. Wang, Y. Liu, A. Alsaedi, et al., Environmental remediation and application of nanoscale zero-valent iron and its composites for the removal of heavy metal ions: a review, Environ. Sci. Technol. 50 (14) (2016) 7290–7304.

[124] T. Pasinszki, M. Krebsz, Synthesis and application of zero-valent iron nanoparticles in water treatment, environmental remediation, catalysis, and their biological effects, Nanomaterials 10 (5) (2020) 917.

[125] Y. Monga, P. Kumar, R.K. Sharma, J. Filip, R.S. Varma, R. Zbořil, M.B. Gawande, Sustainable synthesis of nanoscale zerovalent iron particles for environmental remediation, ChemSusChem 13 (13) (2020) 3288–3305.

[126] H.K. Boparai, M. Joseph, D.M. O'Carroll, Kinetics and thermodynamics of cadmium ion removal by adsorption onto nano zerovalent iron particles, J. Hazard. Mater. *186* (1) (2011) 458–465.

[127] Z. Dongsheng, G. Wenqiang, C. Guozhang, L. Shuai, J. Weizhou, L. Youzhi, Removal of heavy metal lead (II) using nanoscale zero-valent iron with different preservation methods, Adv. Powder Technol. 30 (3) (2019) 581–589.

[128] L. Zhu, L. Tong, N. Zhao, J. Li, Y. Lv, Coupling interaction between porous biochar and nano zero valent iron/nano α-hydroxyl iron oxide improves the remediation efficiency of cadmium in aqueous solution, Chemosphere 219 (2019) 493–503.

[129] G. Kozma, A. Rónavári, Z. Kónya, A. Kukovecz, Environmentally benign synthesis methods of zero-valent iron nanoparticles, ACS Sustainable Chem. Eng. 4 (1) (2016) 291–297.

[130] S. Shad, M.F.A. Belinga-Desaunay-Nault, N. Bashir, I. Lynch, Removal of contaminants from canal water using microwave synthesized zero valent iron nanoparticles, Environ. Sci.: Water Res. Technol. 6 (11) (2020) 3057–3065.

[131] R. Singh, V. Misra, Stabilization of zero-valent iron nanoparticles: role of polymers and surfactants, in: Handbook of Nanoparticles, Springer International Publishing, Switzerland, 2015, pp. 1–19.

[132] C.M. Park, D. Wang, C. Su, Recent developments in engineered nanomaterials for water treatment and environmental remediation, in: Handbook of Nanomaterials for Industrial Applications, Elsevier, 2018, pp. 849–882.

[133] V. Danila, S. Vasarevicius, V. Valskys, Batch removal of Cd (II), Cu (II), Ni (II), and Pb (II) ions using stabilized zero-valent iron nanoparticles, Energy Procedia 147 (2018) 214–219.

[134] K.T.S. Hidalgo, R. Guzmán-Blas, E.O. Ortiz-Quiles, E.R. Fachini, J. Corchado-García, E. Larios, et al., Highly organized nanofiber formation from zero valent iron nanoparticles after cadmium water remediation, RSC Adv. *5* (4) (2015) 2777–2784.

[135] H. Zhu, Y. Jia, X. Wu, H. Wang, Removal of arsenic from water by supported nano zero-valent iron on activated carbon, J. Hazard. Mater. 172 (2–3) (2009) 1591–1596.

[136] Q. Liu, Y. Bei, F. Zhou, Removal of lead (II) from aqueous solution with amino-functionalized nanoscale zero-valent iron, Open Chem. 7 (1) (2009) 79–82.

[137] Y. Su, A.S. Adeleye, Y. Huang, X. Sun, C. Dai, X. Zhou, et al., Simultaneous removal of cadmium and nitrate in aqueous media by nanoscale zerovalent iron (nZVI) and Au doped nZVI particles, Water Res. 63 (2014) 102–111.

[138] C. Yuan, H.L. Lien, Removal of arsenate from aqueous solution using nanoscale iron particles, Water Qual. Res. J. 41 (2) (2006) 210–215.

[139] V.N. Montesinos, N. Quici, E.B. Halac, A.G. Leyva, G. Custo, S. Bengio, et al., Highly efficient removal of Cr (VI) from water with nanoparticulated zerovalent iron: understanding the Fe (III)–Cr (III) passive outer layer structure, Chem. Eng. J. 244 (2014) 569–575.

[140] H. Jabeen, K.C. Kemp, V. Chandra, Synthesis of nano zerovalent iron nanoparticles–graphene composite for the treatment of lead contaminated water, J. Environ. Manage. *130* (2013) 429–435.

[141] X. Lv, J. Xu, G. Jiang, J. Tang, X. Xu, Highly active nanoscale zero-valent iron (nZVI)–Fe3O4 nanocomposites for the removal of chromium (VI) from aqueous solutions, J. Colloid Interface Sci. 369 (1) (2012) 460–469.

[142] X. Lv, X. Xue, G. Jiang, D. Wu, T. Sheng, H. Zhou, X. Xu, Nanoscale zero-valent iron (nZVI) assembled on magnetic Fe3O4/graphene for chromium (VI) removal from aqueous solution, J. Colloid Interface Sci. *417* (2014) 51–59.

[143] X. Zhao, X. Dou, D. Mohan, C.U. Pittman Jr., Y.S. Ok, X. Jin, Antimonate and antimonite adsorption by a polyvinyl alcohol-stabilized granular adsorbent containing nanoscale zero-valent iron, Chem. Eng. J. 247 (2014) 250–257.

[144] W. Jiang, M. Pelaez, D.D. Dionysiou, M.H. Entezari, D. Tsoutsou, K. O'Shea, Chromium (VI) removal by maghemite nanoparticles, Chem. Eng. J. 222 (2013) 527–533.

[145] V.K. Yadav, M.H. Fulekar, Biogenic synthesis of maghemite nanoparticles (γ-Fe2O3) using Tridax leaf extract and its application for removal of fly ash heavy metals (Pb, Cd), Mater. Today Proc. 5 (9) (2018) 20704–20710.

[146] T. Tuutijärvi, J. Lu, M. Sillanpää, G. Chen, As (V) adsorption on maghemite nanoparticles, J. Hazard. Mater. 166 (2–3) (2009) 1415–1420.

[147] H.J. Cui, J.K. Cai, J.W. Shi, B. Yuan, C.L. Ai, M.L. Fu, Fabrication of 3D porous Mn doped α-Fe_2O_3 nanostructures for the removal of heavy metals from wastewater, RSC Adv. 4 (20) (2014) 10176–10179.

[148] T. Madrakian, A. Afkhami, B. Zadpour, M. Ahmadi, New synthetic mercaptoethylamino homopolymer-modified maghemite nanoparticles for effective removal of some heavy metal ions from aqueous solution, J. Ind. Eng. Chem. 21 (2015) 1160–1166.

[149] R.K. Sharma, K. Solanki, R. Dixit, S. Sharma, S. Dutta, Nanoengineered iron oxide based sorbents for separation of various water pollutants: current status, opportunities and future outlook, Environ. Sci.: Water Res. Technol. 7 (2021) 818–860.

[150] R.K. Sharma, S. Dutta, S. Sharma, R. Zboril, R.S. Varma, M.B. Gawande, Fe_3O_4 (iron oxide)-supported nanocatalysts: synthesis, characterization and applications in coupling reactions, Green Chem. 18 (11) (2016) 3184–3209.

[151] M.B. Gawande, Y. Monga, R. Zboril, R.K. Sharma, Silica-decorated magnetic nanocomposites for catalytic applications, Coord. Chem. Rev. 288 (2015) 118–143.

[152] S. Dutta, R.K. Sharma, Sustainable magnetically retrievable nanoadsorbents for selective removal of heavy metal ions from different charged wastewaters, in: Separation Science and Technology, vol. 11, Academic Press, 2019, pp. 371–416.

[153] S. Venkateswarlu, M. Yoon, Surfactant-free green synthesis of Fe 3 O 4 nanoparticles capped with 3, 4-dihydroxyphenethylcarbamodithioate: stable recyclable magnetic nanoparticles for the rapid and efficient removal of Hg (II) ions from water, Dalton Trans. 44 (42) (2015) 18427–18437.

[154] R.K. Sharma, A. Puri, Y. Monga, A. Adholeya, Acetoacetanilide-functionalized Fe_3O_4 nanoparticles for selective and cyclic removal of Pb 2+ ions from different charged wastewaters, J. Mater. Chem. A 2 (32) (2014) 12888–12898.

[155] R.K. Sharma, H. Kumar, A. Kumar, A highly efficient and magnetically retrievable functionalized nano-adsorbent for ultrasonication assisted rapid and selective extraction of Pd 2+ ions from water samples, RSC Adv. 5 (54) (2015) 43371–43380.

[156] L. Zhou, J. Xu, X. Liang, Z. Liu, Adsorption of platinum (IV) and palladium (II) from aqueous solution by magnetic cross-linking chitosan nanoparticles modified with ethylenediamine, J. Hazard. Mater. *182* (1–3) (2010) 518–524.

[157] C. Jia, J. Zhao, L. Lei, X. Kang, R. Lu, C. Chen, et al., Novel magnetically separable anhydride-functionalized Fe 3 O 4@ SiO 2@ PEI-NTDA nanoparticles as effective adsorbents: synthesis, stability and recyclable adsorption performance for heavy metal ions, RSC Adv. 9 (17) (2019) 9533–9545.

[158] M. Naushad, T. Ahamad, K.A. Al-Ghanim, H. Ala'a, G.E. Eldesoky, A.A. Khan, A highly porous nanocomposite (Fe3O4@ BFR) for the removal of toxic Cd (II) ions from aqueous environment: adsorption modelling and regeneration study, Compos. Part B Eng. 172 (2019) 179–185.

[159] R.S. Bangari, V.K. Yadav, J.K. Singh, N. Sinha, Fe3O4-functionalized boron nitride nanosheets as novel adsorbents for removal of arsenic (III) from contaminated water, ACS Omega 5 (18) (2020) 10301–10314.

[160] L. Adlnasab, N. Shekari, A. Maghsodi, Optimization of arsenic removal with Fe3O4@ Al2O3@ Zn-Fe LDH as a new magnetic nano adsorbent using Box-Behnken design, J. Environ. Chem. Eng. 7 (2) (2019) 102974.

[161] S. Dutta, K. Manna, S.K. Srivastava, A.K. Gupta, M.K. Yadav, Hollow polyaniline microsphere/Fe_3O_4 nanocomposite as an effective adsorbent for removal of arsenic from water, Sci. Rep. 10 (1) (2020) 1–14.

[162] N.S. Behbahani, K. Rostamizadeh, M.R. Yaftian, A. Zamani, H. Ahmadi, Covalently modified magnetite nanoparticles with PEG: preparation and characterization as nano-adsorbent for removal of lead from wastewater, J. Environ. Health Sci. Eng. 12 (1) (2014) 1–12.

[163] J. Li, S. Chen, H. Xiao, G. Yao, Y. Gu, Q. Yang, B. Yan, Highly efficient removal of Cr (vi) ions from wastewater by the pomegranate-like magnetic hybrid nano-adsorbent of polydopamine and Fe_3O_4 nanoparticles, New J. Chem. 44 (29) (2020) 12785–12792.

[164] M. Hasanzadeh, F. Farajbakhsh, N. Shadjou, A. Jouyban, Mesoporous (organo) silica decorated with magnetic nanoparticles as a reusable nanoadsorbent for arsenic removal from water samples, Environ. Technol. 36 (1) (2015) 36–44.

[165] S.M. El-Bahy, New iminodiacetate chelating resin-functionalized Fe3O4 nanoparticles: synthesis, characterization, and application for the removal of some noxious metal ions from wastewater, J. Chem. Eng. Data 63 (6) (2018) 2299–2313.

[166] S. Venkateswarlu, M. Yoon, Core–shell ferromagnetic nanorod based on amine polymer composite (Fe3O4@ DAPF) for fast removal of Pb (II) from aqueous solutions, ACS Appl. Mater. Interfaces 7 (45) (2015) 25362–25372.

[167] X. Wei, P.J. Sugumaran, E. Peng, X.L. Liu, J. Ding, Low-field dynamic magnetic separation by self-fabricated magnetic meshes for efficient heavy metal removal, ACS Appl. Mater. Interfaces 9 (42) (2017) 36772–36782.

[168] R. Bhateria, R. Singh, A review on nanotechnological application of magnetic iron oxides for heavy metal removal, J. Water Process Eng. 31 (2019) 100845.

[169] G. Bereket, A.Z. Arog, M.Z. Özel, Removal of Pb (II), Cd (II), Cu (II), and Zn (II) from aqueous solutions by adsorption on bentonite, J. Colloid Interface Sci. 187 (2) (1997) 338–343.

[170] G. Bharath, N. Ponpandian, Hydroxyapatite nanoparticles on dendritic α-Fe_2O_3 hierarchical architectures for a heterogeneous photocatalyst and adsorption of Pb (II) ions from industrial wastewater, RSC Adv. 5 (103) (2015) 84685–84693.

[171] T.S. Sakthivel, S. Das, C.J. Pratt, S. Seal, One-pot synthesis of a ceria–graphene oxide composite for the efficient removal of arsenic species, Nanoscale 9 (10) (2017) 3367–3374.

[172] A.S. Poursani, A. Nilchi, A. Hassani, S. Tabibian, L.A. Amraji, Synthesis of nano-γ-Al 2 O 3/chitosan beads (AlCBs) and continuous heavy metals removal from liquid solution, Int. J. Environ. Sci. Technol. 14 (7) (2017) 1459–1468.

[173] J. Yang, Y. Chu, Z. Li, Y. Zhang, Effective removal of heavy metals by nanosized hydrous zirconia composite hydrogel and adsorption behavior study, Environ. Sci. Pollut. Res. 25 (33) (2018) 33464–33477.

[174] L. Wang, C. Shi, L. Pan, X. Zhang, J.J. Zou, Rational design, synthesis, adsorption principles and applications of metal oxide adsorbents: a review, Nanoscale 12 (8) (2020) 4790–4815.

[175] M. Mon, R. Bruno, J. Ferrando-Soria, D. Armentano, E. Pardo, Metal–organic framework technologies for water remediation: towards a sustainable ecosystem, J. Mater. Chem. A 6 (12) (2018) 4912–4947.

[176] J. Li, X. Wang, G. Zhao, C. Chen, Z. Chai, A. Alsaedi, et al., Metal–organic framework-based materials: superior adsorbents for the capture of toxic and radioactive metal ions, Chem. Soc. Rev. 47 (7) (2018) 2322–2356.

[177] P.A. Kobielska, A.J. Howarth, O.K. Farha, S. Nayak, Metal–organic frameworks for heavy metal removal from water, Coord. Chem. Rev. 358 (2018) 92–107.

[178] D.T. Sun, L. Peng, W.S. Reeder, S.M. Moosavi, D. Tiana, D.K. Britt, et al., Rapid, selective heavy metal removal from water by a metal–organic framework/polydopamine composite, ACS Cent. Sci. 4 (3) (2018) 349–356.

[179] R. Soltani, R. Pelalak, M. Pishnamazi, A. Marjani, A.B. Albadarin, S.M. Sarkar, S. Shirazian, A novel and facile green synthesis method to prepare LDH/MOF nanocomposite for removal of Cd (II) and Pb (II), Sci. Rep. 11 (1) (2021) 1–15.

[180] N. Assaad, G. Sabeh, M. Hmadeh, Defect control in Zr-based metal–organic framework nanoparticles for arsenic removal from water, ACS Appl. Nano Mater. 3 (9) (2020) 8997–9008.

[181] Y. Peng, H. Huang, Y. Zhang, C. Kang, S. Chen, L. Song, et al., A versatile MOF-based trap for heavy metal ion capture and dispersion, Nat. Commun. *9* (1) (2018) 1–9.

[182] G. Boix, J. Troyano, L. Garzon-Tovar, C. Camur, N. Bermejo, A. Yazdi, et al., MOF-beads containing inorganic nanoparticles for the simultaneous removal of multiple heavy metals from water, ACS Appl. Mater. Interfaces 12 (9) (2020) 10554–10562.

[183] H. Shahriyari Far, M. Hasanzadeh, M. Najafi, T.R. Masale Nezhad, M. Rabbani, Efficient removal of Pb (II) and Co (II) ions from aqueous solution with a chromium-based metal–organic framework/activated carbon composites, Ind. Eng. Chem. Res. 60 (11) (2021) 4332–4341.

[184] L. Huang, M. He, B. Chen, B. Hu, Magnetic Zr-MOFs nanocomposites for rapid removal of heavy metal ions and dyes from water, Chemosphere 199 (2018) 435–444.

[185] T. Chowdhury, L. Zhang, J. Zhang, S. Aggarwal, Removal of arsenic (III) from aqueous solution using metal organic framework-graphene oxide nanocomposite, Nanomaterials 8 (12) (2018) 1062.

[186] N. Wang, X.K. Ouyang, L.Y. Yang, A.M. Omer, Fabrication of a magnetic cellulose nanocrystal/metal–organic framework composite for removal of Pb (II) from water, ACS Sustainable Chem. Eng. 5 (11) (2017) 10447–10458.

[187] P. Wang, T. Shen, X. Li, Y. Tang, Y. Li, Magnetic mesoporous calcium carbonate-based nanocomposites for the removal of toxic Pb (II) and Cd (II) ions from water, ACS Appl. Nano Mater. *3* (2) (2020) 1272–1281.

[188] H. Li, H. Liu, J. Zhang, Y. Cheng, C. Zhang, X. Fei, Y. Xian, Platinum nanoparticle encapsulated metal–organic frameworks for colorimetric measurement and facile removal of mercury (II), ACS Appl. Mater. Interfaces 9 (46) (2017) 40716–40725.

[189] Y. Wang, G. Ye, H. Chen, X. Hu, Z. Niu, S. Ma, Functionalized metal–organic framework as a new platform for efficient and selective removal of cadmium (II) from aqueous solution, J. Mater. Chem. A 3 (29) (2015) 15292–15298.

[190] D. Li, X. Tian, Z. Wang, Z. Guan, X. Li, H. Qiao, et al., Multifunctional adsorbent based on metal-organic framework modified bacterial cellulose/chitosan composite aerogel for high efficient removal of heavy metal ion and organic pollutant, Chem. Eng. J. 383 (2020) 123127.

[191] A. Sabir, F. Altaf, R. Batool, M. Shafiq, R.U. Khan, K.I. Jacob, Agricultural waste absorbents for heavy metal removal, in: Green Adsorbents to Remove Metals, Dyes and Boron from Polluted Water, Springer, Cham, 2021, pp. 195–228.

[192] J. Shen, I. Kaur, M.M. Baktash, Z. He, Y. Ni, A combined process of activated carbon adsorption, ion exchange resin treatment and membrane concentration for recovery of dissolved organics in pre-hydrolysis liquor of the kraft-based dissolving pulp production process, Bioresour. Technol. 127 (2013) 59–65.

[193] O.A. Dar, M.A. Malik, M.I.A. Talukdar, A.A. Hashmi, Bionanocomposites in water treatment, in: Bionanocomposites, Elsevier, 2020, pp. 505–518.

[194] S. Das, J. Chakraborty, S. Chatterjee, H. Kumar, Prospects of biosynthesized nanomaterials for the remediation of organic and inorganic environmental contaminants, Environ. Sci. Nano 5 (12) (2018) 2784–2808.

[195] Z. Zia, A. Hartland, M.R. Mucalo, Use of low-cost biopolymers and biopolymeric composite systems for heavy metal removal from water, Int. J. Environment. Sci. Technol. 17 (2020) 4389–4406.

[196] Y. Koriche, M. Darder, P. Aranda, S. Semsari, E. Ruiz-Hitzky, Bionanocomposites based on layered silicates and cationic starch as eco-friendly adsorbents for hexavalent chromium removal, Dalton Trans. 43 (27) (2014) 10512–10520.

[197] R. Eivazzadeh-Keihan, F. Radinekiyan, S. Asgharnasl, A. Maleki, H. Bahreinizad, A natural and eco-friendly magnetic nanobiocomposite based on activated chitosan for heavy metals adsorption and the in-vitro hyperthermia of cancer therapy, J. Mater. Res. Technol. 9 (6) (2020) 12244–12259.

[198] V.M. Alamillo-López, V. Sánchez-Mendieta, O.F. Olea-Mejía, M.G. González-Pedroza, R.A. Morales-Luckie, Efficient removal of heavy metals from aqueous solutions using a bionanocomposite of eggshell/Ag-Fe, Catalysts *10* (7) (2020) 727.

[199] J.D. Sachs, From millennium development goals to sustainable development goals, Lancet 379 (9832) (2012) 2206–2211.

[200] G. Assembly, Sustainable development goals, in: SDGs Transform Our World, vol. 2030, UN, 2015.

[201] A. Srivastava, S. Dutta, S. Ahuja, R.K. Sharma, Green chemistry: key to reducing waste and improving water quality, in: Handbook of Water Purity and Quality, Elsevier, 2021, pp. 359–407.

[202] J.D. Sachs, G. Schmidt-Traub, M. Mazzucato, D. Messner, N. Nakicenovic, J. Rockström, Six transformations to achieve the sustainable development goals, Nat. Sustain. 2 (9) (2019) 805–814.

[203] D. Griggs, M. Stafford-Smith, O. Gaffney, J. Rockström, M.C. Öhman, P. Shyamsundar, et al., Sustainable development goals for people and planet, Nature 495 (7441) (2013) 305–307.

[204] A.Y. Hoekstra, A.K. Chapagain, P.R. Van Oel, Advancing water footprint assessment research: challenges in monitoring progress towards sustainable development goal 6, Water 9 (6) (2017) 438.

[205] J.G. Hering, S. Maag, J.L. Schnoor, A call for synthesis of water research to achieve the sustainable development goals by 2030, Environ. Sci. Technol. 50 (2016) 6122–6123.

Chapter 8

Abatement of pesticides in drinking water by nanoscale carbon materials

Afrah Harafan[a], Reetesh K. Gangwar[b], and Shihabudheen M. Maliyekkal[a,*]

[a]*Department of Civil and Environmental Engineering, Indian Institute of Technology Tirupati, Tirupati, India,* [b]*Department of Physics and CAMOST, Indian Institute of Technology Tirupati, Tirupati, India*

**Corresponding author: e-mail address: shihab@iittp.ac.in*

Introduction

A pesticide is a chemical agent or mixture of chemical agents used to control or destroy pests and weeds, thereby protecting crop plants, animals, and people from damage and diseases [1]. The first recorded use of pesticides was reported in Egypt, dating back to 1550 BCE [2,3]. Although pesticides application to combat pests originated centuries ago, the large-scale use of synthetic organic chemicals (SOCs) as pesticide management systems started in the early 1940s and 1950s [4]. In 1990, more than 3000 pesticide formulations were in use, including 300 insecticides, 165 fungicides, 290 herbicides, and other chemicals [5,6]. Numerous pesticides with diverse chemical nature are available now, with an annual consumption of 2 billion kg worldwide [7]. Fig. 1A shows the chemical structure of a few commonly used pesticides. Fig. 1B shows a comparative chart of pesticides used in India and the rest of the world.

Despite the revolutionary role the SOCs have played in agricultural production, controlling insect vectors, and other beneficial use, their indiscriminate use has resulted in the widespread contamination of surface and groundwater resources [8,9]. The runoff from the agricultural sector is identified as the primary source of contamination of water bodies. To less extent, the release of nonagricultural domestic and industrial chemical compounds to the environment also contributes to pollution [10–12]. Most of these pesticides possess a half-life time ranging from days to years, resulting in bioaccumulation and biomagnification [13]. Due to their persistent nature, more than 5 billion kg of unused and obsolete pesticides paves the way for possible entry into human and animal food chains [7].

The health hazards posed by pesticides have become a massive area of concern for developing and developed countries worldwide. Studies performed on various pesticides based on exposure indicate cytotoxicity, inhibition of cell proliferation, mutation, genotoxicity, fetotoxicity, and decreased fertility in mammals [14]. The recent use of nano pesticides has aggravated the situation by penetrating the blood placental barrier, causing adverse fetal defects [15]. According to the World Health Organization (WHO) and United Nations Environment Programme (UNEP), globally, around 0.2 million deaths and 3 million pesticide poisoning are reported annually, with a majority of cases from developing nations [16].

Nevertheless, the production and uses of SOCs are increasing. According to the estimates, total pesticide consumption in India has increased from 0.15 million kg in 1954 to 0.5 billion kg in 2017 [17]. The data also show that it has a high potential for growth in India because the use of agricultural pesticides is markedly low at 0.5 kg/ha as against 7 kg/ha in the USA and 2.5 kg/ha in Europe [18]. Even though synthetic pesticides have been increasing rapidly, lately, there has been a remarkable shift from hazardous pesticides to green pesticides. However, the high cost of green pesticides limits their use in developing countries, leading to the widespread contamination of persistent pesticides [12].

There is a need to limit the massive application of pesticides to mitigate the contamination of water bodies and the associated health and environmental problems. Adopting efficient agronomic practices, sustainable agroecosystems, integrated pest control practices, using less hazardous and natural pesticides, and other proven scientific agricultural methods are some of the ways to achieve it [17]. Nonetheless, the nonbioactive pesticides that have already reached water bodies need attention due to their potential human toxicity and persistent nature. Diverse technologies such as adsorption,

Separation Science and Technology, Volume 15. https://doi.org/10.1016/B978-0-323-90763-7.00001-9

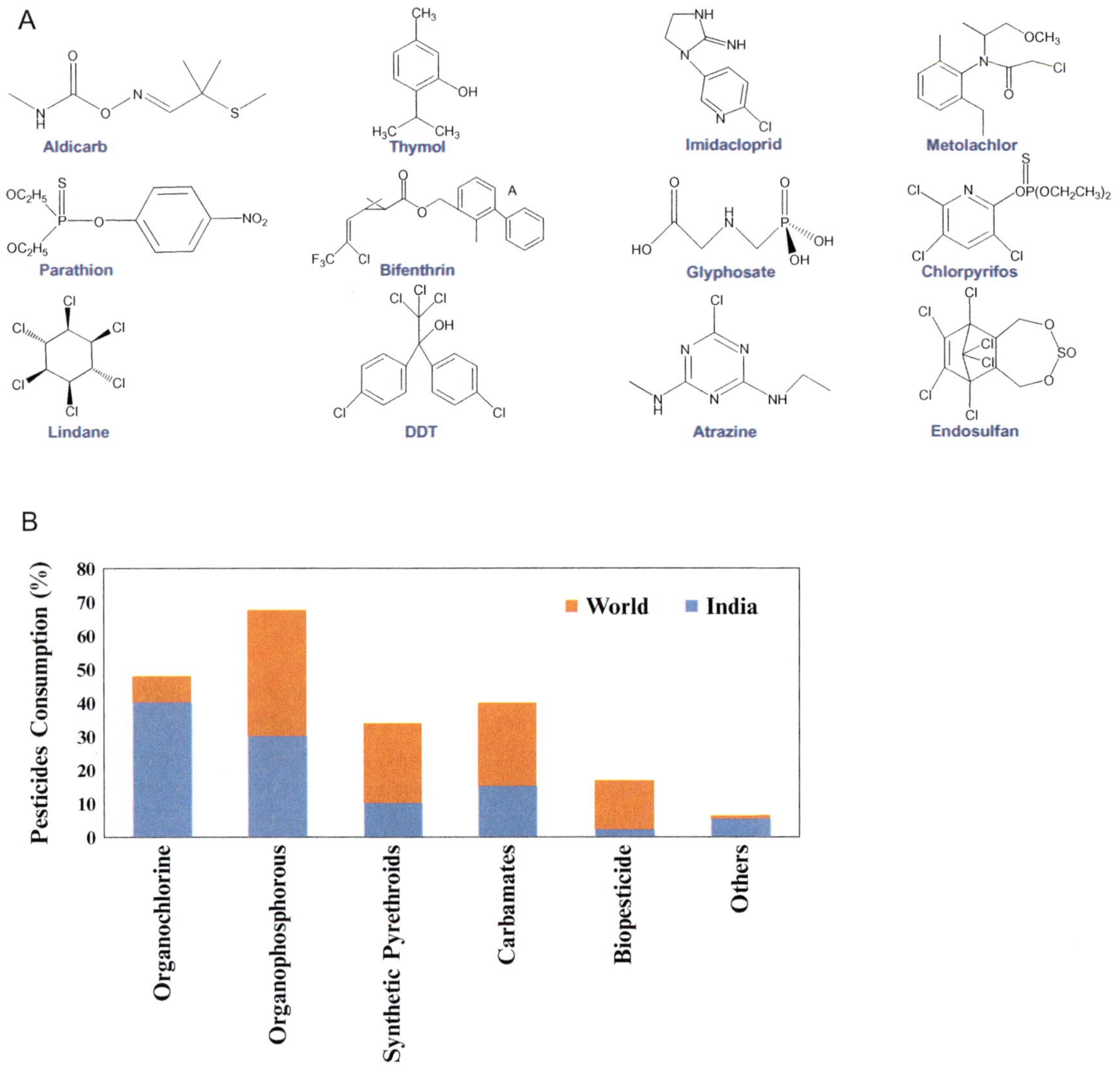

FIG. 1 (A) Chemical structure of a few commonly used pesticides. (B) Group-wise comparison of pesticides usage in India and the rest of the World. *Modified from I.C. Yadav, N.L. Devi, J.H. Syed, Z. Cheng, J. Li, G. Zhang, K.C. Jones, Current status of persistent organic pesticides residues in air, water, and soil, and their possible effect on neighboring countries: A comprehensive review of India, Sci. Total Environ. 511 (2015) 126.*

membrane filtration, advanced oxidation processes (AOPs), and ozonation are employed to improve the water quality and supply pesticide-free water. Even though AOPs such as electrocatalysis, photocatalysis, ozonation, and electro-photocatalysis effectively remove SOCs, its relatively high capital cost is a drawback in utilizing the technology [19]. Membrane technologies can remove organics, inorganics, and solids from water using a semipermeable membrane separating the permeate and concentrate. Regardless of the existence of various kinds of membranes characterized by their pore size, surface charge and hydrophobicity, the risk of fouling, clogging, and concentration polarization minimizes its performance [20]. Although the technique is limited to phase separation, the adsorption process is considered one of the most effective methods, as it complies with environmental quality standards. Moreover, process flexibility, low energy requirements, ease of operation, and low installation, and operation costs have made the process emerge as one of the most preferred treatment techniques in treating pollutants from water matrices [20,21].

Over the years, carbonaceous materials have been preferred over other materials as adsorbent media due to their stability, sustainability, nonspecific action, cost, and efficiency [22,23]. Depending on the organic contaminants structure, carbon nanomaterials (CNM) can interact with SOCs through various forces and eliminate them from waterbodies. Interestingly, most of the modern treatment technologies use the granular activated charcoal bed for the enhanced removal of organic pollutants [24]. Various carbon derivatives are developed to alter their characteristics, such as porosity, surface area, morphology, functionality, and polarity of carbon materials [25]. Recently, nanocarbon materials such as graphene, carbon nanotube (CNT), fullerenes, carbon dots, and carbon nanohorns started gaining interest in water purification due to their unique and favorable physical and chemical properties [26,27]. Apart from nanocarbon, several other nanomaterials, including metal oxides, noble metals, and zerovalent iron, have also exhibited adsorption potential in removing organophosphorus (OP) and organochlorine (OC) pesticides from aqueous media. Anshup and Pradeep have presented an excellent review of the application of noble metal nanoparticles for pesticide removal, and readers may want to read the article for more details [28]. This chapter reviews the literature on various nanoscale carbons with exciting properties and their applications for treating pesticide-laden drinking water. The factors influencing the removal, challenges, prospects, and future development in the area are discussed. A brief outlook concerning the history, source and occurrence of pesticides are also summarized.

History of pesticide use

The term pest is derived from the Greek word "pestis," which means "to annoy." Any organism that disturbs human beings is a pest and generally comprises of bacteria, insects, fungi, viruses and nematodes. Before World War II, crop damage due to locusts accounted for a loss of 20 Million USD worldwide. Later, the bubonic plague (1346–53) and potato blight disease (the 1840s) in Europe led to millions of deaths due to famine starvation [29,30]. As a result, agrochemicals were developed to enhance crop yields. However, their large-scale production and usage have resulted in soil, air and water pollution.

The issue of pesticides contaminations has been prevalent since the origin of agriculture [11,27,31]. Perhaps, the first use of pesticides was recorded around 2500–1500 BCE when Sumerians and Egyptians used sulfur compounds to control insects and mites. Since 1000 BCE, various natural (pyrethrin from chrysanthemum, rotenone from roots of wild vine, nicotine), inorganic elements (arsenic, copper, mercury), and inorganic compounds (copper arsenate, copper sulfate) were extensively used as insecticides and herbicides to protect the crops from foreign attack [32]. The synthesis of dichlorodiphenyl-trichloroethane (DDT), an OC, in the 1940s marked an important event in history. Increased DDT consumption was reported due to its broad-spectrum activity and efficiency [1,27]. During World War II, DDT was extensively used against disease-causing pests, thereby reducing the mortality rate. Parallel to the discovery of OC, OP, and carbaryl compounds, insecticides were also introduced to control weeds, which later dominated the pesticide market [33]. Consequently, the migration of labors from farms to industries during the period 1950–1990s, urged the government and private sectors to encourage the usage of synthetic pesticides by farmers to improve crop productivity [34–36]. The application of these pesticides was reported to be the highest in Europe, and residues were detected in various environmental matrices [1,5]. It was not until the mid-1950s, the first case of DDT poisoning, that these compounds persistence and bioaccumulation were noted and was restricted progressively in many countries [37]. Overconsumption of pesticides had resulted in pesticide resistance genes, which created a pesticide spiral effect [38]. Later, in 1962, Rachel Carson, a marine biologist, discussed the potential ecological damage caused by pesticides in her book "Silent Spring" which initiated standardized pesticide regulations [5].

Due to the risk posed by complex mixtures of pesticides in the environment, many pesticides have been grouped under priority pollutants [39]. Since then, there has been a growing demand for sustainable herbicides, insecticides, fungicides, and other pesticides [40]. Due to the adverse effects posed by synthetic pesticides, many natural pesticides were also exploited [41,42]. Majority of these modern pesticides are less toxic to the environment [41]. For instance, pyrethroids are less toxic to humans and are more photostable than their plant origin, pyrethrin. Currently, imidacloprid and glyphosate are the most widely used insecticide and herbicide, respectively, owing to their fewer toxic properties and durability [43].

Several approaches have also been initiated to reduce the use of pesticides. Advances in genetic engineering have succeeded in producing resistant crop varieties by introducing genetically modified substances to minimize the usage of chemical pesticides [44]. Integrated pest management (IPM) was another initiative proposed in the 1960s to promote agricultural development in a sustainable manner [19]. The use of predatory ants to protect orchards can be regarded as the oldest example of IPM. Lately, usage of pheromones and growth regulators are encouraged to alter organisms' physiology and regulate the populations of pests [45]. The development of such tools is crucial to minimize long term impacts and risks posed in the environment. Table 1 provides an overview of the evolution of pesticide use.

Origin of pesticides: Point and nonpoint sources

Due to the excessive use of pesticides and the inability of the plants to utilize them effectively on target organisms, these chemicals often seeps into freshwater bodies either in their pristine form or metabolized forms through various routes [9] Hence, they pose risks to ecosystems through bioaccumulation or developing resistant genes [13]. As there is a substantial variation in the spatial distribution of pesticides, there is a need to identify their source of origin to predict their fate and transport and mitigate their effects [9,46].

TABLE 1 Important milestones in the history of pesticide use

Year	Milestones
2500 BCE	Sulfur compounds to control mites
1000 BCE	Mercury and arsenic to control body lice Use of predatory ants to protect from caterpillars and beetles
149 BCE–800 CE	Hellebore plant oil to kill rodents and pests Amurca to kill weeds
1649	Nicotine to control lace bugs
1807	Copper sulfate to control bunt disease
1867	Paris green to control potato beetles and vines from pest
1885	Bordeaux Mixture to control downy mildew on grapes Pyrethrum used as a louse powder
1896	Iron sulfate to kill broad leaf weeds
1913	Organic mercury compounds to protect seeds from rust and other diseases
1934	Development of thiram as fungicide
1939	DDT as an insecticide
1943	Synthesis of organophosphorous, parathion
1949	Introduction of herbicide 2,4-D
1958–1960	Synthesis of herbicides atrazine, paraquat, picloram
1970	Imidazole class fungicides
1967	Introduction of the concept IPM
1960	Synthesis of alachlor
1962	Rachel Carson, initiated a movement toward pesticide regulation
1960–1974	Synthesis of first synthetic pesticide pyrethroid
1980–1990	Development of neonicotinoids
1981	A nationwide study on public health
2000s	Development of chitin, pheromones, sterilant, genetically modified organisms

Data from Abubakar et al. Chapter 3—pesticides, history, and classification, in: Natural Remedies for Pest, Disease and Weed Control, 2020, pp. 29–42. Pesticides in the early part of 20th century. *in: A History of Pesticides*,1; E.A. Murphy, M. Aucott, An assessment of the amounts of arsenical pesticides used historically in a geographical area, Sci. Total Environ. 218 (2) (1998) p. 83.

The release of pesticides can occur via a point source or a nonpoint source, among which the agricultural sector is the primary source of freshwater contamination [13,26]. In response to the growing demand for food, there has been a sharp increase in the utilization of pesticides by farmers [13,26,47]. As these pesticides are exposed to the environment, they can infiltrate into the aquatic domain through various matrices [48–50]. Based on seasonal variations and characteristics, pesticides are either drained into surface or groundwater matrix via runoff or leaching [9,18,49]. In water bodies, they often exist as complex mixtures. Hence, exposure to a mix of pesticides may induce increased toxic effects than individual pesticides [13,49]. The translocation of these chemicals into the air matrix may occur either by volatilization or by the influence of atmospheric conditions during the application [18]. Depending on the physicochemical properties, these pesticides can accumulate in the environment, extend across regions and precipitate [13,51]. In addition, pesticides manufacturing industries paves way to contamination by the release poorly treated effluents into the waterbodies [18].

Moreover, improper handling of pesticides has also resulted in pesticides spills, resulting in their migration to the soil, air, and water reservoirs. Fig. 2 shows the various routes through which pesticides accumulate in freshwater bodies. A considerable quantity of agrochemicals (3%–25%) is volatilized during the application, which later deposits back in

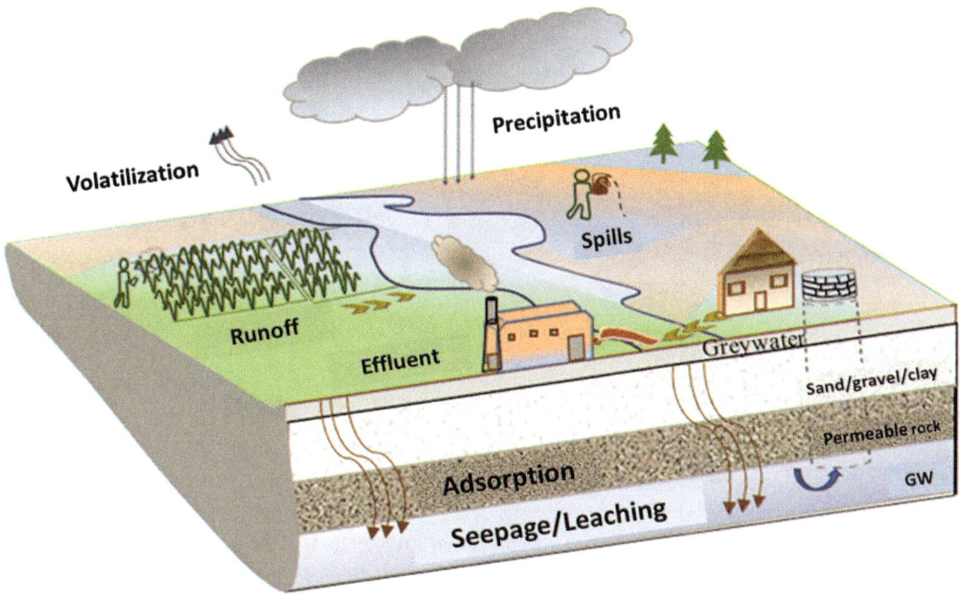

FIG. 2 Fate and transport of pesticides in the environment.

the reservoirs through precipitation [8]. The rainfall washes out these pesticides from the field and discharges them into water bodies. During this process, a portion of these pesticides is either adsorbed or leached into the soil, contaminating groundwater. Consequently, this has led to the detection of pesticide residues in drinking water samples [52].

Occurrence of pesticides in water bodies

The elevated presence of pesticide concentrations in the environment has gained significant importance in the recent years due to their nonselective nature and prolonged toxic effects [12,51,53]. Based on their mobility and partition coefficient, these agrochemicals coexist in various matrices. Pesticide contaminated waterbodies extend across the globe, but the concentrations and types of pesticides can vary based on region and historical use, as shown in Fig. 3 [53–55]. In India, pesticides such as endosulfan (37,560 ng/L), chlorpyrifos (860 ng/L), and methyl parathion (430 ng/L) were reported in surface water samples from river Tapi [56]. Also, persistent organic pollutants (POPs) like lindane (2100–2618 ng/L) and dichloro-diphenyl-trichloroethane (1500 ng/L) were found in water samples collected from the river Ganges [55,57–59]. Despite stringent regulation policies, pesticides were also detected in European waters, among which Spain and Italy reported the highest [7,60]. The high pesticide concentration in Italian surface waters can be attributed to excessive pesticide application to crops that had high intrinsic leaching potential [61]. Fig. 4 shows the percentage of various types of pesticides detected in water worldwide during 2012–2020. It is evident from the data that herbicides usage was more in developed countries, whereas insecticides usage dominated in developing nations [62–65]. In India, insecticides application was the highest due to the ubiquitous presence of stagnant waters, which served as a breeding ground for insects [33]. A study in Ahmedabad revealed high levels of DDT and its metabolites, aldrin and pyrethroids, in human milk samples, which can be as a result of exposure of pesticide-contaminated water [66]. Globally, the highest concentration of insecticides was reported in Costa Rica. The high presence of dimethoate with a concentration of 61.2 μg/L in Costa Rica could be assigned to surface runoff during the initial stages of rainfall [60,64,67]. The other frequently reported pesticides in water bodies were atrazine and its metabolites, metolachlor, chlorpyrifos, tebuconazole, and carbendazim.

Despite many pesticides being banned, their durability pose a severe risk to ecology and humans. In India, the first case of pesticide poisoning was reported in 1958 in Kerala, where more than 100 people died due to the consumption of parathion contaminated wheat flour [68]. Madhya Pradesh and Uttar Pradesh have also recorded several pesticide poisoning cases, placing over 0.1 million people at risk [26,55,58,59]. An approximate of 3 million people are at health risk annually due to OP poisoning worldwide [69].

Exposure routes and impacts on human health

Various studies have linked the increased incidences of diseases in humans to pesticides through direct or indirect exposure. Consequently, the growing use of pesticides in agriculture and food industry has imposed severe human health concerns.

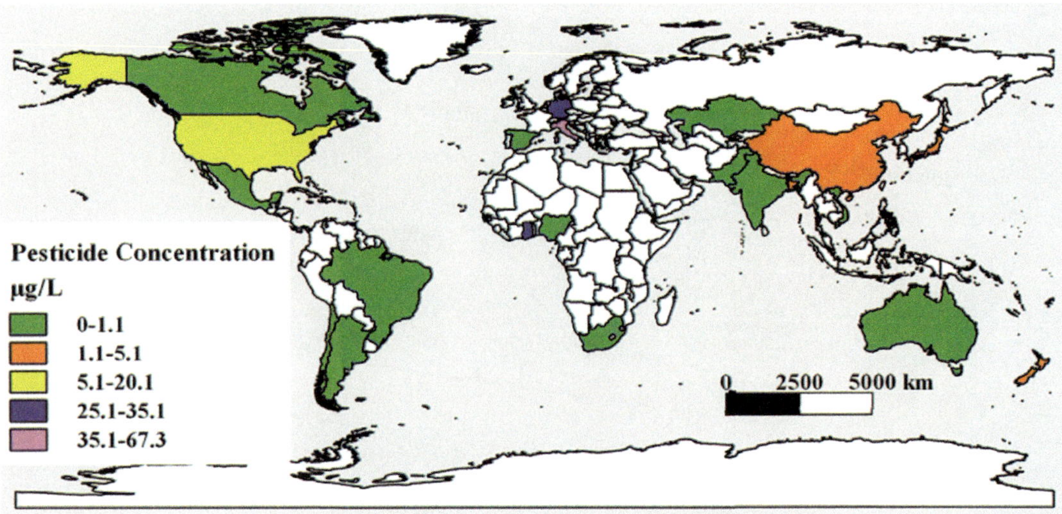

FIG. 3 Concentrations of pesticide residues in the aquatic matrix during the period 2012–2020. *M.V. Barbieri, A. Peris, C. Postigo, A. Moya-Garcés, L.S. Monllor-Alcaraz, M. Rambla-Alegre, E. Eljarrat, M. López de Alda, Evaluation of the occurrence and fate of pesticides in a typical Mediterranean delta ecosystem (Ebro River Delta) and risk assessment for aquatic organisms, Environ. Pollut. (2020) 115813; M.E. Close, B. Humphries, G. Northcott, Outcomes of the first combined national survey of pesticides and emerging organic contaminants (EOCs) in groundwater in New Zealand 2018, Sci. Total Environ. 754 (2021), 142005; R.M. de Souza, D. Seibert, H.B. Quesada, F. de Jesus Bassetti, M.R. Fagundes-Klen, R. Bergamasco, Occurrence, impacts and general aspects of pesticides in surface water: A review, Process Saf. Environ. Prot. 135 (2020) 22–37; T. di Lorenzo, M. Cifoni, B. Fiasca, A. di Cioccio, D.M.P. Galassi, Ecological risk assessment of pesticide mixtures in the alluvial aquifers of Central Italy: toward more realistic scenarios for risk mitigation, Sci. Total Environ. 644 (2018) 161-172; R. Elibariki, M.M. Maguta, Status of pesticides pollution in Tanzania – A review, Chemosphere 178 (2017) 154–164; R. Meffe, I. de Bustamante, Emerging organic contaminants in surface water and groundwater: A first overview of the situation in Italy, Sci. Total Environ. 481 (2014) 280-295; D. Pietrzak, J. Kania, G. Malina, E. Kmiecik, K. WaÛtor, Pesticides from the EU first and second watch lists in the water environment, CLEAN – Soil, Air, Water 47 (7) (2019) 1800376; M.A. Rippy, A. Deletic, J. Black, R. Aryal, J.-L. Lampard, J.Y.-M. Tang, D. McCarthy, P. Kolotelo, J. Sidhu, W. Gernjak, Pesticide occurrence and spatio-temporal variability in urban run-off across Australia, Water Res. 115 (2017) 245–255; N.I. Rousis, R. Bade, L. Bijlsma, E. Zuccato, J. Sancho, v, Hernandez, F. and Castiglioni, S., Monitoring a large number of pesticides and transformation products in water samples from Spain and Italy, Environ. Res. 156 (2017) 31–38; N. Tauchnitz, F. Kurzius, H. Rupp, G. Schmidt, B. Hauser, M. Schrödter, R. Meissner, Assessment of pesticide inputs into surface waters by agricultural and urban sources—A case study in the Querne/ Weida catchment, Central Germany, Environ. Pollut. 267 (2020) 115186; M.S.J. Warne, R.A. Smith, R.D.R. Turner, Analysis of pesticide mixtures discharged to the lagoon of the great barrier reef, Australia, Environ. Pollut. 265 (2020) 114088; Y. Zhou, J. Wu, B. Wang, L. Duan, Y. Zhang, W. Zhao, F. Wang, Q. Sui, Z. Chen, D. Xu, Q. Li, G. Yu, Occurrence, source and ecotoxicological risk assessment of pesticides in surface water of Wujin District (northwest of Taihu Lake), China, Environ. Pollut. 265 (2020) 114953.*

Though the toxicity level of these products varies, continuous exposure can cause serious adverse effects, as shown in Fig. 5. Human exposure to pesticides can be through various active or passive routes. People involved in the manufacturing, transportation, sale and application of pesticides are exposed directly to pesticides. Passive exposure routes include consuming contaminated food, water and inhalation of pesticides aerosols [51]. The contaminated air could also diffuse or drift to nearby areas from the point of release. The mobility depends on the chemicals partitioning with various environmental matrices [9,33]. Besides humans, it was reported that these agrochemicals have a significant effect on lower trophic level organisms. For instance, small concentrations of malathion were suspected of modifying the plankton population composition, consequently affecting tadpoles' growth [70].

Most pesticides are grouped under POPs, soluble in fat, and can accumulate in human tissues and breast milk. Thus many pesticides are banned due to their possible detrimental effects on human health and ecology. Generally, farmers are most vulnerable to its adverse impacts due to continuous exposure. According to a report, most of the pesticide poisoning (73%) in India is contributed by OP compounds [71]. A comparative study of pesticides in adults' blood serum in China, India, America, Slovakia, Mexico, and Saudi Arabia indicated elevated concentrations of various pesticides [71]. Based on epidemiological data, it is also suspected that exposure to pesticides might increase Parkinson's disease risk [72]. Also, most of these pesticides act as endocrine disrupters, affecting the hormone secretions in humans. According to the European Food Safety Authority (EFSA), pesticides can affect thyroid follicular cells with a high-risk factor for hypothyroidism [73]. Chronic effects include genotoxicity, reproductive issues, cancer, and neurological effects [13,74]. Diazinon, an OP pesticide, is known to impart neurotoxicity and mutations in the cell. Similarly, atrazine, an extensively utilized herbicide, is reckoned to be carcinogenic and can also induce muscle degenerations in case of prolonged exposure [75]. Malathion, an insecticide used prominently in vegetable crops, decomposes with temperature to form toxic products.

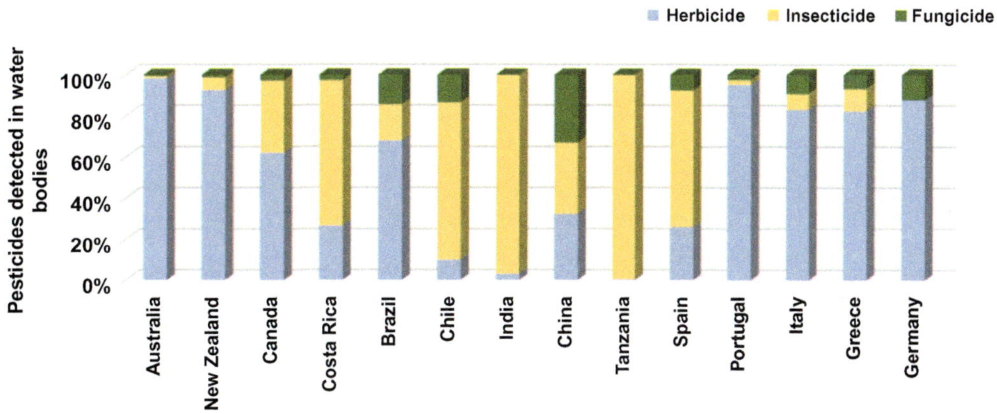

FIG. 4 Percentage of pesticide concentrations detected in various continents during 2012–2020. *M.V. Barbieri, A. Peris, C. Postigo, A. Moya-Garcés, L.S. Monllor-Alcaraz, M. Rambla-Alegre, E. Eljarrat, M. López de Alda, Evaluation of the occurrence and fate of pesticides in a typical Mediterranean delta ecosystem (Ebro River Delta) and risk assessment for aquatic organisms, Environ. Pollut. (2020) 115813; M.E. Close, B. Humphries, G. Northcott, Outcomes of the first combined national survey of pesticides and emerging organic contaminants (EOCs) in groundwater in New Zealand 2018, Sci. Total Environ. 754 (2021), 142005; R.M. de Souza, D. Seibert, H.B. Quesada, F. de Jesus Bassetti, M.R. Fagundes-Klen, R. Bergamasco, Occurrence, impacts and general aspects of pesticides in surface water: A review, Process Saf. Environ. Prot. 135 (2020) 22–37; T. di Lorenzo, M. Cifoni, B. Fiasca, A. di Cioccio, D.M.P. Galassi, Ecological risk assessment of pesticide mixtures in the alluvial aquifers of Central Italy: toward more realistic scenarios for risk mitigation, Sci. Total Environ. 644 (2018) 161–172; R. Elibariki, M.M. Maguta, Status of pesticides pollution in Tanzania – A review, Chemosphere 178 (2017) 154–164; R. Meffe, I. de Bustamante, Emerging organic contaminants in surface water and groundwater: A first overview of the situation in Italy, Sci. Total Environ. 481 (2014) 280–295; D. Pietrzak, J. Kania, G. Malina, E. Kmiecik, K. Wątor, Pesticides from the EU first and second watch lists in the water environment, CLEAN - Soil, Air, Water 47 (7) (2019) 1800376; M.A. Rippy, A. Deletic, J. Black, R. Aryal, J.-L. Lampard, J.Y.-M. Tang, D. McCarthy, P. Kolotelo, J. Sidhu, W. Gernjak, Pesticide occurrence and spatio-temporal variability in urban run-off across Australia, Water Res. 115 (2017) 245–255; N.I. Rousis, R. Bade, L. Bijlsma, E. Zuccato, J. Sancho, v, Hernandez, F. and Castiglioni, S., Monitoring a large number of pesticides and transformation products in water samples from Spain and Italy, Environ. Res. 156 (2017) 31–38; N. Tauchnitz, F. Kurzius, H. Rupp, G. Schmidt, B. Hauser, M. Schrödter, R. Meissner, Assessment of pesticide inputs into surface waters by agricultural and urban sources—A case study in the Querne/ Weida catchment, Central Germany, Environ. Pollut. 267 (2020) 115186; M.S.J. Warne, R.A. Smith, R.D.R. Turner, Analysis of pesticide mixtures discharged to the lagoon of the great barrier reef, Australia, Environ. Pollut. 265 (2020) 114088; Y. Zhou, J. Wu, B. Wang, L. Duan, Y. Zhang, W. Zhao, F. Wang, Q. Sui, Z. Chen, D. Xu, Q. Li, G. Yu, Occurrence, source and ecotoxicological risk assessment of pesticides in surface water of Wujin District (northwest of Taihu Lake), China, Environ. Pollut. 265 (2020) 114953.*

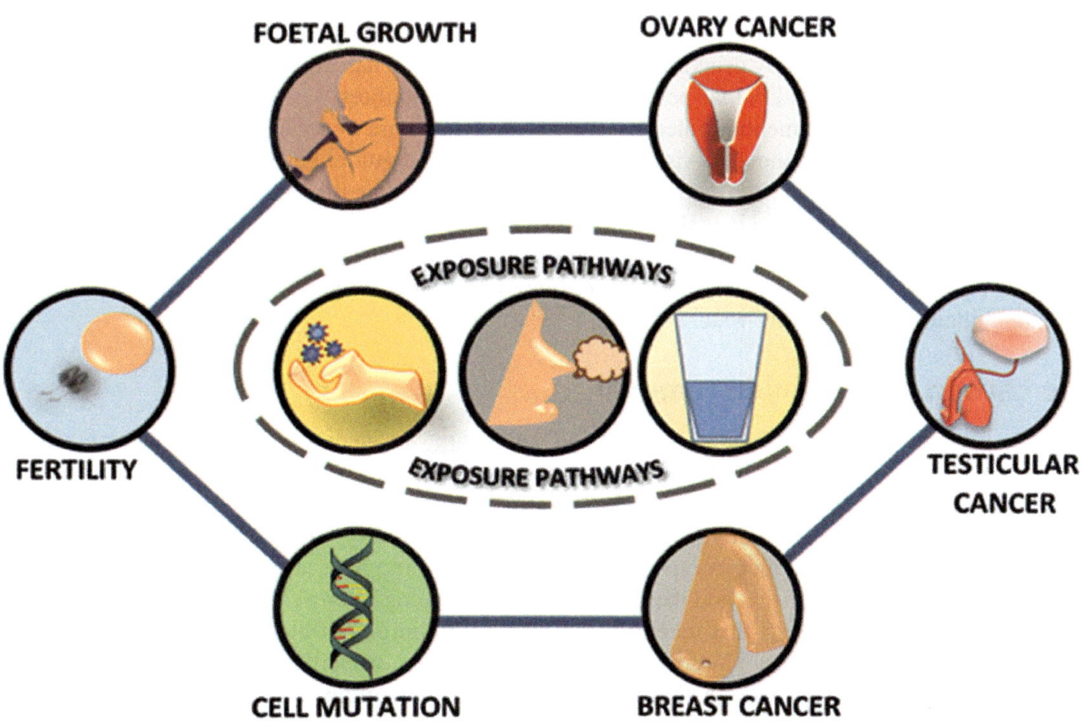

FIG. 5 Exposure pathway and chronic health risk.

TABLE 2 Human health impacts linked to pesticide use.

Pesticide	Health effects	Affected population	References
Carbamate	Lymphoma	Adult	[69]
DDT	Cell mutations	Adult	[71]
Diazinon	Ovarian cancer	Adult	[69]
Glyphosate	Asthma with allergies	Adult	[77]
Malathion	Thyroid cancer	Adult	[78]
Organophosphate	Type 2 diabetes	Adult	[79]
Polychlorinated biphenyl	Breast cancer	Adult	[80]
p,p′-dichlorodiphenyldichloroethylene (DDE)	Testicular germ cell tumor	Adult	[69]
Pyrethroid	Leukemia	Children	[69]
Triazine	Ovary cancer	Adult	[69]
Triazole	Congenital malformation	Infants	[81]

This toxin has a rapid effect on human health and has the affinity to be adsorbed by all organs. Acute exposures may affect the respiratory tract and can cause dizziness and nausea. However, increased exposure can cause a heart attack or even death [76]. Table 2 summarizes various health impacts linked to pesticide usage.

Water quality criteria and standards

Consequences due to the indiscriminate use of pesticides have led to the enactment of regulatory legislation worldwide. The earliest regulation to control the application, consumption, and protect the US environment was set by Federal Insecticide, Fungicide, and Rodenticide Act (FIFRA). During the initial stages, pesticide registration was done based on pesticide effectiveness and their inability to produce undesirable effects [82]. However, the impact of pesticides on biotic systems was not considered in the guideline. Later, based on the pesticide dosage, United States Environmental Protection Agency (USEPA) introduced Maximum Concentration Limits (MCL) for all commonly detected pesticides in drinking water. In European (EU) countries, aquifers were the primary drinking water sources [19]. This has urged the EU legislation to pass the most stringent standards of 0.1 μg/L for individual pesticides and their metabolites with total pesticides concentrations not exceeding 0.5 μg/L. Swedish Environmental Protection Agency and the Swiss Federal office of the environment have also adopted EU standards [75]. Based on oncology risk, the WHO has laid out guidelines for chemicals derived from agricultural activities. These values were set under default assumptions based on a healthy individual (60 kg for adults; 10 kg for children) consuming 1 to 2 L water/day [76]. However, this can vary based on jurisdictions. For instance, the National Health and Medical Research Council (NHMRC) of Australia determine MCL based on health by adopting the average weight of an adult as 70 kg and 2 L consumption of water/day [5]. Table 3 shows the comparison between WHO and USEPA pesticide guidelines for drinking water.

Nanoscale carbon for pesticide removal

Carbon is an attractive material and has a wide variety of applications in various fields, including water purification. Water purification is one of the oldest and most commonly used applications of carbon materials. Carbon has been extensively used as an adsorbent for removing disinfection-byproducts, taste, and odor causing compounds, pesticides, heavy metals, among others due to its abundant availability, cost-effectiveness, high stability, increased surface area, and recyclability [23,83]. Different forms of carbon and their composites are developed to improve adsorption efficacy [84]. Among them, nanoscale carbon such as CNTs, graphene, and their derivatives are gaining interest among the research community. The recent advancement in research suggests that several environmental pollution problems can be resolved or enhanced by using carbon-based nanomaterials. These kinds of materials with exciting properties can transform purification technologies and change our everyday lives. This section presents literature pertaining to the application of nanoscale carbon for the purification of water.

TABLE 3 Guidelines for concentrations of pesticides in drinking water.

Pesticide	WHO Guideline (μg/L)	USEPA (MCL) (μg/L)
Bentazone	25	—
Metolachlor	5	—
Pendimethalin	17	—
Propanil	175	—
Ethylene dibromide	0.4	0.05
Glyphosate	—	700
Pyridate	60	—
Simazine	17	4
Trifluralin	170	—
Aldicarb	10	7
Carbofuran	—	40
Molinate	6	—
Aldrin	0.03	—

Data from D.J. Hamilton, A. Ambrus, R.M. Dieterle, A.S. Felsot, C.A. Harris, P.T. Holland, A. Katayama, N. Kurihara, J. Linders, J. Unsworth, S.-S. Wong, Regulatory limits for pesticide residues in water, Pure Appl. Chem. 75 (8) (2003).

Carbon nanotubes and their derivatives

Since the discovery in 1991, CNTs have found their application in diverse fields such as sensing, catalysis, solid-phase extraction, and water purification. CNTs are synthesized using various techniques, including arc discharge, chemical vapor deposition (CVD), laser ablation, and pyrolysis. This carbon allotrope can be viewed as a rolled form of graphitic sheets with a diameter extending to nanometers and length to centimeters [85]. They are classified based on the number of tubes as single-walled carbon nanotubes (SWCNT) and multiwalled carbon nanotubes (MWCNT). In view of their unique structural properties, they possess strong adsorption capabilities in removing organic contaminants in water through various complex mechanisms [86]. The adsorptive removal of pesticides varies based on the type of CNTs used and their morphology. However, based on Brooks et al. [87] observations, pesticide uptake onto CNT is not restricted to the surface area alone but is also linked to parameters such as metal impurities, carbon content, and pore volume of the CNT bundles. For instance, Dichiara et al. [88] demonstrated a real-time lab-scale column study on removing pesticide diquat dibromide using activated carbon (AC), MWCNT, SWCNT, and SWCNT functionalized with hydroxyl groups (SWCNT-OH). The efficacy of the system was studied by maintaining the flow rate, column dimensions, and varying the feed concentration (C_0) of diquat dibromide. CNT was reported to exhibit an increased affinity to the pesticide than AC due to their open pore structure, which provided easy access to the adsorbate molecules. At higher C_0, an increase in feed molecules led to a faster breakthrough due to quick saturation of active sites in the adsorbent. The addition of hydroxyl functional groups improved the adsorption capacity due to the additional binding property of diquat dibromide to adsorbent by electrostatic interactions. Hence, the uptake capacity followed the order: SWCNT-OH > SWCNT > MWCNT, with SWCNT demonstrating high performance due to their increased surface area.

de Martino et al. [89] investigated 2-methyl-4-chlorophenoxyacetic acid (MCPA) extraction from water using SWCNT and MWCNTs of different diameters. Experiments were conducted at room temperature with an initial concentration of MCPA varying from 2.5–100 mg/L and pH 5–3.2. Based on the results, SWCNT reported higher removal efficiency than MWCNT, with a regeneration efficiency of more than 90% even after 5 cycles. Owing to the acid dissociation constant, pKa (3.07) value of MCPA, and structural properties of CNT, the adsorption mechanism followed a combination of π-π electron-donor interaction and Van der Waals forces of attraction. Also, the recovery of SWCNT using ethanol showed a desorption efficiency of 92%, indicating the potential of SWCNT as an effective media for pesticide removal. Nevertheless, in a batch adsorption study of 4-chloro-2-nitrophenol onto SWCNT and MWCNT with varying initial concentrations from 2–10 mg/L (pH 2–12 and 25°C), MWCNT portrayed higher adsorption efficiency than SWCNTs ranging from 56.96% to

81.28%. This was due to more active sites and interstitial groove spaces in MWCNT with the adsorption mechanism controlled by electron-donor acceptor (EDA) interactions and hydrogen (H) bonding. Furthermore, the influence of adsorbent dosage (0.005–0.5 g) and pH on the system was also investigated. It was observed that the adsorption capacity increased with dosage till 0.1 g due to an increase in available adsorption sites and later remained constant on attaining equilibrium. Also, at $pH < pK_a$, adsorption was considered favorable. At $pH > pK_a$, electrostatic repulsions between phenolate ions and negatively charged CNT lowered the adsorption efficiency [90]. In a study conducted by Dehghani et al. [76], MWCNT showed effective removal of malathion from aqueous solutions at an initial concentration of 6 mg/L within 30min.

To improve the conjugation of CNT with the target pollutant, covalent functionalization (—COOH, —OH, —C=O) and noncovalent functionalization are carried out. Fig. 6 represents some of the CNT derivatives used in pesticide removal. Chemical functionalization of CNTs is carried out using strong acids, incorporating oxygen-rich functional groups onto intrinsic defects. Deng et al. [91] reported increased surface area and pore volume on oxidized MWCNT, enhancing the adsorptive removal of diuron from aqueous solutions. Based on batch studies, it was observed that the interaction of herbicide with the adsorbent was pH-dependent, with higher adsorption toward the neutral and basic conditions. In addition, the interaction of Cu^{2+} ions and the adsorbent was investigated due to their coexistence. The presence of Cu^{2+} ions decreased the adsorption capacity of oxidized MWCNT due to cross-linking effect, thereby decreasing the pore size and restricting the entry of diuron.

Similarly, diquat dibromide herbicide was extracted using oxygen-containing MWCNT (OMWCNT) and its derivatives, OMWCNT-Fe_3O_4, OMWCNT-k-, and carrageenan-Fe_3O_4 with the magnetic property. The experiments were carried out using a varying concentration of the pollutant, maintaining a pH of 6.5 at 25°C. During the initial stages, the adsorption process was rapid due to vacant available sites, which gradually decreased. Based on the studies, OMWCNT was reported as a better adsorbent. The reduced adsorption in magnetized adsorbents can be attributed to filling active sites by Fe_3O_4 particles, thereby decreasing surface area and the hindrance in π-π interaction. The adsorption process was reported to be spontaneous and endothermic, with the OMWCNT system exhibiting a Langmuir adsorption capacity of 58.3 mg/g, with no significant decrease in capacity during regeneration [92]. Comparably, the effect of lead on MWCNT and OMWCNTs with

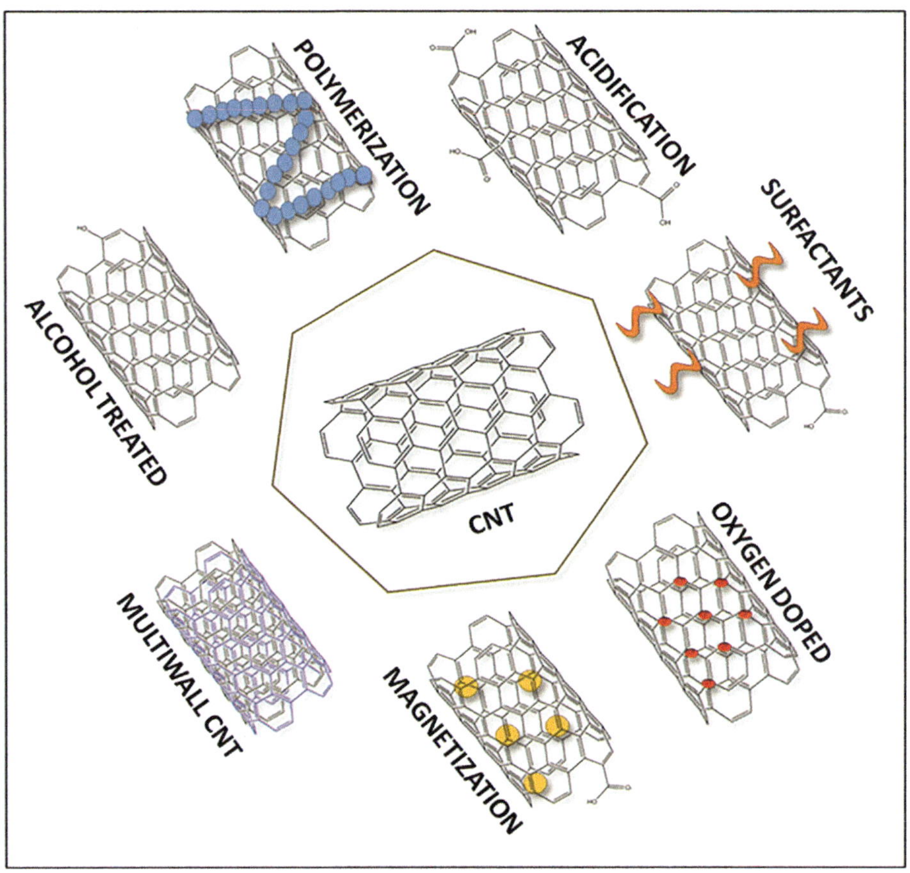

FIG. 6 Summary of various CNTs used in pesticide removal from water.

varying oxygen content was studied by Chen et al. [93] for the adsorption of diuron and dichlobenil from the water. The batch system was maintained at room temperature and pH adjusted to 6, with the concentrations of diuron and dichlobenil ranging from 0.6 to 22 mg/L and 0.6 to 13 mg/L, respectively. They found a positive correlation between the pesticides adsorption capacity and the specific surface area of MWCNTs. As the oxygen content increased, adsorption capacity decreased due to the removal of a proton from carboxylic groups. Besides, the presence of lead suppressed the adsorption of organic pollutants by 10% to 30%, probably due to the hydration and complexation of lead with -COOH group of OMWCNTs. In another study by Ateia and coworkers, atrazine adsorption capacity decreased from 57.8 to 27.6 mg/g while using ozone-assisted regeneration of MWCNT. The decrease in the capacity was due to the formation of ozonation by-products occupying the adsorbent sites. Generally, noncovalent functionalization is carried out by wrapping the CNT using surfactants or polymers. Saleh and coworkers [94] evaluated the performance of polymer grafted CNT for the adsorptive removal of phenol. The batch study was performed at an adsorbent dosage of 250 mg. The experiments were performed by varying the initial concentrations of the pollutants (20–400 mg/L), temperature (20°C–50°C), and pH (2–8). The nanocomposite showed adequate recyclability with a sorption efficiency of 62% after 4 cycles. The system was well described by pseudo-second-order kinetic with a Langmuir adsorption capacity of 261.6 mg/g. A decrease in spontaneity was observed with an increase in temperature. Based on the data, the solution pH and dose of the system played a significant role in the extraction of phenol.

Generally, CNTs show a strong affinity to nonpolar organic compounds. Chen et al. [95] investigated the affinity of CNTs to organic compounds having various properties. It was reported that the adsorption affinity correlated poorly with hydrophobicity but followed the order of nonpolar aliphatic < nonpolar aromatics < nitroaromatics. The authors attributed this trend to the lone pair of electrons on amino groups, which accelerates the adsorption capacity between nitroaromatics and adsorbate, thereby enhancing π-π EDA interaction. In another study, Lin and Xing [96] reported a direct correlation between the number of aromatic rings on polar compounds such as cyclohexanol, phenol, catechol, pyrogallol, 2-phenylphenol, 1-naphthol, and naphthalene and the adsorption affinity of CNTs. The authors also investigated the role of phenol substitution in the polar aromatics-CNT system. They found that the adsorption efficiency of CNTs increased with the number of aromatic rings in the order of cyclohexanol < phenol < 2-phenylphenol < 1-naphthol. Also, the sorption capacity was greatly enhanced with the addition of hydroxyl groups with hydrophobic, electrostatic, H-bonding and π-π as the possible interaction mechanisms. Table 4 summarizes the various CNT used in pesticide removal with their removal efficiency.

CNTs pose a significant challenge in their commercial production due to difficulty in scaling, low yield, and high cost. Deokar et al. [97] synthesized CNT from plastic waste and used it for removing diuron from aqueous solutions. The system exhibited an adsorption capacity of 40.37 mg/g at 303 K, with the experimental data best fitting the Hill Isotherm model. Regardless of its comprehensive favorable properties, CNT utilization is not widely accepted due to its interaction with abiotic and biotic components throughout their lifecycle. Due to its increased cytotoxicity and resistance to degradation, these nanomaterials can be transported easily and can induce harmful effects for animals and humans [105,106]. Also, the initial cost during the production of CNT makes it undesirable for its large-scale use.

Graphene and its derivatives

Due to the constraints posed by CNTs, using them in water purification was often considered challenging. Graphene, sp^2 hybridized carbon allotrope, has received increased attention since its discovery in 2004. Due to its exceptional optical, thermal, and mechanical properties, graphene has many medical and industrial applications [107,108]. Graphene is also used in water treatment owing to its chemical inertness and increased stability. In addition to its excellent specific surface area, the presence of delocalized π-electrons makes it a suitable adsorbent for removing various SOCs from water. For instance, a study investigated the performance of graphene nanosheets (GNS) on the removal of frequently detected pesticides, polyaromatic pesticides (PAHs), from aqueous solutions by choosing naphthalene as a representative adsorbate. Kinetic studies were conducted for naphthalene at an initial concentration of 4.6 mg/L and an adsorbent dosage of 25 mg/L. The adsorption process was observed to be rapid for the first 1 h, reaching equilibrium in 4 h with hydrophobic and π-π interactions governing the process. Based on SEM analysis, it was observed that at lower concentrations of PAH, the graphene nanosheets crumbled, creating wrinkles and grooves, which facilitated the binding of PAH molecules due to their high surface energy. At higher concentrations of the PAH, the formation of wrinkles in GNS became pronounced, which led to stacking of GNS, thereby decreasing the partitioning of adsorbate to the adsorbent [109]. Various modifications such as oxidation, reduction, magnetization, incorporation of nanomaterials, and polymerization have been carried out to improve graphenes stability and dispersibility. These modifications are summarized in Fig. 7.

The earliest graphene synthesis was achieved by mechanical exfoliation and later by CVD. However, the poor yield restricted its usage to laboratory-scale applications. The bulk production of graphene was achieved by chemically oxidizing

TABLE 4 Efficiency of various CNT derivatives used in pesticide removal.

Material	Pesticide	Concentration (μg/L)	Removal Mechanism	Adsorption Capacity (mg/g)/Removal Efficiency (%)	References
MWCNT (in the presence of Pb^{2+})	Diuron and Dichlobenil	60–22,000	H-bonding	23.43–182.72 23.46–96.3	[93]
CNT from plastic waste	Diuron	20,000	Not specified	greater than 70%	[97]
MWCNT Oxidized MWCNT		1130–9070	Hydrophobic interactions,	42.6 49.2	[91]
SWCNT	Atrazine	100	Hydrophobic interactions	4.97 (μg/g) (m^3/mg)	[87]
OMWCNT (0.85%, 2.16%, 7.07%) (In the presence of Cu^{2+}, Cd^{2+} and Pb^{2+})		500–20,000 pH 2–7	Hydrophobic interactions	47.6 (0.85%), 36.1 (2.16%), 20.1 (7.07%).	[98]
Magnetic MWCNT (in the presence of Cu^{2+})		5000	Van der Waals forces, H- bonds, Hydrophobic interactions	42	[99]
MWCNT regenerated with ozone treatment		1000–30,000	π-π interactions	85%–93%	[100]
MWCNT	Isoproturon	Not specified	Electrostatic interactions, H-bonding	16.3	[101]
MWCNT	Malathion	6000	Not specified	100%	[76]
MWCNT	Fenuron	100	π-σ, π-π stacked, Hydrophobic interactions (π-alkyl type)	90%	[102]
OMWCNT	Diquat dibromide	5430–1630	π-π bonding, Electrostatic	~90%	[92]
SWCNT	Diuron Fluridone Norflurazon	100–32,000 10–8000 400–16,000	H-bonding	122.2 126.81 332.02	[103]
MWCNT	Diuron Fluridone Norflurazon	100–32,000 10–8000 400–16,000	H-bonding	27.93–126.82 38.10–183.05 45.96–122.82	[104]

graphite to graphene oxide (GO) [110,111]. The epoxy, hydroxyl, and carboxyl groups present on the chemically exfoliated graphene render a promising material for contaminant remediation. In a study, Catherine et al. [104] investigated various phenolic compounds' adsorption behavior on GO nanoflakes synthesized via the Hummers method. The kinetic studies were performed in centrifuging vials containing 5 mg of GO and 20 mg/L of adsorbate at 150 rpm. The time dependency was also examined, indicating an increasing adsorption rate during the initial 30 min, gradually attaining saturation at 480 min. The isotherm data fitted well for both Freundlich and Langmuir models with the compounds' adsorption capacities varying between 19–30 mg/g. Among the phenolic compounds, phenol portrayed a higher affinity to GO due to H-bonding with oxygen functional groups. The study also reported the exceptional stability of GO in the presence of electrolytes NaCl, $MgCl_2$, and $AlCl_3$.

Generally, GO is converted to reduced graphene oxide (rGO) at higher temperatures. Alternatively, the conversion to rGO can also occur through chemical, hydrothermal or microwave (MW) treatment. Wang and Chen [112] investigated the

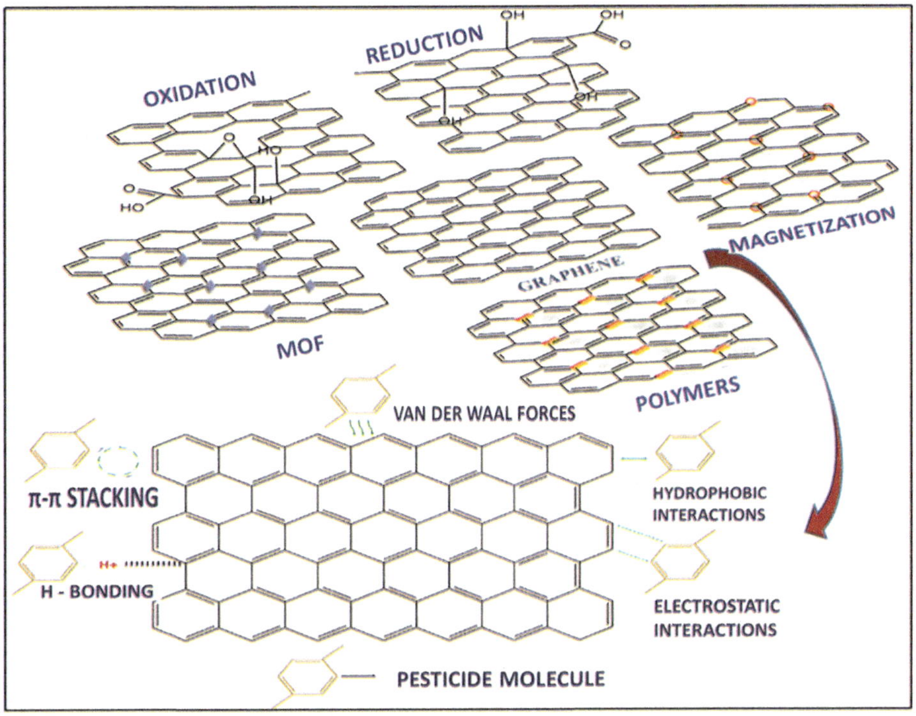

FIG. 7 Graphenic derivatives and possible mechanisms of pesticide removal.

adsorptive and co-adsorptive properties of naphthalene and 1-naphthol onto GO, chemically reduced graphene (CRG), and annealed reduced graphene (ARG) in the presence of cadmium. Single-component adsorption studies were carried out by varying initial concentrations of naphthalene (0.2–20 mg/L) and 1-naphthol (1–50 mg/L), adsorbent dosage (5–200 mg/L for GO; 0.2–10 mg/L for CRG and ARG), and pH (2 − 12). Despite low hydrophobicity, removal of 1-naphthol was significantly more than naphthalene with an adsorption capacity of 282 mg/g at pH 5. This increased removal of 1-naphthol attributed to the presence of hydroxyl groups on the adsorbate surface, imparting significant electronic properties to the material with the sorption performance following the order: CRG > ARG > GO. Despite CRG showing high specific surface area, the affinity of pollutants to adsorbent was not exclusively based on the surface area. In addition, co-adsorption of heavy metals with organic pollutants was investigated. The presence of Cd^{2+} was reported to interact differently with polar and nonpolar aromatics onto graphene-type materials. Decreased uptake of 1-naphthol by ARG was observed due to the filling of active sites by cadmium and facilitating cation interaction thereby restricting H-bonding. Contrastingly, the increased affinity of cadmium ions to GO neutralized the charges by the compression of double-layer thereby enhancing the adsorption capacity of 1-naphthol by 120%. Furthermore, the pH dependency in the decontamination of 1-naphthol using CRG and ARG was also investigated. The adsorption efficiency was found to increase with pH and reached a maximum at $pH \sim pK_a$ as a result of n-π interactions. In another study, rGO showed exceptional ability to remove chlorpyrifos (CP), endosulfan (ES), and malathion in water. At an initial concentration of 2 mg/L, the materials exhibited 800–1200 mg/g adsorption capacity. The study also probed the interference of ions during the adsorption process. The absence of interference and pH independency indicated its potential in the purification of water, as shown in Fig. 8A and B. Furthermore, density functional theory calculations for various configurations of water, pesticide, and graphene were carried out to elucidate the entropic effects. The quantified energies demonstrated the influence of water in the formation of the pesticide-graphene complex [113]. Yu et al. [114] demonstrated the suitability of rGO in the simultaneous removal of phenol and naphthol in a bicomponent system. Batch studies were conducted at ambient conditions with an adsorbent dosage of 0.1 g/L and the adsorbate concentration of 20 mg/L for a duration of 48 h. Kinetic experiments revealed that the rate of adsorption of phenol and naphthol was rapid for the initial 4 h and gradually attained saturation after 6 h. The influence of pH on adsorption capacity was also studied. At this point, adsorption capacity decreased due to repulsion between negatively charged rGO and phenolate ions. The increase in adsorption capacity until $pH < pH_{PHZ}$ can be as a result of π-π interaction mechanisms. The adsorption system showed the Langmuir model as a better model to predict the experimental data with an adsorption capacity of naphthol and phenol as 2.3 and 2.7 mmol/g, respectively. Competitive removal of pollutants onto rGO was also investigated by varying the pH and time. However, the maximum values of phenol and naphthol (1.5 mmol/g for

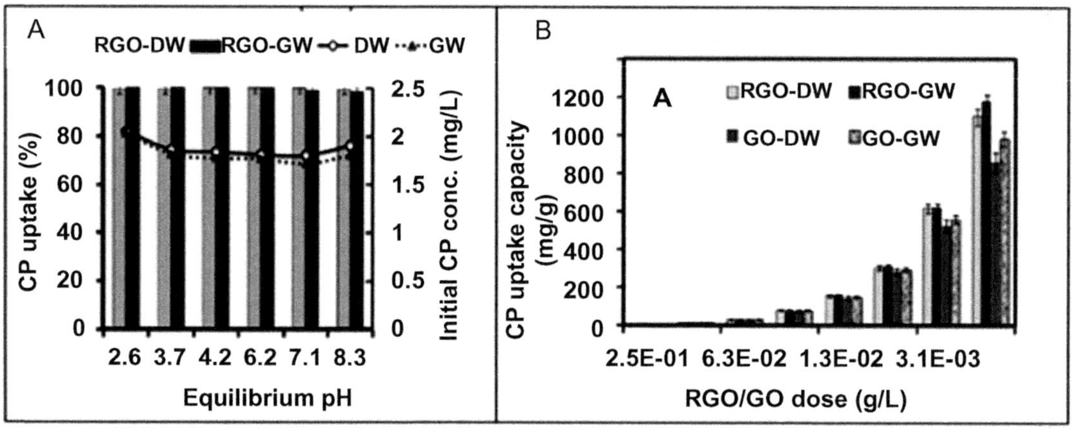

FIG. 8 (A) Effect of pH on adsorption of chlorpyrifos on rGO (B) Effect of rGO/GO dose on the adsorption of chlorpyrifos. *Reproduced with the permission from S.M. Maliyekkal, T.S. Sreeprasad, D. Krishnan, S. Kouser, A.K. Mishra, U.V. Waghmare, T. Pradeep, Graphene: a reusable substrate for unprecedented adsorption of pesticides, Small 9 (2) (2013) 273–283.*

phenol and 1.6 mmol/g for naphthol) were higher than the single-component system suggesting that the adsorption capacities of rGO elevated during the coexistence of pollutants. The hydroxyl groups on naphthol preferentially got adsorbed to rGO leaving the hydrophobic face toward the aqueous media. As a result, this hydrophobic face act as a new active site for phenol adsorption, thereby accelerating the adsorption performance.

As the separation of graphene and its derivatives from water poses a challenge, magnetization and functionalization are carried out to control the movement of nanoparticles in the fluid medium. In a study, magnetically recoverable Fe_3O_4/rGO synthesized using graphite powder and Fe_3O_4 nanopowder as a precursor exhibited a remarkable affinity for the triazine class of pesticides. Batch studies were carried out for varying initial concentrations (2–54 mg/L), pH (3 – 11) and temperature (20°C–35°C). The removal efficiency followed: 93.61%, 91.34%, 88.55%, 81.22%, and 75.24% in the order prometryn > simazine > ametryn > atrazine > simeton, respectively for an initial concentration of 10 mg/L, pH 5 and 25°C. The difference in adsorption affinity was due to the differing S, O, Cl functional groups within the compounds, which influenced the van der Waal forces of attraction and π bonding network. Furthermore, the regeneration studies resulted only in a 0.05% decrease in adsorption capacity following 7 cycles, which validate iron oxide decorated rGO as a potential adsorbent for pesticide removal [115]. A study showed that the adsorption capacity of Fe_3O_4/rGO positively correlated with increasing concentration of glyphosate herbicides. The experimental data well-fitted the Langmuir isotherm model with a maximum adsorption capacity of 70.08 mg/g, with glyphosate linking to the adsorbent via H-bonding and electrostatic interactions. However, the presence of competing ions decreased the adsorption efficiency in the order $PO_4^{3-} < H_2PO_4^{-} < SO_4^{2-} < NO_3$-$^{-} < Cl^{-} < Br^{-}$ due to the complexation of adsorbent with phosphate ions, thereby reducing the available sites for adsorption [116]. Similarly, adsorption of isomeric pesticides endrin and dieldrin was effectively carried out on Fe_3O_4/GO with a removal efficiency of 86% and 74%, respectively, for a contact time of 30 min and an initial concentration of 5 mg/L. The removal rate was high for the acidic condition due to H-bonding between the adsorbate and the hydrophilic groups on the surface of GO along with π-π stacking [117].

Functionalization of GO is employed through covalent (-COOH, -OH, epoxy), noncovalent bonding or via doping to improve its stability and reduce agglomeration in the aquatic matrix. Interestingly, the incorporation of Ag-nanoparticles (Ag-NPs) onto the surface of rGO via oxidation–reduction was reported to be effective in the degradation of halogenated pesticides. The removal followed a two-step mechanism with the degradation of lindane to trichloro-benzene by Ag-NPs followed by adsorption. This isomeric conversion displayed a removal efficiency of >99.9% in 15 min for an initial lindane concentration of 2 mg/L. Moreover, the influence of the common ion effect was examined by the addition of NaCl solution into the adsorption system [118]. A similar mechanism was exhibited by silver nanoparticles immobilized on rGO in the degradation of OC pesticides. The adsorption studies were done in batch reactors in ambient conditions. In this study, ES pesticide was reported to exhibit better removal efficiency when compared with CP due to the interaction between silver ions and chloride ions on ES. The exceptional adsorption capacity of 1534 mg/g of the nanocomposite and its regeneration capability renders it an excellent material for water treatment technologies [119].

In recent years, metal–organic frameworks (MOFs) have gained wide popularity due to their diverse use and ability to integrate with carbon-based materials. Insecticides such as OC, OP, and neonicotinoids are commonly used in developing countries for pest control. In a study by Yang et al. [120], MOFs were used to remove glyphosate. The adsorption studies

TABLE 5 Efficiency of various graphene derivatives used in pesticide removal.

Material	Pesticide	Initial Concentration (mg/L)	Removal Mechanism	Adsorption Capacity (mg/g)/ Removal Efficiency (%)	References
Fe_3O_4/rGO	2,4-D	50–750	Electrostatic, Hydrophobic, π-π effect	67.26	[121]
Magnetic graphene oxide–β cyclodextrin MOF	Neonicotinoid	0.5–100	H- bonding, Hydrophobic, Electrostatic, π-π stacking	14.4%	[122]
Nitrogen doped rGO	Phenol and p-Nitrophenol	20–200	Electrostatic, Hydrophobic, π-π interactions	93%–97%	[123]
GO and rGO	Chlorpyrifos, Endosulfan and Malathion	2	Electrostatic	>100%	[113]
Graphene Quantum Dots (GQD)	Oxamyl	50–150	Electrostatic interactions	91.10–71.10	[124]
GQD + citrate and dodecyl sulfate	2,4,6-Trichloro phenol	65	Hydrophobic, Hydrogen bond and π–π interaction	119	[125]
KOH activated graphene nanosheet	1-Naphthylamine and Phenanthrene	0.2–20 and 0.02–2	π–π interaction	180.7 and 317	[126]

were carried out with 40mg/L glyphosate concentration and 20mg/L adsorbent dosage at ambient conditions. The adsorption followed Langmuir model with an adsorption capacity of 482.2 mg/g in acidic pH due to the interaction of phosphate ions with the oxygen group of adsorbent UiO-67/GO. This hybrid nanocomposite was reported to possess abundant Zr-OH groups on the surface of GO which enabled excellent adsorption capability. Table 5 summarizes various graphene derivatives used in pesticide removal. Despite the key attributes displayed by graphene and its derivatives, an enormous quantity of water is required in the production, which pose difficulty in countries facing water shortage. However, recent researches have been inclined toward the green synthesis of graphene, which projects its potential for impressive growth in future.

Supported nanocarbon materials

Despite the exciting properties showcased by CNMs, there are several concerns regarding the fate and transport of these materials within the environment and their interaction with organic matter. The difficulty in solid–liquid separation and its susceptibility to leaching limits their direct use in water purification. Moreover, nanomaterials increased cost and ion-induced agglomeration can also hamper their practical utility. These limitations can be addressed by immobilizing the nanoparticles on suitable support matrices such as sand, fibers, polymers, and stainless steel.

Graphene sand composite (GSC) was synthesized using cane sugar as the sole carbon source by combustion as follows: 100°C for 30min, 100°C–200°C for 30min, 200°C for 1 h to melt the sugar, 750°C for 3 h to ensure complete graphitization. Beyond 800°C, loss in adhesive properties with the sand surface was observed. Batch studies were carried out with an adsorbent dosage of 100mg and a working volume of 10 mL to check the adsorptive efficiency of the synthesized material. The material showed good affinity to CP (initial concentration: 3 mg/L) with an adsorption capacity of

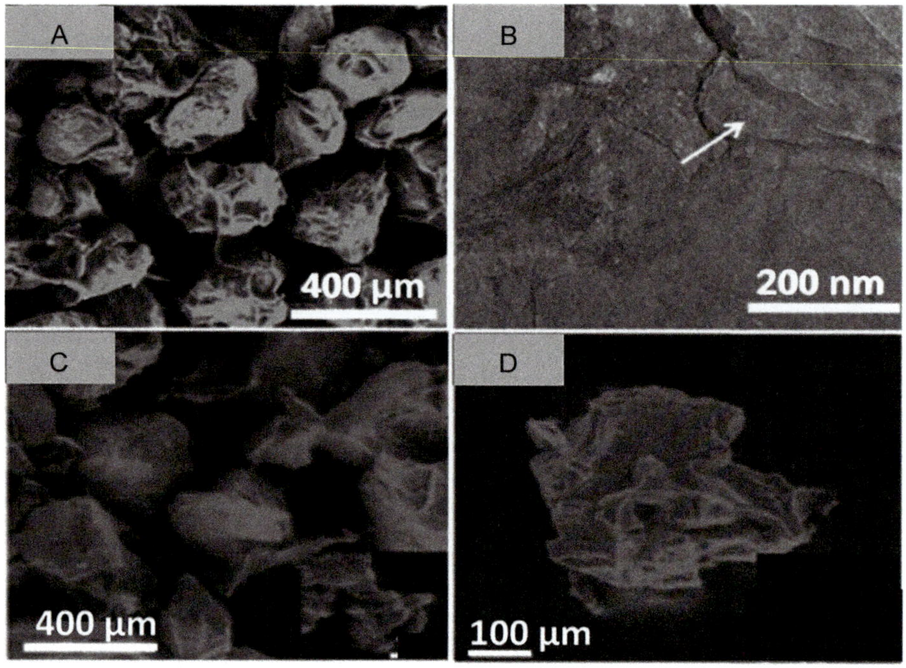

FIG. 9 (A) SEM image of graphene sand composite (B) HRTEM image of extracted graphene sheets (C) and (D) SEM images of graphene sand composite from asphalt (5% loading). *Reproduced with the permission from S.S. Gupta, T.S. Sreeprasad, S.M. Maliyekkal, S.K. Das, T. Pradeep, Graphene from sugar and its application in water purification, ACS Appl. Mater. Interfaces 4 (8) (2012) 4156–4163; T.S. Sreeprasad, S. Gupta, S.M. Maliyekkal, T. Pradeep, Immobilized graphene-based composite from asphalt: facile synthesis and application in water purification, J. Hazard. Mater. 246–247 (2013) 213–220.*

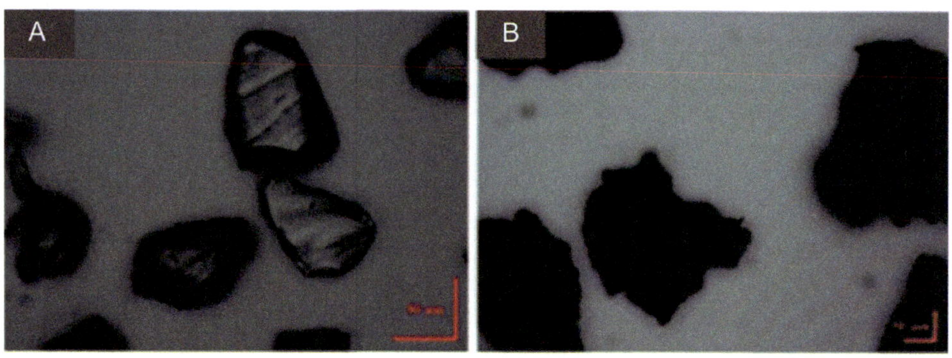

FIG. 10 Photograph of (A) Bare silica (B) Graphene coated silica. *Reproduced with the permission from X. Liu, H. Zhang, Y. Ma, X. Wu, L. Meng, Y. Guo, G. Yu, Y. Liu, Graphene-coated silica as a highly efficient sorbent for residual organophosphorus pesticides in water, J. Mater. Chem. A 1 (5) (2013).*

48 mg/g. Fig. 9A and B depict the scanning electron microscope (SEM) image of GSC composite and high-resolution transmission electron microscope (HRTEM) image of extracted graphene sheets [127]. In another study, GSC was synthesized using asphalt as the carbon source, immobilized on river sand. The carbon loading on the sand particles was varied by changing the proportion of asphalt content. After several iterations, 400°C was chosen as the optimized temperature to get a stable carbon coating onto the sand surface. The morphology of the structure is shown in Fig. 9C and D. The suitability of this material as an adsorbent was determined by the adsorptive removal of CP (4 mg/L) in a batch reactor. Complete removal was observed for an adsorbent dosage of 50 g/L [128]. Graphene coated on silica substrate was synthesized by reacting completely dispersed GO with acid-treated silica, as shown in Fig. 10A and B. The stable composite exhibited increased removal in the order: profenofos > fenthion > pirimiphos-methyl > parathion-methyl > phorate > chlorfenvinphos > methidathion > malathion > isocarbophos > phonamiphos > dimethoate. The adsorption mechanism was influenced by the electron-donating abilities of sulfur, phosphorous and nitrogen atoms in the adsorbate. The effects of dosage (90–170 mg), pH (3–11), and concentration (1–40 mg/L) were also studied using chlorfenvinphos as the model pesticide. The composite

achieved more than 95% removal efficiency at an adsorbent dosage of 7.5 g/L with no significant dependency on pH [129]. In another study, stainless steel wire coated with GO and OH terminated polydimethylsiloxane expressed affinity in removing triazole fungicides with the potential to be reused after aging [130].

Conclusions

The scarcity of clean and affordable water due to global climate change and expanding anthropogenic activities has been a massive area of concern in the past few decades. Pesticides play an inevitable role in the agricultural domain by minimizing pest-related losses. According to estimates, about 30% of crop loss globally is due to pests, and hence their usage is an integral part of pest management to ensure food security. The potential buyers of pesticides are not limited to the agricultural sector alone. It is also used extensively in industrial facilities, highways, and railroads to control the vegetation and grass. Pesticide usage and production have increased manifold over the past decades due to its high demand. Globally, the pesticide market is dominated by Western Europe and the United States (22.7%), and India stands 12th among the south Asian countries.

Pesticides can adsorb, leach, infiltrate, or volatilize and coexist in different matrices based on their physicochemical properties. Their presence in water bodies in pristine form or as a derived product is reported worldwide. Among various water pollutants, pesticides are of significant importance owing to their widespread presence and toxicity. It is proven that these pesticides can pose risks to ecosystems and have devastating effects on human health, plants, and mammals. Several studies have linked the exposure of pesticides to serious health risks, including cancerous and noncancerous diseases. Hence, treating drinking water is necessary to remove dissolved pesticides and meet the citizens' safe drinking water needs. Several treatment technologies exist to address the problem, including adsorption, membrane filtration, and AOPs. However, relatively high capital, operation, and maintenance costs, inability to comply with the stringent regulatory standards, and prolonged treatment time limit their use in many practical situations. Carbon is an attractive material and has a wide variety of applications in various fields, including water purification, due to its abundant availability, chemical inertness, high surface area, and cost-effectiveness. Nanoscale carbons such as nanotubes, graphene, and their composites are promising candidates for purifying pesticides contaminated water due to their excellent adsorption capability and stability. However, due to their nanosize, they are highly susceptible to leaching and may interact with various environmental components causing detrimental effects. Besides, the production cost and low yield limit their use for mass applications such as water purification. Hence, methods to produce nanoscale carbon materials are crucial to realize their application in water purification. Also, immobilizing them on suitable matrices is essential for successfully implementing the material in the field.

Though the removal of pesticides using advanced materials is an attractive solution for treating water, there is a need for a significant reduction in toxic pesticide usage without compromising the growing populations' food demands to address the problem effectively. Using biopesticides as an alternative to conventional pesticides is a sustainable solution. However, higher costs limit their usage in developing countries. Policy interventions to regulate the use of synthetic pesticides by imposing a tax on agrochemicals, encouraging crop rotation, or conducting awareness campaigns are needed to control pesticide use.

Acknowledgment

Authors gratefully acknowledge the Government of India, Ministry of Science and Technology, Department of Science and Technology (DST), Technology Mission Division [Grant no: DST/TM/WTI/WIC/2K17/82(C)] for supporting the work. All the authors also thank IIT Tirupati for the support.

Abbreviations

AC activated carbon
AOPs advanced oxidation processes
ARG annealing reduced graphene
CNM carbon nanomaterial
CP chlorpyrifos
CRG chemically reduced graphene
CVD chemical vapor deposition
DDE p,p'-dichlorodiphenyldichloroethylene

DDT	dichloro-diphenyl-trichloroethane
2,4-D	2,4-dichlorophenoxyacetic acid
EDA	electron donor and acceptor
EFSA	European Food Safety Authority
EPA	Environmental Protection Agency
ES	endosulfan
EU	European Union
FIFRA	Federal Insecticide, Fungicide, and Rodenticide Act
GC-ECD	gas chromatography-electron capture detector
GNS	graphene nanosheet
GO	graphene oxide
GSC	graphene sand composite
H	hydrogen
HDA	hydrogen electron donor acceptor
HRTEM	high-resolution transmission electron microscope
IPM	integrated pest management
MCL	maximum concentration limit
MCPA	2-methyl-4-chlorophenoxyacetic acid
MOFs	metal organic frameworks
MW	microwave
MWCNT	multiwalled carbon nanotube
NHMRC	National Health and Medical Research Council
NP	nanoparticle
OC	organochlorine
OP	organophosphorous
PAH	poly aromatic hydrocarbon
POP	persistent organic pollutants
rGO	reduced graphene oxide
SEM	scanning electron microscope
SOCs	synthetic organic chemicals
SWCNT	single-walled carbon nanotube
UNEP	United Nations Environment Programme
WHO	World Health Organization

References

[1] R.L. Rose, E. Hodgson, R.M. Roe, Chapter 28—Pesticides, Toxicology (1999) 663–697. https://doi.org/10.1016/B978-012473270-4/50087-0ff.

[2] M.M. Amr, Pesticide monitoring and its health problems in Egypt, a third world country, Toxicol. Lett. 107 (1–3) (1999) 1–13, https://doi.org/10.1016/S0378-4274(99)00026-0.

[3] Taylor, L. Eric, A.G. Holley, M. Kirk, Pesticide Development, 2007. https://www.e-education.psu.edu/.

[4] C. Olisah, O.O. Okoh, A.I. Okoh, Global evolution of organochlorine pesticides research in biological and environmental matrices from 1992 to 2018: A bibliometric approach, Emerg. Contam. 5 (2019) 157–167, https://doi.org/10.1016/j.emcon.2019.05.001.

[5] I. Newton, Chapter 15—pesticides and pollutants, in: Population limitation in birds, 1998, pp. 407–447. https://doi.org/10.1016/B978-012517365-0/50015-6. https://www.nhmrc.gov.au/health-advice/guidelines. Last accessed on 20-12-2020.

[6] J.R. Plimmer, Chemistry of pesticides, in: Handbook of Pesticide Toxicology, 2001, pp. 95–107, https://doi.org/10.1016/B978-012426260-7.50005-7.

[7] I.C. Yadav, N.L. Devi, J.H. Syed, Z. Cheng, J. Li, G. Zhang, K.C. Jones, Current status of persistent organic pesticides residues in air, water, and soil, and their possible effect on neighboring countries: A comprehensive review of India, Sci. Total Environ. 511 (2015) 123–137, https://doi.org/10.1016/j.scitotenv.2014.12.041.

[8] W. Aktar, D. Sengupta, A. Chowdhury, Impact of pesticides use in agriculture: their benefits and hazards, Interdiscip. Toxicol. 2 (1) (2009) 1–12, https://doi.org/10.2478/v10102-009-0001-7.

[9] V. Ochoa, B. Maestroni, Chapter 9—pesticides in water, soil, and sediments, in: Integrated Analytical Approaches for Pesticide Management, 2018, pp. 133–147, https://doi.org/10.1016/B978-0-12-816155-5.00009-9.

[10] L. Aicher, M.F. Wilks, From risk assessment to regulation, in: Exposure and Risk Assessment of Pesticide Use in Agriculture, 2021, pp. 3–23, https://doi.org/10.1016/B978-0-12-812466-6.00012-9.

[11] V. Novotny, Diffuse pollution from agriculture — A worldwide outlook, Water Sci. Technol. 39 (3) (1999) 1–13, https://doi.org/10.1016/S0273-1223(99)00027-X.

[12] L. Rani, K. Thapa, N. Kanojia, N. Sharma, S. Singh, A.S. Grewal, A.L. Srivastav, J. Kaushal, An extensive review on the consequences of chemical pesticides on human health and environment, J. Clean. Prod. 283 (2020) 124657. https://doi.org/10.1016/j.jclepro.2020.124657.

[13] R.M. de Souza, D. Seibert, H.B. Quesada, F. de Jesus Bassetti, M.R. Fagundes-Klen, R. Bergamasco, Occurrence, impacts and general aspects of pesticides in surface water: A review, Process Saf. Environ. Prot. 135 (2020) 22–37, https://doi.org/10.1016/j.psep.2019.12.035.

[14] T.A. Hashmi, R. Qureshi, D. Tipre, S. Menon, Investigation of pesticide residues in water, sediments and fish samples from Tapi River, India as a case study and its forensic significance, Environ. Forensics 21 (1) (2020) 1–10, https://doi.org/10.1080/15275922.2019.1693441.

[15] M. Nehra, N. Dilbaghi, G. Marrazza, A. Kaushik, C. Sonne, K.H. Kim, S. Kumar, Emerging nanobiotechnology in agriculture for the management of pesticide residues, J. Hazard. Mater. 401 (2021) 123369. https://doi.org/10.1016/j.jhazmat.2020.123369.

[16] U. Arisekar, R. Jeya Shakila, R. Shalini, G. Jeyasekaran, Pesticides contamination in the Thamirabarani, a perennial river in peninsular India: the first report on ecotoxicological and human health risk assessment, Chemosphere 267 (2021), https://doi.org/10.1016/j.chemosphere.2020.129251.

[17] M.C. Vagi, A.S. Petsas, Recent advances on the removal of priority organochlorine and organophosphorus biorecalcitrant pesticides defined by directive 2013/39/EU from environmental matrices by using advanced oxidation processes: An overview (2007–2018), J. Environ. Chem. Eng. 8 (1) (2020) 102940. https://doi.org/10.1016/J.JECE.2019.102940.

[18] I.A. Saleh, N. Zouari, M.A. Al-Ghouti, Removal of pesticides from water and wastewater: chemical, physical and biological treatment approaches, Environ. Technol. Innov. 19 (2020), 101026. https://doi.org/10.1016/j.eti.2020.101026.

[19] A. Mojiri, J.L. Zhou, B. Robinson, A. Ohashi, N. Ozaki, T. Kindaichi, H. Farraji, M. Vakili, Pesticides in aquatic environments and their removal by adsorption methods, Chemosphere 253 (2020) 126646. https://doi.org/10.1016/j.chemosphere.2020.126646.

[20] C.Y. Yin, M.K. Aroua, W.M.A.W. Daud, Review of modifications of activated carbon for enhancing contaminant uptakes from aqueous solutions, Sep. Purif. Technol. 52 (3) (2007) 403–415, https://doi.org/10.1016/j.seppur.2006.06.009.

[21] İ. Demiral, C. Samdan, H. Demiral, Enrichment of the surface functional groups of activated carbon by modification method, Surf. Interfaces 22 (2020) 100873. https://doi.org/10.1016/j.surfin.2020.100873.

[22] A.L. Cukierman, G.V. Nunell, P.R. Bonelli, Removal of emerging pollutants from water through adsorption onto carbon-based materials, in: Emerging and Nanomaterial Contaminants in Wastewater, 2019, pp. 159–213, https://doi.org/10.1016/B978-0-12-814673-6.00007-3.

[23] R. Gusain, N. Kumar, S.S. Ray, Recent advances in carbon nanomaterial-based adsorbents for water purification, Coord. Chem. Rev. 405 (2020) 213111. https://doi.org/10.1016/j.ccr.2019.213111.

[24] J. Altmann, D. Rehfeld, K. Träder, A. Sperlich, M. Jekel, Combination of granular activated carbon adsorption and deep-bed filtration as a single advanced wastewater treatment step for organic micropollutant and phosphorus removal, Water Res. 92 (2016) 131–139.

[25] A.A.P. Khan, A. Khan, A.M. Asiri, Nanocarbon and its composites for water purification, in: Nanocarbon and its Composites: Preparation, Properties and Applications, 2018, pp. 711–731, https://doi.org/10.1016/B978-0-08-102509-3.00024-9.

[26] W. Lai, Pesticide use and health outcomes: evidence from agricultural water pollution in China, J. Environ. Econ. Manage 86 (2017) 93–120, https://doi.org/10.1016/j.jeem.2017.05.006.

[27] D. Widawsky, S. Rozelle, S. Jin, J. Huang, Pesticide productivity, host-plant resistance and productivity in China, Agric. Econ. 19 (1) (1998) 203–217, https://doi.org/10.1016/S0169-5150(98)00049-8.

[28] T. Pradeep, Anshup., Noble metal nanoparticles for water purification: a critical review, Thin Solid Films 517 (24) (2009) 6441–6478.

[29] G. Matthews, Pesticides in the early part of 20th century, in: A history of pesticides, 2018, p. 1. https://doi.org/10.978/1786394901.

[30] A. Özkara, D. Akyil, M. Konuk, Pesticides, environmental pollution, and health, in: Environmental Health Risk – Hazardous Factors to Living Species, 2016, https://doi.org/10.5772/63094.

[31] D.J. Ecobichon, Pesticide use in developing countries, Toxicology 160 (1) (2001) 27–33, https://doi.org/10.1016/S0300-483X(00)00452-2.

[32] W. Van der Hoek, F. Konradsen, K. Athukorala, T. Wanigadewa, Pesticide poisoning: A major health problem in Sri Lanka, Soc Sci Med 46 (4) (1998) 495–504, https://doi.org/10.1016/S0277-9536(97)00193-7.

[33] R.M.A. Sjerps, P.J.F. Kooij, A. van Loon, A.P. van Wezel, Occurrence of pesticides in Dutch drinking water sources, Chemosphere 235 (2019) 510–518, https://doi.org/10.1016/j.chemosphere.2019.06.207.

[34] D. Barceló, M.-C. Hennion, Chapter 1—pesticides their degradation products:characteristics usage environmental behaviour, in: Techniques and Instrumentation in Analytical Chemistry, 19, 1997, pp. 1–94. https://doi.org/10.1016/S0167-9244(97)80002-9.

[35] I.G. Dubus, J.M. Hollis, C.D. Brown, Pesticides in rainfall in Europe, Environ. Pollut. 110 (2) (2000) 331–344, https://doi.org/10.1016/S0269-7491(99)00295-X.

[36] M.A. Van der Gaag, J. Worm, Persistent pesticides: the need for criteria to control atmospheric transport, Stud. Environ. Sci. 72 (1998) 341–351, https://doi.org/10.1016/S0166-1116(98)80020-4.

[37] Y. Abubakar, H. Tijjani, C. Egbuna, C.O. Adetunji, S. Kala, T.L. Kryeziu, J.C. Ifemeje, K.C. Patrick-Iwuanyanwu, Chapter 3—pesticides, history, and classification, in: Natural Remedies for Pest, Disease and Weed Control, 2020, pp. 29–42. https://doi.org/10.1016/B978-0-12-819304-4.00003-8.

[38] Y. Zhang, W. Cui, W. An, L. Liu, Y. Liang, Y. Zhu, Combination of photoelectrocatalysis and adsorption for removal of bisphenol A over TiO2-graphene hydrogel with 3D network structure, Appl. Catal. Environ. 221 (2018) 36–46, https://doi.org/10.1016/j.apcatb.2017.08.076.

[39] B. Ayoubi-Feiz, M.H. Mashhadizadeh, M. Sheydaei, Preparation of reusable nano N-TiO2/graphene/titanium grid sheet for electrosorption-assisted visible light photoelectrocatalytic degradation of a pesticide: effect of parameters and neural network modeling, J. Electroanal. Chem. 823 (2018) 713–722, https://doi.org/10.1016/j.jelechem.2018.07.020.

[40] M. Younes, H. Galal-Gorchev, Pesticides in drinking water—A case study, Food Chem. Toxicol. 38 (2000) S87–S90. https://doi.org/10.1016/S0278-6915(99)00132-5.

[41] J. Ruberson, H. Nemoto, Y. Hirose, Chapter 11—pesticides and conservation of natural enemies in pest management, in: Conservation Biological Control, 1998, pp. 207–220. https://doi.org/10.1016/B978-012078147-8/50057-8.

[42] J.C. Zadoks, H. Waibel, From pesticides to genetically modified plants: history, economics and politics, NJAS-Wagen. J. Life Sci. 48 (2) (2000) 125–149, https://doi.org/10.1016/S1573-5214(00)80010-X.

[43] D.W. Sparling, Current use pesticides, in: Basics of Ecotoxicology, 2017, pp. 109–152, https://doi.org/10.1016/B978-0-12-801947-4.00005-6.

[44] H. Ghanbarlou, B. Nasernejad, M. Nikbakht Fini, M.E. Simonsen, J. Muff, Synthesis of an iron-graphene based particle electrode for pesticide removal in three-dimensional heterogeneous electro-Fenton water treatment system, Chem. Eng. J. 395 (2020) 125025. https://doi.org/10.1016/j.cej.2020.125025.

[45] M. Agarwal, A. Verma, Modern technologies for pest control: A review, in: Biotechnology in Mining and Metallurgical Industry, 2020, https://doi.org/10.5772/intechopen.93556.

[46] S. Hintze, G. Glauser, D. Hunkeler, Influence of surface water – groundwater interactions on the spatial distribution of pesticide metabolites in groundwater, Sci. Total Environ. 733 (2020) 139109. https://doi.org/10.1016/j.scitotenv.2020.139109.

[47] J.J. Rasmussen, P. Wiberg-Larsen, A. Baattrup-Pedersen, N. Cedergreen, U.S. McKnight, J. Kreuger, D. Jacobsen, E.A. Kristensen, N. Friberg, The legacy of pesticide pollution: An overlooked factor in current risk assessments of freshwater systems, Water Res. 84 (2015) 25–32, https://doi.org/10.1016/j.watres.2015.07.021.

[48] A. Deknock, N. de Troyer, M. Houbraken, L. Dominguez-Granda, I. Nolivos, W. van Echelpoel, M.A.E. Forio, P. Spanoghe, P. Goethals, Distribution of agricultural pesticides in the freshwater environment of the Guayas river basin (Ecuador), Sci. Total Environ. 646 (2019) 996–1008, https://doi.org/10.1016/j.scitotenv.2018.07.185.

[49] M.A. Hassaan, A. el Nemr, Pesticides pollution: classifications, human health impact, extraction and treatment techniques, Egypt. J. Aquat. Res. 46 (3) (2020) 207–220, https://doi.org/10.1016/j.ejar.2020.08.007.

[50] R. Schäfer, P. van den Brink, M. Liess, Impacts of pesticides on freshwater ecosystems, in: Ecological Impacts of Toxic Chemicals, 2011, pp. 111–137.

[51] S.S. Hassan, M.A. Al-Ghouti, M. Abu-Dieyeh, G. McKay, Novel bioadsorbents based on date pits for organophosphorus pesticide remediation from water, J. Environ. Chem. Eng. 8 (1) (2020), https://doi.org/10.1016/j.jece.2019.103593.

[52] H.B. Mathur, S. Johnson, R. Mishra, A. Kumar, B. Singh, Analysis of pesticide residues in bottled water, 2003. www.cseindia.org.

[53] R. Chow, R. Scheidegger, T. Doppler, A. Dietzel, F. Fenicia, C. Stamm, A review of long-term pesticide monitoring studies to assess surface water quality trends, Water Res. X 9 (2020) 100064. https://doi.org/10.1016/j.wroa.2020.100064.

[54] M.A. Rippy, A. Deletic, J. Black, R. Aryal, J.-L. Lampard, J.Y.-M. Tang, D. McCarthy, P. Kolotelo, J. Sidhu, W. Gernjak, Pesticide occurrence and spatio-temporal variability in urban run-off across Australia, Water Res. 115 (2017) 245–255, https://doi.org/10.1016/j.watres.2017.03.010.

[55] L. Zhu, C. Jiang, S. Panthi, S.M. Allard, A.R. Sapkota, A. Sapkota, Impact of high precipitation and temperature events on the distribution of emerging contaminants in surface water in the mid-Atlantic, United States, Sci. Total Environ. 755 (2021) 142552. https://doi.org/10.1016/j.scitotenv.2020.142552.

[56] T. Parween, S. Jan, Perceptive exploitation of pesticides, in: Ecophysiology of Pesticides, 2019, pp. 189–222, https://doi.org/10.1016/b978-0-12-817614-6.00006-8.

[57] S.A. Covert, M.E. Shoda, S.M. Stackpoole, W.W. Stone, Pesticide mixtures show potential toxicity to aquatic life in U.S. streams, water years 2013–2017, Sci. Total Environ. 745 (2020) 141285. https://doi.org/10.1016/j.scitotenv.2020.141285.

[58] D. Pietrzak, J. Kania, G. Malina, E. Kmiecik, K. Wątor, Pesticides from the EU first and second watch lists in the water environment, *CLEAN – Soil, Air*, Water 47 (7) (2019) 1800376, https://doi.org/10.1002/clen.201800376.

[59] Y. Zhou, J. Wu, B. Wang, L. Duan, Y. Zhang, W. Zhao, F. Wang, Q. Sui, Z. Chen, D. Xu, Q. Li, G. Yu, Occurrence, source and ecotoxicological risk assessment of pesticides in surface water of Wujin District (northwest of Taihu Lake), China, Environ. Pollut. 265 (2020) 114953. https://doi.org/10.1016/j.envpol.2020.114953.

[60] W. Fang, Y. Peng, D. Muir, J. Lin, X. Zhang, A critical review of synthetic chemicals in surface waters of the US, the EU and China, Environ. Int. 131 (2019) 104994. https://doi.org/10.1016/j.envint.2019.104994.

[61] L. Fava, M.A. Orrù, S. Scardala, E. Alonzo, M. Fardella, C. Strumia, E. Funari, Pesticides and their metabolites in selected Italian groundwater and surface water used for drinking, Ann. Ist. Super. Sanita 46 (2010) 309–316, https://doi.org/10.4415/ANN_10_03_15.

[62] R. Meffe, I. de Bustamante, Emerging organic contaminants in surface water and groundwater: A first overview of the situation in Italy, Sci. Total Environ. 481 (2014) 280–295, https://doi.org/10.1016/j.scitotenv.2014.02.053.

[63] D. Pietrzak, J. Kania, E. Kmiecik, G. Malina, K. Wątor, Fate of selected neonicotinoid insecticides in soil–water systems: current state of the art and knowledge gaps, Chemosphere 255 (2020) 126981. https://doi.org/10.1016/j.chemosphere.2020.126981.

[64] C. Postigo, A. Ginebreda, M.V. Barbieri, D. Barceló, J. Martín-Alonso, A. de la Cal, M.R. Boleda, N. Otero, R. Carrey, V. Solà, E. Queralt, E. Isla, A. Casanovas, G. Frances, M. López de Alda, Investigative monitoring of pesticide and nitrogen pollution sources in a complex multi-stressed catchment: the lower Llobregat River basin case study (Barcelona, Spain), Sci. Total Environ. 755 (2021) 142377. https://doi.org/10.1016/j.scitotenv.2020.142377.

[65] N.I. Rousis, R. Bade, L. Bijlsma, E. Zuccato, J. Sancho, v, Hernandez, F. and Castiglioni, S., Monitoring a large number of pesticides and transformation products in water samples from Spain and Italy, Environ. Res. 156 (2017) 31–38, https://doi.org/10.1016/j.envres.2017.03.013.

[66] N. Anand, P. Chakraborty, S. Ray, Human exposure to organochlorine, pyrethroid and neonicotinoid pesticides: comparison between urban and semi-urban regions of India, Environ. Pollut. 270 (2020) 116156. https://doi.org/10.1016/j.envpol.2020.116156.

[67] N. Tauchnitz, F. Kurzius, H. Rupp, G. Schmidt, B. Hauser, M. Schrödter, R. Meissner, Assessment of pesticide inputs into surface waters by agricultural and urban sources—A case study in the Querne/Weida catchment, Central Germany, Environ. Pollut. 267 (2020) 115186. https://doi.org/10.1016/j.envpol.2020.115186.

[68] P.C. Abhilash, N. Singh, Pesticide use and application: An Indian scenario, J. Hazard. Mater. 165 (1–3) (2009) 1–12, https://doi.org/10.1016/j.jhazmat.2008.10.061.

[69] C. Bolognesi, F.D. Merlo, Pesticides: human health effects, in: Encyclopedia of Environmental Health, 2019, pp. 118–132, https://doi.org/10.1016/B978-0-12-409548-9.11818-4.

[70] T. Parween, S. Jan, Pesticides and environmental ecology, in: Ecophysiology of Pesticides, 2019, pp. 1–38, https://doi.org/10.1016/b978-0-12-817614-6.00001-9.

[71] J.D. Meeker, B. Hang, J. Gao, J. Luo, K. Wu, T. Dong, K. Zhou, X. He, W. Hu, W. Wu, C. Lu, X. Wang, Y. Xia, Current pesticide profiles in blood serum of adults in Jiangsu Province of China and a comparison with other countries, Environ. Int. 102 (2017) 213–222, https://doi.org/10.1016/j.envint.2017.03.004.

[72] S. Shrestha, C.G. Parks, D.M. Umbach, M. Richards-Barber, J.N. Hofmann, H. Chen, A. Blair, L.E. Beane Freeman, D.P. Sandler, Pesticide use and incident Parkinson's disease in a cohort of farmers and their spouses, Environ. Res. 191 (2020) 110186. https://doi.org/10.1016/j.envres.2020.110186.

[73] M. Requena, A. López-Villén, A.F. Hernández, T. Parrón, Á. Navarro, R. Alarcón, Environmental exposure to pesticides and risk of thyroid diseases, Toxicol. Lett. 315 (2019) 55–63, https://doi.org/10.1016/j.toxlet.2019.08.017.

[74] E. Djangalina, N. Altynova, S. Bakhtiyarova, U. Kapysheva, B. Zhaksymov, E. Shadenova, M. Baizhanov, O. Sapargali, A. Garshin, A. Seisenbayeva, M. Delannoy, S. Jurjanz, E. Khussainova, B. Bekmanov, L. Djansugurova, Comprehensive assessment of unutilized and obsolete pesticides impact on genetic status and health of population of Almaty region, Ecotoxicol. Environ. Saf. 202 (2020) 110905. https://doi.org/10.1016/j.ecoenv.2020.110905.

[75] Z. Li, A. Jennings, Global variations in pesticide regulations and health risk assessment of maximum concentration levels in drinking water, J. Environ. Manage. 212 (2018) 384–394, https://doi.org/10.1016/J.JENVMAN.2017.12.083.

[76] M.H. Dehghani, Z.S. Niasar, M.R. Mehrnia, M. Shayeghi, M.A. Al-Ghouti, B. Heibati, K. Yetilmezsoy, Optimizing the removal of organophosphorus pesticide malathion from water using multi-walled carbon nanotubes, Chem. Eng. J. 310 (2017) 22–32.

[77] P.K. Henneberger, X. Liang, S.J. London, D.M. Umbach, D.P. Sandler, J.A. Hoppin, Exacerbation of symptoms in agricultural pesticide applicators with asthma, Int. Arch. Occup. Environ. Health 87 (4) (2014), https://doi.org/10.1007/s00420-013-0881-x.

[78] V. Kumar, P. Kumar, Pesticides in agriculture and environment: impacts on human health, in: Contaminants in Agriculture and Environment: Health Risks and Remediation, vol. 1, 2019, p. 76. https://doi.org/10.26832/AESA-2019-CAE-0160-07.

[79] L.M. Jaacks, S. Yadav, P. Panuwet, S. Kumar, G.H. Rajacharya, C. Johnson, I. Rawal, D. Mohan, V. Mohan, N. Tandon, D.B. Barr, K.M.V. Narayan, D. Prabhakaran, Metabolite of the pesticide DDT and incident type 2 diabetes in urban India, Environ. Int. 133 (2019) 105089. https://doi.org/10.1016/j.envint.2019.105089.

[80] W. Huang, Y. He, J. Xiao, Y. Huang, A. Li, M. He, K. Wu, Risk of breast cancer and adipose tissue concentrations of polychlorinated biphenyls and organochlorine pesticides: a hospital-based case-control study in Chinese women, Environ. Sci. Pollut. Res. 26 (31) (2019), https://doi.org/10.1007/s11356-019-06404-3.

[81] E. Giavini, E. Menegola, Are azole fungicides a teratogenic risk for human conceptus? Toxicol. Lett. 198 (2) (2010) 106–111, https://doi.org/10.1016/j.toxlet.2010.07.005.

[82] F. Suo, X. You, Y. Ma, Y. Li, Rapid removal of triazine pesticides by P doped biochar and the adsorption mechanism, Chemosphere 235 (2019) 918–925, https://doi.org/10.1016/j.chemosphere.2019.06.158.

[83] J. Jampílek, K. Kráľová, Potential of nanoscale carbon-based materials for remediation of pesticide-contaminated environment, in: Carbon Nanomaterials for Agri-Food and Environmental Applications, 2020, pp. 359–399, https://doi.org/10.1016/B978-0-12-819786-8.00017-7.

[84] N. Rao, R. Singh, L. Bashambu, Carbon-based nanomaterials: synthesis and prospective applications, in: Materials Today: Proceedings, 2020, https://doi.org/10.1016/j.matpr.2020.10.593.

[85] B. Constantin, Carbon nanotubes as a new solid phase extraction sorbent for analysis of environmental pollutants, in: Carbon Nanotubes, IntechOpen, 2010, March, https://doi.org/10.5772/39442.

[86] A.V. Herrera-Herrera, M.Á. González-Curbelo, J. Hernández-Borges, M.Á. Rodríguez-Delgado, Carbon nanotubes applications in separation science: a review, Anal. Chim. Acta 734 (2012) 1–30, https://doi.org/10.1016/j.aca.2012.04.035.

[87] A.J. Brooks, H.N. Lim, J.E. Kilduff, Adsorption uptake of synthetic organic chemicals by carbon nanotubes and activated carbons, Nanotechnology 23 (29) (2012) 294008. https://doi.org/10.1088/0957-4484/23/29/294008.

[88] A.B. Dichiara, S.F. Harlander, R.E. Rogers, Fixed bed adsorption of diquat dibromide from aqueous solution using carbon nanotubes, RSC Adv. 5 (76) (2015), https://doi.org/10.1039/C5RA11167F.

[89] A. de Martino, M. Iorio, B. Xing, R. Capasso, Removal of 4-chloro-2-methylphenoxyacetic acid from water by sorption on carbon nanotubes and metal oxide nanoparticles, RSC Advances 2 (13) (2012), https://doi.org/10.1039/c2ra00038e.

[90] A. Mehrizad, M. Aghaie, P. Gharbani, S. Dastmalchi, M. Monajjemi, K. Zare, Comparison of 4-chloro-2-nitrophenol adsorption on single-walled and multi-walled carbon nanotubes, Iran. J. Environ. Health Sci. Eng. 9 (1) (2012), https://doi.org/10.1186/1735-2746-9-5.

[91] J. Deng, Y. Shao, N. Gao, Y. Deng, C. Tan, S. Zhou, X. Hu, Multiwalled carbon nanotubes as adsorbents for removal of herbicide diuron from aqueous solution, Chem. Eng. J. 193–194 (2012) 339–347, https://doi.org/10.1016/j.cej.2012.04.051.

[92] O. Duman, C. Özcan, T. Gürkan Polat, S. Tunç, Carbon nanotube-based magnetic and non-magnetic adsorbents for the high-efficiency removal of diquat dibromide herbicide from water: OMWCNT, OMWCNT-Fe3O4 and OMWCNT-K-carrageenan-Fe3O4 nanocomposites, Environ. Pollut. 244 (2019) 723–732, https://doi.org/10.1016/j.envpol.2018.10.071.

[93] G.-C. Chen, X.-Q. Shan, Z.-G. Pei, H. Wang, L.-R. Zheng, J. Zhang, Y.-N. Xie, Adsorption of diuron and dichlobenil on multiwalled carbon nanotubes as affected by lead, J. Hazard. Mater. 188 (1) (2011) 156–163, https://doi.org/10.1016/j.jhazmat.2011.01.095.

[94] T.A. Saleh, A. Sarı, M. Tuzen, Carbon nanotubes grafted with poly(trimesoyl, m-phenylenediamine) for enhanced removal of phenol, J. Environ. Manage. 252 (2019) 109660, https://doi.org/10.1016/j.jenvman.2019.109660.

[95] W. Chen, L. Duan, D. Zhu, Adsorption of polar and nonpolar organic chemicals to carbon nanotubes, Environ. Sci. Tech. 41 (24) (2007), https://doi.org/10.1021/es071230h.

[96] D. Lin, B. Xing, Adsorption of phenolic compounds by carbon nanotubes: role of aromaticity and substitution of hydroxyl groups, Environ. Sci. Tech. 42 (19) (2008), https://doi.org/10.1021/es801297u.
[97] S.K. Deokar, G.S. Bajad, P. Bhonde, R.P. Vijayakumar, S.A. Mandavgane, Adsorptive removal of diuron herbicide on carbon nanotubes synthesized from plastic waste, J. Polym. Environ. 25 (2) (2017) 165–175, https://doi.org/10.1007/s10924-016-0794-3.
[98] G.-C. Chen, X.-Q. Shan, Y.-S. Wang, Z.-G. Pei, X.-E. Shen, B. Wen, G. Owens, Effects of copper, lead, and cadmium on the sorption and desorption of atrazine onto and from carbon nanotubes, Environ. Sci. Tech. 42 (22) (2008), https://doi.org/10.1021/es801376w.
[99] W.-W. Tang, G.-M. Zeng, J.-L. Gong, Y. Liu, X.-Y. Wang, Y.-Y. Liu, Z.-F. Liu, L. Chen, X.-R. Zhang, D.-Z. Tu, Simultaneous adsorption of atrazine and cu (II) from wastewater by magnetic multi-walled carbon nanotube, Chem. Eng. J. 211–212 (2012), https://doi.org/10.1016/j.cej.2012.09.102.
[100] M. Ateia, M. Ceccato, A. Budi, E. Ataman, C. Yoshimura, M.S. Johnson, Ozone-assisted regeneration of magnetic carbon nanotubes for removing organic water pollutants, Chem. Eng. J. 335 (2018), https://doi.org/10.1016/j.cej.2017.10.166.
[101] J.L. Sotelo, A.R. Rodríguez, M.M. Mateos, S.D. Hernández, S.A. Torrellas, J.G. Rodríguez, Adsorption of pharmaceutical compounds and an endocrine disruptor from aqueous solutions by carbon materials, J. Environ. Sci. Health B 47 (7) (2012), https://doi.org/10.1080/03601234.2012.668462.
[102] I. Ali, O.M.L. Alharbi, Z.A. ALOthman, A.M. Al-Mohaimeed, A. Alwarthan, Modeling of fenuron pesticide adsorption on CNTs for mechanistic insight and removal in water, Environ. Res. 170 (2019) 389–397, https://doi.org/10.1016/j.envres.2018.12.066.
[103] K. Sun, Z. Zhang, B. Gao, Z. Wang, D. Xu, J. Jin, X. Liu, Adsorption of diuron, fluridone and norflurazon on single-walled and multi-walled carbon nanotubes, Sci. Total Environ. 439 (2012) 1–7, https://doi.org/10.1016/j.scitotenv.2012.08.022.
[104] H.N. Catherine, M.H. Ou, B. Manu, Y. Shih, hsin., Adsorption mechanism of emerging and conventional phenolic compounds on graphene oxide nanoflakes in water, Sci. Total Environ. 635 (2018) 629–638, https://doi.org/10.1016/j.scitotenv.2018.03.389.
[105] C. Liné, C. Larue, E. Flahaut, Carbon nanotubes: impacts and behaviour in the terrestrial ecosystem—A review, Carbon 123 (2017) 767–785, https://doi.org/10.1016/j.carbon.2017.07.089.
[106] B. Sarkar, S. Mandal, Y.F. Tsang, P. Kumar, K.H. Kim, Y.S. Ok, Designer carbon nanotubes for contaminant removal in water and wastewater: A critical review, Science of the Total Environment 612 (2018) 561–581, https://doi.org/10.1016/j.scitotenv.2017.08.132.
[107] G. Ersan, O.G. Apul, F. Perreault, T. Karanfil, Adsorption of organic contaminants by graphene nanosheets: A review, Water Res. 126 (2017) 385–398, https://doi.org/10.1016/j.watres.2017.08.010.
[108] E. Gracia-Espino, F. López-Urías, H. Terrones, Novel nanocarbons for adsorption, in: Novel Carbon Adsorbents, 2012, pp. 3–34, https://doi.org/10.1016/B978-0-08-097744-7.00001-6.
[109] J. Wang, Z. Chen, B. Chen, Adsorption of polycyclic aromatic hydrocarbons by graphene and graphene oxide nanosheets, Environ. Sci. Tech. 48 (9) (2014) 4817–4825, https://doi.org/10.1021/es405227u.
[110] W.S. Hummers Jr., R.E. Offeman, Preparation of graphitic oxide, J. Am. Chem. Soc. 80 (6) (1958) 1339, https://doi.org/10.1021/ja01539a017.
[111] M. Mohandoss, S.S. Gupta, A. Nelleri, T. Pradeep, S.M. Maliyekkal, Solar mediated reduction of graphene oxide, RSC Adv. 7 (2) (2017) 957–963, https://doi.org/10.1039/C6RA24696F.
[112] J. Wang, B. Chen, Adsorption and coadsorption of organic pollutants and a heavy metal by graphene oxide and reduced graphene materials, Chem. Eng. J. 281 (2015) 379–388, https://doi.org/10.1016/j.cej.2015.06.102.
[113] S.M. Maliyekkal, T.S. Sreeprasad, D. Krishnan, S. Kouser, A.K. Mishra, U.V. Waghmare, T. Pradeep, Graphene: a reusable substrate for unprecedented adsorption of pesticides, Small 9 (2) (2013) 273–283, https://doi.org/10.1002/smll.201201125.
[114] S. Yu, X. Wang, W. Yao, J. Wang, Y. Ji, Y. Ai, X. Wang, Macroscopic, spectroscopic, and theoretical investigation for the interaction of phenol and naphthol on reduced graphene oxide, Environ. Sci. Tech. 51 (6) (2017) 3278–3286, https://doi.org/10.1021/acs.est.6b06259.
[115] P.K. Boruah, B. Sharma, N. Hussain, M.R. Das, Magnetically recoverable Fe3O4/graphene nanocomposite towards efficient removal of triazine pesticides from aqueous solution: investigation of the adsorption phenomenon and specific ion effect, Chemosphere 168 (2017) 1058–1067, https://doi.org/10.1016/j.chemosphere.2016.10.103.
[116] Y. Li, C. Zhao, Y. Wen, Y. Wang, Y. Yang, Adsorption performance and mechanism of magnetic reduced graphene oxide in glyphosate contaminated water, Environ. Sci. Pollut. Res. 25 (21) (2018), https://doi.org/10.1007/s11356-018-2282-x.
[117] K. Shrivas, A. Ghosale, N. Nirmalkar, A. Srivastava, S.K. Singh, S.S. Shinde, Removal of endrin and dieldrin isomeric pesticides through stereoselective adsorption behavior on the graphene oxide-magnetic nanoparticles, Environ. Sci. Pollut. Res. 24 (32) (2017), https://doi.org/10.1007/s11356-017-0159-z.
[118] S. Sen Gupta, I. Chakraborty, S.M. Maliyekkal, T. Mark Adit, D.K. Pandey, S.K. Das, T. Pradeep, Simultaneous dehalogenation and removal of persistent halocarbon pesticides from water using graphene nanocomposites: A case study of lindane, ACS Sustain. Chem. Eng. 3 (6) (2015) 1155–1163, https://doi.org/10.1021/acssuschemeng.5b00080.
[119] D. Koushik, S. Sen Gupta, S.M. Maliyekkal, T. Pradeep, Rapid dehalogenation of pesticides and organics at the interface of reduced graphene oxide-silver nanocomposite, J. Hazard. Mater. 308 (2016) 192–198, https://doi.org/10.1016/j.jhazmat.2016.01.004.
[120] Q. Yang, J. Wang, W. Zhang, F. Liu, X. Yue, Y. Liu, M. Yang, Z. Li, J. Wang, Interface engineering of metal organic framework on graphene oxide with enhanced adsorption capacity for organophosphorus pesticide, Chem. Eng. J. 313 (2017) 19–26, https://doi.org/10.1016/j.cej.2016.12.041.
[121] S. Nethaji, A. Sivasamy, Graphene oxide coated with porous iron oxide ribbons for 2, 4-Dichlorophenoxyacetic acid (2, 4-D) removal, Ecotoxicol. Environ. Saf. 138 (2017) 292–297.
[122] G. Liu, L. Li, D. Xu, X. Huang, X. Xu, S. Zheng, Y. Zhang, H. Lin, Metal–organic framework preparation using magnetic graphene oxide–β-cyclodextrin for neonicotinoid pesticide adsorption and removal, Carbohydr. Polym. 175 (2017) 584–591, https://doi.org/10.1016/j.carbpol.2017.06.074.

[123] R. Zhao, Y. Li, J. Ji, Q. Wang, G. Li, T. Wu, B. Zhang, Efficient removal of phenol and p-nitrophenol using nitrogen-doped reduced graphene oxide, Colloids Surf. A Physicochem. Eng. Asp. 125866 (2020), https://doi.org/10.1016/j.colsurfa.2020.125866.

[124] S. Agarwal, N. Sadeghi, I. Tyagi, V.K. Gupta, A. Fakhri, Adsorption of toxic carbamate pesticide oxamyl from liquid phase by newly synthesized and characterized graphene quantum dots nanomaterials, J. Colloid Interface Sci. 478 (2016) 430–438, https://doi.org/10.1016/j.jcis.2016.06.029.

[125] Q. Yao, S. Wang, W. Shi, C. Lu, G. Liu, Graphene quantum dots in two-dimensional confined and hydrophobic space for enhanced adsorption of nonionic organic adsorbates, Ind. Eng. Chem. Res. 56 (2) (2017), https://doi.org/10.1021/acs.iecr.6b02389.

[126] J. Wang, B. Chen, B. Xing, Wrinkles and folds of activated graphene nanosheets as fast and efficient adsorptive sites for hydrophobic organic contaminants, Environ. Sci. Tech. 50 (7) (2016), https://doi.org/10.1021/acs.est.5b04865.

[127] S.S. Gupta, T.S. Sreeprasad, S.M. Maliyekkal, S.K. Das, T. Pradeep, Graphene from sugar and its application in water purification, ACS Appl. Mater. Interfaces 4 (8) (2012) 4156–4163, https://doi.org/10.1021/am300889u.

[128] T.S. Sreeprasad, S.s. Gupta, S.M. Maliyekkal, T. Pradeep, Immobilized graphene-based composite from asphalt: facile synthesis and application in water purification, J. Hazard. Mater. 246–247 (2013) 213–220, https://doi.org/10.1016/j.jhazmat.2012.12.022.

[129] X. Liu, H. Zhang, Y. Ma, X. Wu, L. Meng, Y. Guo, G. Yu, Y. Liu, Graphene-coated silica as a highly efficient sorbent for residual organophosphorus pesticides in water, J. Mater. Chem. A 1 (5) (2013), https://doi.org/10.1039/C2TA00173J.

[130] A. Amiri, M. Baghayeri, F. Karimabadi, F. Ghaemi, B. Maleki, Graphene oxide/polydimethylsiloxane-coated stainless steel mesh for use in solid-phase extraction cartridges and extraction of polycyclic aromatic hydrocarbons, Microchim. Acta 187 (4) (2020) 1–8, https://doi.org/10.1016/j.trac.2019.115772.

Further reading

[131] M.V. Barbieri, A. Peris, C. Postigo, A. Moya-Garcés, L.S. Monllor-Alcaraz, M. Rambla-Alegre, E. Eljarrat, M. López de Alda, Evaluation of the occurrence and fate of pesticides in a typical Mediterranean delta ecosystem (Ebro River Delta) and risk assessment for aquatic organisms, Environ. Pollut. (2020) 115813. https://doi.org/10.1016/j.envpol.2020.115813.

[132] E. Carazo-Rojas, G. Pérez-Rojas, M. Pérez-Villanueva, C. Chinchilla-Soto, J.S. Chin-Pampillo, P. Aguilar-Mora, M. Alpízar-Marín, M. Masís-Mora, C.E. Rodríguez-Rodríguez, Z. Vryzas, Pesticide monitoring and ecotoxicological risk assessment in surface water bodies and sediments of a tropical agro-ecosystem, Environ. Pollut. 241 (2018) 800–809, https://doi.org/10.1016/j.envpol.2018.06.020.

[133] M.E. Close, B. Humphries, G. Northcott, Outcomes of the first combined national survey of pesticides and emerging organic contaminants (EOCs) in groundwater in New Zealand 2018, Sci. Total Environ. 754 (2021) 142005. https://doi.org/10.1016/j.scitotenv.2020.142005.

[134] T. di Lorenzo, M. Cifoni, B. Fiasca, A. di Cioccio, D.M.P. Galassi, Ecological risk assessment of pesticide mixtures in the alluvial aquifers of Central Italy: toward more realistic scenarios for risk mitigation, Sci. Total Environ. 644 (2018) 161–172, https://doi.org/10.1016/j.scitotenv.2018.06.345.

[135] R. Elibariki, M.M. Maguta, Status of pesticides pollution in Tanzania – A review, Chemosphere 178 (2017) 154–164, https://doi.org/10.1016/j.chemosphere.2017.03.036.

[136] D.J. Hamilton, Á. Ambrus, R.M. Dieterle, A.S. Felsot, C.A. Harris, P.T. Holland, A. Katayama, N. Kurihara, J. Linders, J. Unsworth, S.-S. Wong, Regulatory limits for pesticide residues in water, Pure Appl. Chem. 75 (8) (2003), https://doi.org/10.1351/pac200375081123.

[137] J. Kaushal, M. Khatri, S.K. Arya, A treatise on organophosphate pesticide pollution: current strategies and advancements in their environmental degradation and elimination, Ecotoxicol. Environ. Saf. 207 (2021) 111483. https://doi.org/10.1016/j.ecoenv.2020.111483.

[138] E.A. Murphy, M. Aucott, An assessment of the amounts of arsenical pesticides used historically in a geographical area, Sci. Total Environ. 218 (2) (1998) 89–101, https://doi.org/10.1016/S0048-9697(98)00180-6.

[139] M.S.J. Warne, R.A. Smith, R.D.R. Turner, Analysis of pesticide mixtures discharged to the lagoon of the great barrier reef, Australia, Environ. Pollut. 265 (2020) 114088. https://doi.org/10.1016/j.envpol.2020.114088.

[140] Global Pesticide Market, 2020–2024—by Origin, Application, Unorganized and Organized Market and Crops, 2020. https://www.researchandmarkets.com/reports/5062080/global-agricultural-pesticides-market-2020-2024. (Last accessed on 20-03-2021).

Chapter 9

Removing chromium (VI) from contaminated water using a nano-chitosan–coated diatomaceous earth

Johannes A. DeMessie[a,b], George A. Sorial[a], and Endalkachew Sahle-Demessie[c,*]

[a]*Environmental Engineering and Science, College of Engineering and Applied Sciences, University of Cincinnati, Cincinnati, OH, United States,* [b]*William Mason High School, Mason, OH, United States,* [c]*Department of Chemistry, Missouri University of Science & Technology, Rolla, MO, United States*

**Corresponding author: e-mail address: sahle-demessie.endalkachew@epa.gov*

Introduction

There is a global water problem from Cape Town, South Africa, to Flint, Michigan, USA, from rural sub-Saharan Africa to the teeming megacities along India's Ganges river, water crises are risking human health. The struggle to access clean and safe water of sufficient quantity is still a major challenge [1]. Pollution by heavy metals can cause severe human health and ecological problems due to their high toxicity and bioaccumulation [2]. Water pollution is a global issue, with heavy metal pollution representing one of the most serious problems. Therefore, the removal of heavy metals is of particular concern to scientists and engineers because of their harmful effects on many life forms. Dissolved metal contamination in drinking water can pose a serious health risk. The primary sources of the contamination are geological weathering, which provides the background level, industrial processing of ores and metals, the use of metal and metal compounds, such as chromium salts in tanneries, copper compounds in agriculture, and tetraethyl lead as an antiknock agent in gasoline, leaching of heavy metals from domestic wastes and solid waste dumps, and heavy metals in human and animal excretions [2–4]. Most heavy metals are nonbiodegradable and accumulate in living tissues, particularly in human bodies, causing various diseases and significant physiological disorders such as central nervous system damage.

Hexavalent chromium, Cr(VI), compounds are manufactured and widely used in many industries [5,6]. Hexavalent chromium is a toxic form of the element chromium [6]. Some major industrial sources of hexavalent chromium are chromate pigments in dyes, paints, inks, and plastics, chromates added as anticorrosive agents to paints, primers, and other surface coatings, chrome plating by depositing chromium metal onto an item's surface using a solution of chromic acid, particles released during smelting of ferrochromium ore, fume from welding stainless steel or nonferrous chromium alloys, and impurity present in portland cement. Once in the water, chromium mainly occurs in two oxidation states Cr(III) and Cr(VI) and related ion forms depending on pH values, redox potential, and presence of natural reducing agents [7,8]. Hexavalent chromium is a major concern to public health owing to its toxic effects on humans, animals, plants, and microorganisms [9–11]. Risks for human health range from skin irritation to DNA damages and cancer development, depending on dose, exposure level, and duration [12–14]. Remediation strategies commonly used for Cr (VI) removal include physicochemical and biological methods (Fig. 1) [15–17].

Regular operations and several accidents involving contamination by large releases of heavy metals in groundwater and rivers have attracted great concern regarding river water safety [1,2]. Heavy metals are mainly emitted from several industries, including metal plating, battery manufacture, fertilizer, mining, fossil fuels, and tannery. These metal ions can finally accumulate in living bodies, including humans, causing various health problems [3,4]. Generally, physical and chemical methods exist for the removal of metal ions from aqueous solutions. The most well-developed include chemical precipitation, filtration, oxidation, reduction, ion exchange, membrane separation, reverse osmosis, electrocoagulation, and solvent extraction, and adsorption [18].

Separation Science and Technology, Volume 15. https://doi.org/10.1016/B978-0-323-90763-7.00005-6

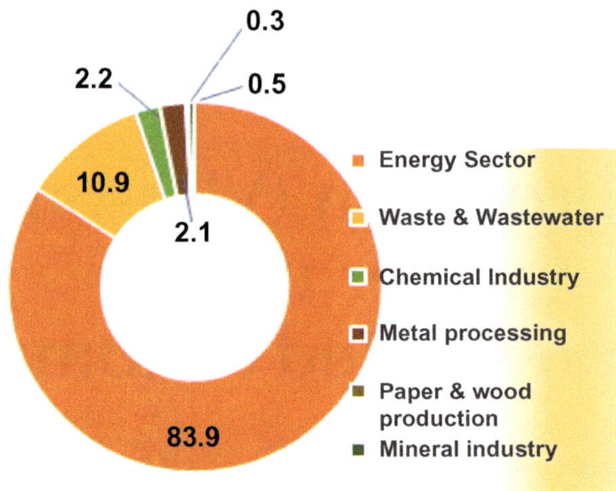

FIG. 1 Percentage contribution of industrial Cr(VI) emission to water pollution. *Data source: M. Tumolo, et al., Chromium pollution in European water, sources, health risk, and remediation strategies: an overview. Int. J. Environ. Res. Public Health 17(15) (2020) 5438.*

Among these, adsorption remains one of the commonest for removing metal ions due to its high efficiency and low cost [19,20]. A variety of materials, both organic and inorganic, have been used as adsorbents. Biomass, chelating resin, fly ash [21,22], activated carbon including granular, powder, fiber, and cloth form [23–26], clays (sepiolite, vermiculite, bentonite, zeolite, etc.) [25,27,28], magnetic carbon nanotubes, inorganic–organic nanocomposite [28], and graphene are widely used as adsorbents in industrial wastewater treatment. However, many problems remain solved in practical applications using common adsorbents due to high cost, low efficiency, difficult recycling, and hazardous metal-bearing contaminated water [22]. Accidental water pollution by toxic heavy metals still cannot be effectively controlled and may cause severe harm to the ecological environment [29]. As a result, developing highly efficient, recoverable adsorbents for the quick removal of heavy-metal pollution from water, especially natural water, remains a significant challenge. The maximum contaminant level of chromium in drinking water allowed by the Environmental Protection Agency (EPA) and the World Health Organization (WHO) is 100 and 50 μg/L, respectively [19]. When Cr(VI) concentrations become elevated, serious health problems can occur [20]. Chromium poisoning can cause kidney damage, anemia, and toxicity symptoms, including impaired kidney function, hypertension, and headache.

Chitosan, a deacetylated derivative of chitin, is a natural macromolecular compound offering the advantages of abundant sources, low cost, biocompatibility, biodegradability, and nontoxicity. In recent years, chitosan has attracted widespread attention as a highly efficient chelating agent removing toxic metal ions from an aqueous solution [22]. However, powdered chitosan shows a tendency to agglomerate or form colloids in water, rendering its active sites not readily available for the removal of metal ions [30]. It is, therefore, necessary to use supporting material to immobilize chitosan and increase the possibility of bonding between its active sites ($-NH_2$, $-OH$) and metal ions. Moreover, coating chitosan on a cost-effective supporting material can reduce chitosan consumption while leaving the overall adsorption capacity unaffected [31]. Removal of arsenic (III) and arsenic(V) from an aqueous medium using chitosan-coated biosorbents have been reported [32]. Hasan et al. [33] investigated the adsorption of Cd(II) from an aqueous solution onto chitosan-coated perlite granules. However, chitosan has a low specific surface area, tends to aggregate and swell, and has poor mechanical properties that reduce its application [34]. Hence, coating chitosan on diatomaceous earth can improve the physical properties of chitosan to enhance its applications.

This study aimed to develop a high surface area novel adsorbent by coating chitosan on diatomaceous earth to remove heavy metals such as Cr(VI) from contaminated water. The adsorbent, which is chitosan-coated diatomaceous earth (CCDE), will be used as a filter material for a point-of-use application in developing countries rural areas and emergency preparedness to produce purified drinking water. Chitosan is chosen due to the low-cost alternative synthetic adsorbent such as carbon or ion exchange. The study included a factorial study in determining the effects of water pH, initial metal concentration, adsorbent mass, contact times, equilibrium adsorption tests, kinetic studies, and testing disposal of used adsorbent.

Experimental and materials used

Chitosan synthesis

Chitosan is a natural organic material, a linear polysaccharide compound of (poly-β-(1 → 4)—linked 2-aminio-2-deoxy D-glucose) commercially produced by partial deacetylation of chitin. Chitin is a substance constituent in the exoskeleton of arthropods and the cell wall of find, which is the second plentiful organic component in nature next to cellulose polymer (Fig. 2). Due to the presence of amino ($-NH_2$) and hydroxyl (–OH) groups in its structure, chitosan is capable of adsorbing heavy metals from aqueous solutions by creating an electronic bond between these active sites and metal ions.

Medium molecular weight chitosan (deacetylated chitin molecular weight ~190,000–310,000 Da) and diatomaceous earth were obtained from Sigma Aldrich Co. (St. Louis, MO, USA). Fig. 3 presents the steps for the synthesis of CCDE that is similar to the literature-reported protocol [35]. The chitosan was obtained by treatment with HCl 1 N and NaOH 2% at 100°C for 2 h. The chitin was placed in a 10 L glass reactor with a mechanical stirrer. The solid/liquid was added in a 1:15 ratio. The mixture was stirred at 500 rpm and heated at 90°C. A 10% (w/v) solution of $NaBH_4$ was added to avoid oxidation and depolymerization. After 90 min the final product was washed with warm deionized water to eliminate NaOH. The product was dried using a vacuum drier at room temperature. Deionized water was used during the experimental studies, and all chemicals used in this study were analytical grade.

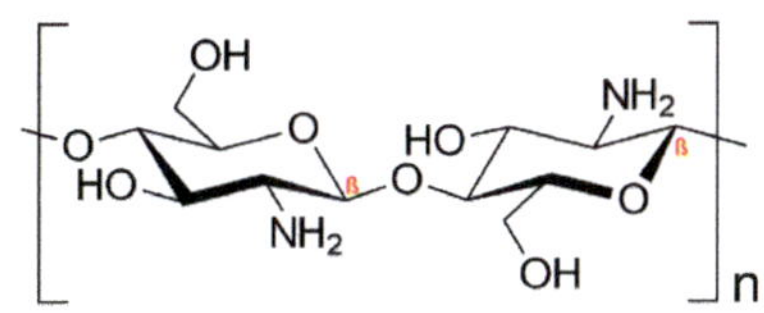

FIG. 2 Chemical structure of chitosan showing glucose amine structure.

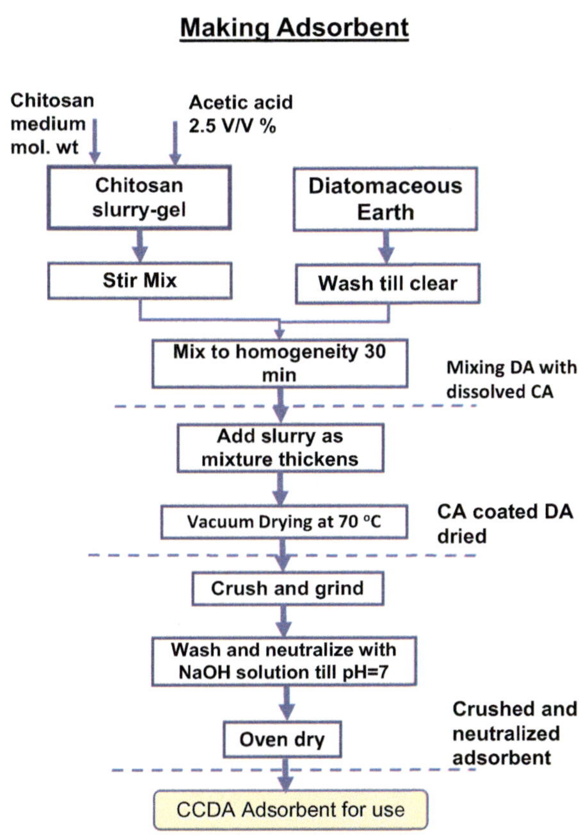

FIG. 3 Schematic summary of chitosan coating to diatomaceous earth and project outline.

Infrared spectroscopy

Changes in the surface functional groups were analyzed for neat chitosan and chitosan-coated on diatomaceous earth. Infrared spectra were measured using an FT-IR Agilent Cary 660 ATR spectrophotometer coupled to a PC with OMNIC analysis software. The films were placed in the holder directly in the IR laser beam. Spectra were recorded at a resolution of 4 cm^{-1}, and 64 scans were accumulated.

Thermogravimetric analysis

A Thermal Analysis D500 thermogravimetric system with a microprocessor-driven temperature control unit and a TA data station was used. The mass of the samples was generally in the range of 5–10 mg. The sample pan was placed in the balance system equipment and the temperature was raised from 30 to 600°C at a heating rate of 10°C per minute. The mass of the sample pan was continuously recorded as a function of temperature and time.

Digital optical microscopy

The microscopic analysis represents a powerful tool for understanding the size, structure, and morphology of adsorbents. Hirox optical microscope is used to characterize chitosan-coated adsorbents. 3D measurement is achieved with a digital microscope by image stacking. The system takes images from the lowest focal plane in the field of view to the highest focal plane using a step motor,. Then it reconstructs these images into a 3-dimensional (3D) model based on contrast to give a 3D color image of the sample. From these 3D models, measurements can be made, but their accuracy is based on the step-motor and depth of field of the lens.

Hexavalent chromium analysis

In this study, inductively coupled plasma optical emission spectrometry (ICP-OES) was used for the determination of chromium concentrations before and after the adsorption treatment. U.S. EPA Method 200.7 has been selected as the standard method for the analysis of total chromium in either drinking water or wastewater for regulatory compliance can be measured ICP-OES using [36]. Standards were prepared in 5% (v/v) HNO_3 from a single elemental solution (Sigma Aldrich Co.). Seven standard solutions were prepared to ensure a broader range of experimental designs. This method offers a lower detection limit for the occurrence monitoring for the program. Water samples from the adsorption studies were collected, and a few drops of 0.15% double distilled HNO_3. A method detection limit of 6 ppb passed the criteria set by the U.S. EPA Method 200.7.

Multilevel factorial design attributes

To determine which is the most important factor affecting the adsorption of hexavalent chromium Cr(VI) is used a multi-level factorial experimental design. The base design includes three experimental factors, initial Cr(VI) concentration, and mass of adsorbent, and one response, removal efficiency.

$$\eta = \frac{C_o - C_e}{C_o} \tag{1}$$

where C_o and C_e are initial and equilibrium aqueous concentrations of chromium.

A multilevel factorial design consisting of ($4 \times 4 \times 3$) with two replicates and 96 runs was created. The design is to be run in 2 blocks. The order of the experiments has been fully randomized. This will protect against the effects of lurking variables. Table 1 below shows the total number of runs for two replicate studies 144, error degrees of freedom: 132.

TABLE 1 Test factors for design of experiment.

sss	Low	High	Levels	Units
Initial Cr(VI) concentration	100.0	1000.0	4	microgram/L
Mass adsorbent	20.0	500.0	4	mg
Solution pH	4.0	8.0	3	[]

Degrees of Freedom refers to the maximum number of logically independent values, which have the freedom to vary in the data sample. The tests were randomized to avoid any systemic errors and cross-contamination during sample preparations or tests.

Based on the final concentrations of Cr(VI), adsorption capacity is determined using equation:

$$q_e = \frac{\text{Mass of Cr(VI) adsorded}}{\text{Mass of chitosan}} \tag{2}$$

where q_e is the adsorption capacity in mg of chromium per gram of adsorbent chitosan.

Results

Morphology of chitosan adsorbent

Chitosan particles were observed under an optical microscope, showed the surface to be rough and porous. Optical images of adsorbents used show particle size and morphology of CCDE with 20 and 30 wt% of DE (Fig. 4). The surface appears to be heterogeneous, which results from the presence of crystalline domains of soft segment of chitosan and DE mixture. The particle size varied from 300 to 500 mm. Scanning electron microscopy (SEM) images of fresh and used chitosan adsorbents showed a well-developed open cell structure, suggesting a high porous surface (Fig. 4).

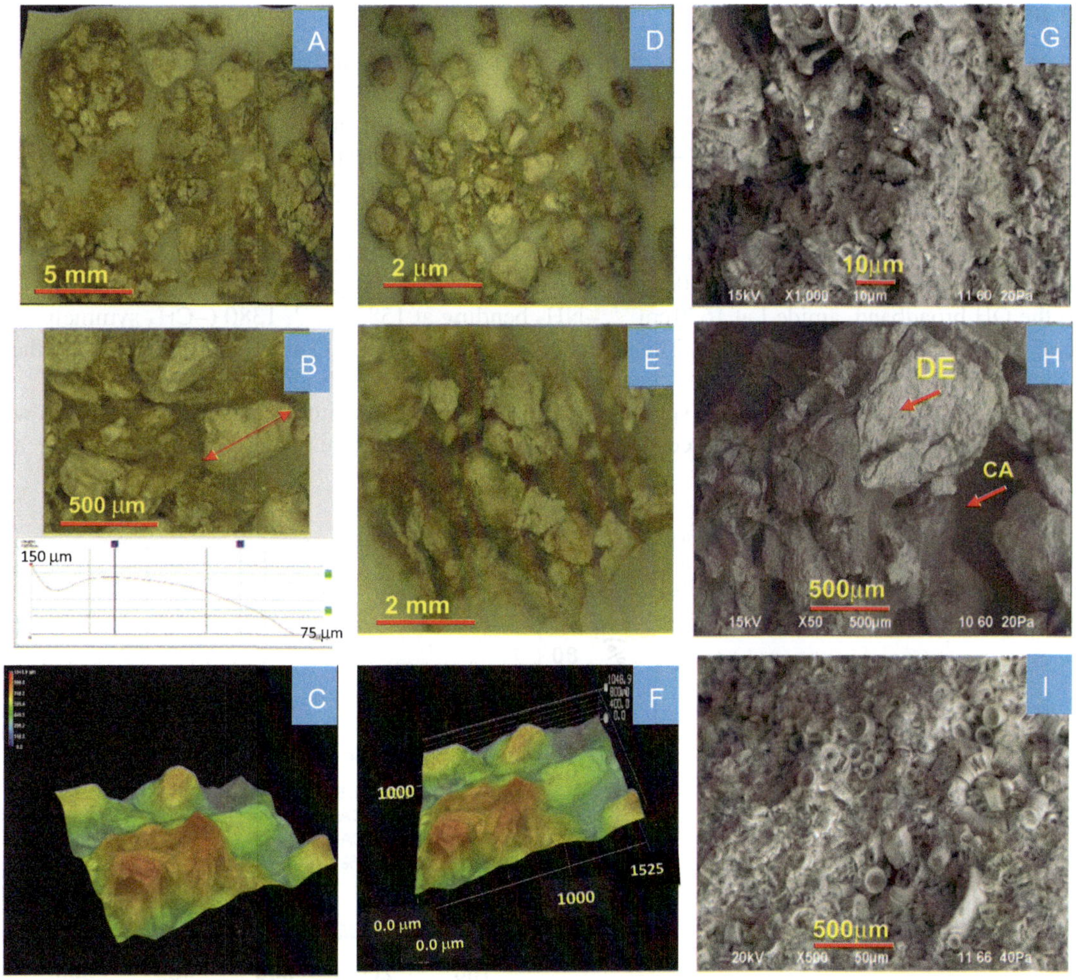

FIG. 4 Optical and scanning electron micrograph images of CCDE used for adsorption of Cr(VI) for 20 wt% chitosan (A), (C) and 30 wt% (D) to (F); (G) and (H) show scanning electron micrograph (SEM) of fresh chitosan at 1000 × and 50 ×; (I) show SEM-X-ray diffraction patterns for chromium in used chitosan adsorbent (DE is diatomaceous earth).

Chitosan coating might have blocked some of the pores on the surface of diatomaceous earth, and the chitosan film was not porous enough to retain the original porosity of diatomaceous earth. However, the coating affords a combination of higher surface area with a favorable chemical functional surface. The higher magnification images show the porous nature of the adsorbent. Energy-dispersive X-ray microanalysis of spent adsorbent was used with SEM for the elemental analysis or chemical characterization of samples. The images of green spots show a mapping of adsorbed chromium on used chitosan.

Thermogravimetric analysis

The thermal decomposition of neat and diatomaceous earth chitosan is shown in Fig. 5. A temperature range of decomposition for chitosan with a similar degree of deacetylation has been reported with a decomposition temperature of 299°C. Kittur et al. concludes that decomposition temperatures decrease with decreasing acetyl units [37]. This reasoning was shown using chitosan with the same number of acetyl groups but different molecular weights. It is generally agreed that a decrease in the degree of polymerization reduces thermal stability.

In chitosan, the decomposition process of the *N*-acetylated compound is overlapped by the *N*-deacetylated unit, thereby increasing the widening process seen at temperatures up to 400°C. In this study, although the molecular weight and degree of deacetylation were different, both chitosan adsorbents had almost the same thermal decomposition profile. The percentage weight loss at 600°C, matches closely with the chitosan fraction of diatomaceous earth coated adsorbent.

Fourier transfer infrared spectroscopy

Infrared spectra (FTIR) were taken for chitosan, diatomaceous earth (DE), and chitosan-coated 20% and 30% DE. The FTIR spectra of the tested adsorbent confirmed the presence of the characteristic absorption bands for chit. Fig. 6 shows that FTIR results of fresh chitosan diatomaceous earth granules have peaks at 3355 and 1025 cm^{-1} which are bands of stretching vibration of O—H, and related to C—C bond, respectively. IR spectra bands for chitosan are observed between 1313 and 1650 cm^{-1} which corresponds to C=N stretching bond, and at 2800 up to 2900 cm^{-1} which is for C—H stretching vibration.

The spectrum of chitosan adsorbent shows the characteristics bands of chitosan reported by other authors [25–27]. Bands of NH stretching modes at 3370 and 3300 cm^{-1} as well as those of CH groups at 2930 and 2871 cm^{-1} appear superimposed to the OH broadband, amide I at 1650 cm^{-1}, $-NH_2$ bending at 1585 cm^{-1}, 1380 ($-CH_3$ symmetric deformation), and 1313 cm^{-1} (amide III and CH_2 wagging). The stretching bonds of the Si—O and Si-O-Si groups and the deformation bonds for these groups correspond to 630 cm^{-1} in chitosan diatomaceous earth granules.

The stretching frequencies of the infrared spectra of the neat and coated chitosan all functional groups were the same. The absorbance peak at 1650 cm^{-1} shows secondary amide in the solid-state. This band is due to the antisymmetric

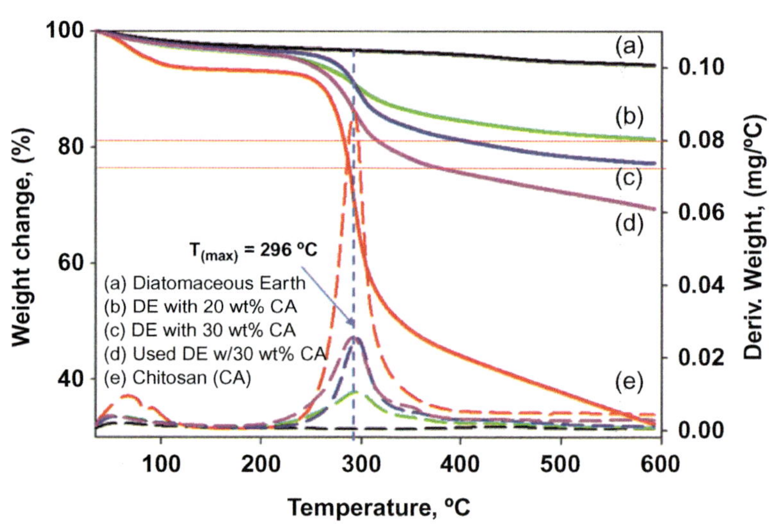

FIG. 5 Thermogravimetric analysis of chitosan and chitosan-coated diatomaceous earth. The *vertical broken line* shows the maximum weight loss rate, the *horizontal broken lines* indicate residual mass of adsorbent with 20 and 30 wt % chitosan.

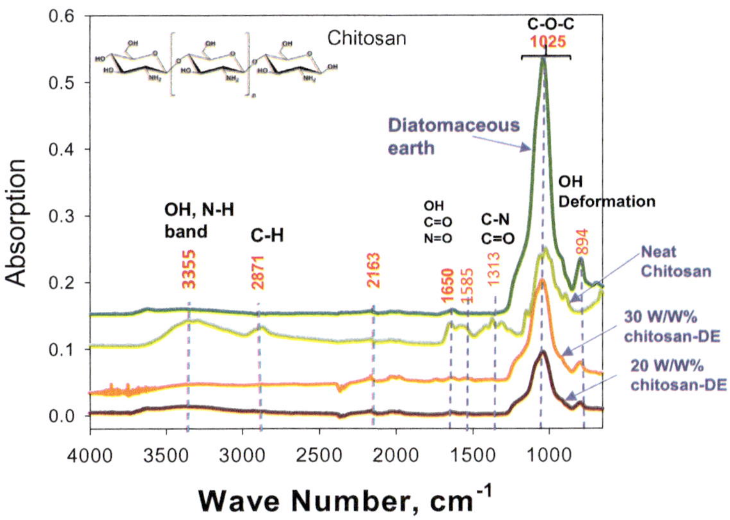

FIG. 6 FTIR spectra of neat and diatomaceous earth coated chitosan.

deformational vibration due to the formation -NH_3^*, which forms the band appearing at 1521 cm^{-1} may be assigned to the symmetric NH_3^* deformational mode. For this study FTIR, the physical and chemical property of the prepared chitosan was compared with the neat chitosan, and it was observed that surface functional groups are identical.

Adsorption studies

Standards were prepared for the calibration of inductively coupled plasma atomic emission spectroscopy showed good linearity. Control tests on samples with no adsorbent at pH of 4 and 8, the DI water, and samples showed no change in chromium concentrations. Samples with diatomaceous earth showed no chromium presence.

The influence of pH is an essential parameter for adsorption of Cr(VI) from the aqueous solution because it affects the solubility of the metal ions, concentrations of counterion on the functional groups of the adsorbent, and the degree of ionization of the adsorbent. Adsorption of Cr(VI) was studied at three levels of 4, 6, and 8 to examine the effect of pH on the Cr removal efficiency. The optimal efficiency was around 4, where a higher pH removal efficiency decline. Metals such as Cr(VI) and arsenic exist as an anion. At lower pH, the amine group on chitosan is protonated to varying degrees.

Effect of initial solution-pH

Solution-pH has a strong effect on the speciation of Cr ions. At lower pH range 2–6, dichromate ($Cr_2O_7^{2-}$) and hydrogen chromate ($HCrO_4^-$) are present, whereas at pH >6 $Cr_2O_4^{2-}$ is the more stable ion form [34]. Generally, the initial pH value of the metal solution significantly affects the adsorption onto solid–liquid interfaces. To determine the optimum solution pH for removal of heavy metal ions by chitosan, the adsorption capacity of Cr(V) ions from the water was studied at various pH values in the range from 4 to 8, as shown in Fig. 7. The amine groups of chitosan are the primary active groups for binding metal ions, and amine can be protonated to NH_3^+ in an acidic medium. Free electron pairs of amino groups can bind with unoccupied orbitals of metal ions to form complexes (Fig. 8).

According to Lewis's acid–base theory, the interactions between chitosan and metal ions where M^{2+} represents a divalent cation. Treating contaminated water with an initial concentration of 1000 mg/L, the removal efficiency of 95% or more is needed to achieve clean drinking water standards as set by World Health Organization (1). This study shows that chitosan adsorbent has achieved this goal (Fig. 8). At lower pH, more amino (−NH_2) groups are inclined to bind with hydrogen ions to form –NH^{3+}. Protonation of available sites (–NH_2) at lower pH was obtained, leading to increased adsorption capacity. This result agrees with an increase in the zeta potential of CCDE with the decrease pH, where the point of zero charges of the CCDE is at pH 5.8 [35]. Thus the surface charge of CCDE strongly determines the bond type created between the Cr(VI) ion and the adsorbent. The widening of -NH_2 stretching band of the FTIR spectra suggests hydrogen bonding for the interaction of Cr(VI) interacts with CCDE with electrostatic adsorption between the positively charged surface and chromate ion [28]. These results are in good agreement with other studies [35,38].

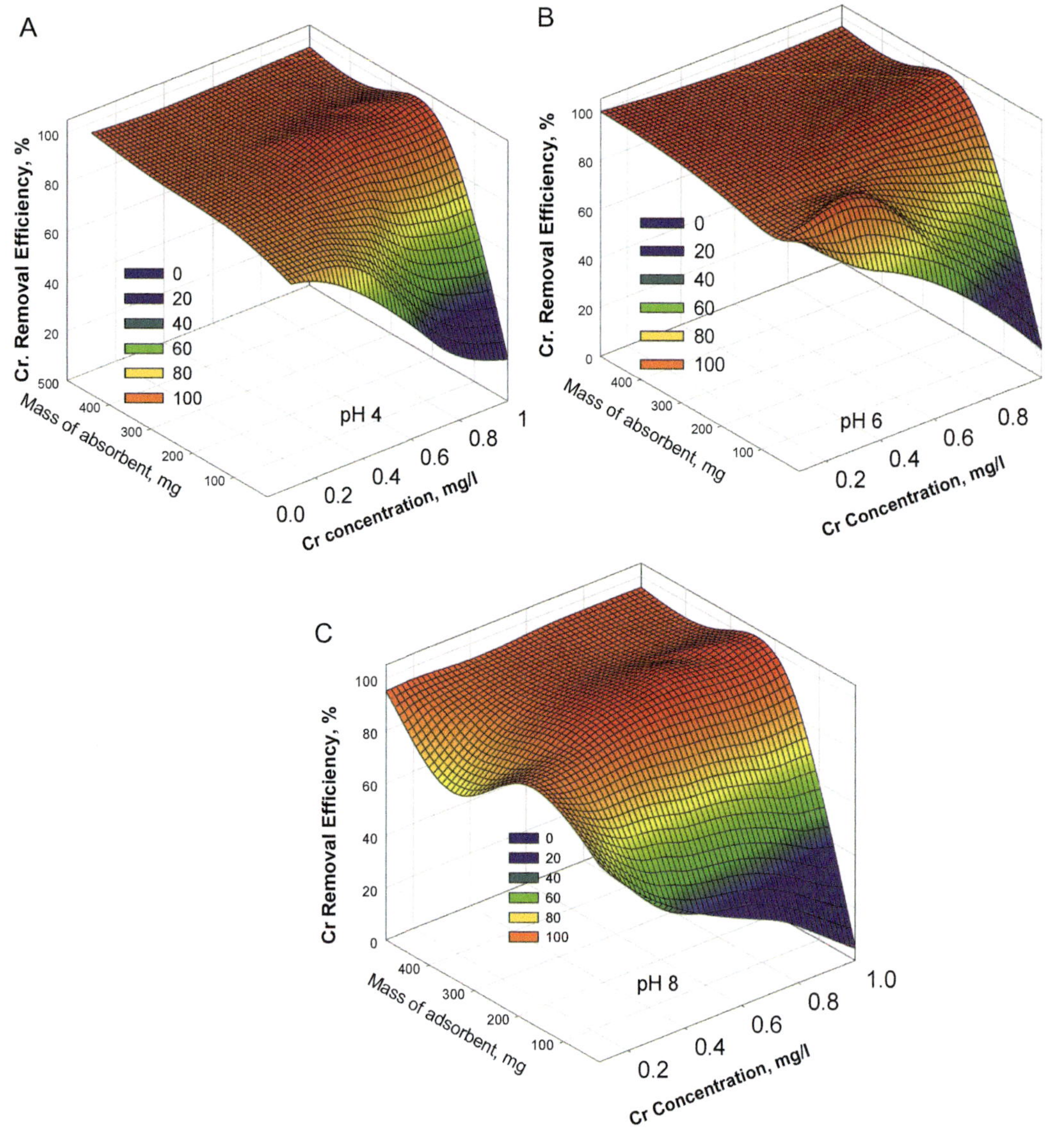

FIG. 7 Effect of mass of adsorbent and initial concentration of chitosan at (A) pH 4, (B) pH 6, and (C) pH 8.

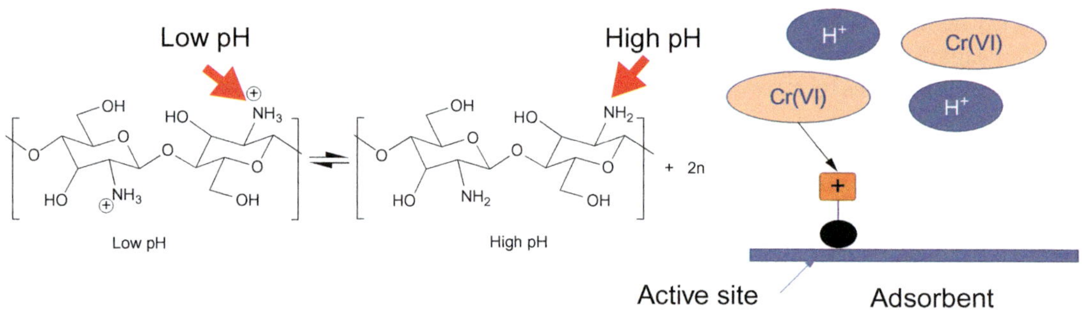

FIG. 8 Effect of pH on chitosan hydration and effect on chromium adsorption.

Effect of initial metal ion concentration

The dependence of Cr(VI) sorption on the adsorbent dose was studied by varying the amount of chitosan from 0.4 to 10 g/L, at four-factor levels. Cr(VI) removal efficiency for all the adsorbents (including control tests were neat diatomaceous earth). The removal efficiency of the chitosan increased with increasing doses, which is expected due to the higher dose of adsorbents in the solution. Hence, the greater the availability of exchangeable sites for the ions, the higher the removal efficiencies. The removal efficiency at 1 mg/L of Cr(VI) CCDE varied from 67%, 93%, 96%, and 98% as chitosan amount increased from 0.4, 3.6, 7.2, and 10 g/L, respectively. The results suggest that after a certain dose of adsorbent, the maximum adsorption reaches on the free ions remaining in solution with a further increase in the dose of adsorbent, there were no improvements in removal efficiencies.

To determine the adsorption capacity of the adsorbent, it is necessary to carry out equilibrium experiments at different initial metal concentrations. As shown in Fig. 7, it was observed that an increase in the initial concentration (10–300 mg/L) of Cr(VI) ions enhanced the corresponding adsorption capacity of CCDE. At lower initial metal ion concentration, the adsorption capacity increased proportionally with the initial concentration, indicating sufficient adsorption sites on the chitosan to bind with the metal ions in solution. Consequently, the adsorption capacity depended on the number of metal ions transported from the solution to the adsorbent surfaces. However, on a further increase of the initial metal ion concentration, the adsorption capacities increased slowly or even remained unchanged, suggesting that the number of active sites on CCDE could be the factor limiting its adsorption capacities. Therefore, the maximum adsorption value of metal ions on CCDE was obtained at a higher concentration of metal ions due to the saturation of available active sites on the chitosan surface.

Adsorption isotherms

Sorption isotherm models are widely used for fitting the data using Langmuir and Freundlich equations to examine the relationship between sorbed (q_e) and aqueous concentration *Ce* at equilibrium. The Langmuir model assumes that the sorption of metal ions occurs on a homogeneous surface by monolayer adsorption without any interaction between adsorbed ions. The equilibrium data were obtained, using increasing Cr(VI) concentrations were varied while the adsorbent mass was kept constant. The sorption samples were tumbled for 48 h before they were centrifuged, filtered, and prepared for ICP-OES analysis.

Equilibrium data are usually evaluated using the Langmuir and Freundlich isotherm models, which can be expressed using the following equations:

$$q_e = \frac{Q_e b C_e}{(1 + bC_e)} \tag{3}$$

C_e (mg/L) is the equilibrium concentration of metal ions in solution, q_e (mg g^{-1}) is the adsorption capacity, q_{max} (mg g^{-1}) is the maximum adsorption capacity, Q_e (Lmg^{-1}) is the Langmuir constant,

$$\frac{C_e}{q_e} = \frac{1}{bQ_o} + \frac{C_e}{Q_o} \tag{4}$$

The constants Q_o and b are the characteristics of the Langmuir equation and can be determined from a linearized form of the Langmuir equation, represented by Eq. (4). Therefore, a plot of C_e/q_e versus C_e gives a straight line of slope $1/Q_e$ and intercept $1/bQ_o$. The constants Q_o of the Langmuir equilibrium constant and the $1/Q_o$ give the theoretical monolayer saturation capacity. The data fit well with the Langmuir isotherm. At the equilibrium concentration of 0.5 mg/L, the uptake of Cr(VI) was 2 mg/L (Fig. 9). The maximum adsorption capacity at pH of 4 was estimated at 18.69 mg/L.

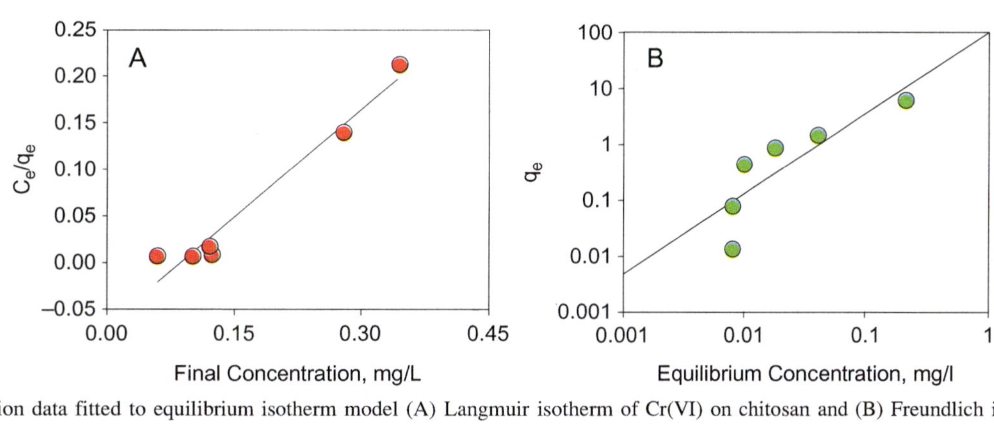

FIG. 9 Adsorption data fitted to equilibrium isotherm model (A) Langmuir isotherm of Cr(VI) on chitosan and (B) Freundlich isotherm of Cr(VI) adsorption on chitosan.

For Freundlich isotherm shown in Eq. (5), Q_o (mg g^{-1}) (L mg^{-1})$^{1/n}$ and n are the Freundlich constants, representing adsorption capacity and adsorption intensity, respectively. The experimental data were fit using C_e/q_e versus C_e curves, as shown in Fig. 8. The R^2 values are considered as a measure of the goodness of fit of experimental data by isotherm models.

Freundlich equation is an empirical equation based on adsorption on a heterogeneous surface that is valid only up to a certain concentration. The equation is represented by

$$q_e = K_d C_e^{1/n} \tag{5}$$

where K_d is the partitioning coefficient, and $1/n$ is the Freundlich constants characteristics of the system, indicating the adsorption capacity and the adsorption intensity, respectively. The K_d term in Freundlich equation implies that the energy of adsorption on a homogeneous surface is independent of surface coverage. Eq. (5) can be simplified by linearized form:

$$\ln q_e = b \ \ln C_e + \ \ln K_d \tag{6}$$

The Langmuir models fit the data better than the Freundlich model for adsorption of Cr(VI) on chitosan (Fig. 9A).

Adsorption kinetics

Fig. 10 shows the amount of metal ions adsorbed onto CCDE versus contact time. Fast initial adsorption was observed in the first 10 min, with adsorption equilibrium being reached after approximately 60 min.

Therefore, it was considered that CCDE could be used as an adsorbent to quickly remove toxic metal ions from polluted water, especially for application in emergency treatment for sudden pollution accidents. The rapid adsorption in the initial stage may be due to the plentiful availability of active sites on the CCDE. At the same time, the rate gradually slows down with decreasing active sites during the adsorption process. The experimental data were fit using two kinetic models: pseudo-first-order model and a pseudo-second-order, which is expressed in the following equation:

$$\log(q_e - q_t) = \log q_e - \left(\frac{k_1}{2.303}\right) t \tag{7}$$

$$\frac{t}{q_t} = \frac{1}{k_2 q_e^2} + \frac{t}{q_e} \tag{8}$$

where q_e (mg g^{-1}) and q_t (mg g^{-1}) represent the amounts of adsorbed metal ions at equilibrium and at any time t (min), respectively. k_1 (min^{-1}) and k_2 (g mg^{-1} min^{-1}) are the rate constants for the pseudo-first-order and pseudo-second-order model, respectively.

The kinetics plots for absorption of metal ions onto CCDE are shown in Figs. 11. According to the correlation coefficients ($R^2 > 0.99$) of the kinetics plots for the pseudo-second-order kinetic models, the adsorption process was the best fit by the pseudo-second-order model and the theoretical values (q_e) approached the experimental values (q_e) closely. The pseudo-second-order kinetic parameters were obtained based on nonlinear assumptions [38–42].

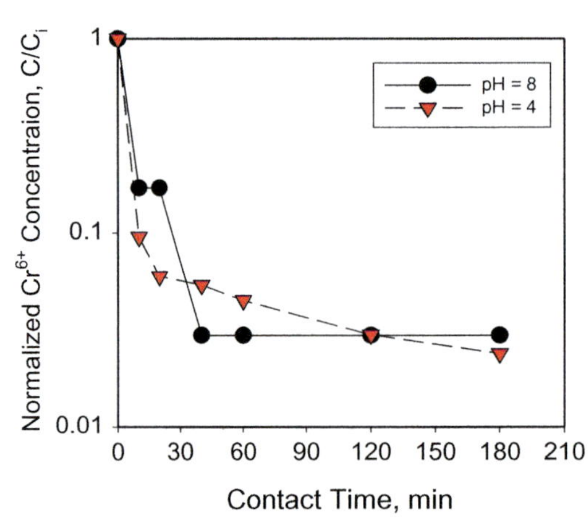

FIG. 10 Adsorption kinetics of Cr(VI) on chitosan adsorbent at pH 4 and 8, initial chromium concentration 0.3 mg/L, and mass of adsorbent 270 mg.

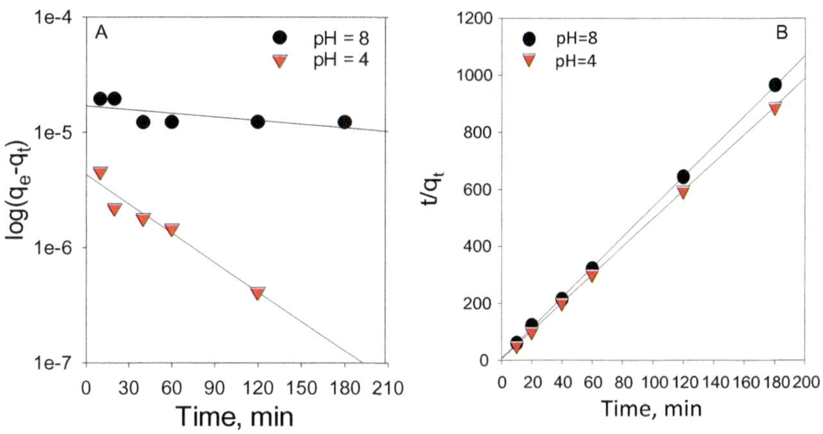

FIG. 11 Data represented based on the (A) First order adsorption kinetics model for Cr(VI) on chitosan (B) second order adsorption kinetics model for Cr(VI) on chitosan.

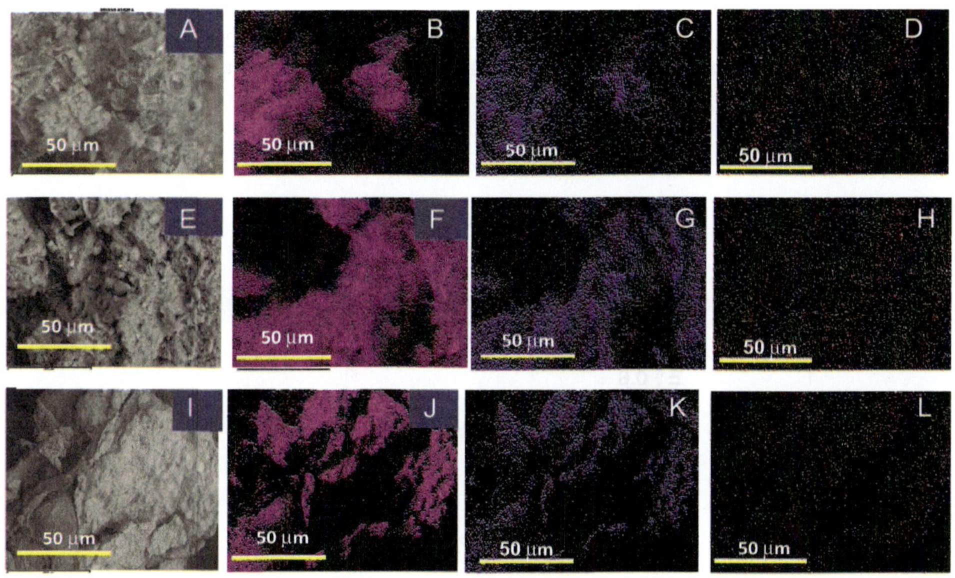

FIG. 12 Analysis of sorption of metal ions on CCDE showing SEM images (A) fresh CCDE 20% CA, (E) fresh CCDE 30% CA, (I) spent CCDE. SEM-XRD maping of silica images (B), (F) and (J), alumina (C), (G) and (K), and iron (D), (H) and (L).

Scanning electron microscopy with X-ray diffraction

The surface textural characterization of granular chitosan-diatomaceous earth adsorbent before and after Cr(VI) adsorption was carried out using a scanning electron microscope at various magnifications. Scanning electron microscope and energy dispersive X-ray microanalysis (SEM–EDX) determined fresh and spent adsorbent's microstructure and elemental composition (Fig. 12). According to the previously published methodology, the elemental analysis and ions distribution mapping on adsorbent samples were performed using the Oxford Instrument energy dispersive X-ray system [30,33]. SEM–EDX generates information about the chemical composition of a sample, including what elements are present as well as their distribution and concentration EDX detected the emission of various elements, including silicon (Si), aluminum (Al), iron (Fe) which are from diatomaceous earth. Analysis of used adsorbent detected all the above elements and uniform distribution of Cr, which indicates that active adsorption sites are uniformly distributed (Fig. 13).

Chitosan adsorbent for point-of-use filter

Point-of-use (POU) water filters are using at individual locations where water is dispensed for consumption. POU systems offer a solution in developing communities as low-cost, low maintenance, and grid-independent systems. There is an

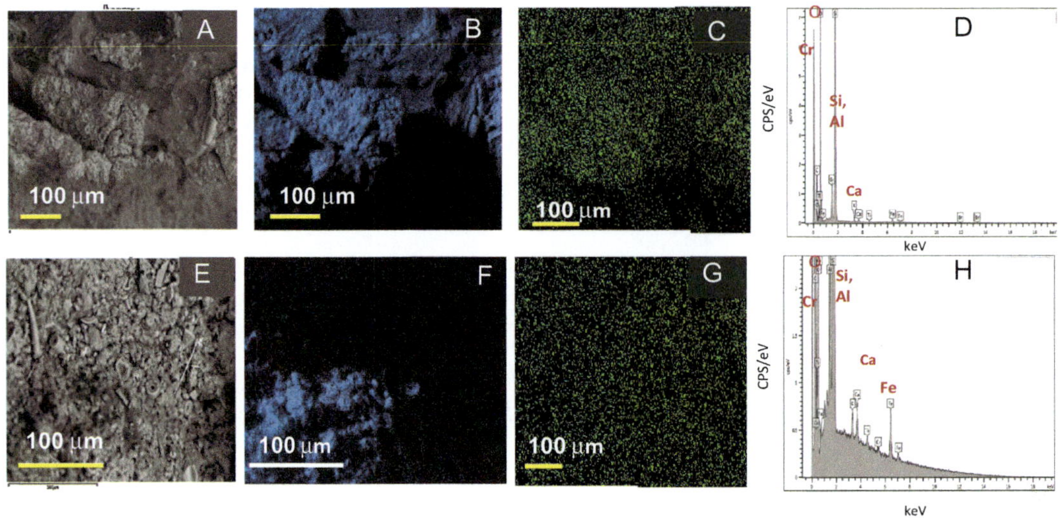

FIG. 13 Scanning electron micrograph of used adsorbent (A) CCDA with 30% CA, (E) CCDA with 20% CA, XRD mapping of elemental analysis of oxygen ions (B) and (F), and adsorbed Cr (C) and (G). Images (D) and (H) showing X-ray diffraction analysis indicating the presence of characteristic diffraction peaks of CCDE and adsorbed Cr.

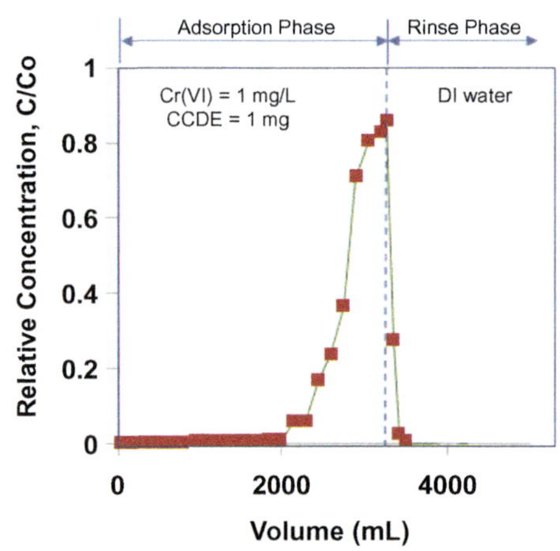

FIG. 14 Two-phase Cr(VI) adsorption followed by rinsing with deionized water showing irreversible adsorption of chromium ions on active sites.

extensive volume of literature on the POU option as an effective alternative for small-scale, nonacute, and high-risk emergency applications [43,44]. Most of these POU applications are effective at microbiological and disease reduction. Although health impact and the reduction of water borne pathogens are the gold standards for POU applications, these matrices do not show their effectiveness with organic and inorganic pollutants [45]. The use of POU or point-of-entry technologies offers a potentially viable and cost-effective method for providing safe water from sources that are contaminated with organic and inorganic chemicals [45].

A multilayered filter is needed to remove the various types of contaminants that are likely to encounter in surface waters. These contaminants include biological, organic, inorganic, and turbidity. For example, a filter that combines sand bed with CCDE adsorbent and clay adsorbent would effectively reduce inorganic pollutants such as heavy metals, organic contaminants that are toxic or affect the taste and odor of water [38–41,46]. Following adsorption, Cr(VI) saturated packed column bed of CCDE was rinsed with de-ionized water as the eluent. DI water was pumped upward into the column (Fig. 14) at room temperature at a fixed flow rate of 0.01 L/min, and effluent samples were collected for analysis. The adsorbent bed with 1 mg of CCDE has the capacity of treating 2 L before the breakthrough. As the Cr(VI) solution is changed to rinse water, there was no detectable leaching of adsorbed Cr(VI). The result shows little or no leaching of adsorbed Cr(VI) on CCDE.

Conclusions

Chitosan-coated on diatomaceous earth was prepared and used as an adsorbent to remove Cr(VI) ions from an aqueous solution. The study indicated that the removal rate of Cr(VI) ions from aqueous solution onto chitosan-coated diatomaceous earth powder was highly effective, reducing Cr(VI) from 1 mg/L to below US Environmental Protection Agency safe drinking water standard of 0.1 mg/L. Approximately 1 mg of chitosan is sufficient to treat 42 L of chromium-contaminated water with an initial concentration of 0.5 mg/L to achieve safe drinking water levels. The Cr(VI) adsorption data on chitosan followed the Langmuir isotherm model, showing a maximum adsorption capacity of 18.69 mg g^{-1} goodness of fitting, R^2 value 0.95. The optimal pH was found to be 4, which suggests that adjusting pH can improved filter performance.

The Cr(VI) adsorption data on chitosan followed the Langmuir isotherm model, showing a maximum adsorption capacity of 18.69 mg g^{-1} goodness of fitting, R^2 value 0.95. The optimal pH was found to be 4, which suggests that adjusting pH can improved filter performance. The studies showed that chitosan could convert to a cost-effective adsorbent to remove metal ions from contaminated waters, such as Cr(VI). The studies showed that chitosan could be converted to a cost-effective adsorbent to remove metal ions from contaminated waters, such as Cr(VI). Adsorption equilibrium was reached after approximately 30 min, and the adsorption kinetics was a good fit by a second-order model for Cr(VI). Leaching tests of used chitosan showed no removal of adsorbed Cr(VI). Leaching tests of used chitosan showed no removal of adsorbed Cr(VI). Considering the low cost of the raw material and convenient preparation route, CCDE can be used as a novel adsorbent for remediation of heavy metal ions from an aqueous solution. Future studies will include the effectiveness of chitosan adsorption on other heavy metal water pollutants such as arsenic and lead, investigating the impacts of other ions and organic matter, and developing point-of-use water treatment filters that work on natural waters on multiple pollutants.

References

[1] T. Thompson, J. Fawell, S. Kunikane, D. Jackson, S. Appleyard, P. Callan, J. Bartram, P. Kingston, S. Water, World Health Organization, *Chemical safety of drinking water: assessing priorities for risk management*, World Health Organization, 2007.

[2] S. Yin, et al., Heavy metal pollution in the surface water of the Yangtze estuary: a 5-year follow-up study, Chemosphere 138 (2015) 718–725.

[3] R. Xiao, et al., Distribution and pollution, toxicity and risk assessment of heavy metals in sediments from urban and rural rivers of the Pearl River delta in southern China, Ecotoxicology 22 (10) (2013) 1564–1575.

[4] A. Abbas, et al., Heavy metal removal from aqueous solution by advanced carbon nanotubes: critical review of adsorption applications, Sep. Purif. Technol. 157 (2016) 141–161.

[5] M. Owlad, et al., Removal of hexavalent chromium-contaminated water and wastewater: a review, Water Air Soil Pollut. 200 (1) (2009) 59–77.

[6] R. Saha, R. Nandi, B. Saha, Sources and toxicity of hexavalent chromium, J. Coord. Chem. 64 (10) (2011) 1782–1806.

[7] C.-Y. Hu, et al., Hexavalent chromium removal from near natural water by copper–iron bimetallic particles, Water Res. 44 (10) (2010) 3101–3108.

[8] A.S. Ellis, T.M. Johnson, T.D. Bullen, Chromium isotopes and the fate of hexavalent chromium in the environment, Science 295 (5562) (2002) 2060–2062.

[9] S.A. Katz, H. Salem, The toxicology of chromium with respect to its chemical speciation: a review, J. Appl. Toxicol. 13 (3) (1993) 217–224.

[10] S. Mishra, R.N. Bharagava, Toxic and genotoxic effects of hexavalent chromium in environment and its bioremediation strategies, J. Environ. Sci. Health C 34 (1) (2016) 1–32.

[11] I. Moffat, et al., Hexavalent chromium in drinking water, J. Am. Water Works Assoc. 110 (5) (2018) E22–E35.

[12] A.K. Patlolla, et al., Oxidative stress, DNA damage, and antioxidant enzyme activity induced by hexavalent chromium in Sprague-Dawley rats, Environ. Toxicol. Inter. J. 24 (1) (2009) 66–73.

[13] C. Pellerin, S.M. Booker, Reflections on hexavalent chromium: health hazards of an industrial heavyweight, Environ. Health Perspect. 108 (9) (2000) A402–A407.

[14] R.M. Sedman, et al., Review of the evidence regarding the carcinogenicity of hexavalent chromium in drinking water, J. Environ. Sci. Health C 24 (1) (2006) 155–182.

[15] E.L. Hawley, et al., Treatment technologies for chromium (VI), Chromium (VI) Handbook 275 (2004) 309.

[16] B.R. James, et al., Oxidation-reduction chemistry of chromium: relevance to the regulation and remediation of chromate-contaminated soils, Soil Sediment Contam. 6 (6) (1997) 569–580.

[17] B.R. James, Peer reviewed: the challenge of remediating chromium-contaminated soil, Environ. Sci. Technol. 30 (6) (1996) 248A–251A.

[18] B. Volesky, Z. Holan, Biosorption of heavy metals, Biotechnol. Prog. 11 (3) (1995) 235–250.

[19] M. Tumolo, et al., Chromium pollution in European water, sources, health risk, and remediation strategies: an overview, Int. J. Environ. Res. Public Health 17 (15) (2020) 5438.

[20] J. Wang, C. Chen, Chitosan-based biosorbents: modification and application for biosorption of heavy metals and radionuclides, Bioresour. Technol. 160 (2014) 129–141.

[21] M.M. Rao, et al., Removal of copper and cadmium from the aqueous solutions by activated carbon derived from Ceiba pentandra hulls, J. Hazard. Mater. 129 (1–3) (2006) 123–129.

[22] T. Nguyen, et al., Applicability of agricultural waste and by-products for adsorptive removal of heavy metals from wastewater, Bioresour. Technol. 148 (2013) 574–585.
[23] J. Febrianto, et al., Equilibrium and kinetic studies in adsorption of heavy metals using biosorbent: a summary of recent studies, J. Hazard. Mater. 162 (2–3) (2009) 616–645.
[24] O. Duman, E. Ayranci, Adsorption characteristics of benzaldehyde, sulphanilic acid, and p-phenolsulfonate from water, acid, or base solutions onto activated carbon cloth, Sep. Sci. Technol. 41 (16) (2006) 3673–3692.
[25] O. Duman, E. Ayranci, Attachment of benzo-crown ethers onto activated carbon cloth to enhance the removal of chromium, cobalt and nickel ions from aqueous solutions by adsorption, J. Hazard. Mater. 176 (1–3) (2010) 231–238.
[26] E. Ayranci, O. Duman, In-situ UV-visible spectroscopic study on the adsorption of some dyes onto activated carbon cloth, Sep. Sci. Technol. 44 (15) (2009) 3735–3752.
[27] O. Duman, S. Tunc, T.G. Polat, Adsorptive removal of triarylmethane dye (basic red 9) from aqueous solution by sepiolite as effective and low-cost adsorbent, Microporous Mesoporous Mater. 210 (2015) 176–184.
[28] O. Duman, S. Tunç, T.G. Polat, Determination of adsorptive properties of expanded vermiculite for the removal of CI basic red 9 from aqueous solution: kinetic, isotherm and thermodynamic studies, Appl. Clay Sci. 109 (2015) 22–32.
[29] X.-J. Zhang, et al., Emergency Drinking Water Treatment during Source Water Pollution Accidents in China: Origin Analysis, Framework and Technologies, ACS Publications, 2011.
[30] V.M. Boddu, et al., Removal of hexavalent chromium from wastewater using a new composite chitosan biosorbent, Environ. Sci. Technol. 37 (19) (2003) 4449–4456.
[31] M.-W. Wan, et al., Copper adsorption through chitosan immobilized on sand to demonstrate the feasibility for in situ soil decontamination, Carbohydr. Polym. 55 (3) (2004) 249–254.
[32] V.M. Boddu, et al., Removal of arsenic (III) and arsenic (V) from aqueous medium using chitosan-coated biosorbent, Water Res. 42 (3) (2008) 633–642.
[33] S. Hasan, et al., Adsorption of divalent cadmium (cd (II)) from aqueous solutions onto chitosan-coated perlite beads, Ind. Eng. Chem. Res. 45 (14) (2006) 5066–5077.
[34] W.W. Ngah, L. Teong, M.M. Hanafiah, Adsorption of dyes and heavy metal ions by chitosan composites: a review, Carbohydr. Polym. 83 (4) (2011) 1446–1456.
[35] S.S. Salih, T.K. Ghosh, Preparation and characterization of chitosan-coated diatomaceous earth for hexavalent chromium removal, Environ. Process. 5 (1) (2018) 23–39.
[36] Juliasih, N.L.G.R., et al., Verification of the determination method of dissolved metal content using ICP-OES and its application for river water in Bandar Lampung City. Jurnal Kimia Sains dan Aplikasi. 24(1): p. 29–36.
[37] F. Kittur, et al., Characterization of chitin, chitosan and their carboxymethyl derivatives by differential scanning calorimetry, Carbohydr. Polym. 49 (2) (2002) 185–193.
[38] X.-J. Hu, et al., Adsorption of chromium (VI) by ethylenediamine-modified cross-linked magnetic chitosan resin: isotherms, kinetics and thermodynamics, J. Hazard. Mater. 185 (1) (2011) 306–314.
[39] S. Al-Asheh, et al., Predictions of binary sorption isotherms for the sorption of heavy metals by pine bark using single isotherm data, Chemosphere 41 (5) (2000) 659–665.
[40] S. Basha, Z. Murthy, B. Jha, Sorption of Hg (II) onto Carica papaya: experimental studies and design of batch sorber, Chem. Eng. J. 147 (2–3) (2009) 226–234.
[41] M. Rinaudo, Chitin and chitosan: properties and applications, Prog. Polym. Sci. 31 (7) (2006) 603–632.
[42] Y.-S. Ho, Second-order kinetic model for the sorption of cadmium onto tree fern: a comparison of linear and non-linear methods, Water Res. 40 (1) (2006) 119–125.
[43] D. Lantagne, T. Clasen, Point-of-use water treatment in emergency response, Waterlines (2012) 30–52.
[44] P.J. Sheffer, et al., Efficacy of new point-of-use water filter for preventing exposure to Legionella and waterborne bacteria, Am. J. Infect. Control 33 (5) (2005) S20–S25.
[45] L.T. Rozelle, Point-of-use and point-of-entry drinking water treatment, J. Am. Water Works Assoc. 79 (10) (1987) 53–59.
[46] W. Jiang, et al., Spherical polystyrene-supported chitosan thin film of fast kinetics and high capacity for copper removal, J. Hazard. Mater. 276 (2014) 295–301.

Chapter 10

Recent trends of carbon nanotubes and chitosan composites for hexavalent chromium removal from aqueous samples

Mian Muhammad Ahson Aslam[a], Hsion-Wen Kuo[a,*], Walter Den[b,*], Muhammad Sultan[c,*], Kashif Rasool[d], and Muhammad Bilal[c]

[a]*Department of Environmental Science and Engineering, Tunghai University, Taichung City, Taiwan, R.O.C,* [b]*Institute for Water Resources Science and Technology, Department of Mathematical, Physical, and Engineering Sciences, Texas A&M University-San Antonio, San Antonio, TX, United States* [c]*Department of Agricultural Engineering, Bahauddin Zakariya University, Multan, Pakistan,* [d]*Qatar Environment and Energy Research Institute, Hamad Bin Khalifa University (HBKU), Qatar Foundation, Doha, Qatar*

[*]*Corresponding authors: e–mail addresses: hwkuo@thu.edu.tw; walter.den@tamusa.edu; muhammadsultan@bzu.edu.pk*

Introduction

Modern life involves using Earth's resources beyond their viability and depleting or exhausting limited resources [1]. The industrial revolution has caused environmental problems by releasing either improperly handled or untreated industrial effluents into water systems, causing a waste buildup of hazardous organic and inorganic pollution in receiving water bodies [2,3]. Many contaminants do not degrade and end up in food chains that affect the ecosystem's service and the health of marine feeds [4,5]. The volume of waste in water systems increased and often exceeded the ability of aquatic systems to self-clean and purify. Such discharges increase with the migration and concentration of the population in urban regions due to modern life styles and the adverse effects of climate change [6–9].

Chromium is one of the most toxic trace elements in water systems, even at parts per billion levels, because of its usage in many industrial segments such as dyeing, nuclear power, leather tanning, photography, electroplating, metallurgy, refractories, and textile industries [6,10]. The strong solubility and diffusiveness of hexavalent chromium (Cr(VI)) species (e.g., chromate, CrO_4^{2-}, and dichromate, $Cr_2O_7^{2-}$) help them penetrate easily across bio-membrane tissue [11–13]. Trivalent chromium (Cr(III)) is another environmentally important Cr species which considered nontoxic at small concentrations but may be harmful at high concentrations [11]. The rise in the number of articles on Cr toxicity published in the previous decade reflects the desire to explain and remediate the Cr contaminations [6]. Several conventional approaches have been examined to remediate Cr from the aqueous solutions. These methods include precipitation [12], ultrafiltration [13], coagulation/flotation [14], electrochemical and reverse osmosis methods [15], solvent extraction [16] and ion-exchange methods [17]. Most of these methods are effective, but they have limitations in terms of high energy requirements, incomplete disposal, high operating costs, complex operations, and the formation of toxic sludge; they also may not be able to treat low concentrations of metal ions in water [18,19]. Table 1 shows the key advantages and disadvantages associated with these methods of removing heavy metal ions from an aqueous medium. Among these techniques, adsorption is widely recognized for its ability to economically eliminate various contaminants from wastewaters [19–21]. Besides, the adsorbent can undergo regeneration by a proper desorption procedure [19].

In recent decades, scientists worldwide have shown great interest in developing and using the different adsorbents to sequestrate Cr(VI) in environmental samples [22–26]. Mostly, the literature focuses on the adsorbents such as functionalized CNTs [27], graphene [28], chitosan [29], silica [30], polypyrrole [31], AC [32], biochar [33], and clay minerals [34]-based adsorbents and nanomaterials in general [6]. Although a wide variety of adsorbents can extract Cr(VI) from the aqueous solutions, the adsorption process coincides with the transference of the adsorbate from a liquid phase to

Separation Science and Technology, Volume 15. https://doi.org/10.1016/B978-0-323-90763-7.00006-8

TABLE 1 Advantages and disadvantages of different techniques used for heavy metals remediation from aqueous solutions.

Method	Advantages	Disadvantages
Adsorption	• Large variety of adsorbents • Ease of operation • Works in wide range of pH • Highly effective in terms of performance • Usually, low cost and economical process • Better recovery and fast adsorption kinetics	• Waste product • Requires modification of adsorbents to enhance their adsorption performance • Selective depending on adsorbents
Chemical precipitation	• Cheap • Removal of most metals	• Mass production of sludge • Disposal problems • Ineffective for trace metal ions • Waste products
Chemical coagulation	• Dewatering • Sludge settlement	• Expensive • Requires many chemicals
Ion-exchange	• Selective • Easy to regenerate for several times	• Expensive • Remove limited metal ions • Occurrence of fouling • Better performance under low concentrations • Highly pH-dependent • Occurrence of free acids results into lower affinity of ion-exchange media • Reactor may clog and saturated rapidly • Cationic exchange resins saturate more rapidly than the anionic exchange resins • Resins may degrade with respect to time and target contaminant • Lake of selectivity in conventional ion-exchangers
Membrane filtration	• Production of reduced solid waste • Every efficient	• Reduced remote performance over time • Production of concentrated sludge • Requires high initial, operational and maintenance costs • Requires high energy and costs • Clogs rapidly • Smaller throughputs • Works effectively only under low flow rates • Not effective under low concentrations
Electrochemical methods	• Pick out metals • Does not consume chemicals • Pure metal ions extraction is possible • Efficient	• High operational and startup cost • High investment cost • Deposition of sludge on electrodes • Requires additional cost of sludge treatment • Passivation of anodes • Requires additional cost of posttreatment process
Oxidation	• A rapid removal of metal ion	• High energy requirement • By-product formation
Biological treatment	• Volume reduction • Active in removing individual metal ions	• Need sophisticated technology • Favorable environment is needed for growth and operation of biological media • Production of biological sludge and its treatment • The complexity of the mechanisms, management, and maintenance

TABLE 1 Advantages and disadvantages of different techniques used for heavy metals remediation from aqueous solutions.—cont'd

Method	Advantages	Disadvantages
		• Time consuming process • Low biodegradability of certain molecules • Several molecules are not bio-degradable • Possibility of sludge bulking and foaming
Photochemical	• Do not produce sludge • Rapid degradation	• Effective only in lab-scale operation • pH-dependent • Requires high of operations which make nonfeasible for small and/or medium scale operations • Smaller throughputs • Complex process of operations • Produce several types of by-products • Requires extremely high Energy and pressure conditions
Flotation	• Effective performance under lower retention time • Suitable for primary settling of effluents • Can combine with physicochemical processes • Vast range of ionic and nonionic collectors	• Requires high initial, installation, operational, energy and maintenance costs • Highly selective method and pH-dependent

the adsorbent surface. The adsorption efficiency is determined by factors such as temperature, pH, amount and structure of adsorbent, the initial concentrations, an ionic strength, and the presence of other pollutants [35–37]. In contrast to the previous review articles published, the objective of the current study is to outline the latest advancements in the field of adsorption using CNTs and chitosan-based adsorbents with current literature leading the trend toward the use of composites. In addition, it provides a one-step resource material of CNTs and chitosan-based adsorbents for Cr(VI) uptake.

Chromium chemistry and toxicity

Steely gray-colored Cr having the atomic number 24 is a stiff and brittle, and lustrous metal generally found in the crystalline solid of earth's crust [38]. It has an atomic weight of 61.9 u, melting point of 1907°C, and boiling point of 2672°C. Cr is known to have many oxidation states, but there are only two species that are of environmental importance, namely Cr(III) and Cr(VI) that differ in their physicochemical properties and toxicity. There are four main kinds of Cr(III) in an aqueous solution, wherein the soluble Cr(III) complex predominates at $pH < 4.0$ and precipitates as $Cr(OH)_3$ within the pH ranging from ~6.0 to ~12.0. Meanwhile, Cr(VI) exist as H_2CrO_4 at $pH < 1.0$, $HCrO_4^-$ and $Cr_2O_7^{2-}$ at pH in between 2.0 to 6.0, and CrO_4^{2-} exist at $pH > 6.0$, as depicted in Fig. 1. The ratio of Cr(VI) to Cr(III) mainly depends on the solution pH, potential of a redox reaction, and kinetic characteristics and easily switch from one oxidation state to another [6]. Moreover, the external factors between the bottom and upper sediment layers (e.g., solar radiation, water complexity, redox gradients) and other sensitive redox elements (i.e., Fe and Mn) can also affect the Cr speciation in the aquatic system [6]. Fig. 2 shows the transition scheme for Cr species within a water column, involving the water-atmosphere and water-sediment interfaces [39]. It is difficult to assess Cr speciation in natural water directly as the sample specimen must be well-preserved during its processing, pretreatment, collection, and determination. The accuracy of the findings requires the implementation of sufficient analytical and speciation methodologies.

The toxicity of Cr(VI) compounds is well-documented as compared to Cr(III) but the latter has been linked to inhibiting immune system activity [40] and destroying red blood cells [41]. Moreover, it has been reported that, in extracellular and intracellular structures, Cr(III) is first converted into Cr(V) and then transformed to Cr(VI) [42,43]. The toxicity of high levels of Cr(III) compounds cannot be overlooked as research has shown the conversion of Cr(III) into Cr(VI) during the chlorination process of wastewater treatment [44]. Hence, removing Cr(III) from different environmental matrices has been

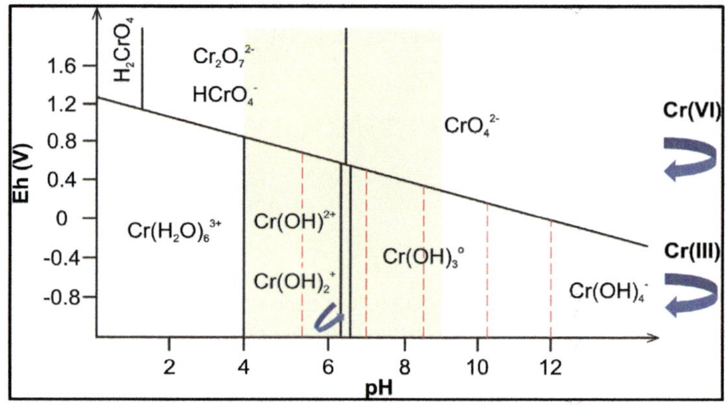

FIG. 1 Speciation profile of Cr in water at different pH range. *Adopted from B. Markiewicz, I. Komorowicz, A. Sajnóg, M. Belter, D. Barałkiewicz, Chromium and its speciation in water samples by HPLC/ICP-MS—technique establishing metrological traceability: a review since 2000. Talanta 132 (2015) 814–28. doi:10.1016/j.talanta.2014.10.002 with the permission of Elsevier.*

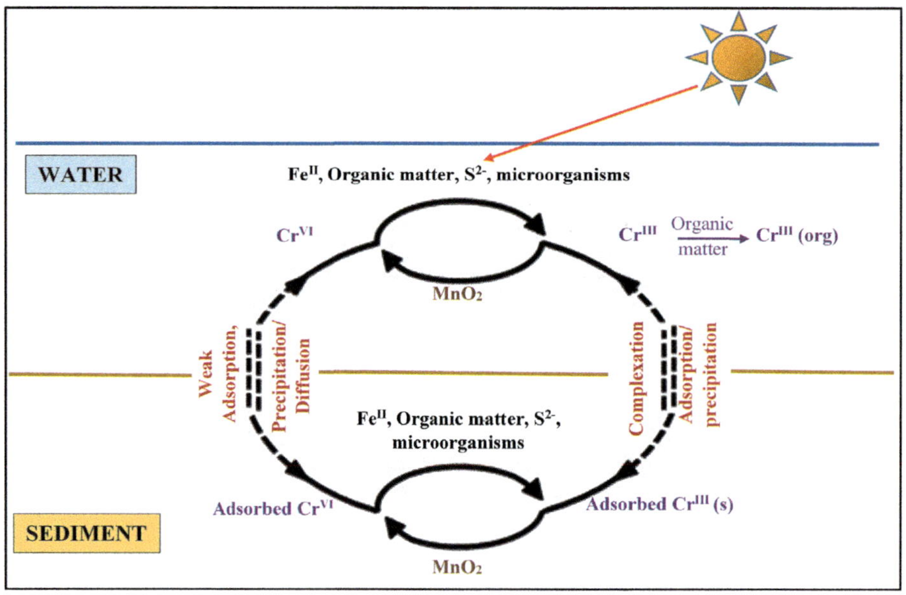

FIG. 2 Cr cycle in the water-sediment system. *Modified from B. Markiewicz, I. Komorowicz, A. Sajnóg, M. Belter, D. Barałkiewicz, Chromium and its speciation in water samples by HPLC/ICP-MS—technique establishing metrological traceability: a review since 2000. Talanta 132 (2015) 814–28. https://doi.org/10.1016/j.talanta.2014.10.002 with the permission of Elsevier.*

a subject of urgency [45,46]. Also, the compounds of Cr(VI) (i.e., $Cr_2O_7^{2-}$, CrO_4^{2-}, and $HCrO_4^-$) are oxidizing agents; they are mobile, soluble, and bioavailable [11]. Cr(VI) species are hazardous to human health as they are related to many adverse health effects such as ash, weak immune system, inflammation of the ears and throat, ulcers, allergic reactions, damage to the kidneys and liver, genetic materials, stomach injuries, even death [30]. The Cr(VI) toxicity to the human body is mainly recognized by the chemical structural similarity between oxyanions of chromium (CrO_4^{2-}) and sulfate ions (SO_4^{2-}), thereby allowing the former to pass through the cell wall and the sulfate channel [47]. A study conducted by Wise and coworkers confirms the genotoxicity of Cr(VI) causing the aneuploidy in human urothelial cells [48]. Studies have also shown that the Cr(VI) concentrations as low as 0.5–3.0 mM can cause DNA damage [49]. Due to the health consequences linked to Cr(VI) toxicity as mentioned above, several research groups have attempted to new methods for improving the extraction of Cr compounds from different substrates.

Release of chromium into the environment

Rocks weathering, soil erosion, and rainwater leaching (natural processes), and industrial discharges such as pigments, leather tanning, mining, electroplating, wood processing stainless steel production, and paints (anthropogenic processes) are the key sources accountable for the Cr input in aquatic environments [6,50,51]. In the metal finishing industry, Cr plays a critical role in surface coating for commercial materials applied in many sectors, including construction, aviation, telecommunications, transport, and jewelry. Cr can improve its appearance and reflectivity, such as color and transparency, abrasion resistance, corrosion resistance, pressure resistance, chemical resistance, electrical resistance, and toughness to alter surface characteristics of the material as it is crucial for subsequent operations [52]. As a priority metal contaminant, Cr typically enters water bodies in different industrial cycles, and the poor storage conditions in many industrial activities cause its deposition in soil and water [53]. According to the U.S. Geological Survey (USGS), worldwide, the annual production of Cr in 2019 amounted to 44.8 Gg with an approximate increase of 9.1 Gg from 2017 and 1.7 Gg from 2018. Furthermore, it has been estimated that the value will remain within the level of 40.2 Gg in 2020, as shown in Fig. 3. Mining activities constitute the primary sources of Cr contamination in countries such as South Africa, Turkey, Kazakhstan, India, and Finland. Approximately 95% of the world's Cr resources are found in South Africa and Kazakhstan, and these two countries have been considered the foremost producers of chromite ore [54]. In Europe, more than 650 galvanizing plants have been installed, with 160 in Germany, 90 in Italy, 72 in Spain, 62 in the UK, and 60 in France [6]. The leather tanning industries producing the perishable hides into leather are releasing Cr(III) globally, accounting for about 20% of total Cr emissions. The use of chromium sulfate ($CrOHSO_4$) has also been reported to permanently stabilize the skin matrix to prevent biodegradation [55]. This industry is prevalent in Asia, Africa, and South America [6]. For example, the Chinese leather tanning industry is responsible for discharging 20% of Cr into the water. The average total between 1990 and 2009 exceeds 500 tons per year [56]. Among the 91 pollutants being released into the surrounding environment, the *E*-PRTR (European Pollutant Discharge and Transfer Register) documented an estimated release of 550 tons of Cr into European waters during 2007–2015 [6].

Chromium Remediation via Adsorption

Cr(VI) was removed from aqueous solutions by adsorption processes because of its dissolution tendency in aqueous solutions and exploited by its physical (physisorption), chemical (chemisorption), and bio-remediation characteristics [24,26,57]. A variety of adsorbents such as synthetic polymers, carbon nanotubes (CNTs), activated carbons (AC), graphene oxide (GO), nanoparticles, biomass, and bio-sorbents have been explored for the confiscation of trace elements, especially Cr(VI) [5,58,59]. The adsorbents can be employed as synthesized, or through a modified by cross-linking, grafting, or impregnating with other advanced composite materials. Several adsorbents are being utilized as scaffolds to create different materials in an adsorbent1@adsorbent2@adsorbent3 fashion, with more adsorption sites necessary for the enhanced sorption capacity. In these structures, the dominant functional groups confine adsorption within the pores and the surface of an adsorbent [11,60]. The fact that composites deliver superior performance cannot be an ignored inference. For instance, the functionality of activated carbon may lead to the loss of carbon structure with the resulting carbon pores spreading from one side to the other of the particle surface. These materials are not ideally suited for adsorption because they mainly serve as sieves [11]. Several recent studies have developed effective adsorbents for Cr removal that display excellent adsorption capacity, high binding kinetics, and selectivity at lower dosages and concentrations. Wide variability in the adsorption rate of several adsorbents has been reported. For example, magnetite particles prepared from the residue of titanium, electro-spun carbon nanofiber, and carbon-coated montmorillonite nanocomposite showed an adsorption capacity of 14 mg/g [61], 100 mg/g [62], and 119 mg/g, respectively [63]. The higher selectivity of magnetite particles prepared from titanium residue in the presence of Cl^-, PO_4^{3-}, SO_4^{2-}, As(III), and As (V) ions collapsed in an aqueous sample containing fluoride [61]. This result indicates that there is still room for research in the production of super adsorbents by leveraging the characteristics of individual material components.

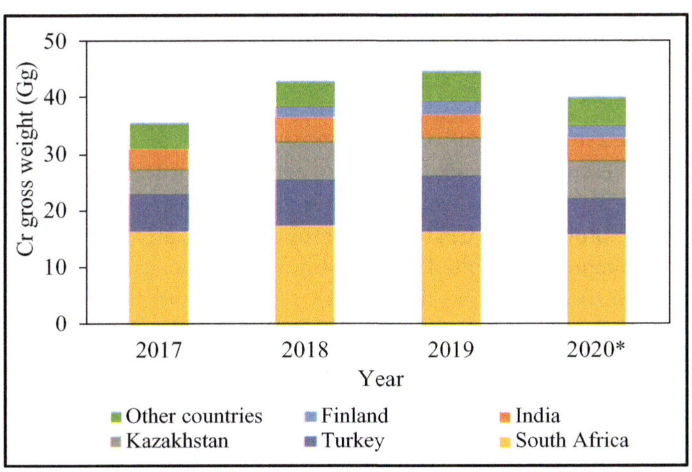

FIG. 3 Major countries and annual Cr mine production. The data obtained from USGS [54] (Note that "*" indicates an estimated data for the year 2020).

The complex interactions occur between Cr(VI) and the functional groups of adsorbents which lead to the reduction of Cr(VI) into Cr(III), making the Cr confiscation incomplete and thus requiring new or improved adsorbents for its complete removal. Four types of Cr(VI) removal routes have been identified with different types of sorbents [64]. First, the adsorption of anionic Cr(VI) species by electrostatic forces of an adsorbent loaded with positively charged functional groups such as NH_3^+ or $COOH_2^+$. The second mechanism involves the adsorption of Cr(VI) onto the surface of the adsorbent and then completely reduced to Cr(III). The third case includes the adsorption of a portion of Cr(VI) onto the adsorbent surface while another portion of Cr(VI) is ultimately reduced to Cr(III), and then Cr(III) is adsorbed on the surface of these materials. In the fourth case, a portion of Cr(VI) is adsorbed, and another portion is reduced to Cr(III) and finally discharged into the solution. Cr(VI) can be transformed either directly or indirectly to Cr(III) [65]. Direct transformation occurs when the Cr(VI) ions are directly linked to sites that provide electrons (O, N, and S) for its reduction to Cr(III). Cr(VI) can also be reduced to Cr(III) by cations such as Fe^{2+} on the adsorbents surface. On the other hand, indirect transformation occurs where the adjacent sites of the adsorbed Cr(VI) donate electrons to facilitate reduction. The adsorption sites are oxidized (i.e., the hydroxyl groups are converted into carboxyl groups) when Cr(VI) reduces to Cr(III), thereby hindering the reusability of the adsorbents [11]. While some studies do not consider the impact of the reduction process [66,67], several studies have illustrated the reduction processes using techniques such as FTIR, EDS, and XPS to verify the presence of Cr(III) species and the oxidation of surface groups [68–71]. In addition to the above four mechanisms, studies have also demonstrated Cr(VI) removal via esterification and the coupling of Cr(VI) ions with catechol [72].

The reduction of Cr(VI) to Cr(III) has also been investigated as a removal mechanism using nanoscale zerovalent iron particles [22,73,74]. The available literature reports different strategies of handling reduced Cr(III). Although some studies have investigated the readsorption of Cr(III) [61,75–77], the focus has mostly been on the transition of Cr(VI) to a less toxic form [22,68,71,78,79]. In addition to the above chemical adsorption processes, there are also physical adsorption processes, particularly for materials having a wider range of surface area and larger pore size such as ACs [32], GOs [80], and zeolites [81,82].

Carbonaceous nanomaterials for Cr removal

Various types of carbonaceous nanomaterials have been investigated to remove heavy metal ions from water and wastewater. Nanomaterials, e.g., activated carbon, carbon fibers, graphene, and CNTs, have played an important role in efficient wastewater treatment processes by effectively removing Cr ions from water samples. Depending on the surface properties and test conditions, different types of adsorption mechanisms on the carbonaceous nanomaterials surfaces have been shown simultaneously [83]. Although each carbon-based adsorbent material shows specific structural and functional properties, all of them have an active surface with common functional groups that provide chemical and physical interactions, attributing to the adsorption of heavy metal ions. A wealth of literature have documented the effect of surface functional groups on separation mechanisms [84]. Different physical and chemical processes, like oxidation, impregnation, and grafting, are used to activate the heteroatoms (O, N, S) contained in the functional groups on the carbon-based nanomaterials surfaces. These processes may synergistically enhance the surface properties such as surface area and pore size distribution [85]. Fig. 4 shows the various adsorption mechanisms including, electrostatic interaction, ion exchange, physical, chemical, precipitation, and surface complexion during the adsorption process. However, in this study, we focus on the application of CNTs-based adsorbents for the uptake of Cr(VI) from aqueous samples.

Carbon nanotubes (CNTs)

CNTs build on a series of fused benzene rings which are crimped into a tubular structure consisting only of carbon atoms. This new sort of artificial nanomaterial has a place with the fullerene family and represents carbon's third allotrope characterized by sp^2 and sp^3 forms of graphite and diamond, respectively [86–88]. Generally, there are two kinds of CNTs based on the number of layers, namely single-walled carbon nanotubes (SWCNTs) and multiwalled carbon nanotubes (MWCNTs), as depicted in Fig. 5 [89]. CNTs are made of graphene layers rolled into a cylindrical shape, and have either capped or open ends. They usually form hexagonal close-packed structure with a diameter in a few nanometers and several micrometers long. SWCNTs with a diameter as small as 0.4–2 nm are products of a single graphene sheet rolled into a cylindrical shape. Similarly, MWCNTs with an outer and inner diameter ranging from 2 to 100 nm and 1 to 3 nm, respectively, consist of two or more graphene sheets incasing with a hollow core in the same way as in SWCNTs [9,88,90–92]. Here, Table 2 compares SWCNTs and MWCNTs based on their properties.

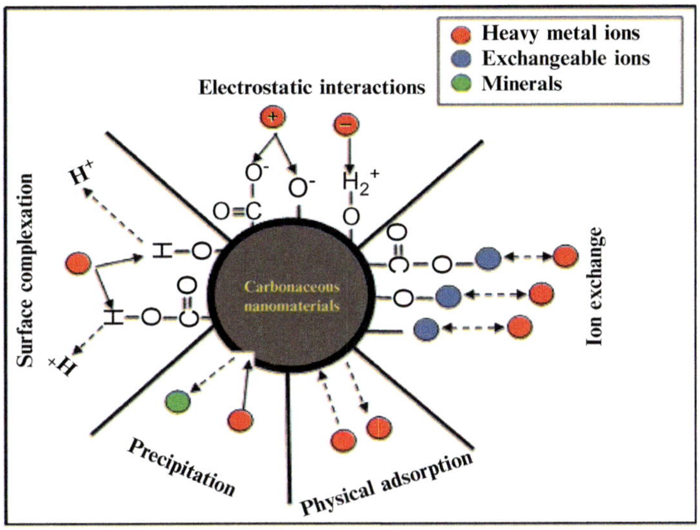

FIG. 4 Illustration of mechanisms (electrostatic interactions, ion exchange, physical adsorption, precipitation, and surface complexation) of heavy metal ions onto carbonaceous adsorbents. *Adopted from X. Yang, Y. Wan, Y. Zheng, F. He, Z. Yu, J. Huang, et al. Surface functional groups of carbon-based adsorbents and their roles in the removal of heavy metals from aqueous solutions: a critical review. Chem. Eng. J. 366 (2019) 608–21. https://doi.org/10.1016/j.cej.2019.02.119 with the permission of Elsevier.*

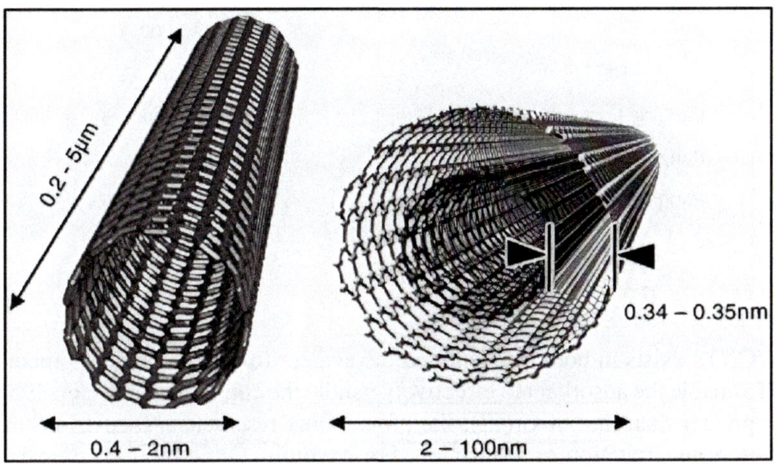

FIG. 5 Types of typical dimensions of CNTs, SWCNTs (*left*), and MWCNTs (*right*). *Adopted from M.M-A. Aslam, H-W. Kuo, W. Den, M. Usman, M. Sultan, H. Ashraf, Functionalized carbon nanotubes (CNTs) for water and wastewater treatment: preparation to application. Sustainability 13 (2021) 5717. https://doi.org/10.3390/su13105717 and W. Cheung, F. Pontoriero, O. Taratula, A.M. Chen, H. He, DNA and carbon nanotubes as medicine. Adv. Drug Deliv. Rev. 62 (2010) 633–649. https://doi.org/10.1016/j.addr.2010.03.0071 with the permission of Elsevier.*

Cr adsorption by modified CNT-based adsorbents

The groundbreaking chemical and physical properties of CNTs, including excellent mechanical strength, high chemical stability, and large surface area, play a fascinating role in the carbon material family [93]. However, a study conducted by Gupta and his coworkers determined that SWCNTs and MWCNTs cannot effectively remove Cr(VI) ions in wastewater. Nevertheless, compared with MWCNTs, the SWCNTs showed higher adsorption capacities under different experimental conditions (contact time, system pH, initial Cr(VI) ion concentration and competitive anions). The Cr(VI) adsorption capacity for SWCNTs and MWCNTs were 13.15 and 1.31 mg/g, respectively. They observed that the key role of competitive anions and pH, such as the competition of sulfate anions inhibits the Cr(VI) ions adsorption on CNTs. Furthermore, both CNTs (SWCNTs and MWCNTs) showed the maximum adsorption capacity occurred at the pH of 2.5 [94]. The Cr(VI) adsorption on the oxidized MWCNTs is also regulated by a solution pH. When the system pH value increases, the Cr(VI) adsorption on the oxidized MWCNTs decreases. During the adsorption process, adsorbed Cr(VI) on the surface of MWCNTs is ultimately reduced to Cr(III), and a part of Cr(III) migrates back in the solution at a low pH value [38]. In terms of selectivity, maximum uptake capacity, pH range, amount of organic solvent, interference, and reusability, the analytical methods defined with MWCNTs are considered superior to the documented Cr separation-concentration methods.

TABLE 2 Comparison between properties of SWCNTs and MWCNT [9].

Properties	SWNTs	MWCNTs
Layer type	Single graphene layer	Multiple graphene layer
Catalyst requirement	Essential during synthesis	No need during synthesis
Bulk or massive production	Difficult	Easy
Purity level	Low	Large
Defect's level	High	Low
Characterization	Easy	Difficult
Manage	Easily twisted	Cannot be twisted easily
Specific gravity (g/cm^3)	About 0.8	<1.8
Elastic modulus (TPa)	About 1.4	Ranging from 0.3 to 1
Strength (GPa)	Ranging from 50 to 500	Ranging from 10 to 60
Electrical conductivity (S/cm)	Ranging from 102 to 106	Ranging from 103 to 105
Electron mobility ($cm^2/V s$)	About 105	Ranging from 104 to 105
Thermal conductivity (W/m K)	About 6000	About 2000
Coefficient of thermal expansion (K^{-1})	$> 1.1 \times 10^{-3}$	About 1.37×10^{-3}
Thermal stability in air (°C)	Ranging from 600 to 800	Ranging from 600 to 800
Resistivity (μΩ cm)	Ranging from 50 to 500	Ranging from 50 to 500
Specific surface area (m^2/g)	Ranging from 400 to 900	Ranging from 200 to 400

The XPS studies revealed that Cr adsorbed by MWCNTs exists in both trivalent and hexavalent forms. The higher preconcentration and solidification capabilities of MWCNTs enable the adsorbent to effectively handle the contaminated waters [95]. It is worth noting that the Cr(VI) toxicity is much greater than that of Cr(III); therefore, some researchers focused on the reduction process of Cr(VI) toxicity, i.e., Cr(III) before precipitation or adsorption. For example, Qiu et al. [96] prepared magnetic carbon nano-adsorbents (MC-N and MC-O) through the carbonization of cellulose and reduction of Fe_3O_4 nanoparticles. They identified the reduction and precipitation methods to extract Cr(VI) by using MC-N and MC-O. After the reaction, the MC-N showed a removal capacity of 327.5 mg/g, marginally higher than that of the MC-O (293.8 mg/g) at the solution pH of 1.0. As a result, they proposed that the mechanisms for Cr(VI) removal were related to the reductive reaction of Cr(VI) and the precipitation of Cr(III).

Covalent and noncovalent functionalization for CNTs modification is a common technique to tailor specific applications. For instance, MWCNTs were modified with nanoscale zero-valent iron (nZVI-MWCNT) strip Cr(VI) from the aqueous samples. The composite material was synthesized by the reductions of ferrous sulfate to deposit nZVI particles on MWCNTs [97]. The magnetically-modified MWCNTs with Fe_3O_4 imparting magnetism in MWCNTs helps recover nanomaterials quickly and prevent aggregation in aqueous solutions. The electronic state of Fe_3O_4 nanoparticles can also enhance Cr(VI) adsorptions. In addition, using metal ion scavengers such as MnO_2 helps improve the CNTs performance for eliminating Cr species from water samples. For example, at pH 2, the Cr(VI) adsorption capacity of MWCNTs modified with MnO_2 and Fe_3O_4 (MnO_2/Fe_3O_4/MWCNTs) was 186 mg/g, an encouraging result attributed to the electrostatic interactions of both MnO_2 and Fe_3O_4 with Cr(VI), combined with cation exchange and surface complex surface mechanisms [98]. Krishna Kumar et al. [99] and Huang et al. [100] reported that their MWCNTs functionalized with an ionic liquid improved the adsorption capacity and effective removal of Cr(VI) ions. Cation, anion-π, and electrostatic interactions occurred, in the presence of ionic liquid, between the quaternary ammonium cations and the surface oxygenated functional groups (carboxyl and hydroxyl). Recently, oxidized MWCNTs modified with polypyrrole (PPy/OMWCNTs) have been proven an effective adsorbent to remove Cr ions from the aqueous solution. PPy/OMWCNTs synthesized by in-situ chemical oxidation polymerization process showed a notable Cr adsorption trend (370.3 mg/g at 45°C). An electrostatic interaction occurred between the protonated site of PPy/OMWCNT and the Cr metal ions [101]. Schwarz et al. [102] showed that CNTs modified by titanium complexes had the advantages of large surface area, complex structure, strong chemical stability, low density, high reusability, and small size, so they can effectively extract Cr(VI) from water samples. Atieh et al. [103] determined that the use of CNTs modified with AC could effectively extract Cr(VI) ions from the wastewater, showing that CNTs coated with AC improved the Cr adsorption capacities. All of these studies point to the direction that CNTs-based modified adsorbent materials can effectively remove heavy metal ions for the water and wastewater remediation.

Mohamed et al. [70] synthesized a novel material by electrospinning of polyacrylonitrile (PAN) and CNTs modified with amino-functionalized titanium dioxide (TiO_2-NH_2), abbreviated as (PAN–CNT/TiO_2-NH_2). The composite adsorbent material showed an excellent Cr(VI) 714.27 mg/g adsorption capacity at an equilibrium time of 30 min, Cr(VI) concentration of 180 mg/L, and solution pH of 2. This study was a typical example of synthetic composite materials in adsorbent1@adsorbent2@adsorbent3@adsorbent4, where both ends of this system (adsorbent 1 and adsorbent 4) were loaded with amino groups that shows high uptake affinity for Cr(VI) ions. Despite the high adsorption capacities, one should not overlook the various technical and economic obstacles with these CNT-based nanomaterials. For example, one of the usual knocks on using nanoparticles for environmental remediation is the reusability and a similar concern exists for CNTs due to their small size (<100 μm). In addition, despite the maturing production that has driven down the unit costs of MWCNTs, the cost–benefit of CNTs-based adsorbent remains unjustified at this early phase of development. These limitations confine the use MWCNTs to niche applications where it is used in a small, controlled manner to remove or extract environmental toxins with commercial resale values [38].

Chitosan

Chitosan is known as the second most abundant natural biopolymer after cellulose. It can be extracted from its precursor called chitin abundant in the exoskeleton of shrimp, crab, and lobster shells. However, the poor solubility and low adsorption performance of chitin limit its large-scale applications. Chemical (alkaline) or enzymatic deacetylation of chitin can eliminate the acetyl groups from chitosan's acetylamino groups to make it more soluble. As shown in Fig. 6 [104], chitosan is a polyaminosaccharide with a glucose-like structure, bounded to β-(1 → 4)-2-amino-2-deoxy-D-glucose complex. It is obvious that the molecular weight of chitosan depends on the commercial production process of the product. Many manufacturing companies can provide different grades of chitosan ranging from 10,000 to 1,000,000 Da, as well as products with different degrees of deacetylation usually greater than 55%.

The degree of deacetylation refers to the mole fraction of deacetylated units found in the polymer backbone. In environmental applications, chitosan is an efficient adsorbent for various contaminants because of the abundant NH_2 and OH groups. However, chitosan showed some limitations like poor mechanical strength, weak stability in the acid regime, weak thermal resistance, and low surface area, which hindered its application in raw form. Therefore, several methods have been

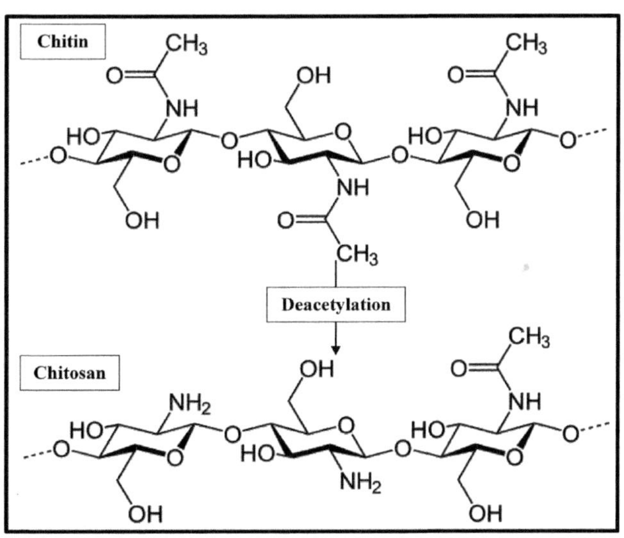

FIG. 6 Chemical structure of Chitin and Chitosan. *Adopted from M. Zuber, K.M. Zia, M. Barikani, Chitin and chitosan based blends, composites and nanocomposites. Adv. Nat. Polym. Springer, 2013, pp. 55–119. https://doi.org/10.1007/978-3-642-20940-6_3 with the permission of Springer.*

developed to modify chitosan, including physical and chemical modifications, to resolve these issues [105–109]. The physical modification includes the processing of chitosan into nanoparticles, membranes, sponges, gel beads, skeletons (shape of honeycomb), and various fibers. In contrast, the crosslinking chitosan or grafting functional groups comes under chemical modifications. Recent research has mainly focused on modifying chitosan chemically with crosslinking and organic ligands [66,110,111], and integrating nanomaterials into its backbone [71,112,113]. Compared with ACs and other bio sorbents, the main advantages of chitosan are its low price and abundance, renewability, antibacterial property, non-toxicity, biocompatibility, biodegradability, macromolecular structure, and hydrophilicity. The presence of many amino and hydroxyl groups is critical to uptake various pollutants with good selectivity. Chitosan and its derivatives have been extensively explored in water and wastewater treatment to eliminate various toxins, such as heavy metal ions, phenols, anions, pesticides, fungicides, dyes, and humic substances.

Cr removal by modified chitosan-based adsorbents

Chitosan materials have displayed a remarkable ability to eliminate various toxic metals, especially Cr species, from aqueous samples [112,114]. Aydin et al. [115] used chitosan flasks to remove Cr(VI) from water samples with a solution pH of 1.5–9.5 and adsorbent dosage of 1.8–24.2 g/dm^3. The results showed a maximum adsorption capacity of Cr(VI) about 22.09 mg/g at an initial Cr(VI) concentration of 30 mg/dm^3, pH of 3, and the adsorbent dose of 13 mg/dm^3. In another study, unmodified and crosslinked chitosan showed the adsorption capacity of 50 and 78 mg Cr(VI)/g, respectively, at the optimum pH of 5 [116]. Rojas and colleagues tested the method of cross-linked chitosan to remove Cr(VI) and Cr(III) from the aqueous samples. The test results indicated that the adsorption capacity of chitosan for Cr(III) was only 6 mg/g, while for Cr(VI), the adsorption capacity was as high as 215 mg/g at the optimum pH value of 4.0. The higher exclusion of anionic Cr under acidic conditions indicates that chitosan has many protonated amino groups under low pH reactions [117]. According to Kahu et al. [71], chitosan's structural modification and functionalization showed excellent adsorption performance for Cr(VI). In their research, the structure was modified by crosslinking chitosan with phosphate, and an ionic solid (ethyl cetyl dimethyl ammonium bromide) was used to functionalize the adsorbent. The modified adsorbent showed a great adsorption potential of 267 mg Cr(VI)/g, showing the synergistic effect of the composite material, and preferred over individual materials [71].

Dima et al. [113] used chitosan micro/nanoparticles (MCH) to extract Cr(VI) synthesized by the gelation method. They found that the crosslinking of the MCH particles with tripolyphosphate (TPP), a nontoxic polyanion, prevented the leaching of chitosan particles, thereby improving the removal efficiency. Li et al. [118] used chitosan nanofibers for Cr(VI) removal synthesized by electrospinning. Compared with nonelectrospun chitosan particles, its adsorption capacity increases from 26.7 to 68.3 mg/g [118]. Although chitosan-based adsorbents show better adsorption capacity, it is difficult to separate the saturated adsorbents after the reaction. Many studies have focused on the development of magnetic adsorbents and showed adequate separation ability and higher potential to treat pollutants. In this regard, Xio et al. [119] reported biochar-chitosan (CS-BC) composite material modified with $FeCl_3$ showed a significant change in adsorption capacity of 26% for Cr(VI), which was higher than the nonmagnetic CS-BC composite material. They observed that the $FeCl_3$ incorporated into the active sites in the composite framework increased the Cr(VI) elimination rate. Hu et al. [120] prepared magnetic chitosan resin (EMCMCR) using glutaraldehyde as a crosslinking agent, followed by condensation with ethylenediamine. After Cr(VI) adsorption, the modified adsorbent material showed an adsorption capacity of 52 mg/g, higher than the unmodified chitosan at the optimum pH of 2 [120].

Chitosan-CNTs composite materials for Cr(VI) removal

According to the literature, both CNTs and chitosan have great potential for environmental applications with excellent adsorption capacity for many toxic metals, including Cr(VI). However, limitations such as poor structural groups and agglomerations of CNTs, heat resistance, poor mechanical strength, weak stability under acidic conditions, and low surface area of chitosan limit their use in actual water and wastewater treatment systems [121]. Combining these two materials can synergistically minimize the limitations linked with each material individually [122]. After these two materials are combined, chitosan endows CNTs with excellent dispersibility and provides the necessary amount of NH_2-based active sites with a high affinity to capture Cr(VI) ions, especially under acidic conditions [123]. On the other hand, CNTs improve the surface area, mechanical strength, heat resistance, and stability of chitosan under different operating conditions [124]. Therefore, chitosan-CNTs composites represent an effective adsorbent due to their robust performance and high adsorption capacity.

Huang et al. [125] prepared a reusable chitosan-modified with MWCNTs (CS-MWCNTs-COOH) to remove aqueous phase Cr(VI). Briefly, the pristine chitosan was dissolved in DIW containing 2% (v/v) acetic acid heated at 40°C, and then MWCNTs-COOH was added into the solution. After 15 min of reaction, glutaraldehyde was added as a crosslinking agent, and the temperature was kept at 40°C for 30 min. Subsequently, 1 L of pure water was added to the solution, followed by adding 0.10 M NaOH. Finally, after cooling under room temperature, the excess water was removed from the solution. The resulting material was oven-dried at 80°C for 12 h before using for adsorption experiments. The results of this study showed that the adsorption process of Cr (VI) was efficient, with an equilibrium time of 30 min and an adsorption capacity of about 164 ± 2 mg/g at the optimum pH (2.0) and temperature (313*K*). According to the literature, chitosan stability, especially in the acidic regime, can be enhanced with the addition of glutaraldehyde as a crosslinking agent [125–127]. The aldehyde groups in glutaraldehyde firstly react with the amino groups of chitosan and making the chain-like structure. Fig. 7 shows the oxygen-containing functional groups of chitosan-CNTs composite material interacts with the –CH_2 group of chitosan. The use of glutaraldehyde as a crosslinking agent is still questionable since the toxicity of glutaraldehyde greatly threatens the aquatic ecosystem [128]. Therefore, incorporating the magnetic nanoparticles can ensure the easy recovery of adsorbents from the treated effluents, thereby resolving the environmental problems associated with such toxic substances [129]. In addition, the introduction of magnetic nanoparticles (Fe_2O_3 or Fe_3O_4) into chitosan-CNT can bind many additional active sites and help form the porous adsorbent structure, thereby improving the adsorption capacity and separation ability of the composite [127]. For example, the removal of Cr(VI) from the water with the utilization of iron oxide/CNTs/chitosan magnetic composite (CLCh/MWCNTs/Fe) showed the adsorption capacity of 450 mg/g owing to a high surface area of 70.9 m^2/g and pore volume of 0.0247 cm^3/g. In contrast, the nonmagnetic composite (CLCh/MWCNTs) showed 49.68 m^2/g of surface area and 0.0172 cm^3/g of the pore volume. These results suggested the important role of Fe_3O_4 nanoparticles in the development of the porous structure of the CLCh/MWCNTs composite [127].

Factors affecting the adsorption capacity of CNTs and chitosan composites

Operational pH

During the adsorption process, the pH of the solution has a considerable impact on the ionization degree of metal ions and their speciation, and the surface properties of the adsorbents. The influence of solution pH on Cr species adsorption onto CNTs and chitosan-based adsorbents revealed that metal speciation stability and adsorbent charge status were pH-dependent. Cr(VI) usually occurs in expected forms of different species (H_2CrO_4 at $pH < 1.0$, $HCrO_4^-$ and $Cr_2O_7^{2-}$ at pH 2.0–6.0, and CrO_4^{2-} when $pH > 6.0$) under different pH conditions [130]. Therefore, CNTs and chitosan-based adsorbents are synthesized according to these predicted species. The acidic environments are usually suitable for extracting Cr(VI) ions. Tables 3 and 4 list the optimal pH to eliminate Cr(VI) ions by CNTs and chitosan-based adsorbents, respectively. For example, Jung et al. [106] studied different carbon-based adsorbents for the adsorption of Cr(VI) and showed that more Cr(VI) is being adsorbed at a pH lower than 4. Hu et al. [163] showed that the effectiveness of oxidized CNTs for adsorbing Cr(VI) is at $pH < 2$. However, these studies have shown that both Cr(III) and Cr(VI) maintain the oxidized MWCNT. Similarly, Ghasemi et al. [137] applied MWCNT/Fe_3O_4 nanoparticles, prepared by combining the magnetite nanoparticles and carboxylate CNTs, for Cr(VI) adsorption. At a pH of 2.0, they attained a Cr(VI) removal efficiency of 88.5% at an equilibrium condition in 45 min. When the solution pH value increased to 11, the removal efficiency decreased to a level of 41.3%. Protonation of oxygen-containing functional groups (OH_2^+, $COOH_2^+$) into CNTs offers a positively charged surface for the adsorption of Cr(VI) anions under acidic circumstances, resulting in increased Cr(VI) adsorption. On the other hand, the increasing pH caused the deprotonation of surface functional groups (−OH, −COOH), making them negatively charged ($-OH^-$, $-COO^-$) and reducing the adsorption rate of Cr(VI) ions due to the increased competition between negatively charged adsorbate and adsorbent. Similar findings were observed when magnetic chitosan-iron(III) hydrogel (MCH-Fe) was evaluated for Cr(VI) adsorption in the pH range of 2.0–8.0. pH at 3.0 resulted in the maximum adsorption of Cr(VI) ions, which were adsorbed by positively charged MCH-Fe via electrostatic and ion-exchange interactions [152]. Chen et al. [107] reported that the Cr(VI) adsorption on chitosan/montmorillonite magnetic (CTS/MMT-Fe_3O_4) microspheres with pH values ranging from 1.0 to 9.0. When the pH was adjusted from 1.0 to 2.0, the adsorption of Cr(VI) increased from 45.84 to 74.20 mg/g, then fell to 20.37 mg/g at pH 9.0. The enhanced removal capacity owing to protonation of amine groups (–NH_3^+) in lower pH ranges was attributed to the electrostatic interaction between the adsorbent and Cr(VI) ions. Kim et al. [164] fabricated a novel chitosan gel functionalized with MWCNTs and polyaniline (CS-MWNT-PAA-PADPA/FG) adsorbent and used it to remove Cr(VI) in pH values ranging from 1.0 to 10.0. They attained the highest limit of Cr(VI) removal at pH 2.0. When the pH increased to 10.0, the rate of Cr(VI) removal declined drastically. Another study also reported the Cr(VI) adsorption on crosslinked chitosan-diethylenetriaminepentaacetic acid (CD) in the pH range of 1.0–11.0, with the

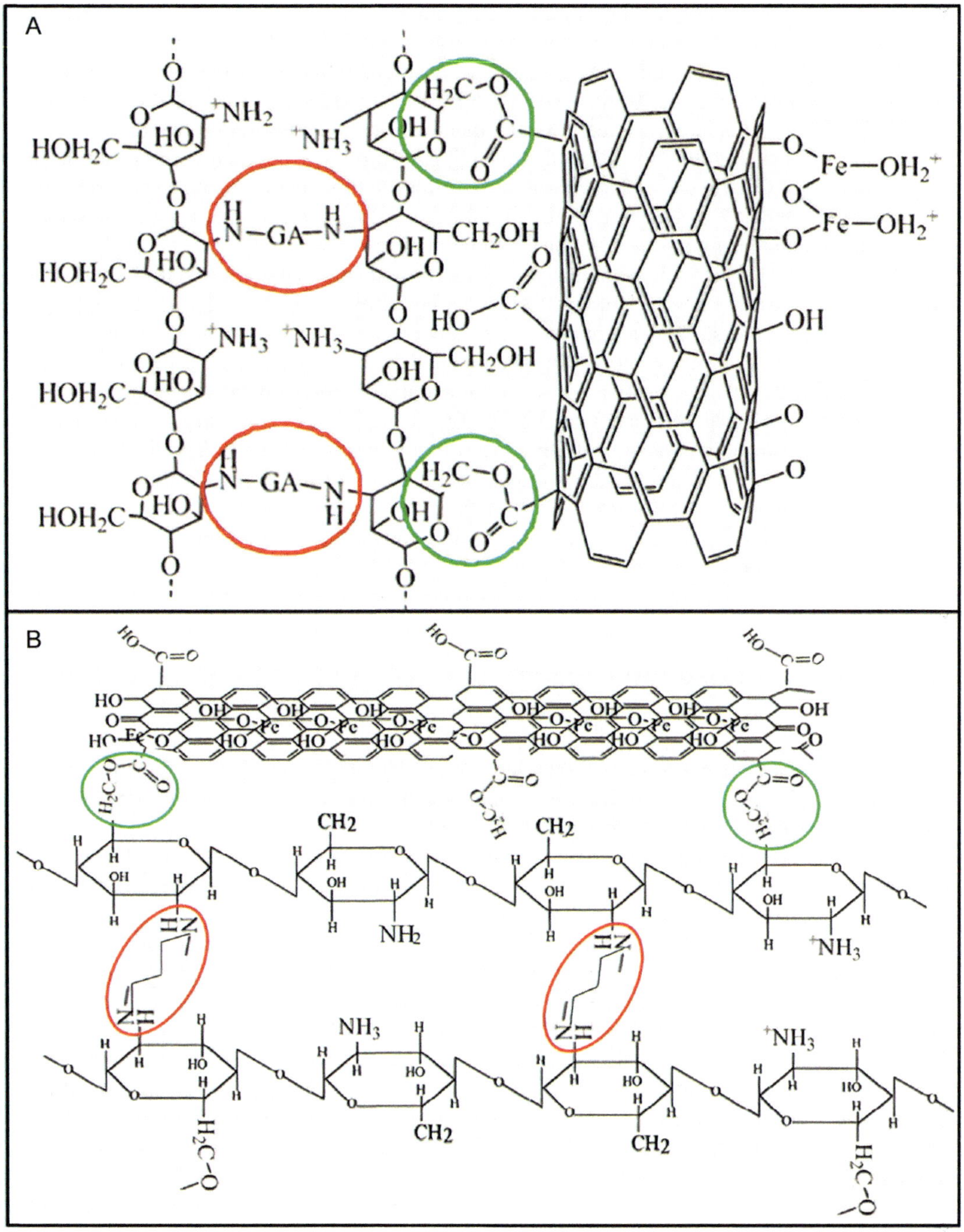

FIG. 7 Representation of glutaraldehyde crosslinking (*red circle*) and interaction of oxygen-containing functional groups of CNTs with chitosan (*green circle*). *Panel (A): Adopted from J. de O. Marques Neto, C.R. Bellato, D. de C. Silva, Iron oxide/carbon nanotubes/chitosan magnetic composite film for chromium species removal. Chemosphere 218 (2019) 391–401. https://doi.org/10.1016/j.chemosphere.2018.11.080. Panel (B): Adopted from M.M.A. Aslam, W. Den, H.-W. Kuo, Encapsulated chitosan-modified magnetic carbon nanotubes for aqueous-phase CrVI uptake. J. Water Process. Eng. 40 (2021) 101793. https://doi.org/10.1016/j.jwpe.2020.101793 with the permission of Elsevier.*

highest adsorption occurring at pH 3.0 [145]. The authors of both investigations concluded that adsorption was strongly pH-dependent. The competition between CrO_4^{2-} and OH^- ions resulted in decreased adsorption capacity at higher pH values.

Measuring the adsorbents' point of zero charge (PZC)—the pH at which the surface of the adsorbent has no net charge—can offer hints to the appropriate operating conditions. The pH_{PZC} of raw CNTs usually lies between pH 4 and 6 [165–167]. Acid treatment of CNTs, however, yields a lower value of pH_{PZC} than those of raw CNTs [165,168–171]. Chitosan generally has a PZC at pH 6.3, and 99% of its amino groups are protonated at pH 4.3.

TABLE 3 Cr(VI) uptake from aqueous solution using carbon nanotube composites.

Adsorbents	Characterization techniques	Experimental conditions	Q_{max}	Isothermal and kinetic model	Nature of adsorption	References
MWCNTs-CTAB (Multiwall carbon nanotubes-cetyl trimethyl ammonium bromide)	SEM, FT-IR, BET, TGA, EDS	IC = AD = 100 mg/L, 5 g/L, CT = 30 min, pH 4.5	27.28	L, Spis	Endothermic	[131]
NMCNTs (Nitrogen doped magnetic carbon nanotubes)	SEM, TEM, HRTEM, XRD, XPS, BET	IC = 40–1000 mg/L, AD = 2.5 g/L, pH 1	970.9	L, PSO	—	[132]
PIL/MWCNT (Polymeric ionic liquids/multiwalled carbon nanotubes)	SEM/EDS, TEM, NMR, FTIR, TGA/DSC	pH 2–3, IC = 10–50 mg/L, AD = 20 mg CT = 2 h, T = room temp.	60%	—	—	[133]
MNP/MWCNTs (Magnetic nanoparticle-multiwalled carbon nanotubes composites)	SEM, TEM, FT-IR, TGA, VSM, XRD	IC = 5 mg/L, AD = 0.4–1 g/L, T = 25–45°C, SR = 150 rpm, pH 2	42.02	L, PSO, IPD	—	[134]
SWCNTs (Single-walled carbon nanotubes)	SEM, TEM, XRD, BET, FT-IR	pH 2.5, CT = 60 min, AD = 5–40 mg/200 cc, IC = 0.2–1.0 mg/L	2.35	L, PSO	—	[94]
MWCNTs (Multiwalled carbon nanotubes)			1.26			
FeOOH/AlOOH/MWCNTs	TEM, SEM, XPS, XRD, FTIR, BET	CT = 300 min, IC = 10–100 mg/L, AD = 0.02 g/20 mL, pH 3	60.6	L, PSO	—	[135]
AAPTS modified MWCNTs (3-(2-aminoethylamino) propyl-trimethoxy silane-multiwalled carbon nanotubes)	FTIR	pH 2.2, IC = 1 mg/L, IC = 40 mg,	9.79	L	—	[136]
MWCNTs/Fe3O4	TEM, XRD	pH 2, T = 25°C, SR = 250 rpm, IC 10 mg/L, CT = 45 min, AD = 0.035 g	76.9	L, PSO	—	[137]
PPy/OMWCNTs NCs (Polypyrrole wrapped oxidized multiwalled carbon nanotubes nanocomposites)	BET, FE-SEM, HR-TEM, ATR-FTIR, XRD, XPS	IC = 50–100 mg/L, AD = 0.5 g/L, CT = 24 h, T = 25°C, SR = 200 rpm, pH 2	294	L, PSO	Endothermic	[101]

Continued

TABLE 3 Cr(VI) uptake from aqueous solution using carbon nanotube composites.—cont'd

Adsorbents	Characterization techniques	Experimental conditions	Q_{max}	Isothermal and kinetic model	Nature of adsorption	References
AC/f-MWCNTs 1% (Activated carbon-functionalized multiwalled carbon nanotubes)	FTIR, SEM, BET	pH 2, AD = 0.2 g/50 mL, IC = 1 × 10–3 M, CT = 24 h, T = 25°C	64	L	—	[138]
AC/f-MWCNTs 3% (Activated carbon-functionalized multiwalled carbon nanotubes)			113.3		—	
AC/f-MWCNTs 5% (Activated carbon-functionalized multiwalled carbon nanotubes)			95		—	
AC/f-CNSs (Activated carbon-functionalized carbon nanospheres)			105.5		—	
IL-oxi-MWCNT (Ionic liquid-oxi-multiwalled carbon nanotubes)	Raman spectroscopy, XPS, FTIR, TEM, SEM, BET, EDX	pH 3, 20 mg/20 mL, AD = 0.15–0.2, CT = 40 min, T = 298–333 K	85.8	L, PSO	Exothermic	[99]
MnO_2/Fe_3O_4/o-MWCNTs (Manganese dioxide/iron oxide/acid oxidized multiwalled carbon nanotube magnetic nanocomposites)	TEM, XRD, FTIR, VSM, BET	IC = 50–30 mg/L, AD = 0.05 g, CT = 150 min T = 298 K, pH 2	186.9	L, PSO, IPD	Endothermic	[98]
DP/MWCNTs (Dodecyl-benzene-sulfonic-acid (DBSA) doped polyaniline/multiwalled carbon nanotubes)	Electron microscope, EDAX, FTIR	IC = 20–140 mg/L, AD = 0.02 g/15 mL, T = 30°C, SR = 200 rpm, pH 2	55.6	F, PSO, IPD	Endothermic	[139]
AA-CNTs (Activated alumina-carbon nanotubes)	SEM, EDAX, BET, FTIR	IC = 1000–2000 mg/L, CT = 1.5–2 h, T = 25–45°C, pH 2, AD = 2.5 g/L	264.5	L and F, PSO	Exothermic	[140]
CeO_2/ACNTs (Ceria nanoparticles supported on aligned carbon nanotubes)	—	pH 3–7, IC = 35.3 mg/L mL, AD = 35 mg/L, T = 25°C	30.2	L, F	—	[141]

Q_{max} = maximum adsorption capacity; IC = initial Cr(VI) concentration; AD = adsorbent dose; CT = contact time; T = solution temperature; SR = stirring rate; L = Langmuir isotherm model; F = Freundlich isotherm model; PSO = Pseudo second order kinetic model; IDP = Intraparticle diffusion model.

TABLE 4 Cr(VI) uptake from aqueous solution using chitosan composites.

Adsorbents	Characterization techniques	Experimental conditions	Q_{max}	Isothermal and kinetic model	Nature of adsorption	References
CS/MWCNTs/Fe beads (Chitosan and maghemite-doped multiwall carbon nanotubes beads)	SEM, FT-IR, XRD, XPS, TGA, EDS	CT = 24 h, IC = 10–50 mg/L, AD = 0.5 g/L, T = 25°C, pH 4	119	L, F, PSO, IPD	Endothermic	[142]
Fe_3O_4@CTS-Cin (Magnetite modified chitosan cinnamaldehyde)	SEM, HR-TEM, FT-IR, EDS, XRD, VSM	pH 2, CT = 80 min, IC = 50–300 mg/L, AD = 0.3 g, T = 303 K	48.1	L, PSO	—	[143]
CTS-Cin (Modified chitosan cinnamaldehyde)		IC = 50–300 mg/L, pH, 3, CT = 35 min, AD = 0.4 g, T = 303 K	61.4			
Chitosan-BC (M) (Chitosan-biochar)	SEM, XPS, XRD, FTIR, BET	T = 30°C, CT = 18 h, pH 3, IC = 0–40 mg/L, AD = 0.5 g/L	30.14	PSO	—	[119]
CCB (Cross-linked chitosan/bentonite composite)	FTIR, BET, XRD, TGA	pH 2, T = 293 K, CT = 240 min, AD = 2 g/L	89.1	L, PSO	—	[144]
Electro spun chitosan nanofibers	FTIR, SEM, XPS	pH 3, T = 298 K, CT = 20 min, AD = 1 g/L	131.6	L, F, PSO	—	[118]
CD (Chitosan cross-linked with diethylenetriaminepentaacetic acid)	TGA, FTIR, XPS, SEM, DSC, EDAX, XRD, ESR,	pH 3.0, T = 303 K, CT = 240 min, AD = 2.0 g/L	192.3	L, PSO	Exothermic	[145]
CS–NZVI-CMβ-CD (Chitosan–nano zero-valent iron–carboxymethyl β-cyclodextrin)	XPS, BET, FTIR, SEM	pH 6, T = 298 K, AD = 3.3	200	L, T, PSO	Endothermic	[146]
CGP (Chitosan grafted polyaniline)	TGA, FTIR, DSC, SEM–EDX, XRD	T = 303 K, CT = 30, AD = 2 g/L, pH 4.2	165.6	L, PSO, ID	Endothermic	[147]
CCGP (Cross linked chitosan grafted polyaniline composite)			179.2			
MCGO-IL (Magnetic chitosan and graphene oxide-ionic liquid)	SEM, FTIR	pH 3–4, T = 303 K, CT = 50 min, AD = 1 g/L	145	L, PSO	—	[148]
CCGO (Magnetic cyclodextrin–chitosan/graphene oxide)	XRD, SEM, FTIR	pH 2, T = 303, AD = 1 g/L	66.7	L, PSO	—	[149]
3ACBZr (3rd generation poly amido amine chitosan beads zirconium)	DSC, XRD, SEM, EDAX, TGA/DTG, FTIR	pH 2–4, T = 303, CT = 120 min, AD = 1 g/L	185	F	Endothermic	[150]
Chitosan	FTIR,	pH 4, IC = 250–250,000 μg/L, AD = 100 mg/L, T = 20 ± 1°C	35.6	L, PSO, IDP	Exothermic	[106]
Magnetic chitosan nanoparticles	XRD, FE-SEM, VSM, TEM, EDS	pH 3, T = 298 K, CT = 100 min, AD = 2 g/L	55.8	L, PSO	Exothermic	[151]

Continued

TABLE 4 Cr(VI) uptake from aqueous solution using chitosan composites.—cont'd

Adsorbents	Characterization techniques	Experimental conditions	Q_{max}	Isothermal and kinetic model	Nature of adsorption	References
MCH–Fe (Magnetic chitosan–iron(III) hydrogel)	SEM, XRD, TG, VSM	pH 3, T = 303 K, CT = 30 min, AD = 1 g/L	144.9	L, PSO	Endothermic	[152]
Zr–CTS (Zirconium–chitosan)	FTIR, SEM, EDS, XRD	pH 5.0, T = 303 K, CT = 300 min, AD = 0.2 g/L	175	L, PSO	Endothermic	[153]
γ-Fe_2O_3-chitosan bead	XPS, FTIR, XRD, SEM, TEM	pH 5.0, T = 298 K, CT = 360 min, AD = 0.5 g/L	106.5	F, PSO	Exothermic	[154]
CTS/MMT–Fe_3O_4 (Chitosan/montmorillonite–Fe_3O_4)	XRD, FTIR, SEM	pH 3, T = 303 K, CT = 90 min, AD = 0.8 g/L	74.2	L, PSO	Endothermic	[107]
TC-chitosan (Thiocarbamoyl chitosan)	SEM, EDX, FTIR	pH 2.0, T = 298 K, CT = 240 min, AD = 1.0 g/L	434.8	L, F, PSO	Endothermic	[155]
ECCR	FTIR, SEM	pH 3, T = 323 K, AD = 1 g/L	112.6	L, PSO	Endothermic	[156]
CCN (Chitosan/clay nanocomposite)	FTIR, XRD, SEM, EDS, TEM	pH 3, T = 308 K, CT = 150 min, AD = 2 g/L	357.1	L, PSO	Endothermic	[157]
sPANIs/Ch-HCl (Chemically synthesized alkyl-substituted polyaniline/chitosan composites)	FTIR, SEM–EDX	pH 4.2, T = 298 K, CT = 240 min AD = 1 g/L	229.8	F, PSO	Endothermic	[158]
EMCMCR (Ethylenediamine-modified cross-linked magnetic chitosan resin)	SEM	pH 2, T = 303 K, CT = 10 min, AD = 5 g/L	52.8	L, T, PSO	Exothermic	[120]
MCAGS (Magnetic chitosan anthranilic acid glutaraldehyde Schiff's base)	FTIR, WAXRD, TGA, SEM	pH 2, T = 318 K, CT = 120 min, AD = 0.5 g/L	58.48	L, PSO	Endothermic	[159]
Ch–Fe (Fe-crosslinked chitosan complex)	—	pH 4.6, T = 298 K, CT = 90 min, AD = 1.25 g/L	295	L-F, PSO	Exothermic	[160]
CLCh/MWCNT/Fe film (Crosslinked chitosan/multiple walled carbon nanotubes doped with magnetic iron oxide)	SEM, XRD, BET, EDS, FTIR, Raman spectroscopy	50–700 mg/L, 0.3 g/L, 25°C, 3 h, pH 4	449.3	L, PSO	Endothermic	[127]
Chitosan/MWCNTs-COOH	SEM, XPS	50–600 mg/L, 1 g/L, 40°C, 24 h, pH 2	163.9	L, PSO	Exothermic	[125]
Cht-MWCNT (Chitosan doped with multiwalled carbon nanotubes)	SEM, FTIR	0–25 mg/L, 1 g/L, 35°C, 2 h, pH 6	26.14	L, PSO	Endothermic	[123]
CS-GO (Chitosan grafted graphene oxide)	SEM, TEM, XRD, FTIR	10–125 mg/L, 0.25 g/L, 27°C, 7 h, pH 2	104.16	L, PSO	Endothermic	[161]
Chi@Fe_3O_4GO (Magnetic chitosan modified with graphene oxide)	SEM, XRD, Raman spectroscopy, FTIR, BET	10–100 mg/L, 0.5 g/L, 22°C, 3 h, pH 2	100.5	F, PSO	Endothermic	[162]

Q_{max} = maximum adsorption capacity; IC = initial Cr(VI) concentration; AD = adsorbent dose; CT = contact time; T = solution temperature; SR = stirring rate; L = Langmuir isotherm model; F = Freundlich isotherm model; L-F = Langmuir–Freundlich model; T = Temkin isotherm model; PSO = Pseudo second order kinetic model; IDP = Intraparticle diffusion model.

The PZC of chitosan-modified CNTs has been reported to be in the range of pH 4–8 [127,172,173]. A negatively charged adsorbent surface appears at $pH > pH_{PZC}$, while a positively charged surface appears at $pH < pH_{PZC}$, resulting in altered adsorption capacities. This is because adsorbing cationic ions becomes easier at $pH > pH_{PZC}$ due to electrostatic interaction with the negatively charged surfaces, whereas anionic ions are less likely to do so. A pH value greater than pH_{PZC} results in oxygen-containing functional groups with a negative charge at the surface for oxidized CNTs, which is conducive to the adsorption of Cr(III) cationic species due to electrostatic interactions. The pH_{PZC} of oxidized CNTs around pH 4 makes them suitable for sorbing Cr (III) from environmental samples at $pH > 4$. On the other hand, Cr(VI) appears as chromate and dichromate in environmental samples due to its high charge/radius ratio. The amino-functionalized oxidized CNTs are the preferred sorbents for these species at pH values lower than pH_{PZC} because their positive surface charges facilitate the sorption of Cr(VI) anions. For example, Marques Neto et al. [127] reported that CLCh/MWCNT/Fe with pH_{PZC} of 5.6 showed maximum removal of Cr(III) at pH 8 (i.e., at $pH > pH_{PZC}$) while maximum Cr(VI) was removed (97%) from solutions with pH 4–5 (i.e., $pH < pH_{PZC}$). CLCh/MWCNT/Fe would therefore be positively charged at $pH < pH_{PZC}$, which would facilitate interaction with Cr(VI) anions. However, negatively charged surfaces of the adsorbent inhibited the attraction of anionic Cr(VI) species at pH values higher than pH_{PZC}. A similar kind of adsorption trend has been reported when chitosan was combined with other adsorbents. For example, zirconium-crosslinked chitosan (Zr-CTS) composite having the PZC of 8.3 can eliminate Cr(VI) ions at pH lower than PZC [153]. The study found that at pH 5.0, the maximum adsorption capacity of 172 mg/g was achieved because the negatively charged Cr(VI) ions were captured by the positively charged surface of the adsorbent at $pH < pH_{PZC}$. Electrostatic attractions and the ligand exchange mechanism also resulted in a greater adsorption capacity for Cr(VI) ions. However, as the pH increased to 8, the repulsion between the OH^- and Cr(VI) ions became more intense, and the adsorption capacity dropped to 24.5 mg/g. Finally, most adsorbents exhibited maximum adsorption in the acidic pH range. It means that the pH of the wastewater needs to be adjusted within that specific range to attain high removal efficiencies. As a result, the electrostatic interaction between Cr(VI) ions and functional groups of adsorbent material should be addressed while designing a wastewater treatment strategy. Nonetheless, the pH of the solution influences the separation of Cr(VI) ions from various kinds of functional groups on the surface of adsorbents, causing problems with adsorbent reusability and limiting their cost windows to more affordable ranges [174,175].

Contact time/reaction time

For all adsorbents, the influence of contact time on adsorption capacity is always investigated to deduce the reaction mechanism that governs the process. Tables 3 and 4 show the ideal contact time for achieving Cr(VI) equilibrium capacity on CNTs and chitosan-based adsorbents, respectively. Most adsorbents follow a similar pattern, adsorbing Cr(VI) ions rapidly at first due to the molar and fixed amount of metal ions at the adsorption site, which causes boundary layer adsorption. The second stage is then the slow adsorption, which is described as diffusion into particles. The final stage is the saturation of adsorption sites, which is called the equilibrium stage. For example, Hossini et al. [176] tested MWCNT functionalized with EDTA and sulfuric acid (f-MWCNT) to study the effect of contact time (0–250 min) on the adsorption of Cr(VI) in aqueous samples with an initial Cr(VI) concentration of 5 mg/L. The results indicated that the maximum adsorption efficiency was 85% within 150 min of contact time. Atieh [103] used carbon nanotubes coated with activated carbon (AC/CNT) and AC alone to remove Cr(VI) ions. After the adsorption process, AC/CNTs outperformed AC, obtaining a maximum removal efficiency of 100% for contact times of 60 and 240 min, respectively, at initial Cr(VI) concentrations of 0.2 and 0.5 mg/L. The contact time of Fe0 nanorods modified with chitosan in porous anodic alumina (PAA) varied from 0 to 400 min. Cr(VI) concentration dramatically decreased within the first 30 min before leveling off, ultimately achieving an equilibrium within 400 min [177]. Another study reported similar results with MWCNT/Fe_3O_4 nanoparticles for adsorbing Cr(VI) ions [137]. In this study, the adsorption rate was rapid in the first 5 min before approaching an equilibrium in 45 min. The contact time also represents the economic output of the process and is related to the kinetics of adsorption. As a result, one can deduce that in an adsorption system, the contact time between the adsorbent and the adsorbate is critical. For example, Bhatt et al. [110] used chitosan cross-linked with trimeric acid (CTMA) for the adsorption of Cr(VI). According to the data, the CTMA attained a maximum adsorption capacity of 129.5 mg/g at an initial Cr(VI) concentration of 336.6 mg/L with a very long period of 882.5 min. The usage of adsorbents in industrial applications may be hindered by such a long equilibration time. In another investigation, the Cr(VI) removal capacity of chitosan/MWCNTs-COOH increased significantly in the first 10 min of contact, notably at Cr (VI) concentrations of 200 mg/L. It then increased steadily until reaching an adsorption equilibrium after about 30 min [125]. Such type of adsorbents with fast kinetics makes their selves the ideals adsorbents for remediation Cr-laden wastewater.

Adsorbent dosage

The appropriate dose of adsorbent must be acquired to optimize the interactions between Cr(VI) ions and the active sites of the adsorbents, as shown in Tables 3 and 4. It was observed that increasing the dosage of MWCNT/Fe_3O_4 nanoparticles for the adsorption of Cr(VI) ions from an optimal dosage of 0.035–0.05 g resulted in no significant change due to nanoparticle aggregation [137]. Polypyrrole-wrapped oxidized MWCNTs nanocomposites (PPy/OMWCNTs NCs) were used for effective adsorption of Cr (VI) at an adsorbent dosage of 0.005–0.1 g [101]. An adsorbent dosage at 0.075 g yielded the maximum adsorption capacity of 370.3 mg/g. Liu et al. [109] studied the effect of nZVI dosage in chitosan/Fe^0-nanoparticles (ECH-CS-nZVI) beads for Cr(VI) removal. They noted that the rate constant of Cr(VI) removal followed by increasing the dosage of nZVI in ECH-CS-nZVI beads. The authors suggested that increasing the dosage of nZVI increased the surface area and reactive sites resulting in an enhanced rate constant [109]. Jung et al. [106] showed that the Cr(VI) confiscation efficiency increased with increasing dosage from 10 to 200 mg/L (i.e., 66–99%, 70.0–96.4%, 11–89%, and 4–69% for MWNTs, PAC, SWNTs, and chitosan, respectively). According to these findings, the number of Cr(VI) adsorption sites was proportionate to the dose supplied to each adsorbent. At doses more than 100 mg/L, however, there was no further increase in adsorption capacity for PAC and chitosan, implying that most of the Cr(VI) ions occupied the active sites. Unlike PAC and chitosan, SWNTs and MWNTs required dosage of more than 100 mg/L to obtain a 95% adsorption capacity. The optimum and most cost-effective dose for water treatment was determined using this dose-dependent Cr(VI) adsorption behavior. It may be concluded that raising the adsorbent dose will increase the number of accessible adsorption sites, resulting in higher removal efficiency. When the adsorbents take up nearly all metal ions in aqueous solutions, however, the number of empty active adsorption sites increases, resulting in a decrease in the adsorbents' adsorption capabilities. For example, the Cr(VI) extraction efficiency of Cr (VI) of CTS/MMT-Fe_3O_4 increased from 22.6% to 99.6%, with an increasing dosage from 10.0 to 70.0 mg [107]. This pattern revealed that as the adsorbent dose increased, more active sites became available. However, Cr (VI) adsorption capability to the adsorbent peaked at 48.3 mg/g at 20 mg and then declined as the adsorbent dose increased. This result suggested that the economy absorbent dose is at 20.0 mg for the removal of Cr (VI) in this system [107].

Interference

According to studies on the impacts of coexisting ions, the developed adsorbents generally perform well in actual water and wastewater treatment systems containing pollutants other than Cr, which may interact and cause competitive adsorption with Cr(VI) binding to the adsorption sites. In an aqueous medium, Cr(VI) ions behave as anions; therefore, the adsorption performance of an adsorbent must be evaluated in the presence of most interfering anions such as F, SO_4^{2-}, NO_3^-, PO_4^{3-}, Cl^-, HCO_3^-, and other cations Zn^{2+}, Ni^{2+}, Cu^{2+}, Cd^{2+}, Pb^{2+}, as shown in Table 5. For example, Bhaumik et al. [101] found that coexisting ions such as Zn^{2+}, Ni^{2+}, Cu^{2+}, F^-, NO_3^-, CO_3^{2-} and SO_4^{2-} had a slight effect on the Cr(VI) adsorption onto PPy/OMWCNTs NCs. Gandhi and Meenakshi [150] showed that the addition of Ca^{2+} and Mg^{2+} ions did not affect the adsorption capacity of poly amidoamine chitosan beads loaded with zirconium (3ACBZr) for Cr(VI) ions uptake. Anions such as SO_4^{2-} and HCO_3^- had only a minor effect on Cr(VI) adsorption, whereas NO_3^- and Cl^- had no discernable effect. In another study, millimeter-sized hydrothermal carbon beads synthesized from chitosan (HCB) could effectively uptake Cr(VI) adsorption in the existence of Cl^-, NO_3^-, HPO_4^{2-} and SO_4^{2-} anions. The results showed that Cl^- and NO_3^- remained little impact on Cr(VI) adsorption as compared to HPO_4^{2-} and SO_4^{2-} which possessed similar ionic radius to $HCrO_4^-$, resulting in competing for the active adsorption sites of HCB [181].

The concentration of background electrolyte can significantly influence the interface potential and the thickness of the double layer, especially the interaction between the adsorbate and the active sites of the adsorbent during the adsorption process. Therefore, this parameter, expressed by ionic strength, also plays a crucial role in the adsorption of metal ions. The background electrolyte concentration affects the type of adsorption process. The electrolyte ions are more sensitive than the inner-sphere complexes because they are on the same plane as the adsorbent's outer sphere. For example, Lv et al. [97] reported that the removal rate of Cr(VI) by nZVI-MWCNT was enhanced by increasing the ionic strength of the solution from 0 to 0.05 M using NaCl, a result attributed to the outer sphere complex mechanism and β plane adsorption. The adsorption of Cr(VI) ions by MWCNT reduced when the ionic strength of the solution was 0.1 M using NaCl. The authors theorized that it was due to Cl^- and Cr(VI) ion competition [169]. Chagas et al. [182] used nanostructured oxide stabilized by chitosan nanocomposite (CT-Fe) for extraction of Cr(VI) from water samples with an ionic concentration of 1 mol/L KCl. The authors found that Cr(VI) adsorption was reduced because of a stronger repulsive interaction between the negative charges of solution ions ($HCrO_4^-$ and Cl^-), in which electrolytic ions can create a barrier on the charged surface with high salt concentration, preventing Cr(VI) ions from being removed. NaCl or $NaNO_3$ were used to evaluate the influence

TABLE 5 Cr(VI) uptake performance of CNTs and chitosan composites in the presence of interfering ions and their reusability.

Adsorbents	Coions	De-sorbents	Reusability	References
MWCNTs-CTAB (Multiwall carbon nanotubes modified with cationic surfactant cetyl trimethyl ammonium bromide)	—	0.1 M KCl	The adsorption efficiency was dropped from 98% to 60% at 25 °C, over the two cycles with a 25% drop in the first cycle followed by a 13% drop in the second cycle	[131]
MNP/MWCNTs (Magnetic nanoparticle-multiwalled carbon nanotubes composites)	Na^+, K^+, Ca^{2+}, Mg^{2+}, Ni^{2+}, Cu^{2+}, Co^{2+}, Pb^{2+}, Cr^{3+}, CO_3^{2-}, PO_4^{3-}, NO_3^{2-}, and Cl^- showed negligible influence on the removal efficiency	0.5 mol/L NaOH	80% of removal efficiency was obtained after the 5th cycle of reuse	[134]
FeOOH/AlOOH/MWCNTs (Binary metal oxyhydroxide, FeOOH and AlOOH decorated oxidized multiwalled carbon nanotubes)	—	0.1, 0.5 and 1.0 M NaOH	Cr(VI) desorption was found to be incomplete, and only 37.9% Cr(VI) was released into the aqueous solution after the third cycle	[135]
AAPTS modified MWCNTs (3-(2-aminoethylamino) propyl-trimethoxy silane-multiwalled carbon nanotubes)	K^+, Na^+, Ca^{2+}, Mg^{2+}, Fe^{3+}, SO_4^{2-}, Cl^-, NO_3^-, and MnO_4^- had no obvious impact on the determination of Cr(VI).	0.8 mol/L HNO_3	The adsorbent can be reused >100 times (tested for column operations)	[136]
IL-oxi-MWCNT (Ionic liquid-oxi-multiwalled carbon nanotubes)	Cu^{2+}, Cd^{2+}, Co^{2+}, Zn^{2+}, Ni^{2+}, or Pb^{2+} influenced the adsorption capacity, however Fe^{2+} and Mn^{2+} did by converting Cr(VI) to Cr(III). In the presence of NO_2^-, Cl^-, SO_3^{2-}, and PO_4^{3-}, Cr(VI) adsorption was hindered whereas a higher concentration of NO_2^-, SO_3^{2-} reduced Cr(*VI*)to Cr(III)	NaOH, KOH, Na_2SO_3, NH_4OH, L-ascorbic acid, $NaNO_2$, $Na_2S_2O_5$	The maximum regeneration capacity of the adsorbent was accomplished using NaOH retaining 99.5% efficiency for 3 cycles	[99]
MnO_2/Fe_3O_4/o-MWCNTs (Manganese dioxide/iron oxide/acid oxidized multiwalled carbon nanotube magnetic nanocomposites)	—	0.1 M NaOH	Adsorption capacity remained up to 85% after five times usage	[98]
DP/MWCNTs (Dodecyl-benzene-sulfonic-acid (DBSA) doped polyaniline/multiwalled carbon nanotubes)	—	DIW, NaOH, NaCl, HCl	DIW and HCl did not show any desorption while 2.9% and 10.5% Cr(VI) were recovered with 0.1 M NaCl and 0.1 M NaOH, respectively	[139]
CMBB (Chitosan modified magnetic bamboo biochar)	The adsorption efficiency affected in the order of $Cl^- < Ca^{2+} < PO_4^{3-} < Fe^{3+} < SO_4^{2-}$. 0.1 MNO_3^- led to a 38.3% reduction in the adsorption capacities	0.01 M $NaNO_3$, HNO_3, Na_2EDTA, $NaHCO_3$, or NaOH	$Na_2 > HNO_3 > NaOH > NaHCO_3 > NaNO_3$. adsorption efficiency remained above 90% after 5 adsorption–desorption cycles	[178]

Continued

TABLE 5 Cr(VI) uptake performance of CNTs and chitosan composites in the presence of interfering ions and their reusability.—cont'd

Adsorbents	Coions	De-sorbents	Reusability	References
PANI@MCTS (polyaniline@ magnetic chitosan nanomaterials)	—	1.0 M NaOH	>90% of removal efficiency was achieved after 4 cycles	[179]
Modified chitosan gel	—	0.5 N NaOH	About 85% of Cr(VI) were observed to be desorbed in the stripping solution	[180]
CD (Chitosan cross-linked with diethylenetriaminepentaacetic acid)	—	0.1 M of NaOH, H_2SO_4, HCl, and HNO_3	About 78.45% of Cr(VI) ions were desorbed in 0.1 N NaOH solution	[145]
MCGO-IL (Magnetic chitosan and graphene oxide-ionic liquid)	—	HCl	A slight loss in adsorption capacity was observed after 4 cycles of adsorption–desorption	[148]
Chitosan	Cl^-, SO_4^{2-} were reported to decrease the Cr(VI) removal effectiveness by >40% and 50%, respectively	—	—	[106]
CLCh/MWCNTs/Fe film (Crosslinked chitosan/multiple walled carbon nanotubes doped with magnetic iron oxide)	—	0.01 NAOH	After 10 cycles the adsorption performance of the composite material remained up to 93%	[127]
chitosan/MWCNTs-COOH	In a solution containing NaCl, KNO_3, KH_2PO_4, and Na_2SO_4 the adsorption capacity was reduced to 76.3, 67.5, 56.1, and 37.5 mg/g, respectively, compared to the adsorption capacity of 101.9 mg/g achieved in the absence of competing anions	0.10 M NaOH	Adsorption performance remained within the range of ~98–100% in the 4th cycles but decreased to ~91% in 5th cycle	[125]
CS/MWCNTs/Fe beads (Chitosan and maghemite-doped multiwall carbon nanotubes beads)	The adsorption capacity was weakened in the order of $PO_4^{3-} > SO_4^{2-} > NO_3^- > Cl^-$	1 M NaOH	During the consecutive five adsorption–desorption cycles, the Cr(VI) removal efficiency remained about 80%	[142]

of ionic strength on Cr(VI) removal by poly(4-vinyl pyridine) decorated magnetic chitosan (VMCP) biopolymer. Increasing salt concentration from 0 to 0.1 M, the suppressed effect of coexisting salt ions (NaCl or $NaNO_3$) on Cr(VI) removal appeared to be insignificant [183].

Adsorption isotherms

Adsorption isotherm is an effective tool characterizing an adsorbate-adsorbent system. Various types of isothermal models have been applied for adsorption, such as Freundlich, Temkin, Langmuir, Redlich–Peterson, and Dubinin–Radushkevich, among others. These isotherm models provide useful information on the distribution of accessible adsorption sites on the adsorbent, adsorption behavior, adsorption system parameters, and the adsorbent's adsorption capacity. Tables 3 and 4 show a summary of the best-fit isotherm model(s) for Cr(VI) adsorption using various CNTs and chitosan composites, respectively.

Dehghani et al. [94] applied D-R isotherms, Langmuir, and Freundlich, to the experimental data of Cr(VI) adsorption onto SWCNTs and MWCNTs from aqueous media. The Langmuir model represented as the best suited for the experimental data. In addition, the author concluded that the mean free energy of adsorption for MWCNTs and SWCNTs calculated by the D-R model were 80.3 and 77.3 kJ/mol, respectively, indicating that the adsorption was ascribed to the chemical adsorption process. Hu et al. [163] used oxidized MWCNTs to adsorb Cr(VI) ions. The Langmuir and Freundlich isotherms models were utilized to fit the experimental data, with the Langmuir adsorption isotherm being the more appropriate model describing the adsorption equilibrium. Huang et al. [132] reported that The adsorption equilibrium data of Cr(VI) onto nitrogen-doped magnetic carbon nanotubes (NMCNTs) was evaluated using Langmuir and Freundlich isotherms. The results showed that the experimental data match the Langmuir model well ($r^2 > 0.99$), implying that the adsorption process is monolayer adsorption. The maximum adsorption capacity of NMCNT was 970.9 mg/g, which is higher than the reported activity of the state-of-the-art carbon-based adsorbents. de la Luz-Asunción et al. [184] explored the nonlinear modeling of Cr(VI) adsorption on carbon-based nanomaterials and found that Langmuir and Freundlich's isotherms are consistent with the equilibrium data with $r^2 > 0.98$ and 0.97, respectively representing mono-layer and heterogeneous surface condition. Recently, commercially available CNTs had been modified with H_2O_2 to enhance the adsorption of Cr(VI) ions. The D-R isotherm revealed that the adsorption isotherm energy ranged from 0.371 to 0.870 kJ/mol, implying that the physical adsorption process was responsible for eliminating of Cr(VI) ions [184]. Hu et al. [120] applied Freundlich, Temkin isotherm, and Langmuir models to study the Cr(VI) adsorption equilibrium by EMCMCR. The results showed that the Temkin and Langmuir models fitted well with the experimental data, and the adsorption process was attributed to the chemical and physical adsorption. Li et al. [118] used chitosan nanofibers for the Cr(VI) ions adsorption. The experimental data followed both the Langmuir and Freundlich isotherms; however, the Freundlich model suited the data somewhat better ($r^2 > 0.98$) than the Langmuir model ($r^2 > 0.97$), showing both mono- and multilayer adsorptions occurred during the adsorption process. Huang et al. [125] showed that the Langmuir isotherm was the better model than the Freundlich model for describing the Cr(VI) adsorption by chitosan/MWCNTs-COOH composites, implying that the energies of the active sites evenly distributed across the surface. Marques Neto et al. [127] applied both the Freundlich and Langmuir isotherms on the adsorption data of Cr(VI) by CLCh/MWCNT/Fe. With an $r^2 > 0.99$, the Langmuir isotherm model provided the best fit, with a maximum adsorption capacity of 449.3 mg/g at 25°C. The adsorption was favorable, as evidenced by the computed value of separation factor (R_L) < 1.0.

Adsorption kinetics

Adsorption kinetics provides valuable information on the possible adsorption mechanisms and their potential rate-limiting step during the adsorption process. It is also an important step to select the optimizing process parameters and the best conditions for a large-scale removal process [185]. Pseudo-first- and second-order, Elovich, and intraparticle diffusion models usually fit well with the adsorption kinetics of metal ions in an aqueous medium, as shown in Tables 3 and 4. For example, Ihsanullah et al. [167] used pseudo-first and second-order and intraparticle diffusion models to characterize the adsorption kinetics of Cr(VI) ions on raw and modified CNTs. The pseudo-second-order kinetic equation resulted in a greater r^2 value than the other two kinetic equations, indicating that the pseudo-second-order kinetic equation effectively described Cr(VI) ion adsorption by raw and modified CNTs and AC. In another investigation, magnetic iron oxide nanoparticle-multiwalled carbon nanotube composites (MNP/MWCNTs) were employed to adsorb Cr(VI) from water samples. The r^2 value obtained from fitting with the pseudo-second-order model was greater than that with the pseudo-first-order models, and the calculated adsorption capacities (q_{ecal}) from the pseudo-second-order model were closer to the experimental values. The authors proposed that valence forces controlled the rate of chemisorption by sharing electrons between Cr(VI) ions and MNP/MWCNTs [134]. Similarly, the kinetic data for Cr(VI) adsorption onto crosslinked chitosan resins matched the pseudo-second-order model with a stronger correlation coefficient [156]. Another study found that Cr(VI) adsorption on CTS/MMT-Fe_3O_4 microspheres followed a pseudo-second-order kinetic model, with the corresponding rate constant decreasing with increasing temperature [107]. Debnath et al. [186] found that the rate constant increased with increasing temperature as the adsorption of Cr(VI) by CGM followed the pseudo-second-order kinetic equation. The adsorption process was controlled by more than one mechanism, with the film diffusion process being the rate-limiting step. Sikder et al. [146] studied that the CS-nZVI-CM-CD beads were used to adsorb Cr(VI) ions. The pseudo-second-order model and the intraparticle diffusion model both explained the adsorption kinetics very well, with the latter being the rate-limiting step.

Thermodynamics

Temperature is another significant element for Cr(VI) ion adsorption since it can alter the solid/liquid interface, adsorbent swellability, and Cr(VI) ion mobility during the adsorption process. In addition to the solution temperature, the thermodynamic characteristics of adsorption reactions are frequently used to determine the nature of the adsorption process, such as spontaneity, unpredictability, endothermic or exothermic. It also determines whether the temperature is favorable or unfavorable to the adsorption process. From Tables 3 and 4, it is evident that adsorption processes of Cr(VI) ions onto CNTs and chitosan-based adsorbent are both exothermic and endothermic. The adsorption capacity of Cr(VI) on the chitosan adsorbent grafted with n-butyl acrylate decreased with increasing temperature [105]. The equilibrium constant dropped with increasing temperature, indicating that adsorption is exothermic and thus more advantageous at lower temperatures. An explanation is that, as the temperature increases, the mobility of Cr(VI) ions also increases, causing the desorption and de-chelation of the adsorbent. Similarly, Bayazit and Kerkez [187] showed that the maximum monolayer adsorption capacity of magnetic MWCNTs (M-MWCNTs) for Cr(VI) ions decreased from 14.28 to 11.27 mg/g with the incremental increases in temperature. Contrarily, Yue et al. [188] showed that the ability of magnetic chitosan microspheres modified with 1,6-hexane diamine (AF-MCTS) to remove Cr(VI) increased as the adsorption temperature rose from 20 to 40°C, indicating that adsorption was endothermic. Similar results were obtained by Marques Neto et al. [127], who reported that the adsorption capacity of CLCh/MWCNTs/Fe film increased from 449.3 to 477.3 mg Cr(VI)/g by increasing the reaction temperature from 25 to 40°C. The expansion of pore size and decrease in mass and electron transfer resistance between Cr(VI) and adsorbent was related to the rise in adsorption capacity with increasing temperature. Another possibility is that, while Cr(VI) species are intensely hydrated, they need the energy to break through the hydration sheath to be adsorbed effectively. As a result, a higher temperature helps metal ion dehydration and promotes the adsorption process.

Regeneration

As previously mentioned, the reusability of an adsorbent can be a determining factor when evaluating the practicality of the adsorbent. Adsorbents showing minimal loss in adsorption efficiency after repeated adsorption–desorption cycles are more valuable than those showing greater capacity but limited reusability. The regeneration of adsorbents may have several advantages, including (i) recovery of pollutants, (ii) reusability of adsorbents in the number of cycles, thereby reducing adsorption and processing costs, (iii) reducing secondary waste, and (iv) determine the adsorption mechanism [189]. According to the scientific literature, the most common reagents used to regenerate CNTs, and chitosan composites are as follows: acids > chelating agents > organic solvents > mixtures of these eluents > alkalis. Alkaline eluents (such as NaOH) have shown the highest desorption performance for the pollutants in the adsorbent with the minimum loss in adsorption efficiency and maximum adsorbents reusability. The combination of two eluents can boost regeneration performance when compared to employing a single eluent. Table 5 shows the adsorption–desorption and reusability performance of CNTs and chitosan composites by using different desorption reagents. For example, Kumar et al.[99] found that the order of tested reagents for regeneration capacity of multiwalled carbon nanotubes functionalized with ionic liquid (IL-oxi-MWCNT) was NaOH (98.0%) > KOH (79.4%) > Na_2SO_3 (65.0%) > NH_4OH (61.0%) > $C_6H_8O_6$ (39.1%) > $NaNO_2$ (28.0%) > $Na_2S_2O_5$ (25.3%). The maximum regeneration capacity of the adsorbent was accomplished using NaOH retaining its performance efficiency for three cycles (99.5%). Gholipour and Hashemipour [190] used 0.03–0.5 M NaOH solution for desorption of Cr(VI) ions from MWCNTs. MWCNTs could be employed for 10 adsorption–desorption cycles at an optimum dose of 0.2 M NaOH as desorption eluent without compromising their adsorption–desorption capabilities, according to the findings. Chen et al. [107] reported that the CTS/MMT-Fe_3O_4 microspheres have a desorption efficiency of 95.8% after three cycles of reuse when using 0.01 M NaOH to desorb Cr(VI) ions. These findings suggested that the microspheres might be regenerated and reused. Debnath et al. [186] reported that, by using the NaOH, the adsorption capacity of CGM remained constant (96%) until the third adsorption–desorption cycle, and it continued to decrease to 85% after the fifth cycle. In a study by Subedi et al. [162], 0.1 M NaOH was preferable over 0.1 M HNO_3 to regenerate magnetic chitosan-based adsorbents loaded with Cr(VI). The performance declined from 77.5% to 75% in four consecutive adsorption–desorption cycles when using NaOH eluent, as compared to 70% to 32.5% using HNO_3. The adsorption performance of a magnetic chitosan-CNT loaded with Cr(VI) exhibited minimal decline after 10 cycles when using NaOH as desorbing agent [127]. Similarly, magnetic chitosan-GO composites lost 4% of their adsorption capacity for Cr(VI) after seven cycles [191]. Aslam et al. [9] reported that the Cr(VI) adsorption capacity of prepared CS/MWCNTs/Fe beads remained up to 80% after five consecutive adsorption–desorption cycles. In conclusion, Cr(VI) ions can be efficiently desorbed by NaOH solution, because the electrostatic connection between the adsorbent and Cr(VI) ions decreases in an alkaline environment, thus encouraging Cr(VI) ion desorption.

Prospects and challenges

Adsorption technology has become an effective and common practice for mitigating Cr(VI) pollution in aqueous solutions. CNTs and chitosan composites possess excellent physical and chemical properties and have drawn extensive research for further improving the removal efficiency of heavy metal ions in water. Still, there is plenty of room for elucidating the science and the applications. An example would be the uncertainties as to whether these types of adsorbents would perform as well in actual samples (e.g., industrial wastewater, seawater) as it does to extract Cr(VI) from deionized water. Marques Neto et al. [127] used ChCL/MWCNTs/Fe film to adsorb the Cr(VI) ions from the water sample. Although the composite material showed promising adsorption capacity (450 mg Cr(VI)/g) in a short equilibrium time (60 min), the applicability in a complex water system containing multiple contaminants is still questionable. Another obstacle is the difficulty to reuse these adsorbents to meet the cost-effective adsorption process. For example, Samuel et al. [161,192] pointed out that the performance of carbonaceous and chitosan materials was greatly reduced after each cycle of Cr(VI) adsorption. Once the adsorbents have been saturated with hazardous Cr ions, they can be injected directly into the environment without any postprocessing. Although these types of adsorbents may present effective options for eliminating specific pollutants (e.g., Cr(VI) ions), once used, they need to be disposed of, leading to concerns over secondary pollution at the disposal sites.

Further, the cost-effectiveness of any treatment method is always a prioritized criterion. Although chitosan is inexpensive and a readily available biomaterial, the costs associated with its downstream processing and modification on CNTs is still in the infancy phase. There has not been any pilot-scale investigation documented to justify the cost effectiveness of equipment design, operation and regeneration of metal-loaded adsorbents for reuse. Successful regeneration step and minimal loss in the adsorption capacity after multiple desorption is critical to ensuring the adsorption process is economical and environmentally friendly. From an economic perspective, the cost of CNTs production, operations, and applications needs to be reduced and optimized. Despite ample works being done at the research phase, much work remains to be done to move these green research methodologies to mass-production of CNTs for industrial water and wastewater treatment. The modification of carbonaceous nanoparticles with biomaterials (especially chitosan) provides a new opportunity for large-scale production of green, renewable, and cost-effective products to eliminate toxic Cr ions in wastewater streams. However, the toxicity and biocompatibility of the CNTs should be thoroughly tested for ecosystems and humans. The combination of these two materials (CNTs and chitosan) can improve the usefulness of these materials in real-life applications to address these shortcomings in an orderly manner.

The Life Cycle Assessment (LCA) approach has received little attention in the area of nanotechnology. Presently, only two reviews have been published on this topic [193,194]. A 2-day workshop of European and American experts convened that the ISO framework of LCA is suitable for nanomaterials [195]. Interestingly, the workshop report also asserts that by examining the various phases of LCA, there are several issues need to be clarified specifically for applications in the field of nanotechnology. For example, defining functional units properly and sufficiently in the Goals and Scope Definition phase; accessing comprehensive and appropriate life cycle inventory data in the Inventory Analysis phase; and developing and incorporating characterization factors for nano-specific impacts in the subsequent Impact Assessment phase. In the literature search, articles on "life-cycle assessment" and "chitosan- and carbon-based adsorbents" were identified, producing eight different studies all dealing with carbon nanotubes [196–203]. Unfortunately, no studies covering chitosan-based adsorbents have been identified. Out of the eight studies identified here, five use a weight unit of a specific nanomaterial as a functional unit. This unit is only relevant when analyzing the manufacturing process of a specific nanomaterial, not when comparing its functionality to traditional materials. Among the eight studies, only four included the entire life cycle (cradle-to-grave), while all the others omitted the end-of-life treatment and the use phases. The inventories used in these studies are not comprehensive since no emissions data is included. Nanoparticles released indoors and outdoors are not characterized in any way in the impact analysis. Therefore, it is undoubtedly vital to undertake a life cycle assessment of such materials in order to overcome the challenges that may come with the use of such materials for water and wastewater treatment applications.

Conclusions

The increasing frequency and intensity of water pollution by heavy metals and the inability of natural water bodies to self-purify have caused adverse consequences such as bioaccumulation and chemical toxicity for aquatic and human lives. This chapter reviews some of the main CNTs and chitosan-based adsorbents used to confiscate Cr(VI) ions from aqueous solutions and provides an updated resource for researchers studying adsorptive removal of Cr(VI). Many of the studies reported promising results using CNTs and chitosan-based materials to remove Cr(VI) ions from water samples owing to properties complementary to each other—CNTs' high surface area and porosity facilitating rapid adsorption of Cr(VI) ions and chitosan's abundance in the amine functional groups for chemically capturing Cr(VI) ions. Other advantages, such as the nontoxicity, biocompatibility, hydrophilicity, and high affinity of chitosan for Cr(VI) ions, have attracted great attention from researchers. Therefore, CNTs and chitosan-based hybrid materials are actively being explored, with their mechanistic approach involving redox and adsorption technologies to provide green solutions to water purification. The scientific literature we reviewed point to the following conclusions: (i) Cr adsorption onto the composite adsorbents is highly pH-dependent; (ii) the adsorbent dosage in relation to feed Cr(VI) concentration is an important parameter for efficient Cr adsorption; (iii) the Langmuir and Freundlich isotherm models are both satisfactory in describing the adsorption process, with the Langmuir model being favored in most cases; (iv) the pseudo-second-order model is the most suitable equation describing the adsorption kinetics, while the intraparticle diffusion model describes well the multilinearity of porous, film diffusion, and equilibrium steps of adsorption mechanisms; (v) most studies showed the endothermic nature of adsorption but several reported the process as an exothermic one; (vi) alkaline reagents such as sodium hydroxide sufficiently regenerate the adsorbents with limited loss in the adsorption capacity. Despite these positive attributes, the applicability of CNTs/chitosan adsorbents warrant more research endeavors because (i) the effectiveness of Cr uptake in complex wastewater or contaminated water remains unclear; (ii) most of the studies were conducted in batch mode, bringing questions to the applicability in continuous-flow operations called upon in typical industrial-scale facilities; (iii) the cost effectiveness in both the production and large-scale operations are largely unknown at this early phase.

Acknowledgments

This study was done at the Department of Environmental Science and Engineering, Tunghai University, Taiwan. The authors gratefully acknowledge the funding support by the Research Council of Texas A&M University-San Antonio, and the Ministry of Science and Technology of R.O.C. (Taiwan), 107-2221-E-029-001-MY2.

Author contributions

Conceptualization, W.D. and M.M.A.A.; resources, W.D. and H.-W.K.; writing—original draft preparation, M.M.A.A.; writing—review and editing, H.-W.K., W.D., M.S., K.R.; visualization, M.B.; supervision, H.-W.K.; All authors have given approval to the final version of the manuscript.

Conflicts of interest

The authors declare no conflict of interest.

References

[1] J. Kitzes, M. Wackernagel, J. Loh, A. Peller, S. Goldfinger, D. Cheng, et al., Shrink and share: humanity's present and future ecological footprint, Philos. Trans. R. Soc. B Biol. Sci. 363 (2008) 467–475, https://doi.org/10.1098/rstb.2007.2164.

[2] P.F.M. Verdonschot, B.M. Spears, C.K. Feld, S. Brucet, H. Keizer-Vlek, A. Borja, et al., A comparative review of recovery processes in rivers, lakes, estuarine and coastal waters, Hydrobiologia 704 (2013) 453–474, https://doi.org/10.1007/s10750-012-1294-7.

[3] M. Aslam, Z. Khan, M. Sultan, Y. Niaz, M. Mahmood, M. Shoaib, et al., Performance evaluation of trickling filter-based wastewater treatment system utilizing cotton sticks as filter media, Pol. J. Environ. Stud. 26 (2017) 1955–1962, https://doi.org/10.15244/pjoes/69443.

[4] W. Jin, H. Du, S. Zheng, Y. Zhang, Electrochemical processes for the environmental remediation of toxic Cr(VI): a review, Electrochim. Acta 191 (2016) 1044–1055, https://doi.org/10.1016/j.electacta.2016.01.130.

[5] V.N. Thekkudan, V.K. Vaidyanathan, S.K. Ponnusamy, C. Charles, S. Sundar, D. Vishnu, et al., Review on nanoadsorbents: a solution for heavy metal removal from wastewater, IET Nanobiotechnol. 11 (2017) 213–224, https://doi.org/10.1049/iet-nbt.2015.0114.

[6] J.C. Almeida, C.E.D. Cardoso, D.S. Tavares, R. Freitas, T. Trindade, C. Vale, et al., Chromium removal from contaminated waters using nanomaterials—a review, TrAC Trends Anal. Chem. 118 (2019) 277–291, https://doi.org/10.1016/j.trac.2019.05.005.

[7] I. Ali, C. Peng, Z.M. Khan, I. Naz, M. Sultan, M. Ali, et al., Overview of microbes based fabricated biogenic nanoparticles for water and wastewater treatment, J. Environ. Manag. 230 (2019) 128–150, https://doi.org/10.1016/j.jenvman.2018.09.073.

[8] I. Ali, C. Peng, I. Naz, Z.M. Khan, M. Sultan, T. Islam, et al., Phytogenic magnetic nanoparticles for wastewater treatment: a review, RSC Adv. 7 (2017) 40158–40178, https://doi.org/10.1039/C7RA04738J.

[9] M.M.-A. Aslam, H.-W. Kuo, W. Den, M. Usman, M. Sultan, H. Ashraf, Functionalized carbon nanotubes (CNTs) for water and wastewater treatment: preparation to application, Sustainability 13 (2021) 5717, https://doi.org/10.3390/su13105717.

[10] M.K. Aroua, F.M. Zuki, N.M. Sulaiman, Removal of chromium ions from aqueous solutions by polymer-enhanced ultrafiltration, J. Hazard. Mater. 147 (2007) 752–758, https://doi.org/10.1016/j.jhazmat.2007.01.120.

[11] V.E. Pakade, N.T. Tavengwa, L.M. Madikizela, Recent advances in hexavalent chromium removal from aqueous solutions by adsorptive methods, RSC Adv. 9 (2019) 26142–26164, https://doi.org/10.1039/C9RA05188K.

[12] J.O. Esalah, M.E. Weber, J.H. Vera, Removal of lead from aqueous solutions by precipitation with sodium di-(n-octyl) phosphinate, Sep. Purif. Technol. 18 (1999) 25–36.

[13] L. Yurlova, A. Kryvoruchko, B. Kornilovich, Removal of Ni(II) ions from wastewater by micellar-enhanced ultrafiltration, Desalination 144 (2002) 255–260, https://doi.org/10.1016/S0011-9164(02)00321-1.

[14] Y. Xu, J. Chen, R. Chen, P. Yu, S. Guo, X. Wang, Adsorption and reduction of chromium(VI) from aqueous solution using polypyrrole/calcium rectorite composite adsorbent, Water Res. 160 (2019) 148–157, https://doi.org/10.1016/j.watres.2019.05.055.

[15] M. Bagherzadeh, Z. Hadizadeh, Z. Akrami, Z.S. Ghahfarokhi, Electrochemical detection of Cr(III) and Cr(VI) in solution by using ZrO2 modified magnetic nanoparticles by redox probes, Mater. Sci. Semicond. Process. 131 (2021) 105840, https://doi.org/10.1016/j.mssp.2021.105840.

[16] Y. Shi, D. Xiong, Y. Zhao, T. Li, K. Zhang, J. Fan, Highly efficient extraction/separation of Cr (VI) by a new family of hydrophobic deep eutectic solvents, Chemosphere 241 (2020) 125082, https://doi.org/10.1016/j.chemosphere.2019.125082.

[17] L. Ren, L.-J. Yan, Y.-H. Bai, Y. Liu, P. Lv, Y.-X. Wang, et al., Effects of loading methods and oxidation degree of support on the tar reforming activity of char-supported Ni catalyst using toluene as a model compound, Fuel Process. Technol. 201 (2020) 106347, https://doi.org/10.1016/j.fuproc.2020.106347.

[18] M. Marcos-Hernández, R.A. Arrieta, K. Ventura, J. Hernández, C.D. Powell, A.J. Atkinson, et al., Superparamagnetic nanoadsorbents for the removal of trace as(III) in drinking water, Environ. Adv. 4 (2021) 100046, https://doi.org/10.1016/j.envadv.2021.100046.

[19] W.S. Chai, J.Y. Cheun, P.S. Kumar, M. Mubashir, Z. Majeed, F. Banat, et al., A review on conventional and novel materials towards heavy metal adsorption in wastewater treatment application, J. Clean. Prod. 296 (2021) 126589, https://doi.org/10.1016/j.jclepro.2021.126589.

[20] S. Cao, R. Du, M. Niu, B. Li, N. Ren, Y. Peng, Integrated anaerobic ammonium oxidization with partial denitrification process for advanced nitrogen removal from high-strength wastewater, Bioresour. Technol. 221 (2016) 37–46, https://doi.org/10.1016/j.biortech.2016.08.082.

[21] F.-F. Li, W.-R. Cui, W. Jiang, C.-R. Zhang, R.-P. Liang, J.-D. Qiu, Stable sp carbon-conjugated covalent organic framework for detection and efficient adsorption of uranium from radioactive wastewater, J. Hazard. Mater. 392 (2020) 122333, https://doi.org/10.1016/j.jhazmat.2020.122333.

[22] S. Mortazavian, H. An, D. Chun, J. Moon, Activated carbon impregnated by zero-valent iron nanoparticles (AC/nZVI) optimized for simultaneous adsorption and reduction of aqueous hexavalent chromium: material characterizations and kinetic studies, Chem. Eng. J. 353 (2018) 781–795, https://doi.org/10.1016/j.cej.2018.07.170.

[23] L. Zhou, R. Li, G. Zhang, D. Wang, D. Cai, Z. Wu, Zero-valent iron nanoparticles supported by functionalized waste rock wool for efficient removal of hexavalent chromium, Chem. Eng. J. 339 (2018) 85–96, https://doi.org/10.1016/j.cej.2018.01.132.

[24] D. Pradhan, L.B. Sukla, M. Sawyer, P.K.S.M. Rahman, Recent bioreduction of hexavalent chromium in wastewater treatment: a review, J. Ind. Eng. Chem. 55 (2017) 1–20, https://doi.org/10.1016/j.jiec.2017.06.040.

[25] M.N. Sahmoune, K. Louhab, A. Boukhiar, Advanced biosorbents materials for removal of chromium from water and wastewaters, Environ. Prog. Sustain. Energy 30 (2011) 284–293, https://doi.org/10.1002/ep.10473.

[26] P. Miretzky, A.F. Cirelli, Cr(VI) and Cr(III) removal from aqueous solution by raw and modified lignocellulosic materials: a review, J. Hazard. Mater. 180 (2010) 1–19, https://doi.org/10.1016/j.jhazmat.2010.04.060.

[27] A.A. Ihsanullah, A.M. Al-Amer, T. Laoui, M.J. Al-Marri, M.S. Nasser, et al., Heavy metal removal from aqueous solution by advanced carbon nanotubes: critical review of adsorption applications, Sep. Purif. Technol. 157 (2016) 141–161, https://doi.org/10.1016/j.seppur.2015.11.039.

[28] J. Xu, Z. Cao, Y. Zhang, Z. Yuan, Z. Lou, X. Xu, et al., A review of functionalized carbon nanotubes and graphene for heavy metal adsorption from water: preparation, application, and mechanism, Chemosphere 195 (2018) 351–364, https://doi.org/10.1016/j.chemosphere.2017.12.061.

[29] L. Ren, J. Xu, Y. Zhang, J. Zhou, D. Chen, Z. Chang, Preparation and characterization of porous chitosan microspheres and adsorption performance for hexavalent chromium, Int. J. Biol. Macromol. 135 (2019) 898–906, https://doi.org/10.1016/j.ijbiomac.2019.06.007.

[30] M.K. Dinker, P.S. Kulkarni, Recent advances in silica-based materials for the removal of hexavalent chromium: a review, J. Chem. Eng. Data 60 (2015) 2521–2540, https://doi.org/10.1021/acs.jced.5b00292.

[31] H.N. Muhammad Ekramul Mahmud, A.K.O. Huq, R. Yahya, binti., The removal of heavy metal ions from wastewater/aqueous solution using polypyrrole-based adsorbents: a review, RSC Adv. 6 (2016) 14778–14791, https://doi.org/10.1039/C5RA24358K.

[32] B. Tu, R. Wen, K. Wang, Y. Cheng, Y. Deng, W. Cao, et al., Efficient removal of aqueous hexavalent chromium by activated carbon derived from Bermuda grass, J. Colloid Interface Sci. 560 (2020) 649–658, https://doi.org/10.1016/j.jcis.2019.10.103.

[33] A. Shakya, A. Núñez-Delgado, T. Agarwal, Biochar synthesis from sweet lime peel for hexavalent chromium remediation from aqueous solution, J. Environ. Manag. 251 (2019) 109570, https://doi.org/10.1016/j.jenvman.2019.109570.

[34] M.K. Uddin, A review on the adsorption of heavy metals by clay minerals, with special focus on the past decade, Chem. Eng. J. 308 (2017) 438–462, https://doi.org/10.1016/j.cej.2016.09.029.

[35] J.P. Valle, B. Gonzalez, J. Schulz, D. Salinas, U. Romero, D.F. Gonzalez, et al., Sorption of Cr(III) and Cr(VI) to K 2 Mn 4 O 9 nanomaterial a study of the effect of pH, time, temperature and interferences, Microchem. J. 133 (2017) 614–621, https://doi.org/10.1016/j.microc.2017.04.021.

[36] Dubey S, Upadhyay SN, Sharma YC. Optimization of removal of Cr by γ-alumina nano-adsorbent using response surface methodology. Ecol. Eng. 2016;97:272–83. https://doi.org/10.1016/j.ecoleng.2016.10.005.

[37] S. Babel, Low-cost adsorbents for heavy metals uptake from contaminated water: a review, J. Hazard. Mater. 97 (2003) 219–243, https://doi.org/10.1016/S0304-3894(02)00263-7.

[38] M.B. Arain, I. Ali, E. Yilmaz, M. Soylak, Nanomaterial's based chromium speciation in environmental samples: a review, TrAC Trends Anal. Chem. 103 (2018) 44–55, https://doi.org/10.1016/j.trac.2018.03.014.

[39] B. Markiewicz, I. Komorowicz, A. Sajnóg, M. Belter, D. Barałkiewicz, Chromium and its speciation in water samples by HPLC/ICP-MS—technique establishing metrological traceability: a review since 2000, Talanta 132 (2015) 814–828, https://doi.org/10.1016/j.talanta.2014.10.002.

[40] G.-R.R. Bernardo, R.-M.J. Rene, A.-D. la Torre Ma Catalina, Chromium (III) uptake by agro-waste biosorbents: chemical characterization, sorption–desorption studies, and mechanism, J. Hazard. Mater. 170 (2009) 845–854, https://doi.org/10.1016/j.jhazmat.2009.05.046.

[41] M. Suwalsky, R. Castro, F. Villena, C.P. Sotomayor, Cr(III) exerts stronger structural effects than Cr(VI) on the human erythrocyte membrane and molecular models, J. Inorg. Biochem. 102 (2008) 842–849, https://doi.org/10.1016/j.jinorgbio.2007.11.020.

[42] L.E. Wu, A. Levina, H.H. Harris, Z. Cai, B. Lai, S. Vogt, et al., Carcinogenic chromium(VI) compounds formed by intracellular oxidation of chromium(III) dietary supplements by adipocytes, Angew. Chem. Int. Ed. 55 (2016) 1742–1745, https://doi.org/10.1002/anie.201509065.

[43] H.A. Headlam, P.A. Lay, Spectroscopic characterization of genotoxic chromium(V) peptide complexes: oxidation of chromium(III) triglycine, tetraglycine and pentaglycine complexes, J. Inorg. Biochem. 162 (2016) 227–237, https://doi.org/10.1016/j.jinorgbio.2016.06.015.

[44] D.R. Lindsay, K.J. Farley, R.F. Carbonaro, Oxidation of CrIII to CrVI during chlorination of drinking water, J. Environ. Monit. 14 (2012) 1789, https://doi.org/10.1039/c2em00012a.

[45] S.E. Gomez-Gonzalez, G.G. Carbajal-Arizaga, R. Manriquez-Gonzalez, W. De la Cruz-Hernandez, S. Gomez-Salazar, Trivalent chromium removal from aqueous solutions by a sol–gel synthesized silica adsorbent functionalized with sulphonic acid groups, Mater. Res. Bull. 59 (2014) 394–404, https://doi.org/10.1016/j.materresbull.2014.07.035.

[46] E. Agrafioti, G. Bouras, D. Kalderis, E. Diamadopoulos, Biochar production by sewage sludge pyrolysis, J. Anal. Appl. Pyrolysis 101 (2013) 72–78, https://doi.org/10.1016/j.jaap.2013.02.010.

[47] C. Cervantes, J. Campos-García, Reduction and efflux of chromate by Bacteria, in: D.H. Nies, S. Silver (Eds.), Mol. Microbiol. Heavy Met, Springer Berlin Heidelberg, Berlin, Heidelberg, Heidelberg, 2007, pp. 407–419, https://doi.org/10.1007/7171_2006_087.

[48] S.S. Wise, A.L. Holmes, L. Liou, R.M. Adam, J.P. Wise Sr., Hexavalent chromium induces chromosome instability in human urothelial cells, Toxicol. Appl. Pharmacol. 296 (2016) 54–60, https://doi.org/10.1016/j.taap.2016.02.015.

[49] W. Liu, F. Chaspoul, C. Botta, M. De Méo, P. Gallice, Bioenergetics and DNA alteration of normal human fibroblasts by hexavalent chromium, Environ. Toxicol. Pharmacol. 29 (2010) 58–63, https://doi.org/10.1016/j.etap.2009.10.001.

[50] Y. Gong, J. Tang, D. Zhao, Application of iron sulfide particles for groundwater and soil remediation: a review, Water Res. 89 (2016) 309–320, https://doi.org/10.1016/j.watres.2015.11.063.

[51] P.Z. Ray, H.J. Shipley, Inorganic nano-adsorbents for the removal of heavy metals and arsenic: a review, RSC Adv. 5 (2015) 29885–29907, https://doi.org/10.1039/C5RA02714D.

[52] G.M. Naja, B. Volesky, Toxicity and sources of Pb, Cd, Hg, Cr, As, and radionuclides in the environment, in: Handb. Adv. Ind. Hazard. wastes Manag., CRC Press, 2017, pp. 855–903.

[53] M.F. Bergamini, D.P. dos Santos, M.V.B. Zanoni, Development of a voltammetric sensor for chromium(VI) determination in wastewater sample, Sensors Actuators B Chem. 123 (2007) 902–908, https://doi.org/10.1016/j.snb.2006.10.062.

[54] USGS mineral commodity summaries, U.S. Geological Survey (2021).

[55] A.A. Belay, Impacts of chromium from tannery effluent and evaluation of alternative treatment options, J. Environ. Prot. (Irvine, Calif) 01 (2010) 53–58, https://doi.org/10.4236/jep.2010.11007.

[56] H. Cheng, T. Zhou, Q. Li, L. Lu, C. Lin, Anthropogenic chromium emissions in China from 1990 to 2009, PLoS One 9 (2014) e87753, https://doi.org/10.1371/journal.pone.0087753.

[57] T.A. Kurniawan, G.Y.S. Chan, W.-H. Lo, S. Babel, Physico–chemical treatment techniques for wastewater laden with heavy metals, Chem. Eng. J. 118 (2006) 83–98, https://doi.org/10.1016/j.cej.2006.01.015.

[58] M. Hua, S. Zhang, B. Pan, W. Zhang, L. Lv, Q. Zhang, Heavy metal removal from water/wastewater by nanosized metal oxides: a review, J. Hazard. Mater. 211–212 (2012) 317–331, https://doi.org/10.1016/j.jhazmat.2011.10.016.

[59] F. Fu, Q. Wang, Removal of heavy metal ions from wastewaters: a review, J. Environ. Manag. 92 (2011) 407–418, https://doi.org/10.1016/j.jenvman.2010.11.011.

[60] N. Li, Y. Tian, J. Zhao, J. Zhang, J. Zhang, W. Zuo, et al., Efficient removal of chromium from water by Mn3O4 @ZnO/Mn3O4 composite under simulated sunlight irradiation: synergy of photocatalytic reduction and adsorption, Appl. Catal. B Environ. 214 (2017) 126–136, https://doi.org/10.1016/j.apcatb.2017.05.041.

[61] F. Ren, G. Li, Z. Zhang, X. Zhang, H. Fan, C. Zhou, et al., A single-layer janus membrane with dual gradient conical micropore arrays for self-driving fog collection, J. Mater. Chem. A 5 (2017) 18403.

[62] J. Wei, C. Tu, G. Yuan, D. Bi, L. Xiao, B.K.G. Theng, et al., Carbon-coated montmorillonite nanocomposite for the removal of chromium(VI) from aqueous solutions, J. Hazard. Mater. 368 (2019) 541–549, https://doi.org/10.1016/j.jhazmat.2019.01.080.

[63] Z. Yuan, X. Cheng, L. Zhong, R. Wu, Y. Zheng, Preparation, characterization and performance of an electrospun carbon nanofiber mat applied in hexavalent chromium removal from aqueous solution, J. Environ. Sci. 77 (2019) 75–84, https://doi.org/10.1016/j.jes.2018.06.016.

[64] B. Saha, C. Orvig, Biosorbents for hexavalent chromium elimination from industrial and municipal effluents, Coord. Chem. Rev. 254 (2010) 2959–2972, https://doi.org/10.1016/j.ccr.2010.06.005.

[65] D. Park, C.K. Ahn, Y.M. Kim, Y.-S. Yun, J.M. Park, Enhanced abiotic reduction of Cr(VI) in a soil slurry system by natural biomaterial addition, J. Hazard. Mater. 160 (2008) 422–427, https://doi.org/10.1016/j.jhazmat.2008.03.044.

[66] M.R. Gopal Reddi, T. Gomathi, M. Saranya, P.N. Sudha, Adsorption and kinetic studies on the removal of chromium and copper onto chitosan-g-maliec anhydride-g-ethylene dimethacrylate, Int. J. Biol. Macromol. 104 (2017) 1578–1585, https://doi.org/10.1016/j.ijbiomac.2017.01.142.

[67] H. Panda, N. Tiadi, M. Mohanty, C.R. Mohanty, Studies on adsorption behavior of an industrial waste for removal of chromium from aqueous solution, S. Afr. J. Chem. Eng. 23 (2017) 132–138, https://doi.org/10.1016/j.sajce.2017.05.002.

[68] J. Preethi, S.M. Prabhu, S. Meenakshi, Effective adsorption of hexavalent chromium using biopolymer assisted oxyhydroxide materials from aqueous solution, React. Funct. Polym. 117 (2017) 16–24, https://doi.org/10.1016/j.reactfunctpolym.2017.05.006.

[69] L. Hlungwane, E.L. Viljoen, V.E. Pakade, Macadamia nutshells-derived activated carbon and attapulgite clay combination for synergistic removal of Cr(VI) and Cr(III), Adsorpt. Sci. Technol. 36 (2018) 713–731, https://doi.org/10.1177/0263617417719552.

[70] A. Mohamed, W.S. Nasser, T.A. Osman, M.S. Toprak, M. Muhammed, A. Uheida, Removal of chromium (VI) from aqueous solutions using surface modified composite nanofibers, J. Colloid Interface Sci. 505 (2017) 682–691, https://doi.org/10.1016/j.jcis.2017.06.066.

[71] S.S. Kahu, A. Shekhawat, D. Saravanan, R.M. Jugade, Two fold modified chitosan for enhanced adsorption of hexavalent chromium from simulated wastewater and industrial effluents, Carbohydr. Polym. 146 (2016) 264–273, https://doi.org/10.1016/j.carbpol.2016.03.041.

[72] A. Nakajima, Y. Baba, Mechanism of hexavalent chromium adsorption by persimmon tannin gel, Water Res. 38 (2004) 2859–2864, https://doi.org/10.1016/j.watres.2004.04.005.

[73] M. Gheju, Hexavalent chromium reduction with zero-valent Iron (ZVI) in aquatic systems, Water Air Soil Pollut. 222 (2011) 103–148, https://doi.org/10.1007/s11270-011-0812-y.

[74] X. Lv, Y. Zhang, W. Fu, J. Cao, J. Zhang, H. Ma, et al., Zero-valent iron nanoparticles embedded into reduced graphene oxide-alginate beads for efficient chromium (VI) removal, J. Colloid Interface Sci. 506 (2017) 633–643, https://doi.org/10.1016/j.jcis.2017.07.024.

[75] I.A. Bhatti, N. Ahmad, N. Iqbal, M. Zahid, M. Iqbal, Chromium adsorption using waste tire and conditions optimization by response surface methodology, J. Environ. Chem. Eng. 5 (2017) 2740–2751, https://doi.org/10.1016/j.jece.2017.04.051.

[76] A.B. Albadarin, S. Solomon, T.A. Kurniawan, C. Mangwandi, G. Walker, Single, simultaneous and consecutive biosorption of Cr(VI) and Orange II onto chemically modified masau stones, J. Environ. Manag. 204 (2017) 365–374, https://doi.org/10.1016/j.jenvman.2017.08.042.

[77] B. Zhang, R. Hu, D. Sun, T. Wu, Y. Li, Fabrication of chitosan/magnetite-graphene oxide composites as a novel bioadsorbent for adsorption and detoxification of Cr (VI) from aqueous solution, Sci. Rep. 8 (2018) 1–12.

[78] M. Omidvar Borna, M. Pirsaheb, M. Vosoughi Niri, R. Khosravi Mashizie, B. Kakavandi, M.R. Zare, et al., Batch and column studies for the adsorption of chromium(VI) on low-cost hibiscus Cannabinus kenaf, a green adsorbent, J. Taiwan Inst. Chem. Eng. 68 (2016) 80–89, https://doi.org/10.1016/j.jtice.2016.09.022.

[79] S. Periyasamy, V. Gopalakannan, N. Viswanathan, Fabrication of magnetic particles imprinted cellulose based biocomposites for chromium(VI) removal, Carbohydr. Polym. 174 (2017) 352–359. https://doi.org/10.1016/j.carbpol.2017.06.029.

[80] R. Huang, X. Ma, X. Li, L. Guo, X. Xie, M. Zhang, et al., A novel ion-imprinted polymer based on graphene oxide-mesoporous silica nanosheet for fast and efficient removal of chromium (VI) from aqueous solution, J. Colloid Interface Sci. 514 (2018) 544–553, https://doi.org/10.1016/j.jcis.2017.12.065.

[81] M.R. Adam, N.M. Salleh, M.H.D. Othman, T. Matsuura, M.H. Ali, M.H. Puteh, et al., The adsorptive removal of chromium (VI) in aqueous solution by novel natural zeolite based hollow fibre ceramic membrane, J. Environ. Manag. 224 (2018) 252–262, https://doi.org/10.1016/j.jenvman.2018.07.043.

[82] L. Aljerf, High-efficiency extraction of bromocresol purple dye and heavy metals as chromium from industrial effluent by adsorption onto a modified surface of zeolite: kinetics and equilibrium study, J. Environ. Manag. 225 (2018) 120–132, https://doi.org/10.1016/j.jenvman.2018.07.048.

[83] Y. Wang, C. Pan, W. Chu, A. Vipin, L. Sun, Environmental remediation applications of carbon nanotubes and graphene oxide: adsorption and catalysis, Nanomaterials 9 (2019) 439, https://doi.org/10.3390/nano9030439.

[84] V.K. Gupta, O. Moradi, I. Tyagi, S. Agarwal, H. Sadegh, R. Shahryari-Ghoshekandi, et al., Study on the removal of heavy metal ions from industry waste by carbon nanotubes: effect of the surface modification: a review, Crit. Rev. Environ. Sci. Technol. 46 (2016) 93–118, https://doi.org/10.1080/10643389.2015.1061874.

[85] X. Yang, Y. Wan, Y. Zheng, F. He, Z. Yu, J. Huang, et al., Surface functional groups of carbon-based adsorbents and their roles in the removal of heavy metals from aqueous solutions: a critical review, Chem. Eng. J. 366 (2019) 608–621, https://doi.org/10.1016/j.cej.2019.02.119.

[86] X. Liu, Y. Ji, Y. Zhang, H. Zhang, M. Liu, Oxidized multiwalled carbon nanotubes as a novel solid-phase microextraction fiber for determination of phenols in aqueous samples, J. Chromatogr. A 1165 (2007) 10–17.

[87] R. Hirlekar, M. Yamagar, H. Garse, M. Vij, V. Kadam, Carbon nanotubes and its applications: a review, Asian J. Pharm. Clin. Res. 2 (2009) 17–27.

[88] B.G.P. Singh, C. Baburao, V. Pispati, H. Pathipati, N. Muthy, S.R.V. Prassana, et al., Carbon nanotubes. A novel drug delivery system, Int. J. Res. Pharm. Chem. 2 (2012) 523–532.

[89] C. Lam, J.T. James, R. McCluskey, S. Arepalli, R.L. Hunter, A review of carbon nanotube toxicity and assessment of potential occupational and environmental health risks, Crit. Rev. Toxicol. 36 (2006) 189–217.

[90] E. Bekyarova, Y. Ni, E.B. Malarkey, V. Montana, J.L. McWilliams, R.C. Haddon, et al., Applications of carbon nanotubes in biotechnology and biomedicine, J. Biomed. Nanotechnol. 1 (2005) 3–17.

[91] H. He, L.A. Pham-Huy, P. Dramou, D. Xiao, P. Zuo, C. Pham-Huy, Carbon nanotubes: applications in pharmacy and medicine, Biomed. Res. Int. 2013 (2013).

[92] W. Cheung, F. Pontoriero, O. Taratula, A.M. Chen, H. He, DNA and carbon nanotubes as medicine, Adv. Drug Deliv. Rev. 62 (2010) 633–649, https://doi.org/10.1016/j.addr.2010.03.007.

[93] S.H. Chen, P.P. Jin, G. Schumacher, N. Wanderka, Microstructure and interface characterization of a cast Mg2B2O5 whisker reinforced AZ91D magnesium alloy composite, Compos. Sci. Technol. 70 (2010) 123–129, https://doi.org/10.1016/j.compscitech.2009.09.015.

[94] M.H. Dehghani, M.M. Taher, A.K. Bajpai, B. Heibati, I. Tyagi, M. Asif, et al., Removal of noxious Cr (VI) ions using single-walled carbon nanotubes and multi-walled carbon nanotubes, Chem. Eng. J. 279 (2015) 344–352, https://doi.org/10.1016/j.cej.2015.04.151.

[95] J. Hu, C. Chen, X. Zhu, X. Wang, Removal of chromium from aqueous solution by using oxidized multiwalled carbon nanotubes, J. Hazard. Mater. 162 (2009) 1542–1550, https://doi.org/10.1016/j.jhazmat.2008.06.058.

[96] B. Qiu, H. Gu, X. Yan, J. Guo, Y. Wang, D. Sun, et al., Cellulose derived magnetic mesoporous carbon nanocomposites with enhanced hexavalent chromium removal, J. Mater. Chem. A 2 (2014) 17454–17462.

[97] X. Lv, J. Xu, G. Jiang, J. Tang, X. Xu, Highly active nanoscale zero-valent iron (nZVI)–Fe3O4 nanocomposites for the removal of chromium(VI) from aqueous solutions, J. Colloid Interface Sci. 369 (2012) 460–469. https://doi.org/10.1016/j.jcis.2011.11.049.

[98] C. Luo, Z. Tian, B. Yang, L. Zhang, S. Yan, Manganese dioxide/iron oxide/acid oxidized multi-walled carbon nanotube magnetic nanocomposite for enhanced hexavalent chromium removal, Chem. Eng. J. 234 (2013) 256–265, https://doi.org/10.1016/j.cej.2013.08.084.

[99] A.S. Krishna Kumar, S.-J. Jiang, W.-L. Tseng, Effective adsorption of chromium(vi)/Cr(iii) from aqueous solution using ionic liquid functionalized multiwalled carbon nanotubes as a super sorbent, J. Mater. Chem. A 3 (2015) 7044–7057, https://doi.org/10.1039/C4TA06948J.

[100] Z. Huang, X. Wang, D. Yang, Adsorption of Cr(VI) in wastewater using magnetic multi-wall carbon nanotubes, Water Sci. Eng. 8 (2015) 226–232, https://doi.org/10.1016/j.wse.2015.01.009.

[101] M. Bhaumik, S. Agarwal, V.K. Gupta, A. Maity, Enhanced removal of Cr(VI) from aqueous solutions using polypyrrole wrapped oxidized MWCNTs nanocomposites adsorbent, J. Colloid Interface Sci. 470 (2016) 257–267, https://doi.org/10.1016/j.jcis.2016.02.054.

[102] A.P. Schwarz, C. Pérez Bergmann, S.B. Fagan, Carbon nanotubes functionalized with titanium complexes for hexavalent chromium adsorption: an ab initio approach, Comput. Theor. Chem. 1113 (2017) 110–119, https://doi.org/10.1016/j.comptc.2017.05.017.

[103] M.A. Atieh, Removal of chromium (VI) from polluted water using carbon nanotubes supported with activated carbon, Procedia Environ. Sci. 4 (2011) 281–293.

[104] M. Zuber, K.M. Zia, M. Barikani, Chitin and chitosan based blends, composites and nanocomposites, Adv. Nat. Polym. Springer (2013) 55–119, https://doi.org/10.1007/978-3-642-20940-6_3.

[105] A. Santhana Krishna Kumar, C. Uday Kumar, V. Rajesh, N. Rajesh, Microwave assisted preparation of n-butylacrylate grafted chitosan and its application for Cr(VI) adsorption, Int. J. Biol. Macromol. 66 (2014) 135–143, https://doi.org/10.1016/j.ijbiomac.2014.02.007.

[106] C. Jung, J. Heo, J. Han, N. Her, S.-J. Lee, J. Oh, et al., Hexavalent chromium removal by various adsorbents: powdered activated carbon, chitosan, and single/multi-walled carbon nanotubes, Sep. Purif. Technol. 106 (2013) 63–71, https://doi.org/10.1016/j.seppur.2012.12.028.

[107] D. Chen, W. Li, Y. Wu, Q. Zhu, Z. Lu, G. Du, Preparation and characterization of chitosan/montmorillonite magnetic microspheres and its application for the removal of Cr (VI), Chem. Eng. J. 221 (2013) 8–15, https://doi.org/10.1016/j.cej.2013.01.089.

[108] C. Shen, H. Chen, S. Wu, Y. Wen, L. Li, Z. Jiang, et al., Highly efficient detoxification of Cr(VI) by chitosan–Fe(III) complex: process and mechanism studies, J. Hazard. Mater. 244–245 (2013) 689–697, https://doi.org/10.1016/j.jhazmat.2012.10.061.

[109] T. Liu, Z.-L. Wang, L. Zhao, X. Yang, Enhanced chitosan/Fe0-nanoparticles beads for hexavalent chromium removal from wastewater, Chem. Eng. J. 189–190 (2012) 196–202, https://doi.org/10.1016/j.cej.2012.02.056.

[110] R. Bhatt, B. Sreedhar, P. Padmaja, Chitosan supramolecularly cross linked with trimesic acid – facile synthesis, characterization and evaluation of adsorption potential for chromium(VI), Int. J. Biol. Macromol. 104 (2017) 1254–1266, https://doi.org/10.1016/j.ijbiomac.2017.06.067.

[111] S. Hena, Removal of chromium hexavalent ion from aqueous solutions using biopolymer chitosan coated with poly 3-methyl thiophene polymer, J. Hazard. Mater. 181 (2010) 474–479, https://doi.org/10.1016/j.jhazmat.2010.05.037.

[112] S. Gokila, T. Gomathi, P.N. Sudha, S. Anil, Removal of the heavy metal ion chromiuim(VI) using chitosan and alginate nanocomposites, Int. J. Biol. Macromol. 104 (2017) 1459–1468, https://doi.org/10.1016/j.ijbiomac.2017.05.117.

[113] J.B. Dima, C. Sequeiros, N.E. Zaritzky, Hexavalent chromium removal in contaminated water using reticulated chitosan micro/nanoparticles from seafood processing wastes, Chemosphere 141 (2015) 100–111, https://doi.org/10.1016/j.chemosphere.2015.06.030.

[114] G.N. Kousalya, M. Rajiv Gandhi, S. Meenakshi, Sorption of chromium(VI) using modified forms of chitosan beads, Int. J. Biol. Macromol. 47 (2010) 308–315, https://doi.org/10.1016/j.ijbiomac.2010.03.010.

[115] Y.A. Aydın, N.D. Aksoy, Adsorption of chromium on chitosan: optimization, kinetics and thermodynamics, Chem. Eng. J. 151 (2009) 188–194, https://doi.org/10.1016/j.cej.2009.02.010.

[116] R. Schmuhl, H.M. Krieg, K. Keizer, Adsorption of cu (II) and Cr (VI) ions by chitosan: kinetics and equilibrium studies, Water SA 27 (2001) 1–8.

[117] G. Rojas, J. Silva, J.A. Flores, A. Rodriguez, M. Ly, H. Maldonado, Adsorption of chromium onto cross-linked chitosan, Sep. Purif. Technol. 44 (2005) 31–36, https://doi.org/10.1016/j.seppur.2004.11.013.

[118] L. Li, Y. Li, L. Cao, C. Yang, Enhanced chromium (VI) adsorption using nanosized chitosan fibers tailored by electrospinning, Carbohydr. Polym. 125 (2015) 206–213, https://doi.org/10.1016/j.carbpol.2015.02.037.

[119] F. Xiao, J. Cheng, W. Cao, C. Yang, J. Chen, Z. Luo, Removal of heavy metals from aqueous solution using chitosan-combined magnetic biochars, J. Colloid Interface Sci. 540 (2019) 579–584, https://doi.org/10.1016/j.jcis.2019.01.068.

[120] X. Hu, J. Wang, Y. Liu, X. Li, G. Zeng, Z. Bao, et al., Adsorption of chromium (VI) by ethylenediamine-modified cross-linked magnetic chitosan resin: isotherms, kinetics and thermodynamics, J. Hazard. Mater. 185 (2011) 306–314, https://doi.org/10.1016/j.jhazmat.2010.09.034.

[121] S.S. Fiyadh, M.A. AlSaadi, W.Z. Jaafar, M.K. AlOmar, S.S. Fayaed, N.S. Mohd, et al., Review on heavy metal adsorption processes by carbon nanotubes, J. Clean. Prod. 230 (2019) 783–793, https://doi.org/10.1016/j.jclepro.2019.05.154.

[122] J. Dou, D. Gan, Q. Huang, M. Liu, J. Chen, F. Deng, et al., Functionalization of carbon nanotubes with chitosan based on MALI multicomponent reaction for Cu2+ removal, Int. J. Biol. Macromol. 136 (2019) 476–485, https://doi.org/10.1016/j.ijbiomac.2019.06.112.

[123] Ş. Parlayıcı, E. Pehlivan, Removal of chromium(VI) from aqueous solution using chitosan doped with carbon nanotubes, Mater. Today Proc. 18 (2019) 1978–1985, https://doi.org/10.1016/j.matpr.2019.06.689.

[124] H.Y. Zhu, R. Jiang, L. Xiao, G.M. Zeng, Preparation, characterization, adsorption kinetics and thermodynamics of novel magnetic chitosan enwrapping nanosized γ-Fe2O3 and multi-walled carbon nanotubes with enhanced adsorption properties for methyl orange, Bioresour. Technol. 101 (2010) 5063–5069. https://doi.org/10.1016/j.biortech.2010.01.107.

[125] Y. Huang, X. Lee, F.C. Macazo, M. Grattieri, R. Cai, S.D. Minteer, Fast and efficient removal of chromium (VI) anionic species by a reusable chitosan-modified multi-walled carbon nanotube composite, Chem. Eng. J. 339 (2018) 259–267. https://doi.org/10.1016/j.cej.2018.01.133.

[126] Z. Zhou, S. Lin, T. Yue, T.-C. Lee, Adsorption of food dyes from aqueous solution by glutaraldehyde cross-linked magnetic chitosan nanoparticles, J. Food Eng. 126 (2014) 133–141, https://doi.org/10.1016/j.jfoodeng.2013.11.014.

[127] J.d.O. Marques Neto, C.R. Bellato, D.d.C. Silva, Iron oxide/carbon nanotubes/chitosan magnetic composite film for chromium species removal, Chemosphere 218 (2019) 391–401, https://doi.org/10.1016/j.chemosphere.2018.11.080.

[128] L.L. Sano, A.M. Krueger, P.F. Landrum, Chronic toxicity of glutaraldehyde: differential sensitivity of three freshwater organisms, Aquat. Toxicol. 71 (2005) 283–296, https://doi.org/10.1016/j.aquatox.2004.12.001.

[129] L. Fan, C. Luo, M. Sun, X. Li, H. Qiu, Highly selective adsorption of lead ions by water-dispersible magnetic chitosan/graphene oxide composites, Colloids Surf. B: Biointerfaces 103 (2013) 523–529, https://doi.org/10.1016/j.colsurfb.2012.11.006.

[130] H. Wang, X. Yuan, Y. Wu, X. Chen, L. Leng, H. Wang, et al., Facile synthesis of polypyrrole decorated reduced graphene oxide–Fe3O4 magnetic composites and its application for the Cr(VI) removal, Chem. Eng. J. 262 (2015) 597–606, https://doi.org/10.1016/j.cej.2014.10.020.

[131] T. Dokmaji, T. Ibrahim, M. Khamis, M. Abouleish, I. Alam, Chemically modified nanoparticles usage for removal of chromium from sewer water, Environ. Nanotechnol. Monit. Manag. 14 (2020) 100319, https://doi.org/10.1016/j.enmm.2020.100319.

[132] J. Huang, Y. Cao, B. Qin, G. Zhong, J. Zhang, H. Yu, et al., Highly efficient and acid-corrosion resistant nitrogen doped magnetic carbon nanotubes for the hexavalent chromium removal with subsequent reutilization, Chem. Eng. J. 361 (2019) 547–558, https://doi.org/10.1016/j.cej.2018.12.081.

[133] M. Matandabuzo, P.A. Ajibade, Vinyl pyridinium polymeric ionic liquid functionalized carbon nanotube composites as adsorbent for chromium(VI) in aqueous solution, J. Mol. Liq. 296 (2019) 111778, https://doi.org/10.1016/j.molliq.2019.111778.

[134] W. Lu, J. Li, Y. Sheng, X. Zhang, J. You, L. Chen, One-pot synthesis of magnetic iron oxide nanoparticle-multiwalled carbon nanotube composites for enhanced removal of Cr(VI) from aqueous solution, J. Colloid Interface Sci. 505 (2017) 1134–1146, https://doi.org/10.1016/j.jcis.2017.07.013.

[135] M.A. Barakat, A.M. Al-Ansari, R. Kumar, Synthesis and characterization of Fe – Al binary oxyhydroxides/MWCNTs nanocomposite for the removal of Cr(VI) from aqueous solution, J. Taiwan Inst. Chem. Eng. 63 (2016) 303–311, https://doi.org/10.1016/j.jtice.2016.03.019.

[136] H. Peng, N. Zhang, M. He, B. Chen, B. Hu, Simultaneous speciation analysis of inorganic arsenic, chromium and selenium in environmental waters by 3-(2-aminoethylamino) propyltrimethoxysilane modified multi-wall carbon nanotubes packed microcolumn solid phase extraction and ICP-MS, Talanta 131 (2015) 266–272. https://doi.org/10.1016/j.talanta.2014.07.054.

[137] R. Ghasemi, T. Sayahi, S. Tourani, M. Kavianimehr, Modified magnetite nanoparticles for hexavalent chromium removal from water, J. Dispers. Sci. Technol. 37 (2016) 1303–1314, https://doi.org/10.1080/01932691.2015.1090906.

[138] S. Parlayici, V. Eskizeybek, A. Avcı, E. Pehlivan, Removal of chromium (VI) using activated carbon-supported-functionalized carbon nanotubes, J. Nanostructure Chem. 5 (2015) 255–263, https://doi.org/10.1007/s40097-015-0156-z.

[139] R. Kumar, M.O. Ansari, M.A. Barakat, DBSA doped polyaniline/multi-walled carbon nanotubes composite for high efficiency removal of Cr(VI) from aqueous solution, Chem. Eng. J. 228 (2013) 748–755. https://doi.org/10.1016/j.cej.2013.05.024.

[140] N. Sankararamakrishnan, M. Jaiswal, N. Verma, Composite nanofloral clusters of carbon nanotubes and activated alumina: An efficient sorbent for heavy metal removal, Chem. Eng. J. 235 (2014) 1–9, https://doi.org/10.1016/j.cej.2013.08.070.

[141] Z.-C. Di, J. Ding, X.-J. Peng, Y.-H. Li, Z.-K. Luan, J. Liang, Chromium adsorption by aligned carbon nanotubes supported ceria nanoparticles, Chemosphere 62 (2006) 861–865, https://doi.org/10.1016/j.chemosphere.2004.06.044.

[142] M.M.A. Aslam, W. Den, H.-W. Kuo, Encapsulated chitosan-modified magnetic carbon nanotubes for aqueous-phase CrVI uptake, J. Water Process. Eng. 40 (2021) 101793, https://doi.org/10.1016/j.jwpe.2020.101793.

[143] T.E. Khalil, A.F. Elhusseiny, A. El-dissouky, N.M. Ibrahim, Functionalized chitosan nanocomposites for removal of toxic Cr (VI) from aqueous solution, React. Funct. Polym. 146 (2020) 104407, https://doi.org/10.1016/j.reactfunctpolym.2019.104407.

[144] Q. Liu, B. Yang, L. Zhang, R. Huang, Adsorptive removal of Cr(VI) from aqueous solutions by cross-linked chitosan/bentonite composite, Korean J. Chem. Eng. 32 (2015) 1314–1322, https://doi.org/10.1007/s11814-014-0339-1.

[145] R. Bhatt, B. Sreedhar, P. Padmaja, Adsorption of chromium from aqueous solutions using crosslinked chitosan–diethylenetriaminepentaacetic acid, Int. J. Biol. Macromol. 74 (2015) 458–466, https://doi.org/10.1016/j.ijbiomac.2014.12.041.

[146] M.T. Sikder, Y. Mihara, M.S. Islam, T. Saito, S. Tanaka, M. Kurasaki, Preparation and characterization of chitosan–caboxymethyl-β-cyclodextrin entrapped nanozero-valent iron composite for cu (II) and Cr (IV) removal from wastewater, Chem. Eng. J. 236 (2014) 378–387, https://doi.org/10.1016/j.cej.2013.09.093.

[147] R. Karthik, S. Meenakshi, Facile synthesis of cross linked-chitosan–grafted-polyaniline composite and its Cr(VI) uptake studies, Int. J. Biol. Macromol. 67 (2014) 210–219, https://doi.org/10.1016/j.ijbiomac.2014.03.035.

[148] L. Li, C. Luo, X. Li, H. Duan, X. Wang, Preparation of magnetic ionic liquid/chitosan/graphene oxide composite and application for water treatment, Int. J. Biol. Macromol. 66 (2014) 172–178, https://doi.org/10.1016/j.ijbiomac.2014.02.031.

[149] L. Li, L. Fan, M. Sun, H. Qiu, X. Li, H. Duan, et al., Adsorbent for chromium removal based on graphene oxide functionalized with magnetic cyclodextrin–chitosan, Colloids Surf. B: Biointerfaces 107 (2013) 76–83, https://doi.org/10.1016/j.colsurfb.2013.01.074.

[150] M. Rajiv Gandhi, S. Meenakshi, Preparation of amino terminated polyamidoamine functionalized chitosan beads and its Cr(VI) uptake studies, Carbohydr. Polym. 91 (2013) 631–637, https://doi.org/10.1016/j.carbpol.2012.08.028.
[151] N.N. Thinh, P.T.B. Hanh, L.T.T. Ha, L.N. Anh, T.V. Hoang, V.D. Hoang, et al., Magnetic chitosan nanoparticles for removal of Cr(VI) from aqueous solution, Mater. Sci. Eng. C 33 (2013) 1214–1218, https://doi.org/10.1016/j.msec.2012.12.013.
[152] Z. Yu, X. Zhang, Y. Huang, Magnetic chitosan–Iron(III) hydrogel as a fast and reusable adsorbent for chromium(VI) removal, Ind. Eng. Chem. Res. 52 (2013) 11956–11966, https://doi.org/10.1021/ie400781n.
[153] L. Zhang, W. Xia, B. Teng, X. Liu, W. Zhang, Zirconium cross-linked chitosan composite: preparation, characterization and application in adsorption of Cr(VI), Chem. Eng. J. 229 (2013) 1–8, https://doi.org/10.1016/j.cej.2013.05.102.
[154] H. Jiang, P. Chen, S. Luo, X. Tu, Q. Cao, M. Shu, Synthesis of novel nanocomposite Fe3O4/ZrO2/chitosan and its application for removal of nitrate and phosphate, Appl. Surf. Sci. 284 (2013) 942–949, https://doi.org/10.1016/j.apsusc.2013.04.013.
[155] D. Chauhan, M. Jaiswal, N. Sankararamakrishnan, Removal of cadmium and hexavalent chromium from electroplating waste water using thiocarbamoyl chitosan, Carbohydr. Polym. 88 (2012) 670–675, https://doi.org/10.1016/j.carbpol.2012.01.014.
[156] Z. Wu, S. Li, J. Wan, Y. Wang, Cr(VI) adsorption on an improved synthesised cross-linked chitosan resin, J. Mol. Liq. 170 (2012) 25–29, https://doi.org/10.1016/j.molliq.2012.03.016.
[157] S. Pandey, S.B. Mishra, Organic–inorganic hybrid of chitosan/organoclay bionanocomposites for hexavalent chromium uptake, J. Colloid Interface Sci. 361 (2011) 509–520, https://doi.org/10.1016/j.jcis.2011.05.031.
[158] A.G. Yavuz, E. Dincturk-Atalay, A. Uygun, F. Gode, E. Aslan, A comparison study of adsorption of Cr(VI) from aqueous solutions onto alkyl-substituted polyaniline/chitosan composites, Desalination 279 (2011) 325–331, https://doi.org/10.1016/j.desal.2011.06.034.
[159] Y.G. Abou El-Reash, M. Otto, I.M. Kenawy, A.M. Ouf, Adsorption of Cr(VI) and as(V) ions by modified magnetic chitosan chelating resin, Int. J. Biol. Macromol. 49 (2011) 513–522, https://doi.org/10.1016/j.ijbiomac.2011.06.001.
[160] A.C. Zimmermann, A. Mecabô, T. Fagundes, C.A. Rodrigues, Adsorption of Cr(VI) using Fe-crosslinked chitosan complex (Ch-Fe), J. Hazard. Mater. 179 (2010) 192–196, https://doi.org/10.1016/j.jhazmat.2010.02.078.
[161] M.S. Samuel, J. Bhattacharya, S. Raj, N. Santhanam, H. Singh, N.D. Pradeep Singh, Efficient removal of chromium(VI) from aqueous solution using chitosan grafted graphene oxide (CS-GO) nanocomposite, Int. J. Biol. Macromol. 121 (2019) 285–292, https://doi.org/10.1016/j.ijbiomac.2018.09.170.
[162] N. Subedi, A. Lähde, E. Abu-Danso, J. Iqbal, A. Bhatnagar, A comparative study of magnetic chitosan (chi@Fe3O4) and graphene oxide modified magnetic chitosan (chi@Fe3O4GO) nanocomposites for efficient removal of Cr(VI) from water, Int. J. Biol. Macromol. 137 (2019) 948–959, https://doi.org/10.1016/j.ijbiomac.2019.06.151.
[163] J. Hu, S.W. Wang, D.D. Shao, Y.H. Dong, J.X. Li, X.K. Wang, Adsorption and reduction of chromium(VI) from aqueous solution by multiwalled carbon nanotubes, Open Environ. Pollut. Toxicol. J. 1 (2009) 66–73, https://doi.org/10.2174/1876397900901010066.
[164] M.K. Kim, K. Shanmuga Sundaram, G. Anantha Iyengar, K.-P. Lee, A novel chitosan functional gel included with multiwall carbon nanotube and substituted polyaniline as adsorbent for efficient removal of chromium ion, Chem. Eng. J. 267 (2015) 51–64, https://doi.org/10.1016/j.cej.2014.12.091.
[165] A. Stafiej, K. Pyrzynska, Adsorption of heavy metal ions with carbon nanotubes, Sep. Purif. Technol. 58 (2007) 49–52. https://doi.org/10.1016/j.seppur.2007.07.008.
[166] C. Lu, C. Liu, Removal of nickel(II) from aqueous solution by carbon nanotubes, J. Chem. Technol. Biotechnol. 81 (2006) 1932–1940, https://doi.org/10.1002/jctb.1626.
[167] A.-K.F.A. Ihsanullah, B. Abu-Sharkh, A.M. Abulkibash, M.I. Qureshi, T. Laoui, et al., Effect of acid modification on adsorption of hexavalent chromium (Cr(VI)) from aqueous solution by activated carbon and carbon nanotubes, Desalin. Water Treat. 57 (2016) 7232–7244, https://doi.org/10.1080/19443994.2015.1021847.
[168] W.-W. Tang, G.-M. Zeng, J.-L. Gong, Y. Liu, X.-Y. Wang, Y.-Y. Liu, et al., Simultaneous adsorption of atrazine and cu (II) from wastewater by magnetic multi-walled carbon nanotube, Chem. Eng. J. 211–212 (2012) 470–478. https://doi.org/10.1016/j.cej.2012.09.102.
[169] K. Pillay, E.M. Cukrowska, N.J. Coville, Multi-walled carbon nanotubes as adsorbents for the removal of parts per billion levels of hexavalent chromium from aqueous solution, J. Hazard. Mater. 166 (2009) 1067–1075, https://doi.org/10.1016/j.jhazmat.2008.12.011.
[170] J.P. Ruparelia, S.P. Duttagupta, A.K. Chatterjee, S. Mukherji, Potential of carbon nanomaterials for removal of heavy metals from water, Desalination 232 (2008) 145–156. https://doi.org/10.1016/j.desal.2007.08.023.
[171] Z. Gao, T.J. Bandosz, Z. Zhao, M. Han, J. Qiu, Investigation of factors affecting adsorption of transition metals on oxidized carbon nanotubes, J. Hazard. Mater. 167 (2009) 357–365. https://doi.org/10.1016/j.jhazmat.2009.01.050.
[172] Marques Neto J. de O, Bellato CR, Milagres JL, Pessoa KD, Alvarenga ES de. Preparation and evaluation of chitosan beads immobilized with Iron(III) for the removal of as(III) and as(V) from water. J. Braz. Chem. Soc. 2013;24:121–32. https://doi.org/10.1590/S0103-50532013000100017.
[173] P. Udaybhaskar, L. Iyengar, A.V.S.P. Rao, Hexavalent chromium interaction with chitosan, J. Appl. Polym. Sci. 39 (1990) 739–747, https://doi.org/10.1002/app.1990.070390322.
[174] M. Mohammed, J. Syeda, K. Wasan, E. Wasan, An overview of chitosan nanoparticles and its application in non-parenteral drug delivery, Pharmaceutics 9 (2017) 53, https://doi.org/10.3390/pharmaceutics9040053.
[175] V. Chandra, J. Park, Y. Chun, J.W. Lee, I.-C. Hwang, K.S. Kim, Water-dispersible magnetite-reduced graphene oxide composites for arsenic removal, ACS Nano 4 (2010) 3979–3986, https://doi.org/10.1021/nn1008897.
[176] H. Hossini, A. Rezaee, S.O. Rastegar, S. Hashemi, M. Safari, Equilibrium and kinetic studies of chromium adsorption from wastewater by functionalized multi-wall carbon nanotubes, React. Kinet. Mech. Catal. 112 (2014) 371–382.

[177] L. Sun, Z. Yuan, W. Gong, L. Zhang, Z. Xu, G. Su, et al., The mechanism study of trace Cr(VI) removal from water using Fe0 nanorods modified with chitosan in porous anodic alumina, Appl. Surf. Sci. 328 (2015) 606–613, https://doi.org/10.1016/j.apsusc.2014.12.094.
[178] H. Zhang, R. Xiao, R. Li, A. Ali, A. Chen, Z. Zhang, Enhanced aqueous Cr(VI) removal using chitosan-modified magnetic biochars derived from bamboo residues, Chemosphere 261 (2020) 127694, https://doi.org/10.1016/j.chemosphere.2020.127694.
[179] C. Lei, C. Wang, W. Chen, M. He, B. Huang, Polyaniline@magnetic chitosan nanomaterials for highly efficient simultaneous adsorption and in-situ chemical reduction of hexavalent chromium: removal efficacy and mechanisms, Sci. Total Environ. 733 (2020) 139316, https://doi.org/10.1016/j.scitotenv.2020.139316.
[180] S.M. Anush, B. Vishalakshi, Modified chitosan gel incorporated with magnetic nanoparticle for removal of cu(II) and Cr(VI) from aqueous solution, Int. J. Biol. Macromol. 133 (2019) 1051–1062, https://doi.org/10.1016/j.ijbiomac.2019.04.179.
[181] R. Yi, W. Cai, C. Dang, B. Han, L. Liu, J. Fan, Mild hydrothermal preparation of millimeter-sized carbon beads from chitosan with significantly improved adsorption stability for Cr(VI), Chem. Eng. Res. Des. 156 (2020) 43–53, https://doi.org/10.1016/j.cherd.2020.01.026.
[182] P.M.B. Chagas, L.B. de Carvalho, A.A. Caetano, F.G.E. Nogueira, A.D. Corrêa, I. Guimarães, do R., Nanostructured oxide stabilized by chitosan: hybrid composite as an adsorbent for the removal of chromium (VI), J. Environ. Chem. Eng. 6 (2018) 1008–1019, https://doi.org/10.1016/j.jece.2018.01.026.
[183] C. Zheng, H. Zheng, Y. Sun, B. Xu, Y. Wang, X. Zheng, et al., Simultaneous adsorption and reduction of hexavalent chromium on the poly(4-vinyl pyridine) decorated magnetic chitosan biopolymer in aqueous solution, Bioresour. Technol. 293 (2019) 122038, https://doi.org/10.1016/j.biortech.2019.122038.
[184] M. de la Luz-Asunción, E.E. Pérez-Ramírez, A.L. Martínez-Hernández, V.M. Castano, V. Sánchez-Mendieta, C. Velasco-Santos, Non-linear modeling of kinetic and equilibrium data for the adsorption of hexavalent chromium by carbon nanomaterials: dimension and functionalization, Chin. J. Chem. Eng. 27 (2019) 912–919, https://doi.org/10.1016/j.cjche.2018.08.024.
[185] J. Febrianto, A.N. Kosasih, J. Sunarso, Y.-H. Ju, N. Indraswati, S. Ismadji, Equilibrium and kinetic studies in adsorption of heavy metals using biosorbent: a summary of recent studies, J. Hazard. Mater. 162 (2009) 616–645, https://doi.org/10.1016/j.jhazmat.2008.06.042.
[186] S. Debnath, A. Maity, K. Pillay, Magnetic chitosan–GO nanocomposite: synthesis, characterization and batch adsorber design for Cr(VI) removal, J. Environ. Chem. Eng. 2 (2014) 963–973, https://doi.org/10.1016/j.jece.2014.03.012.
[187] Ş.S. Bayazit, Ö. Kerkez, Hexavalent chromium adsorption on superparamagnetic multi-wall carbon nanotubes and activated carbon composites, Chem. Eng. Res. Des. 92 (2014) 2725–2733, https://doi.org/10.1016/j.cherd.2014.02.007.
[188] R. Yue, Q. Chen, S. Li, X. Zhang, Y. Huang, P. Feng, One-step synthesis of 1,6-hexanediamine modified magnetic chitosan microspheres for fast and efficient removal of toxic hexavalent chromium, Sci. Rep. 8 (2018) 11024, https://doi.org/10.1038/s41598-018-29499-z.
[189] D.H.K. Reddy, S.-M. Lee, Application of magnetic chitosan composites for the removal of toxic metal and dyes from aqueous solutions, Adv. Colloid Interf. Sci. 201–202 (2013) 68–93, https://doi.org/10.1016/j.cis.2013.10.002.
[190] M. Gholipour, H. Hashemipour, Evaluation of multi-walled carbon nanotubes performance in adsorption and desorption of hexavalent chromium, Chem. Ind. Chem. Eng. Q. 18 (2012) 509–523, https://doi.org/10.2298/CICEQ111104025G.
[191] L. Zhang, H. Luo, P. Liu, W. Fang, J. Geng, A novel modified graphene oxide/chitosan composite used as an adsorbent for Cr(VI) in aqueous solutions, Int. J. Biol. Macromol. 87 (2016) 586–596. https://doi.org/10.1016/j.ijbiomac.2016.03.027.
[192] M.S. Samuel, V. Subramaniyan, J. Bhattacharya, C. Parthiban, S. Chand, N.D.P. Singh, A GO-CS@MOF [Zn(BDC)(DMF)] material for the adsorption of chromium(VI) ions from aqueous solution, Compos. Part B Eng. 152 (2018) 116–125, https://doi.org/10.1016/j.compositesb.2018.06.034.
[193] C. Bauer, J. Buchgeister, R. Hischier, W.R. Poganietz, L. Schebek, J. Warsen, Towards a framework for life cycle thinking in the assessment of nanotechnology, J. Clean. Prod. 16 (2008) 910–926, https://doi.org/10.1016/j.jclepro.2007.04.022.
[194] T. Fleischer, A. Grunwald, Making nanotechnology developments sustainable. A role for technology assessment? J. Clean. Prod. 16 (2008) 889–898, https://doi.org/10.1016/j.jclepro.2007.04.018.
[195] W. Klöpffer, M.A. Curran, P. Frankl, R. Heijungs, A. Köhler, S.I. Olsen, Nanotechnology and life cycle assessment: a systems approach to nanotechnology and the environment, Woodrow Wilson. Int. Cent. Sch. (2007).
[196] M.L. Healy, A. Tanwani, J.A. Isaacs, Economic and Environmental Tradeoffs in SWNT Production, NSTI-Nanotech, Nano Sci Technol Institute, Bost (MA, USA), 2006.
[197] M. Steinfeldt, Nanotechnology and Sustainability: Prospective Assessment of a Future Key Technology, IÖW, 2003.
[198] A. Singh, H.H. Lou, R.W. Pike, A. Agboola, X. Li, J.R. Hopper, et al., Environmental impact assessment for potential continuous processes for the production of carbon nanotubes, Am. J. Environ. Sci. 4 (2008) 522.
[199] D. Kushnir, B.A. Sandén, Energy requirements of carbon nanoparticle production, J. Ind. Ecol. 12 (2008) 360–375.
[200] V. Khanna, B.R. Bakshi, L.J. Lee, Life cycle energy analysis and environmental life cycle assessment of carbon nanofibers production, in: Proc. 2007 IEEE Int. Symp. Electron. Environ, IEEE, 2007, pp. 128–133.
[201] A.L. Roes, L.B. Tabak, L. Shen, E. Nieuwlaar, M.K. Patel, Influence of using nanoobjects as filler on functionality-based energy use of nanocomposites, J. Nanopart. Res. 12 (2010) 2011–2028, https://doi.org/10.1007/s11051-009-9819-3.
[202] V. Khanna, B.R. Bakshi, Carbon nanofiber polymer composites: evaluation of life cycle energy use, Environ. Sci. Technol. 43 (2009) 2078–2084, https://doi.org/10.1021/es802101x.
[203] R. Hischier, T. Walser, Life cycle assessment of engineered nanomaterials: state of the art and strategies to overcome existing gaps, Sci. Total Environ. 425 (2012) 271–282. https://doi.org/10.1016/j.scitotenv.2012.03.001.

Chapter 11

Applications and limitations of graphene oxide for remediating contaminants of emerging concern in wastewater

Swagotom Sarkar[a] and Satinder Ahuja[b,*]

[a]*Bangladesh Reference Institute for Chemical Measurements (BRiCM), Dhaka, Bangladesh,* [b]*Ahuja Consulting, Myrtle Beach, SC, United States*

[*]*Corresponding author: e-mail address: sutahuja@atmc.net*

Introduction

Fresh water comprises only 3% of the total water available to us. Of that, only 0.06% is easily accessible for use by nearly 7 billion people [1]. According to the United Nations (UN), this situation is not likely to improve much because water shortages will be faced by an estimated 2.7 billion people by 2025 [2]. The limited water resources are being contaminated mostly by human activities [3]. These activities include extensive use and disposal of common pollutants containing chemicals, fertilizers, and other hazardous wastes [4]. Groundwater resources can also be polluted by the microbial release of arsenic into groundwater [5]. Arsenic ores, deposited in soil by natural processes, can be decomposed by microorganisms present in groundwater. Consequently, arsenic can be released into groundwater. Conventional pollutants include pathogens, heavy metals (e.g., As and Pb), pesticides, and dyes [6].

Recently, a new type of pollutant was found in water resources; these pollutants are called contaminants of emerging concern (CECs) because of their toxic effects [7]. CECs-contaminated drinking water shows the following adverse effects on humans [8,9]: cancer, reproductive problems, and digestive problems. They can have a significant effect on the economy as well. The existing analytical methods and regulations are not effective in finding CECs in wastewater matrices. Since CECs (both from contamination and their degradation products) are often present in the range of a few ng/L to mg/L in waste streams, their detection requires a high level of sensitivity and selectivity [7]. To remove legacy pollutants from water resources, wastewater treatment plants use various methodologies [10,11] such as coagulation [12,13], flocculation [12,13], chlorination [14], ion exchange [15], adsorption by activated carbon [16], and reverse osmosis [17]. Unfortunately, CECs are difficult to remove by conventional treatments [7]. Conventional water treatment plants are unable to remove CECs because of the inability of microorganisms to metabolize CECs as a source of carbon, and CECs may even inhibit the microorganisms. There are some CECs, i.e., ibuprofen, naproxen, sulfamethoxazole, and iopromide which can be eliminated during a primary-stage adsorption process. But other CECs including different types of antibiotics and antiinflammatory compounds can be 30 to 75% removed by biological treatments. This means that 25 to 70% of those antibiotics and antiinflammatory compounds remain as contaminants in the treated water. Furthermore, wastewater treatment plants (WWTPs) cannot remove contaminants such as iclofenac properly [7]. Glassmeyer et al. [18], in their 2017 survey on the influent and effluent samples from 29 drinking water treatment plants (DWTPs), concluded that many CECs are incompletely removed during treatment. As a result, they can pollute drinking water sources. We need to find alternate ways to remove CECs.

Nanomaterials can offer solutions to water pollution because they have unique physiochemical properties. Much research has been done and is still going on, using a variety of nanomaterials which act as nanoadsorbents, nanomembranes, nanophotocatalysts, and nanoreactors [19]. Of existing nanomaterials, graphene oxide (GO)-based water purification technology could offer the ultimate solution for drinking water and wastewater treatment [20]. GO exhibits some exceptional properties like a large, specific surface area, nanometer-sized pores, sunlight absorption, supercharged carrier mobility, outstanding electrical conductivity, high thermal conductivity, ideal mechanical strength, high adsorption capacity, and

Separation Science and Technology, Volume 15. https://doi.org/10.1016/B978-0-323-90763-7.00012-3

exceptional optical properties. Also, GO can work as a photocatalyst by coupling with semiconductor nanomaterials like TiO_2 which can destroy toxic pollutants. Furthermore, it can be functionalized with hydroxyl and epoxy groups, mainly at the basal plane, and carboxylic acid groups at the edges. So, it becomes hydrophilic because of the presence of oxygen-containing functional groups that help to adsorb polar contaminants. The hydrophobic properties of graphene favor the high adsorption of nonpolar organic contaminants.

Water pollution by CEC materials

The progress of synthetic chemistry has allowed us to produce anthropogenic chemicals exponentially. For instance, in 2000, the annual production was 400 million tons—100 times more than it was in 1930 [7]. At the present time, anthropogenic emissions are more than 540 million tons per year [21]. It is of great concern that these emissions are increasing at a growth rate at 5% per year. Consequently, humans and some forms of aquatic life are being contaminated by unregulated chemicals. The following CECs are of great concern: agricultural pesticides, poly- and per-fluorinated compounds, disinfection by-products, UV-filtering chemicals, manufactured nanomaterials, pharmaceuticals, antibiotic-resistant bacteria, personal care and household products, micro- and nano-plastics. In 2014, Sauvé et al. [22] defined CECs as follows: "naturally occurring or manufactured chemicals or materials which have been found in water whose toxicity is likely to significantly alter the metabolism of a living being." These potential CECs should be labeled "emerging" as long as there is a scarcity of information in the scientific literature about them or there is poorly documented evidence of potential problems caused by them. Some examples of CECs are given in Table 1. Nanotechnology, especially GO and modified forms of GO nanomaterials can offer better ways to remove CECs from water [49–77,129].

Several studies [78–80] have reported detection of CECs in water samples from surface water, groundwater, WWTPs, and DWTPs. These CECs originate from different sources such as landfills, industries, domestic and hospitals wastage or effluents. CECs from these sources pass through air, plants, animals, and/or microorganisms and contaminate water sources. Overall effects of CECs on human beings are as follows: prolonged illness, breast cancer, infertility, pregnancy complications, growth impairment in human embryonic kidney cells, developmental disorders, and increase in cholesterol levels and weight. Acute effects are allergy, fatal growths, abortion, fetal malformation, and high glucose level [7–9]. Aquatic lives are also affected as follows: feminization in fish and reduced reproduction in brown trout. Other ill effects are bleaching and death of coral reefs, shellfish poisoning, damage to the digestive tract and internal organs of seabirds, seals, dolphins, whales, reptiles, and zooplankton [7–9].

Limitations of conventional CEC removal methods

In 2020, Ferreiro and colleagues [50] performed biological treatment on the following contaminants: acetaminophen, caffeine, valsartan, sulfamethoxazole, trimethoprim, methylparaben, acesulfame, losartan-free acid, genistein, eprosartan mesylate, ketoprofen, irbesartan, diclofenac, telmisartan, norfloxacin, 2-hydroxybenzothiazole, progesterone, bezafibrate, perfluorosulfonamide, diuron, carbamazepine, testosterone, ciprofloxacin, amitriptyline, phenytoin, propranolol, and isoproturon. They found that when the concentration of CECs was more than 1500 ng/L, the removal efficiency was higher than 80%. On further investigation, they found that removal rates of some of the CECs (irbesartan, carbamazepine, diuron, and phenytoin) were below 20%. This was due to the different biodegradability and concentration of CECs. They tested CEC removal by an ultrafiltration treatment methodology as well and they were able to remove certain CECs like amitriptyline, caffeine, telmisartan, losartan, norfloxacin, sulfamethoxazole up to 30%. There was only one exception: amitriptyline was about 63% removed. The lower removal rates were attributed to fouling and clogged adsorbed material on the filtration membrane.

Godoy et al. (2015) [130] reported that atenolol (ATL) is not effectively removed during wastewater treatment and has a long half-life (up to 166 days). It may be found in various environmental water samples (varying from undetected to 11 μg/L). A number of researchers [49,131,132] have shown that various types of pharmaceutical-based emerging contaminants like carbamazepine (CBZ) and gemfibrozil (GEM) are not removed efficiently by granular activated carbon (GAC) and by ozone treatment. The properties like high water solubility and/or poor degradability of these contaminants were the main reasons for poor removal by common tertiary treatment technologies such as GAC and ozone. Activated carbon (AC) has these issues as well: regeneration of used activated carbon is expensive and it is not easy. The regeneration process is not suitable for industrial-scale treatment. Moreover, contaminant removal efficiency rate by regenerated AC is less than 40% compared to its intact state value. Advanced wastewater treatment processes (advanced oxidation, adsorption using activated carbon, ozonization, nanofiltration membrane, and reverse osmosis technologies) offer high removal rates (above 99%) for targeted pharmaceutical compounds. Most of these methods are not cost-effective, and chemical reaction methods can produce unexpected by-products.

TABLE 1 Sources and effects of some CECs.

CEC	Example	Sources	Effects
Antibiotic-resistant bacteria and genes [23–48]	*Escherichia coli* NDM1 (New Delhi metallo-β-lactamase-1) [25]	Soil, agricultural settings, sewage, household waste, hospital waste, water treatment, and drinking water	Antibiotic-resistant bacteria and genes
Pesticides [49–118]	Chlorophenol [61]	Agricultural settings, surface water	Abnormalities in cellular bilayer phospholipids, cancer
Pesticides [49–118]	Chlorophenoxy herbicide and its transformation products [106,107]	Surface water and groundwater	Possible DNA damage Carcinogenic
Pharmaceutical [49–118]	Antibiotic: Chlortetracycline (CTC) [82]	Aquatic environment	Growth of antibiotic-resistant bacteria (tetracycline-resistant enterococci)
Pharmaceutical [49–118]	Metformin [71]	Surface water	Endocrine disruption
Pharmaceutical [49–118]	17α-Ethinyl estradiol (EE2) [108,109]	Human excretion, wastewater treatment plants, surface water	Endocrine disruption
Pharmaceutical [49–118]	Acetaminophen [110–112]	Human excretion, wastewater treatment plants, surface water	Endocrine disruption, liver damage
PFAS [49–118]	PFOA (perfluorooctanoic acid) [52]	Food packaging, stain- and water-repellent fabrics and coatings, fire-fighting foams, surface water	Reproductive, liver, and adverse immunological effects in animals, and immunotoxicological and neurodevelopmental issues in children
Disinfection by-products [119,120]	Monochloroacetic acid (MCAA), dichloroacetic acid (DCAA), trichloroacetic acid (TCAA) [120]	Wastewater and drinking water treatment plant during disinfection process	Cancer
Dioxane [121,122]	1,4 Dioxane [121,122]	Solvent stabilizer of 1,1,1-trichloroethane, production of pharmaceuticals, paper, dyes, and adhesives; surface water	Degenerative diseases in renal, nasal and hepatic organs; cancer, chromosome damage in bone marrow and liver.
UV filter compounds [123]	Benzophenone [123]	Sunscreen lotion, personal care products	Endocrine disruption
Algal toxins [124–126]	Microcystin-LR (MC-LR) [124–126]	Release of MC-LR in surface or lake water by cyanobacteria: *Microcystis aeruginosa*	Tumor promoter, liver toxins, fish kills
Micro- (less than 5 mm) and nano-plastic (1–100 nm) [127,128]	Polyethylene [127,128]	Release of micro- and nano-plastics by physical degradation, biodegradation, and photodegradation of large plastic wastes into aquatic bodies	Hindering alimentary canal function, interfering with the endocrine system, ultimately inhibiting body growth, resulting in early mortality
Miscellaneous [49–118]	Polybrominated biphenyls (PBBs) (an ingredient of plastic that makes it flame retardant) [116]	Release of PBBs from plastic into surface water and wastewater	Reproductive and developmental toxicity, affecting the liver, thyroid, and immune systems

Wang and colleagues in 2019 [133] noted that acesulfame-K (ACE) is the most persistent sweetener, with removal rates as low as 40% by sewage treatment when applying conventional mechanical and biological treatments. According to the Criegee mechanism [121], ozone can attack the C=C double bond easily, so ACE can be oxidized by ozonation, but the ratio between ACE and ozone is more restrictive. Accordingly, the ozonization process is not as effective. The removal of ACE by granular activated carbon is not effective either, because ACE has a high octanol–water partition coefficient [133].

In 2017, Li et al. [82] found that a great fraction of antibiotics such as CTC is not completely absorbed or metabolized; therefore, from 30 to 90% of an administered dose is excreted into the environmental water. The removal of CTC from aqueous systems, including an activated sludge process, membrane separation, ozonation, chlorination, photocatalysis, and reverse osmosis, is not efficient.

Zhang and colleagues [61] in 2018 noted that a large number of methods have been developed to remove pesticides, e.g., chlorophenols, from water: adsorption, biological degradation, and electrochemical degradation. However, one of the limitations of the adsorption process is that it cannot convert pesticides into less toxic compounds. Biological treatment suffers from the drawbacks of slow reaction rates and the need for strict control of suitable pHs and temperatures.

Applications of CEC removal with graphene oxide nanomaterials

As mentioned previously, graphene is a modern form of carbon nanomaterials derived from graphite. Its two-dimensional structure, coupled with its interesting structural (large specific surface area), electrical, optical, and mechanical characteristics, make it favorable for use in many fields [20]. Graphene can be an ideal adsorbent for water treatment because it has high theoretical specific surface area (SSA) ($2630\,m^2/g$) and offers two basal planes of its single layer available for contaminant adsorption. GO is the functionalization product of graphene sheets. The surface of GO contains different functional groups (carboxyl–COOH, carbonyl C=O, hydroxyl–OH groups), which allow it to be used as a potential sorbent for organic pollutants, i.e., dye molecules, and inorganic pollutants, e.g., arsenic. GO-based composites exhibit not only high adsorption capacity for metal ions and organic solvents, but also efficient release of adsorbents offering excellent recyclability [20]. GO has been used in wastewater treatment for the removal of conventional pollutants like heavy metal ions, toxic organic solvents, dyes, viruses, bacteria, and radionuclides. Recently, GO has also been applied for CEC removal from wastewater. Some of the applications are covered below.

1. **Antibiotic-resistant bacteria and gene removal**: GO can be considered a potential tool for removing antibiotic-resistant genes (ARGs) and antibiotic-resistant bacteria (ARBs) from wastewater. In 2020, Li and colleagues [25] synthesized nitrogen-doped reduced graphene oxide (NRGO)-wrapped $Bi_2O_2CO_3$ microspheres (NGWMs) for enhanced inactivation of multidrug-resistant *E. coli* NDM-1 and degradation of the plasmid-encoded ARGs (bla_{NDM-1}) in secondary effluents. NGWMs showed an adsorption capacity for *E. coli* NDM-1 bacteria in a secondary effluent of $3.08 \pm 0.27 \times 10^8$ colony-forming unit (CFU)/mg catalyst which was higher than adsorption by $Bi_2O_2CO_3$ microspheres and GO-wrapped microspheres. NGWM also demonstrated excellent inactivation of ARG (bla_{NDM-1}). The enhancement of inactivation and destruction of ARB and ARG were facilitated by the presence of N doping, reactive oxygen species generation, broad-range light absorption, and decreased electron transfer time. Various researchers [23–48] believe that the destruction or removal of ARGs and ARBs by GO depends on the following factors: oxidative stress induction, number and composition of functional groups, high-energy adsorption sites, conjugate pi region sites, lateral dimension and concentration of GO, DNA length, and base composition of ARGs. The toxicity mechanism of GO against bacterial cells is still not completely clear. In real-time applications, GO could adsorb sugars, fats, and proteins in contaminated water while interacting with bacteria and genes. This could lead to a decrease in the removal efficiency.
2. **Pesticide removal**: Maliyekkal et al. [53] studied the adsorption of pesticides named chlorpyrifos, endosulfan, and malathion by GO and reduced GO. They proposed that graphene forms a graphene-water-pesticide (G–W–P) complex with pesticides in the presence of water; this G–W–P complex is the result of electrostatic interaction among graphene, water, and pesticides that precipitate. Consequently, pesticides are removed from contaminated water. It suggests that pesticide removal by graphene in the absence of water is unlikely. This means that water plays an important role in removing these pesticides from water. Graphene-coated silica (GCS). as a highly efficient sorbent, was used for removal of residual organophosphorus pesticides (OPPs) from water. The study shows that the mechanism of adsorption of OPPs on GCS is based on the electron-donating abilities of P, S, and N atoms and the strong π-bonding network of benzene rings.

   Graphene is also a subject for chemical modifications [87] such as functionalization with functional groups or decorated with metallic nanoparticles that can improve adsorptive capacity, selectivity, and other properties, e.g., magnetic property. Wu et al. (2011) [90] synthesized a magnetic iron–graphene ($G\text{-}Fe_3O_4$) nanocomposite and used that magnetic nanocomposite to remove carbamate pesticides (metolcarb, carbofuran, pirimicarb, isoprocarb, and diethofencarb) from water samples. Shrivas and colleagues [89] developed in 2017 a magnetic GO-Fe_3O_4 nanocomposite for the removal of endrin and dieldrin pesticides. They found higher removal efficiencies when the GO-Fe_3O_4 nanocomposite dose was 30 mg. Other optimized parameters were a contact time of 30 min, and pH 4.0 at room temperature. They obtained 86 and 74% removal efficiency for endrin and dieldrin, respectively. It appears that exo-positioned oxygen atoms of an endrin molecule help produce strong interactions with a GO-Fe3O_4 nanocomposite. In contrast, dieldrin's oxygen atom is at the endo-position that does not favor a strong interaction. That is why endrin is removed at a higher rate than dieldrin,

Graphene and GO can be composited with semiconductors for developing a photocatalyst with synergistic photocatalytic effects. In 2018, Zhang and colleagues [61] conducted a study on the synthesis of $Cu_2(OH)PO_4$/rGO nanocomposites for developing a photocatalyst with enhanced photocatalytic activity for degradation of the pesticide 2,4-dichlorophenol 2,4-DCP. They demonstrated that the pesticide's degradation rate is 87.1% after irradiation with infrared (IR) light for 6 h using a nanocomposite of $Cu_2(OH)PO_4$ and reduced GO $Cu_2(OH)PO_4$/rGO. The optimum amount of GO was 0.5 wt%. The introduction of an optimum amount of GO and light absorption in the IR region increased electron–hole pair generation; holes are responsible for the degradation of 2,4-DCP.

They introduced H_2O_2 as an electron acceptor which improved electron–hole pair separation and further generated powerful reactive species HO• radicals via the photo-Fenton reaction, which can further oxidize 2,4-DCP during photocatalytic degradation. This means hydrogen peroxide maximized the photocatalytic degradation of 2,4-DCP. Typically, the rate constant of $Cu_2(OH)PO_4$/rGO and H_2O_2 was more than 6.25 times higher than that of only $Cu_2(OH)PO_4$/rGO, and almost 10 times greater than the value for pure $Cu_2(OH)PO_4$. $Cu_2(OH)PO_4$/rGO did not degrade even after four cycles, showing that this composite was stable. It appears that no obvious cytotoxicity was induced in human umbilical vein endothelial cells, even at concentrations up to 90 mg mL^{-1}, which shows that $Cu_2(OH)PO_4$/rGO nanocomposites have good biosecurity. It can also be concluded that this unique material can be an excellent candidate for removal of different types of pesticides. One of the issues regarding GO adsorption is that adsorption does not make pesticides less toxic. Further treatment is required before dumping used GO adsorbent.

3. **Antibiotic removal**: Li et al. [82] prepared a graphene oxide/titanium dioxide (GO/TiO_2) nanocomposite for eliminating CTCs from contaminated water. The maximum adsorption capacity was 261.10 mg/g at 298 K. Ligand exchange, pi–pi electron donor–acceptor interaction, hydrogen bonding, and cation–pi bonding between GO/TiO_2 and CTC were responsible for this adsorption. Anirudhan and colleagues [83] degraded a ciprofloxacin hydrochloride aqueous solution by visible-light photocatalysis using a nano zinc oxide–incorporated graphene oxide/nanocellulose composites (ZnO-GO/NC). They synthesized GO, ZnO-GO, ZnO-GO/NC with band gaps of 1.77, 2.40, and 2.79 eV, respectively. This means they turned the band gap from the UV region to the visible region by incorporating ZnO and nanocellulose. A nanocomposite made of ZnO-GO/NC absorbs visible light and degrades ciprofloxacin up to 98% at pH 6.0. They also showed that the composite was reusable up to 5 consecutive times, retaining its significant efficiency. A number of researchers [37–40, 96] used graphene (surface area 49.4 m^2/g), graphene (surface area 49.4 m^2/g), noncovalent functionalized GO (surface area 237.4 m^2/g), Fe/Cu@graphene (surface area 108.6 m^2/g), and Fe_3O_4@graphene (surface area: not available), respectively, to remove amoxicillin (conc. 0.2 p.p.m.), cephalexin (0.2 p.p.m.), ciprofloxacin (10 p.p.m.), tetracycline (100 p.p.m.), and oxy-tetracycline (1 p.p.m.), respectively. They achieved removal efficiency of 100, 100, 96.2, 100, and 96.7%, respectively. It is believed that oxygenated functional groups of GO contribute to the removal of these contaminants from water. Although it looks promising for removing antibiotics by using graphene oxide nanomaterials, there are still research gaps that need to be filled. The process, which involves CTC removal by GO, could not recycle GO efficiently after the adsorption of CTC by GO/TiO_2 because the interaction between CTC and GO/TiO_2 were strong.
4. **Removal of Pharmaceuticals**: Table 2 shows removal of some pharmaceutical contaminants. Khalil et al. [49] synthesized porous graphene (PG) and used it for the adsorption of some emerging contaminants (ECs) like atenolol (ATL), carbamazepine (CBZ), ciprofloxacin (CIP), diclofenac (DCF), gemfibrozil (GEM) and ibuprofen (IBP) from water and wastewater. They found that adsorption capacity was more than 100 mg-EC/g-PG for some of these ECs. Interestingly, when the EC concentration was trace in nature, removal efficiency was higher than 99.9% with a PG dose of 100 mg/L. Adsorption occurred due to the presence of electrostatics, pi–pi interactions, hydrogen bonding and van der Walls forces. PG was used for up to four cycles with good adsorption properties. However, the removal efficiency was not very good when there were interfering contaminants. So they suggested the use of a higher dose of PG. Al-Khateeb et al. [66] used graphene nanoplates (GNPs) of surface area 635.2 m^2/g to remove acetaminophen (APAP) and aspirin (ASA) of concentration 20 p.p.m. They achieved 97.4 and 81% removal efficiency with an adsorption capacity of 12.7 and 17 mg/g, respectively. The adsorption is mainly due to the pi–pi electron donor–acceptor interaction between the ASA and APAP

TABLE 2 Removal of pharmaceutical contaminants by GO and its modified forms.

Pharmaceutical	Adsorbent	Removal efficiency	References
Atenolol, carbamazepine, ciprofloxacin, diclofenac, gemfibrozil, and ibuprofen	Porous grapheme (PG)	greater than 99.9% at a low dose of PG (100 mg/L)	[49]
Acetaminophen, aspirin	Graphene nanoplates	97.4%; 81%	[66]
Diclofenac	Graphene	97%	[67–69]
Ibuprofen	Graphene	95.5%	[67]
Salicylic acid	Functionalized graphene	55%	[70]
Metformin	Graphene oxide	80%	[71]
Carbamazepine	GO/βBi_2O_3/TiO_2/$Bi_2Ti_2O_7$	greater than 99%	[72]

molecules and GNPs. Similarly, graphene (surface area 890 m^2/g) used as adsorbent for removal of diclofenac at 10 p.p.m. level produced a removal efficiency of 97% [67,68]. A number of researchers [67,69–72] used reduced graphene oxide, graphene, functionalized graphene, graphene oxide, GO/beta-Bi_2O_3/TiO_2/$Bi_2Ti_2O_7$ to remove sodium diclofenac, ibuprofen, salicylic acid, metformin, carbamazepine at 20–200 ppm, 10 ppm, 50 ppm, 10 ppm, and 20 μmol, respectively. They achieved removal efficiency of >80% for all of them except salicylic acid which showed an efficiency of 55%. It should be noted that GO is a negatively charged molecule with oxygenated functional groups; it repels negatively charged organic pharmaceutical molecules [73,74].

5. **PFAS removal**: Per- and polyfluoroalkyl substances (PFASs) have been described as "forever chemicals" because they are resistant to degradation due to their strong carbon–fluorine bonds [94,97–100]. Sorption of perfluorooctanoic acid (PFOA) was evaluated by graphene oxide GO, an iron oxide–modified reduced-GO composite (FeG), and an activated-carbon (C)/clay/alumina–based adsorbent, RemBindTM (RemB) [52]. The sorption by FeG and RemB (90%) is much greater than GO (60%) alone. Nonelectrostatic forces are responsible for PFOA adsorption. Synthetic photocatalyst like ZnO/reduced-GO nanocomposites improved PFOA removal substantially (56.6% for ZnO vs. 90.9% for ZnO/rGO nanocomposites) [96]. The heterojunctions between ZnO and reduced-GO in the composites are responsible for the enhancement of the photo-degradation by ZnO/rGO. An electrode made of carbon nanotubes (CNTs) and 20% graphene was fabricated [75]. It showed PFOA and perfluorooctanesulfonic acid (PFOS) adsorption by CNT-20% graphene composite electrode 0.86×10^{-3} and 1.19×10^{-3} $g\,\mu g^{-1}\,h^{-1}$, respectively; these values were higher than those obtained with powdered CNT-20% graphene (0.89×10^{-4} and 0.94×10^{-4} $g\,\mu g^{-1}\,h^{-1}$, respectively). In addition, the PFOA and PFOS adsorption capacity of the CNT-20% graphene electrode was 491.9 and 555.8 mg/g, respectively, which was 205 and 88.2 fold higher than that of powdered CNT-20% graphene. Electromigration of PFAS anions and assembling properties of those anions on an electrode surface are most likely responsible for the PFAS adsorption. Graphene oxide and Mg^{2+} composite were used to adsorb PFOS in solution [76]. A removal efficiency of 98% was found; the adsorption equilibrium reached within 2 min. These results were mainly attributed to the bridging of Mg^{2+} with GO and the interaction of Mg^{2+} with PFOS that maximized the adsorption. Degradation of perfluorooctanoic acid (PFOA) was investigated by the persulfate process over a ZnO-reduced graphene oxide (ZnO-rGO) catalyst [77]. The results showed that the O3/UV/ZnO-rGO/$S_2O_8{}^{2-}$ system exhibited over 99% efficiency. However, PFAS sorption by only graphene oxide is not efficient because GO and PFOA and PFOS all exist as negatively charged ions. Repulsive forces exist between the adsorbent and the adsorbate.
6. **Disinfection by-products (DBP) removal**: Magnetic graphene oxide (MGO) adsorbent was synthesized to eliminate natural organic matter (NOM) from contaminated water sample [119]. NOM works as a precursor of disinfection by-products such as trihalomethanes (THMs), haloacetic acids (HAAs), chloral hydrate, and haloacetonitriles (HANs). The contaminated water was treated with MGO adsorbent. This significantly decreased DBP precursor NOM formation. The process is pH sensitive and does not work well at a pH higher than 5.4. At a pH lower than 5.4, MGO and functional groups (carboxyl and phenolic) of NOM become positively and negatively charged, so adsorption

increases. In contrast, at a pH higher than 5.4, both the adsorbent and NOM are negatively charged. As a result, repulsion increases and adsorption decreases. It was pointed out that hydrogen bonds, Lewis acid–base, and π-π interactions were responsible for NOM adsorption by MGO. This process removed hydrophobic (humic and fulvic acid) compounds more efficiently than hydrophilic and nitrogenous compounds. Another advantage of MGO adsorbent is that it can be regenerated by 20% ethanol (v/v) and reused up to 5 consecutive times with good adsorption efficiency. Although this looks promising, it can't stop formation of HANs disinfections. GO was synthesized to remove three different haloacetic acid (monochloroacetic acid (MCAA), dichloroacetic acid (DCAA) and trichloroacetic acid (TCAA)) from contaminated drinking water [120]. The adsorption efficiency reached 88% at lower pH values and within 2.25 min after five applications of GO adsorbent to contaminated drinking water. This efficiency was enough to decrease the total haloacetic acid concentration from 436 to 52.5 μg/L. This meets the US drinking water standard (HAAs $\leq$60 μg/L). The rapid hydrogen bond formation between haloacetic acid and GO at a pH value lower than 3.9 contributes to this fast adsorption process. At a pH value lower than 3.9, GO and HAAs become positive and negative in charge, respectively, and adsorption increases.

7. **Remediation of 1,4-dioxane**: Diego Cortés-Arriagada and colleagues [121] used the density functional theory to find the effectiveness of pristine graphene, oxidized graphene oxide and silicon-doped graphene (SiG) to filter out 1,4-dioxane from contaminated water. They obtained the adsorption energy of about 0.7 and 0.9 eV for removing 1,4-dioxane by oxidized graphene oxide and SiG, respectively. Higher 1,4-dioxane adsorption efficiency by SiG than the oxidized GO may be due to low covalent interaction between hydroxyl and carboxyl functional groups on GO and dioxane. In case of 1,4-dioxane adsorption by SiG, there was covalent interaction between SiG and dioxane. As the covalent interaction is stronger than noncovalent interaction, SiG provides higher adsorption efficiency than oxidized GO. Interestingly, the presence of water increased the adsorption energy of about 0.1 eV for the SiG adsorbent. This means that SiG adsorbent is stable in water which is sufficient for removal of 1,4-dioxane from contaminated water. Finally, the adsorption efficiency can be rated as follows: SiG > GO > pristine graphene. There is a need of more experimental studies with other contaminants [122].
8. **UV-filtering compounds (benzophenones)**: Medina et al. in 2020 [123] synthesized magnetite nanocomposite (Fe_3O_4@SiO_2–3-aminopropyltrietoxysilane@GO) and employed it for the extraction of five benzophenones (benzophenone-1, benzophenone-3, 4-hydroxybenzophenone, ben-zophenone-6 and benzophenone-8) in aqueous samples by dispersive micro-solid phase extraction, combining the magnetic properties of magnetite nanoparticles with an excellent sorption capacity of graphene oxide via hydrophobic interactions with the analytes. Using selected conditions (pH in 3% NaCl w/v: 4; sample volume: 100 mL, type of agitation: ultrasound + mechanical; eluent: methanol) large sample volumes can be extracted by using a small amount of sorbent (20 mg), The whole process is simple and fast and can be used for benzophenone removal from wastewater as well.
9. **Algal toxins removal**: Pavagadhi and colleagues [126] synthesized graphene oxide and used it to remove algal toxins such as microcystin-LR (MC-LR) and microcystin-RR (MC-RR), from water. Later, they compared adsorption capacity of synthesized GO and commercially available activated carbon. Experimental adsorption capacity of GO was 1700 and 1878 μg/g for removing MC-LR and MC-RR, respectively. In contrast, MC-LR and MC-RR adsorption capacities were 1481.7 and 1034.1 μg/g, respectively, for commercially available activated carbon. The maximum adsorption by GO was observed at pH 5. It appears that at pH 5, the ion pairing effect between MC-LR and MC-RR and GO decreased molecular dimensions of MC-LR and MC-RR: hydrogen bond formation and led to higher adsorption by GO. The removal efficiency by GO was more than 90% within 5 min for concentration ranges 500, 700, and 900 μg/L. GO was efficient and can be reused up to 10 times, the removal efficiency should be tested for other microcystins in the presence of factors such as natural organic matter, etc. A cyclodextrin-functionalized magnetic composite of graphene and porous silica (G-Fe_2O_3-γ-CD) was studied for the selective removal of MC-LR from water in the presence of different types of NOM and metal ions [124]. The removal capacity of G-Fe_2O_3-γ-CD for MC-LR removal was 160 mg/g. The high removal capacity of G-Fe_2O_3-γ-CD was attributed to the high loading capacity of graphene for cyclodextrin on its flat surface, which contributed to MC-LR removal. G-Fe_2O_3-γ-CD provided several advantages for MC-LR removal: (1) G-Fe_2O_3-γ-CD was dispersible in water, leading to a better interaction between MC-LR and cyclodextrin; (2) the presence of Fe_2O_3 provided a magnetic separation of adsorbed MC-LR from the adsorbent, and (3) the removal capacity of G-Fe_2O_3-γ-CD for MC-LR removal was minimally affected in the presence of various metal ions and NOM. However, removal capacity decreased around 20% after the use of the third cycle. They proposed it could be because of the decrease of cyclodextrin above the graphene surface, aggregation of the adsorbent material, or inefficient removal of MC-LR during the regeneration procedure (magnetic + ethanol washing). Recently, Kumar and colleagues [125] synthesized rGO-coated sand and used it to adsorb MC-LR from contaminated

water and achieved 91% removal efficiency. They concluded that functional groups like C–O–C, pi–pi interactions, and electrostatic interactions were most likely responsible for the enhancement of MC-LR adsorption.

10. **Microplastics (MP) removal**: In 2020, Sun and colleagues [127] synthesized a sponge by mixing chitin and graphene oxide under pressure. They applied this sponge to adsorb different types of microplastics from water. They found adsorption efficiencies of about 89.8, 72.4, and 88.9% for neat polystyrene, carboxylate-modified polystyrene, and amine-modified polystyrene at pHs from 6 to 8. These adsorption efficiencies were found after the sponge had been used three consecutive times; this means that the sponge is reusable. The adsorption was the result of intraparticle diffusion; electrostatic, hydrogen bond, and pi–pi interactions between chitin/GO and microplastics.Yuan et al. [128] used porous three-dimensional reduced grapheme oxide 3D RGO) as an adsorbent to remove the microscale polystyrene (PS) in water. They found a maximum adsorption capacity of 617.28 mg/g at pH 6.0; microplastics initial concentration; $C_0 = 600$ mg/L; equilibrium time, $t = 120$ min; and temperature, $T = 26°C$. This shows that the adsorption of PS microplastics is due to π–π interaction between a carbon ring and a benzene ring of 3D RGO and PS microplastics, respectively. This approach worked well with a real water sample as well.

Table 3 provides examples of CEC removal technologies based on GO.

Advantages of graphene/graphene oxide for water purification

GO-based composites exhibit not only high adsorption capacity for metal ions and organic solvents, but also offer excellent recyclability because of the efficient release of adsorbents [134]. GO has large functionalized active sites. The oxygen-containing functional groups like hydroxyl, carboxyl, carbonyl, and epoxy on GO surfaces, help to absorb the heavy metal ions and improve the capabilities to remove the pollutants from wastewater [134,135]. GO offers high antibacterial properties that could be used for disinfection, high desalination ability, high energy efficiency, and high permeability and selectivity. It does not produce harmful by-products. Furthermore, it is chemically and physically stable, thin, light and strong, making transport and handling easy. Photocatalytic and adsorption properties of graphene are high because of its high carrier mobility, large surface area, good optical transparency, high transmittance rates, and strong absorption and mechanical flexibility [136]. High elasticity and unusual magnetic properties of graphene help achieve its usefulness in water treatment [137]. Synthesis of GO from the oxidation of graphite is basically easier and offers the potential of large-scale production and biocompatibility [138].

Limitations/challenges

Cost-effective and large-scale manufacturing of GO nanomaterials is not yet well-established [136]. The production cost of GO nanomaterials is high (the current price of 1 g of graphene oxide powder is equal to 147 Euros [138]). In general, GO is produced by the oxidation reaction of graphene. However, the oxidation of graphite introduces impurities in a GO skeleton [139]. Another challenge is the use of the GO-based nanomembrane for treatment of unclean water because their pores may block after use, and the efficiency starts to decrease [19]. In the case of GO nano-adsorbents, the reusability is the main drawback. There is still a need to synthesize nanomaterials that can efficiently remove the pollutants from wastewater. Furthermore, the recovery process of these is not easy. There is little research on nanomaterial release into the environment, travel into water, and existence in water [140]. These types of materials have some adverse toxicologic effects [141,142]. The reported studies show that nanomaterials may cause some health issues. For example, graphene, GO, and reduced GO elicit toxic effects both in vitro and in vivo.

Conclusions

Sophisticated separation science and technology methods must be employed to address the pressing worldwide concerns for clean drinking water and safe irrigation water for food crops. GO has great potential in water purification technology. It can remove conventional wastewater contaminants and thereby prevent contamination of drinking water resources. The unique physicochemical properties of GO help with the remediation of a variety of contaminants. More research is necessary to address the current limitations of GO to fabricate cost-effective, multifunctional, and integrated GO-based water purification technology for efficient removal of CECs from contaminated water. Additional research on scale-up and design optimization of this technology for industrial applications is also necessary.

TABLE 3 Characterization and removal technologies based on graphene oxide.

CECs	Examples	Characterization of CECs	Remediation by graphene oxide and its modified form
Antibiotic-resistant bacteria and genes [23–48]	*Escherichia coli*-NDM1 (New Delhi metallo-β-lactamase-1) [25]	Plate assay method	NRGO (nitrogen-doped reduced graphene oxide) wrapped $Bi_2O_2CO_3$ microspheres (NGWM) [25]
Pesticide [49–118]	Chlorophenol [61]	UV–Visible spectroscopy	$Cu_2(OH)PO_4$/rGO nanocomposites [61]
Pesticide [49–118]	Chlorophenoxy herbicide and its transformation products [106,107]	HPLC	Graphene nano sheet [107]
Pharmaceutical [49–118]	Antibiotic: chlortetracycline (CTC) [82]	UV–Visible spectroscopy	Graphene oxide/TiO_2 [82]
Pharmaceutical [49–118]	Metformin [71]	UV–Visible spectroscopy	Graphene oxide [71]
Pharmaceutical [49–118]	17α-Ethinyl estradiol (EE2) [108,109]	HPLC	Reduced graphene oxide [109]
Pharmaceutical[49–118]	Acetaminophen [110–112]	HPLC–MS	Coal fly-ash decorated with a graphene oxide – tungsten oxide nanorods with adsorbed lead nanocomposite (CFA/GO/WO3NRs + Pb^{2+}) [112]
PFAS [49–118]	PFOA (perfluorooctanoic acid) [52]	LC–MS/MS	Fe-oxide-modified reduced-GO (FeG) [52]
Disinfection by-products [119,120]	Monochloroacetic acid (MCAA), dichloroacetic acid (DCAA), trichloroacetic acid (TCAA) [120]	GC-ECD	Graphene oxide [120]
Dioxane [121,122]	1,4 Dioxane [121,122]	GC–MS/MS	Silicon-doped graphene oxide [121]
UV filter compounds [123]	Benzophenone [123]	LC–MS/MS and SPME-GC–MS/MS	Fe_3O_4@SiO_2-APTES@GO composite [123]
Algal toxins [124–126]	Microcystin-LR (MC-LR) [124–126]	LC–MS/MS	Graphene oxide [126]
Micro- (less than 5 mm) and nanoplastic (1–100 nm) [127,128]	Polyethylene [127,128]	Pyrolysis GC–MS, micro-Raman spectroscopy, FT-infrared (IR) microscopy, fluorescence microscopy	Chitin:GO sponge, 3D reduced GO [127,128]
Miscellaneous [49–118]	Polybrominated biphenyls (PBBs) (an ingredient of plastic that makes it flame retardant) [116]	HPLC, ICICI, GCMS, LCMS	Rapid oxidation by reduced GO/TiO2 composite [116]

Acknowledgment

Swagotom Sarkar acknowledges the valuable suggestions provided by Dr. Satinder Ahuja.

References

[1] Nationalgeographic. 2020. https://www.nationalgeographic.com/environment/freshwater/freshwater-crisis, (Accessed date: June 26, 2020).
[2] S. Ahuja, Overview of advances in water purification techniques, Adv. Water Purif. Tech. (2019) 1–15.
[3] S. Ahuja, Environmental perspective on sustainability and water reclamation, ACS Symp. Ser. (2015) 87–103.
[4] J. Singh, P. Yadav, A. Pal, V. Mishra, Water pollutants: origin and status. *Sensors in water pollutants monitoring*, Role Mater. (2019) 5–20.
[5] S. Ahuja, Delineation of a major worldwide problem of arsenic-contaminated groundwater, Handb. Water Purity Qual. (2009) 17–38.
[6] A. Ahamad, S. Madhav, A. Singh, A. Kumar, P. Singh, Types of water pollutants: conventional and emerging. Sensors in water pollutants monitoring, Role Mater. (2019) 21–41.
[7] T. Nawaz, S. Sengupta, Contaminants of emerging concern: occurrence, fate, and remediation, Adv. Water Purif. Tech. (2019) 67–114.
[8] O. Gustafsson, L. Manukyan, S. Gustafsson, G. Tummala, S. Zaman, A. Begum, M. Alfasane, K. Siddique-e-Rabbani, A. Mihranyan, Scalable and sustainable Total pathogen removal filter paper for point-of-use drinking water purification in Bangladesh, ACS Sustain. Chem. Eng. 7 (17) (2019) 14373–14383.
[9] M. Lei, L. Zhang, J. Lei, L. Zong, J. Li, Z. Wu, Z. Wang, Overview of emerging contaminants and associated human health effects, Biomed. Res. Int. 2015 (2015) 1–12.
[10] S. Ahuja, Overview: evaluating water quality to prevent future disasters, Sep. Sci. Technol. (2019) 1–12.
[11] S. Ahuja, Green chemistry and other novel solutions to water pollution: overview, ACS Symp. Ser. (2013) 1–14.
[12] J. Bratby, Coagulation and flocculation in water and wastewater treatment - second edition, Water Intelligence Online 5 (2015) 9781780402321.
[13] The International Water Association. 2020. https://www.iwapublishing.com/news/coagulation-and-flocculation-water-and-wastewater-treatment (Access date: August 29, 2020).
[14] Central for Disease Control and Prevention. 2020. https://www.cdc.gov/healthywater/drinking/public/chlorine-disinfection.html (Accessed August 29, 2020).
[15] A. Ting, Green approach: microbes for removal of dyes and metals via ion binding, Appl. Ion Exchange Mater. Environ. (2019) 1–23.
[16] P. Jin, X. Jin, X. Wang, Y. Feng, C., X., Biological activated carbon treatment process for advanced water and wastewater treatment, Biomass Now - Cultivation and Utilization (2013).
[17] D. Trishitman, A. Cassano, A. Basile, N. Rastogi, Reverse osmosis for industrial wastewater treatment, Curr. Trends Fut. Develop. (Bio-)Membr. (2020) 207–228.
[18] S. Glassmeyer, E. Furlong, D. Kolpin, A. Batt, R. Benson, J. Boone, O. Conerly, M. Donohue, D. King, M. Kostich, H. Mash, S. Pfaller, K. Schenck, J. Simmons, E. Varughese, S. Vesper, E. Villegas, V. Wilson, Nationwide reconnaissance of contaminants of emerging concern in source and treated drinking waters of the United States, Sci. Total Environ. 581-582 (2017) 909–922.
[19] A. Yaqoob, T. Parveen, K. Umar, M. Mohamad Ibrahim, Role of nanomaterials in the treatment of wastewater: a review, Water 12 (2) (2020) 495.
[20] M.A. Naushad, New Generation Material Graphene: Applications in Water Technology, first ed., Springer, Cham, 2019. XVII-471.
[21] Food and Agriculture Organisation. 2020. http://www.fao.org/3/y3557e/y3557e11.html (Accessed date: August 29, 2020).
[22] S. Sauvé, M. Desrosiers, A review of what is an emerging contaminant, Chem. Cent. J. 8 (1) (2014).
[23] W. Han, Z. Wu, Y. Li, Y. Wang, Graphene family nanomaterials (GFNs)—promising materials for antimicrobial coating and film: a review, Chem. Eng. J. 358 (2019) 1022–1037.
[24] Z. Bytesnikova, L. Richtera, K. Smerkova, V. Adam, Graphene oxide as a tool for antibiotic-resistant gene removal: a review, Environ. Sci. Pollut. Res. 26 (20) (2019) 20148–20163.
[25] D. Li, P. Yu, X. Zhou, J. Kim, Y. Zhang, P. Alvarez, Hierarchical $Bi_2O_2CO_3$ wrapped with modified graphene oxide for adsorption-enhanced photocatalytic inactivation of antibiotic resistant bacteria and resistance genes, Water Res. 184 (2020) 116157.
[26] S. Ma, S. Zhan, Y. Jia, Q. Zhou, Highly efficient antibacterial and Pb(II) removal effects of Ag-$CoFe_2O_4$-GO nanocomposite, ACS Appl. Mater. Interfaces 7 (19) (2015) 10576–10586.
[27] H. Zheng, R. Ma, M. Gao, X. Tian, Y. Li, L. Zeng, R. Li, Antibacterial applications of graphene oxides: structure-activity relationships, molecular initiating events and biosafety, Sci. Bull. 63 (2) (2018) 133–142.
[28] M. Farid, S. Jeong, D. Seo, R. Ahmed, C. Lau, N. Gali, Z. Ning, A. An, Mechanistic insight into thein vitrotoxicity of graphene oxide against biofilm forming bacteria using laser-induced breakdown spectroscopy, Nanoscale 10 (9) (2018) 4475–4487.
[29] S. Liu, T. Zeng, M. Hofmann, E. Burcombe, J. Wei, R. Jiang, J. Kong, Y. Chen, Antibacterial activity of graphite, graphite oxide, graphene oxide, and reduced graphene oxide: membrane and oxidative stress, ACS Nano 5 (9) (2011) 6971–6980.
[30] J. Qiu, D. Wang, H. Geng, J. Guo, S. Qian, X. Liu, How oxygen-containing groups on graphene influence the antibacterial behaviors, Adv. Mater. Interfaces 4 (15) (2017) 1700228.
[31] W. Yu, S. Zhan, Z. Shen, Q. Zhou, D. Yang, Efficient removal mechanism for antibiotic resistance genes from aquatic environments by graphene oxide Nanosheet, Chem. Eng. J. 313 (2017) 836–846.
[32] J. Liu, Adsorption of DNA onto gold nanoparticles and graphene oxide: surface science and applications, Phys. Chem. Chem. Phys. 14 (30) (2012) 10485.

[33] D. Li, S. Zeng, M. He, A. Gu, Water disinfection byproducts induce antibiotic resistance-role of environmental pollutants in resistance phenomena, Environ. Sci. Technol. 50 (6) (2016) 3193–3201.

[34] J. Zhu, A. Fleming, A. Orendt, C. Burrows, Ph-dependent equilibrium between 5-guanidinohydantoin and iminoallantoin affects nucleotide insertion opposite the DNA lesion, J. Organomet. Chem. 81 (2) (2015) 351–359.

[35] E. Hashemi, O. Akhavan, M. Shamsara, S. Valimehr, R. Rahighi, DNA and RNA extractions from eukaryotic and prokaryotic cells by graphene Nanoplatelets, RSC Adv. 4 (105) (2014) 60720–60728.

[36] S. Banerjee, J. Wilson, J. Shim, M. Shankla, E. Corbin, A. Aksimentiev, R. Bashir, Slowing DNA transport using graphene-DNA interactions, Adv. Funct. Mater. 25 (6) (2014) 936–946.

[37] B. Zheng, C. Wang, C. Wu, X. Zhou, M. Lin, X. Wu, X. Xin, X. Chen, L. Xu, H. Liu, J. Zheng, J. Zhang, S. Guo, Nuclease activity and cytotoxicity enhancement of the DNA intercalators via graphene oxide, J. Phys. Chem. C 116 (29) (2012) 15839–15846.

[38] O. Akhavan, E. Ghaderi, A. Akhavan, Size-dependent genotoxicity of graphene nanoplatelets in human stem cells, Biomaterials 33 (32) (2012) 8017–8025.

[39] O. Akhavan, E. Ghaderi, R. Rahighi, Toward single-DNA electrochemical biosensing by graphene nanowalls, ACS Nano 6 (4) (2012) 2904–2916.

[40] O. Akhavan, E. Ghaderi, H. Emamy, F. Akhavan, Genotoxicity of graphene nanoribbons in human mesenchymal stem cells, Carbon 54 (2013) 419–431.

[41] Z. Xu, Y. Bando, L. Liu, W. Wang, X. Bai, D. Golberg, Electrical conductivity, chemistry, and bonding alternations under graphene oxide to graphene transition as revealed by in situ TEM, ACS Nano 5 (6) (2011) 4401–4406.

[42] M. Das, R. Nasani, M. Saha, S. Mobin, S. Mukhopadhyay, Nickel(II) complexes with a flexible piperazinyl moiety: studies on DNA and protein binding and catecholase like properties, Dalton Trans. 44 (5) (2015) 2299–2310.

[43] D. Frigon, B. Biswal, A. Mazza, L. Masson, R. Gehr, Biological and physicochemical wastewater treatment processes reduce the prevalence of virulent Escherichia Coli, Appl. Environ. Microbiol. 79 (3) (2012) 835–844.

[44] J. Li, B. Shao, J. Shen, S. Wang, Y. Wu, Occurrence of chloramphenicol-resistance genes as environmental pollutants from swine feedlots, Environ. Sci. Technol. 47 (6) (2013) 2892–2897.

[45] Q. Yuan, M. Guo, J. Yang, Fate of antibiotic resistant bacteria and genes during wastewater chlorination: implication for antibiotic resistance control, PLoS One 10 (3) (2015) e0119403.

[46] M. Guo, G. Zhang, Graphene oxide in the water environment could affect tetracycline-antibiotic resistance, Chemosphere 183 (2017) 197–203.

[47] S. Rodriguez-Mozaz, S. Chamorro, E. Marti, B. Huerta, M. Gros, A. Sànchez-Melsió, C. Borrego, D. Barceló, J. Balcázar, Occurrence of antibiotics and antibiotic resistance genes in hospital and urban wastewaters and their impact on the receiving river, Water Res. 69 (2015) 234–242.

[48] W. Zou, X. Li, Z. Lai, X. Zhang, X. Hu, Q. Zhou, Graphene oxide inhibits antibiotic uptake and antibiotic resistance gene propagation, ACS Appl. Mater. Interfaces 8 (48) (2016) 33165–33174.

[49] A. Khalil, F. Memon, T. Tabish, D. Salmon, S. Zhang, D. Butler, Nanostructured porous graphene for efficient removal of emerging contaminants (pharmaceuticals) from water, Chem. Eng. J. 398 (2020) 125440.

[50] C. Ferreiro, I. Gómez-Motos, J. Lombraña, A. de Luis, N. Villota, O. Ros, N. Etxebarria, Contaminants of emerging concern removal in An effluent of wastewater treatment plant under biological and continuous mode ultrafiltration treatment, Sustainability 12 (2) (2020) 725.

[51] A.J. Hernandez-Maldonado, L. Blaney, Contaminants of Emerging Concern in Water and Wastewater: Advanced Treatment Process, 1st ed., Elsevier, Oxford, United Kingdom, 2019, pp. 1–434.

[52] S. Lath, D. Navarro, D. Losic, A. Kumar, M. McLaughlin, Sorptive remediation of perfluorooctanoic acid (PFOA) using mixed mineral and graphene/carbon-based materials, Environ. Chem. 15 (8) (2018) 472.

[53] S. Maliyekkal, T. Sreeprasad, D. Krishnan, S. Kouser, A. Mishra, U. Waghmare, T. Pradeep, Graphene: a reusable substrate for unprecedented adsorption of pesticides, Small 9 (2) (2012) 273–283.

[54] S. Sen Gupta, I. Chakraborty, S. Maliyekkal, T. Adit Mark, D. Pandey, S. Das, T. Pradeep, Simultaneous dehalogenation and removal of persistent halocarbon pesticides from water using graphene nanocomposites: a case study of lindane, ACS Sustain. Chem. Eng. 3 (6) (2015) 1155–1163.

[55] Z. Pei, L. Li, L. Sun, S. Zhang, X. Shan, S. Yang, B. Wen, Adsorption characteristics of 1,2,4-trichlorobenzene, 2,4,6-trichlorophenol, 2-naphthol and naphthalene on graphene and graphene oxide, Carbon 51 (2013) 156–163.

[56] J. Björk, F. Hanke, C. Palma, P. Samori, M. Cecchini, M. Persson, Adsorption of aromatic and anti-aromatic systems on graphene through $\Pi-\Pi$ stacking, J. Phys. Chem. Lett. 1 (23) (2010) 3407–3412.

[57] C. Zhang, R. Zhang, Y. Ma, W. Guan, X. Wu, X. Liu, H. Li, Y. Du, C. Pan, Preparation of cellulose/graphene composite and its applications for triazine pesticides adsorption from water, ACS Sustain. Chem. Eng. 3 (3) (2015) 396–405.

[58] M. Cruz, C. Gomez, C. Duran-Valle, L. Pastrana-Martínez, J. Faria, A. Silva, M. Faraldos, A. Bahamonde, Bare TiO2 and graphene oxide TiO2 photocatalysts on the degradation of selected pesticides and influence of the water matrix, Appl. Surf. Sci. 416 (2017) 1013–1021.

[59] A. Keihan, R. Hosseinzadeh, M. Farhadian, H. Kooshki, G. Hosseinzadeh, Solvothermal preparation of ag nanoparticle and graphene co-loaded TiO2 for the photocatalytic degradation of Paraoxon pesticide under visible light irradiation, RSC Adv. 6 (87) (2016) 83673–83687.

[60] V. Gupta, T. Eren, N. Atar, M. Yola, C. Parlak, H. Karimi-Maleh, $CoFe_2O_4$@TiO_2 decorated reduced graphene oxide nanocomposite for photocatalytic degradation of chlorpyrifos, J. Mol. Liq. 208 (2015) 122–129.

[61] C. Zhang, Z. Du, R. Zhou, P. Xu, X. Dong, Y. Fu, Q. Wang, C. Su, L. Yan, Z. Gu, $Cu_2(OH)PO_4$/reduced graphene oxide nanocomposites for enhanced photocatalytic degradation of 2,4-dichlorophenol under infrared light irradiation, RSC Adv. 8 (7) (2018) 3611–3618.

[62] B. Peng, L. Chen, C. Que, K. Yang, F. Deng, X. Deng, G. Shi, G. Xu, M. Wu, Adsorption of antibiotics on graphene and biochar in aqueous solutions induced by Π-Π interactions, Sci. Rep. 6 (1) (2016).

[63] F. Wang, B. Yang, H. Wang, Q. Song, F. Tan, Y. Cao, Removal of ciprofloxacin from aqueous solution by a magnetic chitosan grafted graphene oxide composite, J. Mol. Liq. 222 (2016) 188–194.
[64] P. Tabrizian, W. Ma, A. Bakr, M. Rahaman, Ph-sensitive and magnetically separable Fe/cu bimetallic nanoparticles supported by graphene oxide (GO) for high-efficiency removal of tetracyclines, J. Colloid Interface Sci. 534 (2019) 549–562.
[65] Y. Zhang, Z. Jiao, Y. Hu, S. Lv, H. Fan, Y. Zeng, J. Hu, M. Wang, Removal of tetracycline and oxytetracycline from water by magnetic Fe_3O_4@ graphene, Environ. Sci. Pollut. Res. 24 (3) (2016) 2987–2995.
[66] L. Al-Khateeb, S. Almotiry, M. Salam, Adsorption of pharmaceutical pollutants onto graphene nanoplatelets, Chem. Eng. J. 248 (2014) 191–199.
[67] L. Rizzo, A. Fiorentino, M. Grassi, D. Attanasio, M. Guida, Advanced treatment of urban wastewater by sand filtration and graphene adsorption for wastewater reuse: effect on a mixture of pharmaceuticals and toxicity, J. Environ. Chem. Eng. 3 (1) (2015) 122–128.
[68] B. Hiew, L. Lee, X. Lee, S. Gan, S. Thangalazhy-Gopakumar, S. Lim, G. Pan, T. Yang, Adsorptive removal of diclofenac by graphene oxide: optimization, equilibrium, kinetic and thermodynamic studies, J. Taiwan Inst. Chem. Eng. 98 (2019) 150–162.
[69] I. Jauris, C. Matos, C. Saucier, E. Lima, A. Zarbin, S. Fagan, F. Machado, I. Zanella, Adsorption of sodium diclofenac on graphene: a combined experimental and theoretical study, Phys. Chem. Chem. Phys. 18 (3) (2016) 1526–1536.
[70] X. Lee, N. Chemmangattuvalappil, L. Lee, Adsorptive removal of salicylic acid from aqueous solutions using new graphene-based Nanosorbents, Chem. Eng. Trans. 45 (2015) 1387–1392.
[71] S. Zhu, Y. Liu, S. Liu, G. Zeng, L. Jiang, X. Tan, L. Zhou, W. Zeng, T. Li, C. Yang, Adsorption of emerging contaminant metformin using graphene oxide, Chemosphere 9179 (2017) 20–28.
[72] G. Lee, K. Chu, Y. Al-Hamadani, C. Park, M. Jang, J. Heo, N. Her, D. Kim, Y. Yoon, Fabrication of graphene-oxide/B-Bi_2O_3/TiO_2/$Bi_2Ti_2O_7$ heterojuncted nanocomposite and its sonocatalytic degradation for selected pharmaceuticals, Chemosphere 212 (2018) 723–733.
[73] A. Khan, J. Wang, J. Li, X. Wang, Z. Chen, A. Alsaedi, T. Hayat, Y. Chen, X. Wang, The role of graphene oxide and graphene oxide-based nanomaterials in the removal of pharmaceuticals from aqueous media: a review, Environ. Sci. Pollut. Res. 24 (9) (2017) 7938–7958.
[74] S. Fanourakis, J. Peña-Bahamonde, P. Bandara, D. Rodrigues, Nano-based adsorbent and photocatalyst use for pharmaceutical contaminant removal during indirect potable water reuse, Npj Clean Water 3 (1) (2020).
[75] Z. Niu, Y. Wang, H. Lin, F. Jin, Y. Li, J. Niu, Electrochemically enhanced removal of perfluorinated compounds (Pfcs) from aqueous solution by CNTs-graphene composite electrode, Chem. Eng. J. 328 (2017) 228–235.
[76] C. Zhao, J. Fan, D. Chen, Y. Xu, T. Wang, Microfluidics-generated graphene oxide microspheres and their application to removal of perfluorooctane sulfonate from polluted water, Nano Res. 9 (3) (2016) 866–875.
[77] D. Wu, X. Li, J. Zhang, W. Chen, P. Lu, Y. Tang, L. Li, Efficient PFOA degradation by persulfate-assisted photocatalytic ozonation, Sep. Purif. Technol. 207 (2018) 255–261.
[78] T. Deblonde, C. Cossu-Leguille, P. Hartemann, Emerging pollutants in wastewater: a review of the literature, Int. J. Hyg. Environ. Health 214 (6) (2011) 442–448.
[79] M. Clara, G. Windhofer, W. Hartl, K. Braun, M. Simon, O. Gans, C. Scheffknecht, A. Chovanec, Occurrence of phthalates in surface runoff, untreated and treated wastewater and fate during wastewater treatment, Chemosphere 78 (9) (2010) 1078–1084.
[80] D. Lapworth, N. Baran, M. Stuart, R. Ward, Emerging organic contaminants in groundwater: a review of sources, fate and occurrence, Environ. Pollut. 163 (2012) 287–303.
[81] J. Dutta, A. Mala, Removal of antibiotic from the water environment by the adsorption technologies: a review, Water Sci. Technol. (2020).
[82] Z. Li, M. Qi, C. Tu, W. Wang, J. Chen, A. Wang, Highly efficient removal of chlorotetracycline from aqueous solution using graphene oxide/TiO_2 composite: properties and mechanism, Appl. Surf. Sci. 425 (2017) 765–775.
[83] T. Anirudhan, J. Deepa, Nano-zinc oxide incorporated graphene oxide/nanocellulose composite for the adsorption and photo catalytic degradation of ciprofloxacin hydrochloride from aqueous solutions, J. Colloid Interface Sci. 490 (2017) 343–356.
[84] X. Liu, H. Zhang, Y. Ma, X. Wu, L. Meng, Y. Guo, G. Yu, Y. Liu, Graphene-coated silica as a highly efficient sorbent for residual organophosphorus pesticides in water, J. Mater. Chem. A 1 (5) (2013) 1875–1884.
[85] Z. Ren, Y. Lan, Y. Wang, Aligned Carbon Nanotubes, Springer Berlin, Berlin, 2013.
[86] Y. Zhu, S. Murali, W. Cai, X. Li, J. Suk, J. Potts, R. Ruoff, Graphene and graphene oxide: synthesis, properties, and applications, Adv. Mater. 22 (35) (2010) 3906–3924.
[87] T. Lazarević-Pašti, V. Anićijević, M. Baljozović, D. Anićijević, S. Gutić, V. Vasić, N. Skorodumova, I. Pašti, The impact of the structure of graphene-based materials on the removal of organophosphorus pesticides from water, Environ. Sci. Nano 5 (6) (2018) 1482–1494.
[88] S. Wang, H. Sun, H. Ang, M. Tadé, Adsorptive remediation of environmental pollutants using novel graphene-based nanomaterials, Chem. Eng. J. 226 (2013) 336–347.
[89] K. Shrivas, A. Ghosale, N. Nirmalkar, A. Srivastava, S. Singh, S. Shinde, Removal of endrin and dieldrin isomeric pesticides through stereoselective adsorption behavior on the graphene oxide-magnetic nanoparticles, Environ. Sci. Pollut. Res. 24 (32) (2017) 24980–24988.
[90] Q. Wu, G. Zhao, C. Feng, C. Wang, Z. Wang, Preparation of a graphene-based magnetic nanocomposite for the extraction of carbamate pesticides from environmental water samples, J. Chromatogr. A 1218 (44) (2011) 7936–7942.
[91] D. Koushik, S. Sen Gupta, S. Maliyekkal, T. Pradeep, Rapid dehalogenation of pesticides and organics at the interface of reduced graphene oxide–silver nanocomposite, J. Hazard. Mater. 308 (2016) 192–198.
[92] M. Sarno, M. Casa, C. Cirillo, P. Ciambelli, Complete removal of persistent pesticide using reduced graphene oxide – silver nanocomposite, Chem. Eng. Trans. 60 (2017) 151–156.
[93] H. Liu, T. Qian, D. Zhao, Reductive immobilization of perrhenate in soil and groundwater using starch-stabilized ZVI nanoparticles, Chin. Sci. Bull. 58 (2) (2012) 275–281.

[94] The New York Times. 2020. https://www.nytimes.com/2020/09/23/parenting/pregnancy/pfas-toxins-chemicals.html?smid=em-share (accessed: October 16, 2020).
[95] M. Lv, L. Yan, C. Liu, C. Su, Q. Zhou, X. Zhang, Y. Lan, Y. Zheng, L. Lai, X. Liu, Z. Ye, Non-covalent functionalized graphene oxide (GO) adsorbent with An organic gelator for co-adsorption of dye, endocrine-disruptor, pharmaceutical and metal ion, Chem. Eng. J. 349 (2018) 791–799.
[96] C. Ong, A. Mohammad, L. Ng, E. Mahmoudi, S. Azizkhani, N. Hayati Hairom, Solar photocatalytic and surface enhancement of ZnO/RGO nanocomposite: degradation of perfluorooctanoic acid and dye, Process. Saf. Environ. Prot. 112 (2017) 298–307.
[97] N. Kotlarz, J. McCord, D. Collier, C.S. Lea, M. Strynar, A.B. Lindstrom, A.A. Wilkie, J.Y. Islam, K. Matney, P. Tarte, M.E. Polera, K. Burdette, J. DeWitt, K. May, R.C. Smart, D.R.U. Knappe, J.A. Hoppin, Measurement of novel, drinking water-associated PFAS in blood from adults and children in Wilmington, North Carolina, Environ. Health Perspect. 128 (7) (2020) 77005.
[98] C.F. Kwiatkowski, D.Q. Andrews, L.S. Birnbaum, T.A. Bruton, J.C. DeWitt, D.R.U. Knappe, M.V. Maffini, M.F. Miller, K.E. Pelch, A. Reade, A. Soehl, X. Trier, M. Venier, C.C. Wagner, Z. Wang, A. Blum, Scientific basis for managing PFAS as a chemical class, Environ. Sci. Technol. Lett. 7 (8) (2020) 532–543.
[99] T.C. Guillette, J. McCord, M. Guillette, M.E. Polera, K.T. Rachels, C. Morgeson, N. Kotlarz, D.R.U. Knappe, B.J. Reading, M. Strynar, S.M. Belcher, Elevated levels of per- and polyfluoroalkyl substances in cape fear river striped bass (*Morone saxatilis*) are associated with biomarkers of altered immune and liver function, Environ. Int. 136 (2020) 105358.
[100] Z.R. Hopkins, M. Sun, J.R. DeWitt, D.R.U. Knappe, Recently detected drinking water contaminants: GenX and other per- and polyfluoroalkyl ether acids, J. AWWA 110 (7) (2018) 13–28.
[101] N. Saleh, A. Khalid, Y. Tian, C. Ayres, I. Sabaraya, J. Pietari, D. Hanigan, I. Chowdhury, O. Apul, Removal of poly- and per-fluoroalkyl substances from aqueous systems by nano-enabled water treatment strategies, Environ. Sci. Water Res. Technol. 5 (2) (2019) 198–208.
[102] S. Mahpishanian, H. Sereshti, M. Baghdadi, Superparamagnetic core–shells anchored onto graphene oxide grafted with phenylethyl amine as a nano-adsorbent for extraction and enrichment of organophosphorus pesticides from fruit, vegetable and water samples, J. Chromatogr. A 1406 (2015) 48–58.
[103] K. Yang, B. Xing, Adsorption of organic compounds by carbon nanomaterials in aqueous phase: Polanyi theory and its application, Chem. Rev. 110 (10) (2010) 5989–6008.
[104] W. Chen, L. Duan, D. Zhu, Adsorption of polar and nonpolar organic chemicals to carbon nanotubes, Environ. Sci. Technol. 41 (24) (2007) 8295–8300.
[105] S. Smith, D. Rodrigues, Carbon-based nanomaterials for removal of chemical and biological contaminants from water: a review of mechanisms and applications, Carbon 91 (2015) 122–143.
[106] S. Benfeito, T. Silva, J. Garrido, P. Andrade, M. Sottomayor, F. Borges, E. Garrido, Effects of chlorophenoxy herbicides and their main transformation products on DNA damage and acetylcholinesterase activity, Biomed. Res. Int. 2014 (2014) 1–10.
[107] R. Kamaraj, A. Pandiarajan, M. Gandhi, A. Shibayama, S. Vasudevan, Eco-friendly and easily prepared graphene nanosheets for safe drinking water: removal of chlorophenoxyacetic acid herbicides, ChemistrySelect 2 (1) (2017) 342–355.
[108] L. Joseph, L. Boateng, J. Flora, Y. Park, A. Son, M. Badawy, Y. Yoon, Removal of bisphenol a and 17A-Ethinyl estradiol by combined coagulation and adsorption using carbon nanomaterials and powdered activated carbon, Sep. Purif. Technol. 107 (2013) 37–47.
[109] F. Wang, W. Sha, X. Wang, Y. Shang, L. Hou, Y. Li, Adsorption of 17A-Ethinyl estradiol and bisphenol a to graphene-based materials: effects of configuration of adsorbates and the presence of cationic surfactant, Adsorpt. Sci. Technol. 2021 (2021) 1–12.
[110] https://www.health.state.mn.us/communities/environment/water/contaminants/index.html#a.
[111] A. Macías-García, J. García-Sanz-Calcedo, J. Carrasco-Amador, R. Segura-Cruz, Adsorption of paracetamol in hospital wastewater through activated carbon filters, Sustainability 11 (9) (2019) 2672.
[112] E. Umejuru, E. Prabakaran, K. Pillay, Coal fly ash decorated with graphene oxide–tungsten oxide nanocomposite for rapid removal of Pb^{2+} ions and reuse of spent adsorbent for photocatalytic degradation of acetaminophen, ACS Omega 6 (17) (2021) 11155–11172.
[113] W. Evans, P. Davies, C. McRae, The occurrence of methyl, ethyl, propyl, and butyl parabens in the urban rivers and stormwaters of Sydney, Australia, Environ. Sci. Water Res. Technol. 2 (4) (2016) 733–742.
[114] D. Husein, Facile one-pot synthesis of porous N-doped graphene based NiO composite for parabens removal from wastewater and its reusability, Desalin. Water Treat. 166 (2019) 211–221.
[115] Y. Wang, X. Wei, Y. Qi, H. Huang, Efficient removal of bisphenol-a from water and wastewater by Fe2O3-modified graphene oxide, Chemosphere 263 (2021) 127563.
[116] B. Yao, Z. Luo, D. Zhi, D. Hou, L. Luo, S. Du, Y. Zhou, Current progress in degradation and removal methods of polybrominated diphenyl ethers from water and soil: a review, J. Hazard. Mater. 403 (2021) 123674.
[117] E. Pan, H. Sun, Q. Xu, Q. Zhang, L. Liu, X. Chen, Y. Xu, Polycyclic aromatic hydrocarbons concentrations in drinking water in villages along the Huai River in China and their association with high Cancer incidence in local population, Biomed. Res. Int. 2015 (2015) 1–10.
[118] A. Adeola, P. Forbes, Advances in water treatment technologies for removal of polycyclic aromatic hydrocarbons: existing concepts, emerging trends, and future prospects, Water Environ. Res. 93 (3) (2020) 343–359.
[119] Z. Liu, X. Wang, Z. Luo, M. Huo, J. Wu, H. Huo, W. Yang, Removing of disinfection by-product precursors from surface water by using magnetic graphene oxide, PLoS One 10 (12) (2015) e0143819.
[120] Z. Liu, Z. Zhang, J. Peng, J. Wu, Y. Huo, Rapid removal of trace haloacetic acids from drinking water by a continuous adsorption process using graphene oxide, Environ. Technol. (2020) 1–7.

[121] D. Cortés-Arriagada, S. Miranda-Rojas, D. Ortega, A. Toro-Labbé, Oxidized and Si-doped graphene: emerging adsorbents for removal of dioxane, Phys. Chem. Chem. Phys. 19 (27) (2017) 17587–17597.

[122] A.C. McElroy, M.R. Hyman, D.R.U. Knappe, 1,4-Dioxane in drinking water: emerging for forty years and still unregulated, Curr. Opin. Environ. Sci. Health 7 (2019) 117–125.

[123] A. Medina, F. Casado-Carmona, Á. López-Lorente, S. Cárdenas, Magnetic graphene oxide composite for the microextraction and determination of benzophenones in water samples, Nanomaterials 10 (1) (2020) 168.

[124] A. Sinha, N. Jana, Separation of microcystin-LR by cyclodextrin-functionalized magnetic composite of colloidal graphene and porous silica, ACS Appl. Mater. Interfaces 7 (18) (2015) 9911–9919.

[125] P. Kumar, J. Pérez, M. Cledon, S. Brar, S. Duy, S. Sauvé, É. Knystautas, Removal of microcystin-LR and other water pollutants using sand coated with bio-optimized carbon submicron particles: graphene oxide and reduced graphene oxide, Chem. Eng. J. 397 (2020) 125398.

[126] S. Pavagadhi, A. Tang, M. Sathishkumar, K. Loh, R. Balasubramanian, Removal of microcystin-LR and microcystin-RR by graphene oxide: adsorption and kinetic experiments, Water Res. 47 (13) (2013) 4621–4629.

[127] C. Sun, Z. Wang, L. Chen, F. Li, Fabrication of robust and compressive chitin and graphene oxide sponges for removal of microplastics with different functional groups, Chem. Eng. J. 393 (2020) 124796.

[128] F. Yuan, L. Yue, H. Zhao, H. Wu, Study on the adsorption of polystyrene microplastics by three-dimensional reduced graphene oxide, Water Sci. Technol. 81 (10) (2020) 2163–2175.

[129] S. Ahuja, Trace and Ultratrace Analysis by HPLC, first ed., Wiley, New York, 1991, pp. 1–432.

[130] A. Godoy, F. Kummrow, P. Pamplin, Occurrence, ecotoxicological effects and risk assessment of antihypertensive pharmaceutical residues in the aquatic environment—a review, Chemosphere 138 (2015) 281–291.

[131] K. Betts, Technology solutions: keeping drugs out of drinking water, Environ. Sci. Technol. 36 (19) (2002) 377A–378A.

[132] T. Heberer, Tracking persistent pharmaceutical residues from municipal sewage to drinking water, J. Hydrol. 266 (3–4) (2002) 175–189.

[133] Wang, Z.; Nguyen Song Thuy Thuy, G.; Srivastava, V.; Ambat, I.; Sillanpää, M. Photocatalytic degradation of an artificial sweetener (Acesulfame-K) from synthetic wastewater under UV-LED controlled illumination. Process. Saf. Environ. Prot. 2019, 123, 206–214.

[134] A. Abu-Nada, G. McKay, A. Abdala, Recent advances in applications of hybrid graphene materials for metals removal from wastewater, Nanomaterials 10 (3) (2020) 595.

[135] H. Chi, K. Murali, T. Li, S. Thomas, Recent advances in graphene based photoresponsive materials, Prog. Nat. Sci. Mater. Int. 29 (6) (2019) 603–611.

[136] P. Jayakaran, G. Nirmala, L. Govindarajan, Qualitative and quantitative analysis of graphene-based adsorbents in wastewater treatment, Int. J. Chem. Eng. 2019 (2019) 1–17.

[137] J. Feng, Y. Ye, M. Xiao, G. Wu, Y. Ke, Synthetic routes of the reduced graphene oxide, Chem. Pap. 74 (11) (2020) 3767–3783.

[138] Sigma Aldrich. 2020. https://www.sigmaaldrich.com/catalog/product/aldrich/796034?lang=en&region=BD (Accessed September 04, 2020).

[139] P. Brisebois, M. Siaj, Harvesting graphene oxide—years 1859 to 2019: a review of its structure, synthesis, properties and exfoliation, J. Mater. Chem. C 8 (5) (2020) 1517–1547.

[140] M. Bundschuh, J. Filser, S. Lüderwald, M. McKee, G. Metreveli, G. Schaumann, R. Schulz, S. Wagner, Nanoparticles in the environment: where do we come from, where do we go to? Environ. Sci. Eur. 30 (1) (2018).

[141] S. Richardson, T. Ternes, Water analysis: emerging contaminants and current issues, Anal. Chem. 90 (1) (2017) 398–428.

[142] X. Guo, N. Mei, Assessment of the toxic potential of graphene family nanomaterials, J. Food Drug Anal. 22 (1) (2014) 105–115.

Chapter 12

Nanoadsorption: An innovational and compelling approach toward fluoride remediation from water

Sana Khalid[a] and Rabia Nazir[b,*]

[a]*College of Earth and Environmental Sciences, University of the Punjab, Lahore, Pakistan,* [b]*Pakistan Council of Scientific and Industrial Research Laboratories Complex, Lahore, Pakistan*

**Corresponding author: e-mail address: rabiapcsir@yahoo.com*

Introduction

Water is the most vital resource for all life forms present on this planet. Out of the total volume of water present on Earth, only 2.5% is present in the form of freshwater resources. A small portion of freshwater reserves, that is 0.01%, is available for human use [1]. The rapid population growth, extensive urbanization, reckless development of industries, intense agricultural practices, and their resultant environmental pollution, all are exerting a drastic pressure on water resources. The complex anthropogenic activities are responsible for introduction of diverse organic and inorganic chemicals into the environment that deteriorates both water quantity and quality and makes it very difficult to assure the provision of sustainable supply of water to humans to fulfill their basic daily needs.

Fluoride is the anionic state of fluorine. Geological formations are the predominant source of fluoride in groundwater. It is also present in sedimentary rocks, minerals, clays and soils. The fluoride contamination and its related health problems, is a significant environmental challenge being faced by most of the countries in the world including Pakistan [2–4], Iran [5], China [6], Japan [7], Turkey [8], Kenya [9], Canada [10], Mexico [11], Algeria [12], Norway [13], Brazil [14], Korea [15], and South Africa [16], and hence affecting more than 200 million individuals from 25 countries globally [17]. According to World Health Organization (WHO), the allowable concentration of fluoride in drinking water is 1.5 mg/L [18]. The presence of fluoride in drinking water at concentrations greater than the permissible limits causes severe health concerns such as skeletal and dental fluorosis [19], thyroid dysfunctioning [20], neurological problems [21–23], liver disease, and disorders of some other body organs [24].

The availability of safe and clean drinking water is becoming a major hurdle and an economic burden globally. In order to fulfill the ever-increasing requirements of potable water for the growing population, it is extremely crucial to devise appropriate water purification approaches. Enormous purification methods and technologies involving the basic principles of membrane filtration, electrodialysis, precipitation, adsorption, and ion exchange have been proposed and implemented to check their success for the removal of excess of fluoride from drinking water and industrial effluents [25–28]. However, among these techniques, adsorption is considered the best method in terms of its low-cost and convenient method for operation (Fig. 1). In this regard, an ample variety of materials have been tested to remove excess fluoride from water by adsorption and chemical precipitation processes, for example, activated and amorphous alumina [29], lignite [30], activated carbon [31,32], activated magnesia [33], metal oxides [29,34,35], zeolites [36–38], quick lime [39], iron-coated sand [35,40], calcite [41,42], hydrocalcite [43], soil [44], cement [45], red bricks powder and pavestones [46,47], modified fly ash [48], dual-layer-hydroxides [49,50], bone char [51], and agricultural waste products [32,52,53].

However, the minimal limit for fluoride removal by majority of the adsorbents is >2 mg/L with low sorption potential and high equilibrium time and most of these materials are effective only at extreme pH values, for example, activated carbon which maximize fluoride decontamination at pH <3. Some ultrafine materials-based adsorbents like zirconium

Separation Science and Technology, Volume 15. https://doi.org/10.1016/B978-0-323-90763-7.00003-2

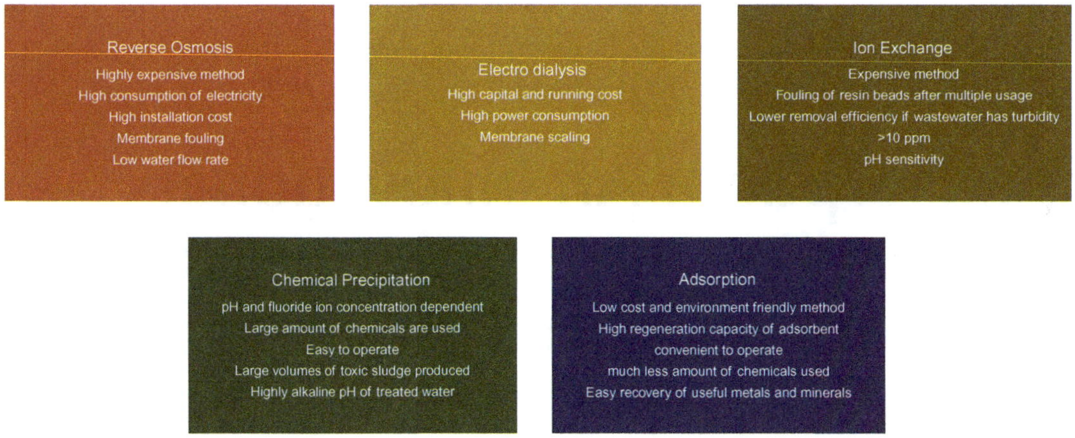

FIG. 1 Primacy of adsorption process over other water purification techniques.

oxide also present the problem of metal leaching into the treated water. Thus, these materials are not favorable for drinking water purification system without any physical or chemical modification [54,55].

In addition, most of the defluoridation materials reported in literature have their particle size in few micrometers or even bigger. Recently, nanomaterials have gained paramount attention and are being extensively used as excellent adsorbents owing to their properties such as large surface area, short diffusion rate, and enhanced optical, chemical, electronic and magnetic characteristics as compared to their bulk counterparts [56–58]. Several metal elements, geopolymer, biopolymer and carbon-based nanomaterials, and their applications in defluoridation of aqueous solutions are highlighted in this chapter.

Metal oxides, hydroxides and carbonates–based nanomaterials

Most of the metal ions have excellent affinity toward fluoride; thereof their nanoparticles (NPs) and composites have been extensively used as a potential candidate to remove excessive fluoride from aqueous solutions (Table 1).

Calcium-based salts, i.e., calcium oxide and calcium carbonate are found useful for the removal of fluoride from the system either by adsorption or precipitation [68]. A study performed to investigate the sorption efficiency of colloidal calcium oxide (particle diameter <15 nm) NPs for fluoride from aqueous solution showed reduction of fluoride up to 98% within 30 min at adsorbent concentration of 0.6 gL^{-1} for baseline concentration of fluoride up to 100 mg/L. The sorption data was best explained by linearized Dubinin–Radushkevich isotherm and pseudo-first-order kinetic models. The mechanism of adsorption was attributed to replacement of fluoride ion by hydroxide group from calcium hydroxide ($Ca(OH)_2$), that is readily formed upon calcium oxide dissolution in water [55] as given in Eq. (1).

$$CaO_{(s)} + H_2O_{(l)} + F^-_{(aq)} \rightarrow Ca(OH)F_{(aq)} + OH^-_{(aq)} \quad (1)$$

Direct use of calcium hydroxide nanorods for the defluoridation have resulted in much higher adsorption capacity, i.e., 450 mg/g in just 45 min at much smaller amounts of the adsorbent (0.12 gL^{-1}) which might be attributed to their much smaller size (4.4 nm), high surface area (367 m^2/g) and pore volume (0.5 mL/g). The same CaF_2 coprecipitation phenomenon was suggested as main removal mechanism. The adsorbent was also found effective in real setting as it was able to remove 91.7% fluoride from the acidic electroplating industrial waste but fluoride scavenging was highly impacted by the presence of competitive ions ($S_2O_3^{-2}$,NO_3^-, PO_4^{-3}, HCO_3^-) at the optimized lower concentrations of adsorbent, i.e., (0.12 g/L) [69].

In order to increase the fluoride uptake, Dhillon et al. [70] fabricated a series of low-cost calcium-based nanocomposites, i.e., cerium (IV)-calcium (II) by using different molar concentrations and calcinations temperatures. The sorbent with molar concentration of 3:1 (Ce:Ca) having particle size and surface area of ~7 nm and 349.9 m^2g^{-1}, respectively, and prepared at calcination temperature of 600 °C was found to be most effective with adsorption capacity of 83.33 mg/g. X-ray spectroscopic studies showed uniform distribution of fluoride on the composite surface exhibiting adsorption to be the major removal mechanism for fluoride ion. Much higher adsorption capacities of fluoride, i.e., 754.36 mg/g were achieved within just 10 min at room temperature and pH 7 by use of calcium carbonate NPs (prepared by coprecipitation approach) having the size of 14.4 nm. Owing to its smaller size the NPs were able to achieve 98–100%

TABLE 1 Metal-based nanoadsorbents and their fluoride removal potential.

Nanoadsorbent	Defluoridation capacity (mg/g)	pH	Surface area (m^2/g)	Adsorption and kinetic models	References
Al/Fe/Ce ternary metal oxide	178	7.0	114	Pseudo-first-order kinetics model, two-site Langmuir isotherm	[56]
Nano magnesium oxide	14	10.0–11.0	92.46	Pseudo-second-order kinetics model, Freundlich isotherm	[59]
Thermally activated dolomite powder	243.25	7.0	5.92	Pseudo-second-order kinetic model, Freundlich isotherm	[60]
Ferromagnetic iron oxide	3.65	–	16.5±2.5	Cluster model	[61]
Ion exchanger functionalized zirconia	134.7	3.5	17.37	Intraparticle diffusion and Pseudo-second-order kinetics model	[62]
Nano goethite	59	5.5–6.0	—	Pseudo-second-order kinetics model, Freundlich isotherm	[63]
Nano alumina	14	6.15	151.7	Pseudo-second-order kinetic model, Langmuir isotherm	[64]
Synthetic titania (201)	5	5.0	31.9	Pseudo-second-order kinetics model, Langmuir isotherm	[65]
Iron (II, III)oxide NPs decorated magnesium oxide	10.96	6.0	—	Pseudo-second-order kinetics model, Langmuir isotherm	[66]
Non thermal plasma functionalized alumina/ cerium oxide	37	3.0–10.0	266.1	Pseudo-second-order and Redlich–Peterson models	[67]

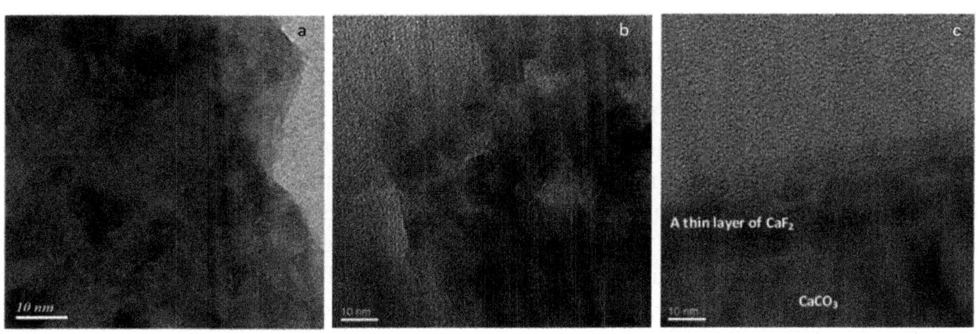

FIG. 2 Removal of fluoride using calcite by adsorption and surface precipitation. HRTEM images of single calcite (A) before adsorption, (B) after reaction with 10 mgL^{-1} F and (C) 40 mgL^{-1} F.

defluoridation of real drinking water samples [71]. X-ray photoelectron spectroscopic (XPS) and high-resolution transmission electron microscopy (HRTEM) studies revealed that calcite NPs react with fluoride to form CaF_2 on the surface at concentrations higher than the 40 mg/L. The studies supported by zeta potential changes suggested that removal of fluoride by calcium carbonate occurs through adsorption (prominent at lower concentrations of fluoride) followed by surface precipitation of calcium fluoride which increases with increasing concentration of fluoride, i.e., 5–40 mg/L [68]. The Fig. 2 represents the changes in calcite surface after adsorption.

Magnesium oxide NPs, in several studies, have been praised for their high removal efficiency for fluoride [59,72]. The most prominent features of MgO with respect to environmental remediation lies in its nontoxic nature, high mechanical stability in water, and ability to efficiently sequester anions especially fluoride from aqueous media. Its reactivity is not solely attributed to large surface area and large fractions of structural defects but low co-ordination site also play equally important role [73]. Furthermore, in contrast to calcium oxide (as discussed earlier), MgO NPs' fluoride scavenging activity is only greatly hampered by presence of carbonate [74] and phosphate as counter anion while for others (Cl^-, NO_3^-, SO_4^{-2}) no considerable impact was observed [75]. Even the presence of Cr(VI) doesn't hamper the uptake capacity of MgO [74]. The comparison of fluoride uptake activity among various magnesium-based NPs follows the order: $Mg(OH)_2$ (2.1 mg/g) < activated MgO (2.3 mg/g) < MgO (9.0 mg/g) < MgO/sand (19.5 mg/g) which depicted that among its counter parts MgO NPs (particle size 17–60 nm) have comparatively high adsorption rates which are enhanced when they are supported on sand particles [75] or loaded onto ϒ-alumina (Particle size 5–30 nm; adsorption efficiency of 95.75%) [76] probably due to increase in surface area.

In addition to increasing the surface area by introduction of support material, decrease in the particle size of MgO NPs (3–7 nm) or enhancing porosity as in case of porous magnesia nanoplates enables quick reduction in fluoride ion concentration as depicted by improved adsorption efficacy of adsorbents, i.e., 20.66 and 185.5 mg/g, respectively. Both of these have been found to work effectively over a pH range of 6.5–8.5 [77,78]. As was the case with CaO, hydroxyl exchange followed by formation of MgF_2 has been suggested to be adsorption mechanism for fluoride removal through magnesium oxide [75,79] as given in Eq. (2).

$$MgO_{(s)} + H_2O_{(l)} + F^-{}_{(aq)} \rightarrow Mg(OH)F_{(aq)} + OH^-{}_{(aq)} \quad (2)$$

However, in the work presented by Jin et al. [77] besides hydroxyl functional group exchange, fluoride adsorption was attributed to the formation of surface carbonates as a result of interaction of atmospheric carbon dioxide with MgO. In case of magnesium oxide (5–25 nm) NPs loaded ϒ-alumina [76], spectroscopic studies revealed that fluoride anion attaches on to the surface of alumina by replacing hydroxide group without Al-O-Al bonds cleavage. As the fluoride concentration increase, Al-O-Al bridge breaks itself to attach more fluoride anions under the strong electronegativity effect caused by fluoride [80]. The same adsorption mechanism goes for nano alumina.

The fluoride adsorption potential of the binary systems like binary metal oxide (Mg (II)/La (III)) and magnesium oxide/magnesium carbonate nano adsorbents was also investigated and it turned out that the later having flower like morphology is more effective candidate for removing fluoride from water with an efficiency of 83%–90% and uptake capacity of 300 mg/g at room temperature and neutral pH. In the former case, needle like shape material, that was achieved from periclase mineral and lanthanum nitrate by using hydrothermal process, resulted in only 63.11 mg/g adsorption capacity [81,82]. In case of mix oxide and carbonate system, the adsorption mechanism was subject to hydroxyl and carbonate groups exchange on the adsorbent surface. The data of study was well fitted to Freundlich isotherm and pseudo-second-order kinetic models [82].

Adsorption of fluoride was also impacted considerably by varying adsorbent's synthesis experimental conditions as shown by a study conducted to investigate fluoride removal efficiency of cerium (IV)-iron (III) binary metal oxide from water. These Ce-Fe binary oxide NPs were prepared by using different Ce:Fe ratios (3:1, 1:1, 1:3), pH (8.5, 10, 11) and calcination temperature (300 °C, 450 °C, 600 °C). The results of adsorption study indicated that higher Ce to Fe ratio (3:1) and pH 10 is much more feasible for fluoride uptake. The former was attributed to formation of NPs with higher BET surface area while later favors formation of metal oxides which is not possible at lower pH values. In case of calcination, metal hydroxides are decomposed resulting in reduction of active sites and consequently limiting fluoride uptake. Adsorption mechanism in this case as well was attributed to ion exchange between fluoride and Ce-OH groups of the sorbent. Interference of counter anions is also suggested to impact the fluoride removal based on ions affinity to the adsorbent which in turn is strongly dependent on the anions' charge to radius (z/r) ratio; both of which obeys the order: $CO_3^{2-} > HCO_3^- > SO_4^{2-} > NO_3^- > Cl^-$ [83].

Owing to very small size of NPs, adsorption experiments suffer from difficulty in separation of NPs from water treatment matrices that make their practical applications very limited with respect to decontamination of the drinking water. However, magnetic NPs address the shortcomings presented by nonmagnetic adsorbents and have been found very effective to be applied in the field of preconcentrations and decontamination of environmental pollutants.

Iron and its oxides, hence being very famous materials in water treatment studies [84] due to their ease in removal and hence are exploited for their fluoride scavenging activity. Ferromagnetic γ-Fe_2O_3 NPs having size of 5–20 nm showed rapid adsorption of fluoride in just 15 min resulting in 95% removal (~3.65 mg/g). The adsorption is highly pH dependent and decreases considerably as pH rises above 4.5. The adsorption mechanism suggests direct bonding of fluoride with the iron atom which passes through various transitional states, i.e., $Fe_2(OH)_6(H_2O)_4$, $Fe_2(OH)_5(H_2O)_4F$, and $Fe_2(OH)_4(H_2O)_4F_2$ clusters as affirmed by computational studies and verified by FT-IR [61]. Cluster modeling affirms that structural transformation of iron oxide from γ-FeOOH to magnetic γ-Fe_3O_4 by ethanol treatment facilitated high adsorption of fluoride which enhance from 10.9 to 62.3 mg/g with remarkable increase in surface area (i.e., 12.6–150.4 m^2/g). The studies also revealed that change in pH during synthesis and heat treatment of γ-FeOOH influence the final structure of ethanol treated γ-Fe_3O_4, which hence determines the adsorption potential. It was observed that heat treatment and increase in pH above 9.0 had negative relation with the sorption capacity of the adsorbent [85].

In an essence to increase the adsorption potential of magnetic nanomaterials, researchers opted for amalgamation of iron with other metals and materials or opted for various treatments. Surface treatment and modification of Fe_3O_4 NPs (~20 nm in diameter) with tetraethylenepantamine appreciably increased removal of some inorganic pollutants including fluoride from water. The sorption efficacy of NPs was reported to be 163.9 mg/g in the pH range of 2–11 [86]. Iron oxide loaded Al_2O_3 nanocomposite (300 nm; qe = 4.82 mg/g) compared to zero-valent NPs (200 nm; qe = 1.47 mg/g) registered an increase in fluoride uptake capacity [87]. Further enhancement in absorption was observed by doping of sulfate into Fe_3O_4/Al_2O_3 nanocomposite (15–20 nm; qe = 70.4 mg/g). In later case, displacement of SO_4^{-2} by F^- ion and development of an inner sphere fluoride complex at Al^{3+} center is suggested to be responsible for fluoride removal (Eqs. (3) and (4).

$$\equiv Al-OH_2{}^+{}_{(s)} + F^-{}_{(aq)} \leftrightarrow \equiv Al-F_{(s)} + H_2O_{(aq)} \quad (3)$$

$$\equiv Al-OH_{(s)} + F^-{}_{(aq)} \leftrightarrow \equiv Al-F_{(s)} + OH^-{}_{(aq)} \quad (4)$$

Elemental analysis depicted a significant reduction in atomic ratio of sulfur (4.7%–1.74%) with consequent elevation in atomic ratio of fluoride (0%–9.89%) demonstrating a successful anionic replacement of SO_4^{-2} with F^- [88].

Significant rise in fluoride uptake potential was observed for $Fe_3O_4@Al_2(OH)_3$ core shell NPs prepared by coprecipitation method. The adsorption was directly dependent on the ratio of $Fe_3O_4:Al_2(OH)_3$ which follows the order: 2:3 (51.28 mg/g), 2:5 (88.49 mg/g), and 2:6 (91.74 mg/g) [89]. Similar impact of ratio was also observed in case of second-line ferrihydrite (FeO(OH))@bayerite (i.e., α-$Al(OH)_3$) nanocomposite which at ratio of 1/20 demonstrated highest adsorption potential as compared to 1/5 and 1/2.5 M ratios. Under optimized conditions, 124.8 mg of F was taken up by the nanocomposite which was suggested to occur due to 2-tiered structure of the composite consisting of large microplates (200 nm) and small NPs (50 nm). Spectroscopic analysis exhibited that fluoride sorption occurs prominently by its substitution with sulfate anions present at sorbent surface [90]. Loading of silver into the Fe_3O_4 also play a prominent role in defluoridation that follows the order $Fe^0 < Fe_3O_4 <$ Fe:Ag (1:3) < Fe:Ag (1:1) < Fe:Ag (3:1) [91].

Changes in molar ratio in ternary metal oxides, analogous to binary metal nanocomposites, have predominant effect on adsorption capacity (as illustrated in Y-Al-Zr laminated composite prepared by coprecipitation approach, Fig. 3A). In addition to that, individual metal and synergistic effects between metals also play considerable role in defluoridation that are much intense in case of ternary metals as compared to binary-metals as depicted by Fig. 3B. Defluoridation mechanism (Fig. 3C) in case of Y-Al-Zr laminated composite was attributed to ion exchange, mesoporous diffusion, and electrostatic interactions. In spite of high surface area, i.e., 256.6 m^2/g and very small pore diameter (2.2 nm), the material was able to adsorb 18.9 g F^-/g of adsorbent [116] in contrast to barium (Ba) metal loaded calcium aluminate powdered (qe = 125 mg/g) nano matrix having small surface area (25.667 m^2/g) with particle size of 40–45 nm [130]. In both cases, the data fit well to

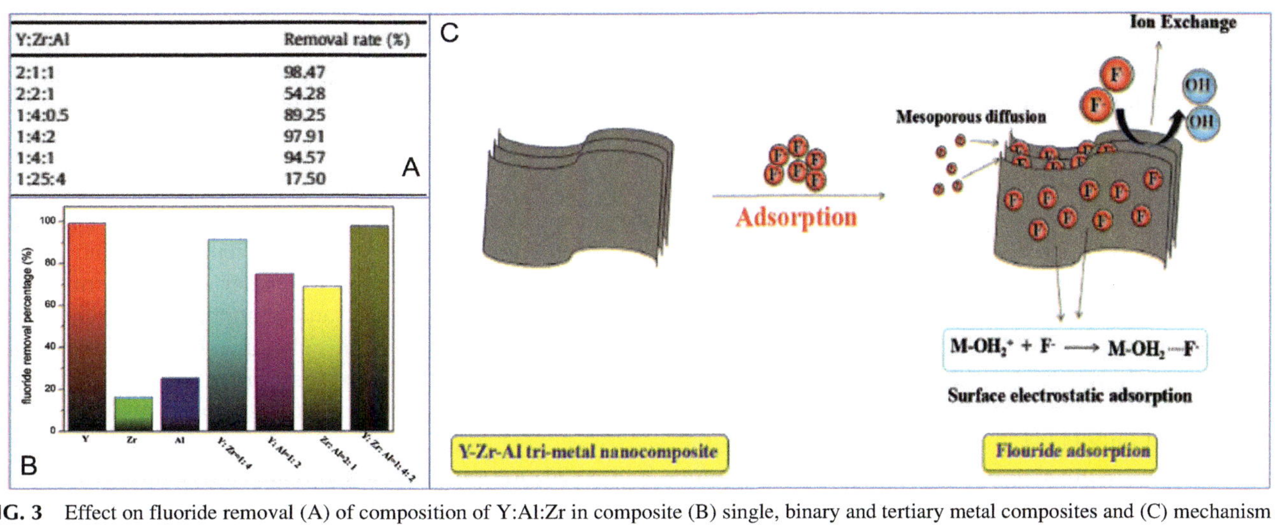

FIG. 3 Effect on fluoride removal (A) of composition of Y:Al:Zr in composite (B) single, binary and tertiary metal composites and (C) mechanism of defluoridation by using Y-Al-Zr composite [116].

pseudo-second order kinetic model but in case of adsorption models, Langmuir best explain the adsorption phenomena in case of Y based composite while for Ba based composite Freundlich isotherm holds good [116,130].

Multifunctional double layered hydroxide materials (LDHs) with co-occurrence of divalent and trivalent ions, large number of functional active sites and large interlayer spacing makes them a prominent material that can be employed as an excellent ion exchanging resins, adsorbent and magnetic composites [112]. These LDHs have great affinity for a number of anions, which is greatly dependent on charge and ionic radius, and for monovalent anions, the affinity order is as follows: $OH^- > F^- > Cl^- > Br^- > NO_3^- > I^-$. The competitive adsorptive capacity of calcinated $Mg/Al/CO_3$ layered double hydroxide for fluoride and phosphate anions was studied by Cai et al. [100]. The spectroscopic results have exhibited a layered morphology along with particles clusters having lateral size of approximately 50–100 nm. Adsorption of fluoride and phosphate decrease significantly in the presence of each other. The sorption data followed the Langmuir isotherm and pseudo-second order kinetic model. The adsorption efficacy of nanocomposite for phosphate and fluoride was 1.79 and 1.94 mmol/g, respectively, in single unit system, which were slightly higher than their respective sorption efficiencies as 1.42 and 0.63 mmol/g for phosphate and fluoride in binary unit system [100].

Plant extracts have vast utilization in green synthesis processes, and serve the purpose of reducing and stabilizing agents in the fabrication of metallic NPs [94,99,134]. The source of plant extracts also greatly influences the NPs characteristics and hence adsorption activity. Generally, these plant based materials depicted low adsorption potential as was the case of *Moringa oleifera* leaves' aqueous extract mediated α-Fe_2O_3NPs with surface area and particle size of 99.79 m^2/g and <100 nm, respectively, which resulted in 1.40 mg fluoride uptake/gram of adsorbent [150]. On other hand, rod-shaped Ag NPs (particle size, 28 nm) biosynthesized by using *Xanthium strumarium* leaves aqueous extract was just able to adsorb fluoride up to 1.5 ppm from simulated wastewater samples [115]. The hybrid nanoadsorbent prepared by grapefruit red (*Citrus paradisi*) peel extract mediated Ag-MgO NPs integration with nanohydroxyapatite and chitosan polymer showed almost similar fluoride uptake potential (qe = 6.86 mg/g) [96]. In an effort to enhance the adsorption and aid separation Mondal and Purkait [132] impregnated clove extract assisted iron NPs on to pH responsive polymer-based membrane comprising of polyethylene glycol and humic acid. The nanocomposite exhibited enhanced fluoride rejection under acidic conditions (pH 4.0) and was reported as 65.8%, 78.56%, and 84.4% against 5, 10, and 20 ppm feed concentrations, respectively. The membrane resulted in high fluoride rejection at low pH (4.0) and high pH (12) which in former case was mainly attributed to presence of strong electrostatic interactions while in later case deprotonation of polymer-based membrane results in pore size shrinkage that enables capturing of fluoride ions [132].

Bio-fabricated TiO_2 NPs synthesized by using bacterial strain "*Bacillus NARW11*" was subjected to calcination at different temperatures (600 °C, 700 °C, and 900°C) to yield various phases of titanium dioxide NPs, i.e., anatase, rutile, and mix of anatase and rutile. Among these, only anatase NPs demonstrated some affinity for fluoride (qe = 0.85 mg/g) [156]. Bio-synthesized hybrid membranes have also been tried to enhance their effectiveness toward removal of Fluoride. For the purpose, bacteria "*Bacillus licheniformis*" mediated alumina-titania nanocomposite was decorated on thermoplastic polyurethane (PU) nanofibrils. The nanomembranes have been able to remove 1.9 mg F^-/g from water [155].

Biominerals and biopolymer–based nanoadsorbents

Biosorption has always been one of the most efficient, simple and cheap methods used for the decontamination of pollutants from water. In recent times, biopolymer-based nanocomposites have gained paramount attention to be used as a potential defluoridation agent for water purification. Table 2 enlists some biomaterials with their defluoridation potential.

Hydroxyapatite-based nanoadsorbents

Hydroxyapatite (HA), a well-recognized member of calcium phosphate family, is highly biocompatible and nontoxic in nature. These merits make it highly suitable to be used as an adsorbent for quick and efficient removal of fluoride from water. The synthetic protocols and raw material usage influence adsorption capacity of the material in concern. The rod like shaped HA NPs (200 nm) prepared by chemical method by use of calcium hydroxide and orthophosphoric acid resulted in sorption capacity of 1845 mg F^-/kg [152] as compared to the one prepared from calcium nitrate and ammonium dihydrogen phosphate, i.e., 1296 mg F^-/kg [153]. On other hand, ultralong HA nanowires (length = 40–60 μm, average diameter = 70–100 nm, surface area = 77.05 m^2/g) prepared by sonochemical method leads to adsorption of 40.56 mg F^-/g of adsorbent only [114].

Generally, removal of fluoride through hydroxyl group follows adsorption and anion exchange mechanisms resulting in the formation of stable fluorapatite ($Ca_5(PO_4)_3F$) and mixed fluoridated $Ca_5(PO_4)_3(OH,F)$ [114] as indicated by presence of characteristic O-H...F bond in FTIR spectra [114,152] and reduction in surface hydroxyl group in case of HA wires from 70.5% to 66.2% after adsorption owing to replacement of hydroxyl groups with fluoride [114]. pH seems to play vital role in

TABLE 2 Biomaterials-based nanoadsorbents and their defluoridation capacity.

Nanoadsorbent	Defluoridation capacity (mg/g)	pH	Fluoride removal mechanism	Adsorption and kinetic models	References
Synthetic hydroxyapatite NPs	1.845	3.0	Adsorption, Ion exchange	Pore diffusion, Pseudo-second-order kinetics, Langmuir isotherm	[152]
Nano hydroxyapatite fabricated chitin	2.84	≤5.9	Adsorption, Ion exchange	Pore diffusion, Pseudo-second-order kinetics, Langmuir isotherm	[154]
Binary (La/Zr) metal hydrous oxide NPs decorated biosorbent	32.5–37.2	3.0	Adsorption, Ligand exchange	Intraparticle diffusion, Pseudo-second-order kinetics, Langmuir isotherm	[149]
Zirconium tungstophosphate modified chitosan	2.025–2.142	3.0	Adsorption, Ion exchange, Surface complexation	Langmuir isotherm	[164]
Magnetic iron (II, III) mixed metal oxide entrapped polypyrrole nanocomposite	17.63–22.31	6.5–7.5	Adsorption, Ion exchange, Surface complexation	Pseudo-second-order kinetics, Freundlich, Freundlich–Langmuir isotherm	[98]

defluoridation process as it is responsible for altering the charge on surface of HA and establishing electrostatic interaction between the hydroxyl group and fluoride ion [114,152]. Defluoridation process seems to be hampered by the presence of counter ions like HCO_3^-, CO_3^{2-}, and PO_4^{3-} resulting in appreciable decrease in desorption capacity while Cl^-, NO_3^-, and SO_4^{2-} show nonsignificant impact [114,152].

The role of waste materials' derived HA as a de-fluorinating agent has also been exploited. Cattle bones sourced char ($104\,m^2/g$ surface area of and 11.1 nm pore diameter), containing HA as a chief constituent, resulted in low uptake of fluoride ($qe = 7.74\,mg/g$) [51]. Defluoridation also occurs in these materials through anion exchange between fluoride and hydroxyl and phosphate functional groups contained in bone char. In addition to that, the existence of strong electrostatic interactions between surface charges of bone char and fluoride ions in the aqueous media facilitates fluoride sorption. The later process can be further explained by Eqs. (5) and (6) [92];

$$\equiv Ca-OH_2^+ + F^- \leftrightarrow \equiv Ca-OH_2-F \quad (5)$$

$$\equiv POH_2^+ + F^- \leftrightarrow \equiv POH_2F \quad (6)$$

In a defluoridation study carried on HA NPs, derived from microwave irradiated phosphogypsum industry waste, resulted in adsorption capacity of 40.8 mg/g and the data was best fitted to Langmuir–Freundlich isotherm and pseudo-first-order kinetic models. Results of the study illustrated that defluoridation mechanism was highly dependent on pH and adsorbent performs best at pH levels below 6. Computational studies carried out explicated that pH has direct impact on the embedding calcium triangles which are dissolved at lower pH exposing the apatite hydroxyl ions to protonate. Hence, the phenomenon of fluoride adsorption is mainly governed by electrostatic interactions and hydrogen bonding between negatively charged fluoride ions and positively charged adsorbent surface. Furthermore, the adsorption of fluoride ions is accompanied by replacement leading to formation of Fluorapatite (Fig. 4A) and consequently discharge of hydroxyl ions which is illustrated by change in pH of the solution as the course of reaction proceeds (Fig. 4B) [177]. Similar formation of fluorapatite was observed on the surface of shell when defluoridation potential of egg shells was exploited. Simple procedure by use of egg shell resulted in removal of 47.9 mg of fluoride/g sorbent [169].

Chitosan-based nanoadsorbents

Chitin is the second most abundantly occurring polysaccharide in nature after cellulose. Chitosan is a biodegradable polymer of poly[(1,4)-β-linked 2-amino-2-deoxy-D-glucose] or (polyglucosamine), which is basically produced by the deacetylation of chitin. The adsorption potential of various nanomaterials is reported to enhance by introducing either chitin or

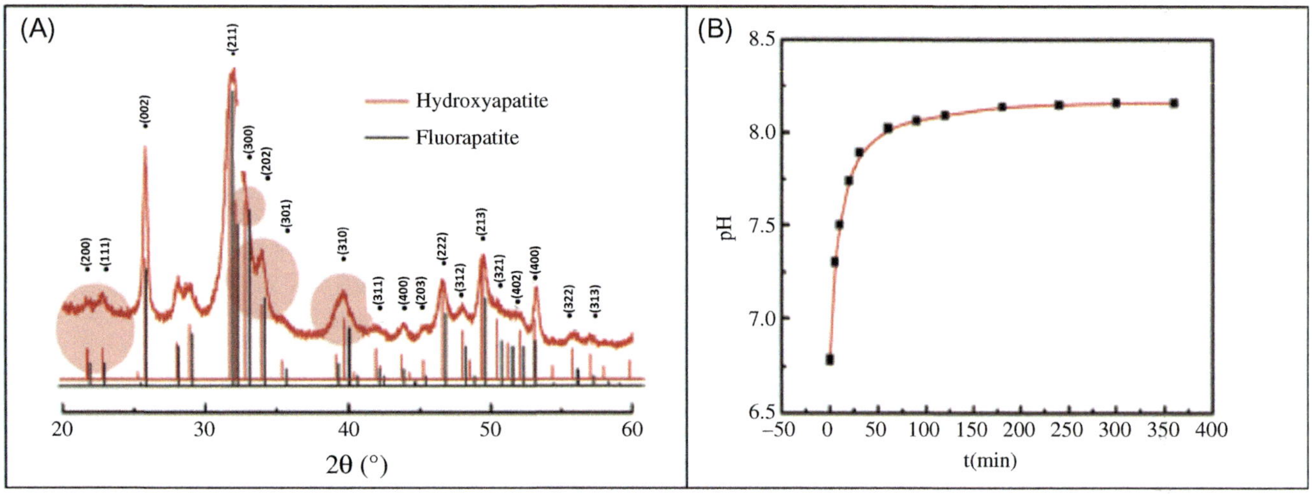

FIG. 4 (A) Powder XRD spectra of HA before and after fluoride adsorption (B) increase in pH of solution during the course of adsorption with respect to contact time [177].

chitosan in it. A hybrid nanocomposite made up of nano HA fabricated chitosan has been found to remove more fluoride, i.e., 1845 mg/kg as compared to the independent nano HA and chitosan with adsorptive capacities of 1296 and 52 mg/kg of adsorbent, respectively [153]. Likewise, mix metal composite of Fe/Mn/Al oxyhydrous oxide (20–30 nm) incorporated chitosan matrix obtained by utilizing laterite clay and steel industry waste enabled 47.6% fluoride removal in just 1 min, while the Fe/Mn/Al oxyhydrous oxide without chitosan resulted in 22.6% removal [101].

The comparative analysis of effect of metal ion doping into the chitosan matrix influenced adsorption potential to minute level. The adsorption potential (mg/g) of aluminum impregnated chitosan, iron impregnated chitosan and Al-Fe doped chitosan hydrogel follows the order: 31.16 (Al-CS-Fe) < 30.69 (Al-CS) < 25.65 (Fe-CS). There are three most plausible mechanisms governing behind defluoridation by Al-chitosan-Fe(III) nanocomposite, i.e., (i) fluoride anion replacement by anion of amine salt (Eq. 6); (ii) electrostatic interaction arising from positively charged active sites on adsorbent surface and negatively charged F^- ion (Eqs. 7 and 8) and (iii) ligand or ion replacement between fluoride and metal center (Eqs. 9 and 10).

$$NH_3{}^+Cl^- + F^- \leftrightarrow NH_3{}^+F^- + Cl^- \tag{6}$$

$$-NH_2 + H^+ \leftrightarrow NH_3{}^+ \ldots F^- \tag{7}$$

$$-OH + H^+ + F^- \leftrightarrow OH^+ \ldots F^- \tag{8}$$

$$M-OH_2 + F^- \leftrightarrow M-F + H_2O \tag{9}$$

$$M-OH + F^- \leftrightarrow M-F + OH^- \tag{10}$$

The prepared adsorbent had the advantage of increase in sorption potential after regeneration owing to enhancement in the number of active functional sites for fluoride binding after treatment with strong alkali [129]. Another study performed by Prabhu and Meenakshi [142] also dealt with the effect of metal ions doping into the chitosan-based matrix. The glutaraldehyde crosslinked Zr(IV) uncoated chitosan-silica gel nanocomposite (244 nm) synthesized for fluoride detoxification from aqueous media portrayed better removal properties (4.35 mg/g) as compared to undoped nanocomposite, i.e., silica gel/chitosan (3.40 mg/g). The adsorption data was best described by the intraparticle diffusion model and adsorption process pseudo-second-order kinetics.

A bio sorbent composed of zirconium hydroxide NPs supported on pomelo peel displayed maximum adsorption capacity of 19.2 mg F^-/g in single component system and 16.4 mg/g in the presence of coexisting anions. The results of study indicated that fluoride adsorption was mainly dominated by zirconium hydrous oxide NPs dispersed in pomelo peel and the electrostatic interactions had been found to play major role in fluoride adsorption mechanism. The composite exhibited a significant chemical stability for fluoride uptake over a broad range of pH without any noticeable organic matter dissolution and zirconium leaching (Yali [111]).

Cellulose-based nanoadsorbents

Cellulose is the most abundant biopolymer in nature and has been extensively tested for the detoxification of several organic and inorganic pollutants from the environment. Likewise, other nanoadsorbents discussed earlier, adsorption in these cellulose-based adsorbents is through ion exchange. Zirconium doped carboxymethyl cellulose nanocomposite exhibited removal of fluoride (47.7 mg/g) under acidic pH and room temperature through hydroxide, carboxyl and Zr^{4+} active sites [168]. Incorporation of metallic dopants into the cellulose matrix to result in cellulose-supported Fe_3O_4NPs doped cerium (IV)-zinc (II) double hydrous oxide enabled much improved adsorption (167.62 mg of fluoride/g sorbent dose) owing to vital role played by multiple mechanisms, i.e., electrostatic interactions, ion exchange, and surface complexation. The later can lead to formation of outer or inner sphere complex depending on the fact that that fluoride ion is exchanged by another anion (Eq. 11) or a ligand (Eq. 12).

$$[X(OH)_6]^+Cl^- + F^- \rightarrow [X(OH)_6]^+F^- + Cl^- \text{ (Outer sphere complex)} \tag{11}$$

$$X(OH)_3 + F^- \rightarrow X(OH)_2F^- + OH^- \text{ (Inner sphere complex) } \{X \text{ is } Zn, Ce, Fe\} \tag{12}$$

The concept of fluoride adsorption could be further explained by "theory of hard soft acid base," according to which the metal ions are regarded as hard acids as they contain Zn^{2+}, Fe^{3+}, H^+ and Ce^{3+}, and tend to bind with a hard base like fluoride ion because of its high electronegativity and low polarization properties [93].

A hybrid composite consisting of nano hydroxyapatite decorated cotton cellulose with surface area of 76.257 m^2/g was synthesized through *in situ* hybridization process and the nanocomposite was capable of adsorbing 4.20 mg of fluoride/g sorbent dose. The adsorption data was well described by Freundlich and Langmuir isotherms and sorption process followed the pseudo-second-order kinetics. Adsorption and ion exchange are the most plausible removal mechanisms for fluoride by nanocomposite [174]. Surface fictionalization of the adsorbents also has considerable effect on the uptake capacity of the adsorbent. A study carried out on cellulosic nanofibers, derived from paper pulp waste, supports the fact that nanofibers in pristine state recorded to have negligible defluoridation capacity which considerably increase after surface modification with positively charged quaternary ammonium functional group [148]. This was in accordance with fluoride decontamination capacity reported for modified chitosan (3–22.4 mg/g) [173].

Alginate-based nanoadsorbents

Alginate, a biopolymer, derived from brown seaweed is a polyuronide type polysaccharide made up of varied numbers of β-D-mannuronic acid and α-L-guluronic acid units that are interlinked with the help of β-1-4 and α-1-4 bonding. The number and distribution sequence of these units along the polysaccharide chain vary a lot according to the source of material used. The fabrication of alginate-based nanocomposite for water purification is becoming popular owing to its unique functional properties such as biocompatibility, nonimmunogenicity, nontoxicity, and excellent biodegradability.

Amalgamation of alginate with other biomaterials that themselves have reported high affinity toward fluoride seems to upgrade the adsorption potential of both materials. Alginate encapsulated HA registered an increase in fluoride adsorption capacity (3.87 mg/g) which was higher than the independent biomaterials, i.e., 0.68 mg/g for HA and 1.29 mg/g for calcium alginate in the same contact time, i.e., 30 min. The morphological attributes of the nanocomposite (i.e., porous structure and irregular surface) (Fig. 5A) that affect the uptake of fluoride experienced considerable changes after fluoride adherence and diffusion (Fig. 5B). As reported previously, the fluoride uptake in nanocomposite is governed by adsorption followed by electrostatic interaction and ion exchange (Fig. 5C) [139].

Doping of metal ions in alginate alter its defluoridation capacity considerably. Alginate impregnation with lanthanum, aluminum, Fe_2O_3, and Zirconium resulted in fluoride uptake of 197.2 [172],75.2 [120], 58.24 [104] and 32.79 mg/g [144], respectively. Incorporation of binary metal oxides like Ce-Al metal oxides reduce adsorption potential to 1.438 mg/g [157] which enhance considerably as Al is replaced by Zr (i.e., Ce-Zr@alginate) resulting in uptake of 137.6 mg F^-/g [103]. In case of ternary metals' immobilization onto alginate, the selection of metals plays imperative role in adsorption of fluoride. Doping of Mg/Al/Ce [166] and Mg/Al/La [167] leads to almost comparable defluoridation capacity, i.e., 26.12 and 30.96 mg/g which is substantially increased (333 mg of fluoride/g) when oxides of Fe/La/Ni [145] were embedded in alginate beads.

The plausible mechanisms suggested for fluoride removal by alginate-based nanocomposites are ion exchange and electrostatic interaction [145,157].

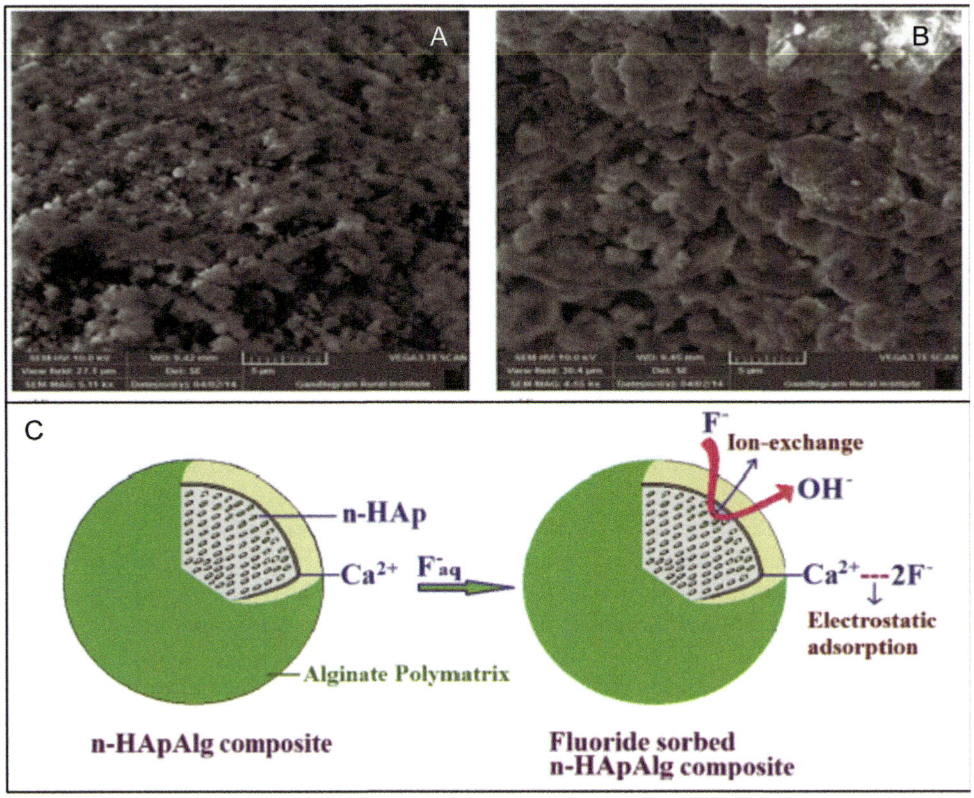

FIG. 5 Alginate encapsulated HA nanocomposite SEM image (A) before adsorption, (B) after adsorption and (C) its mechanism of fluoride ion removal [139].

Other biopolymer-based nanoadsorbents

Collagen, animal skin derived widespread naturally occurring biopolymer, possess hydrophilic characteristics making it water insoluble. The collagen molecule consists of three polypeptide strands with triple helix that are aggregated with the help of hydrogen bonding to structure a collagen fiber [110]. The large numbers of functional groups present in collagen have strong affinity toward metal ions and this merit has attracted the use of collagen fiber as an excellent supporting medium for metals ion to construct versatile NPs and composites. Zr (IV) metal fabricated collagen biopolymer with excellent mechanical stability was used for detoxification of fluoride from water. Under slight acidic conditions (pH 5.0) fluoride uptake potential of 2.29 mmol/g was observed, which gradually dropped down to 1.19 mmol/g under alkaline pH (10.0). The fluoride adsorption on composite surface occurred in single layer and data was best fitted to Langmuir isotherm and pseudo-first-order kinetic models [126]. Another organic biopolymer, polypyrrole alone [119] or its composite [98,118] have been studied for decontamination of fluoride anion from aqueous solutions. The hydrated stannous oxide anchored polypyrrole nanoadsorbent with surface area of 65.758 m^2/g has been used for effective remediation of fluoride. The maximum sorption capability of composite was reported as 26.16–28.99 mg/g under pH range of 3.5–8.5. The hydrated stannous oxide present on nanocomposite played a pivotal role in fluoride adsorption whereas polypyrrole contributed in controlling the point of zero charge of adsorbent [140]. A cerium fabricated fibrous protein nanocomposite with particle size 18.6–21.28 nm has been investigated for its potential to remove phosphate, arsenate and fluoride in batch experiments. The fabrication of cerium metal ion on fibrous protein not only enhances the point of zero charge up to pH value of 9.0, but also improved the adsorptive capacity of nanoadsorbent by increasing its surface electrostatic attraction for these anionic species. The composite was capable of adsorbing fluoride up to 17.5 mg/g at pH value 3.0. The ligand replacement and hydroxide group present on the surface of composite both played an eminent role in the fluoride adsorption [107].

Some others biomaterials-based composites such as magnetic iron (II, III) oxide NPs entrapped carboxylated chitosan matrix [131], silver NPs decorated biopolymer scaffold [123] and bimetal (Al^{3+}, Zr^{+4}) oxide doped pectin beads [179] have also been used for effective remediation of fluoride from water.

Carbonaceous nanomaterials

Activated carbon and biochar-based nanoadsorbents

Biochar is a solid carbonaceous material resulted from thermal decomposition of agricultural waste and by-products in a closed environment with little or no supply of oxygen [105,162]. These carbon-based materials have a long history of utilization for cheap and efficient detoxification of environmental contaminants from water. Some carbon-based nanocomposites along with their defluoridation capacity are listed in Table 3.

Granular activated carbon (GAC) with defluoridation capacity of 1.99 mg/g is managed to be enhanced five times (9.98 mg/g) when activated carbon is doped with 0.05 M Lanthanum oxy-hydrous oxide. Adsorption potential of the prepared nanocomposites is highly dependent on the loading concentration of La (0.01–0.2 M). At lower concentration of doping, i.e., up to 0.05 M the sorption capacity improve but after that increase in doping concentration leads to reduction in fluoride uptake owing to obstruction of active sites. Adsorption on La (III) decorated granular activated carbon was facilitated by fluoride ion was attraction of F^- by La^{3+} and displacement of hydroxyl group with fluoride leading to formation of lanthanum fluoride as presented in Fig. 6A [163]. Incorporation of Zirconium moiety into active carbon also enhances its adsorption capacity (5.4 mg/g) by facilitating fluoride ion attachment with protonated hydroxyl group (H_2O^+) leading to removal of water molecule along with formation of zirconium fluoride through series of fluorinated zirconium species, i.e., mono-, di-, tri-, tetra-, and penta fluoro zirconium species (Fig. 6B) [135]. Similarly, diffusion of hematite and magnetite in Douglas fir derived biochar and boehmite loading into "*Ipomoea carnea*" weed stem-based charcoal enhanced fluoride abatement capacity of 9 and 46 mg/g, respectively [108,147].

Carbon nanotubes-based nanoadsorbents

Carbon nanotubes (CN), single walled (SWCNT) and multiwalled (MWCNT), have gained attention of researchers owing to their high surface area and porous structure. These CNT have been extensively exploited in environmental applications especially water and wastewater treatment [128].

Remediating potential of SWCNT (length 10 μm, diameter 1–2 nm, surface area 700 m^2/g) and MWCNT (length 10 μm, diameter 10–30 nm, surface area 270 m^2/g) was exploited with respect to their efficacy toward fluoride. It was observed that owing to higher surface area and smaller diameter former (2.8 mg/g) performed slightly better than the

TABLE 3 Carbon-based nanocomposites and their defluoridation capacity.

Nanoadsorbent	Defluoridation capacity (mg/g)	pH	Fluoride removal mechanism	Adsorption kinetics	References
Lanthanum immobilized *sargassum*-derived biochar	94.34	3.0–9.0	Ion exchange, Surface complexation, electrostatic interactions	Intraparticle surface diffusion and Pseudo-second-order kinetic models, Langmuir isotherm	[175]
Aluminum/Iron mixed metal hydroxide impregnated Graphene oxide	27.8	4.5–5.5	Adsorption, Ion exchange,	Intraparticle surface diffusion and Pseudo-second-order kinetic models, Langmuir isotherm	[117]
Zr (IV) doped porphyrin crosslinked modified iron (II, III) oxide nanocomposite	Not reported, 92% fluoride removal	5.5	Ion exchange	Not reported	[141]
Aluminum oxide doped carbon nanotubes	9.6	3.0–9.0	Adsorption, Ion exchange	Freundlich isotherm	[125]
Alumina-iron (II, III) oxide- expanded graphite nanosandwich	2.19	2.0–10.0	Ion exchange	Intraparticle diffusion and Pseudo-second-order kinetic models, Langmuir isotherm	[170]

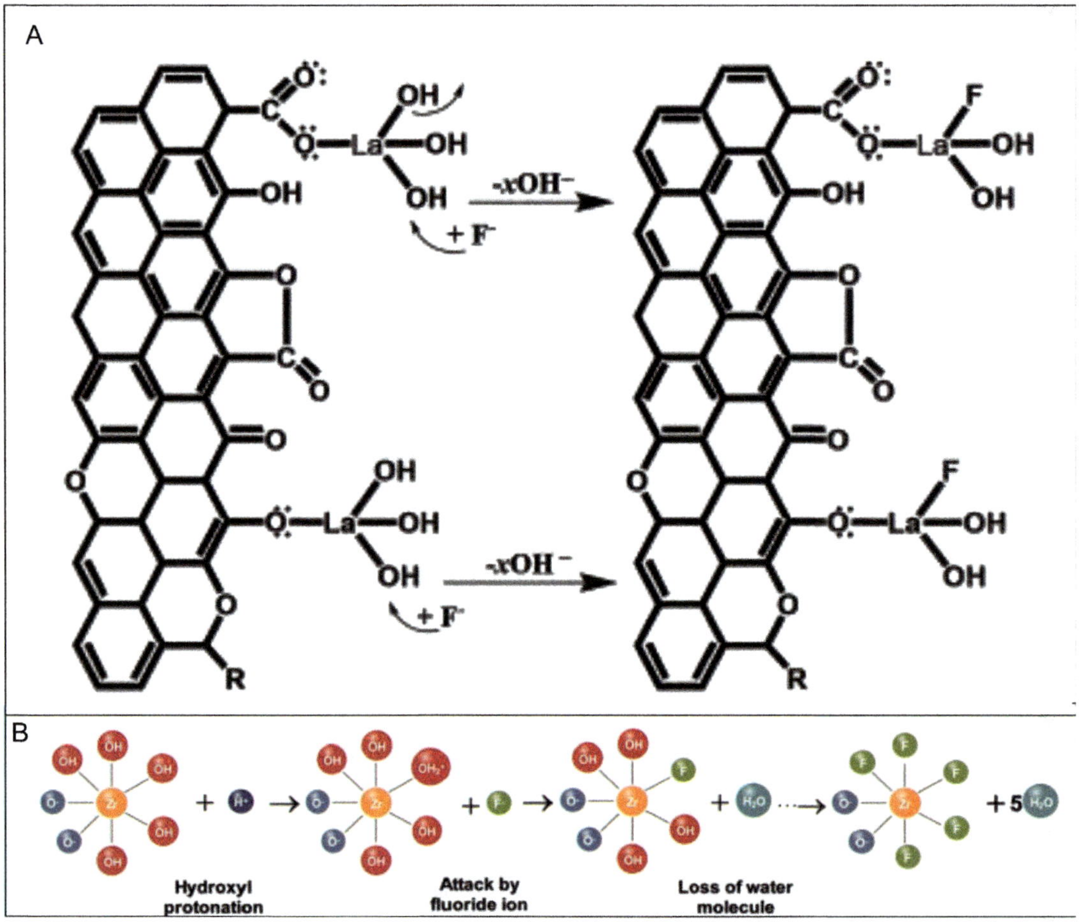

FIG. 6 Defluoridation mechanism on activated carbon doped with (A) Lanthanum and (B) zirconium [135,163].

MWCNT (2.4 mg/g) [106]. SWCNT (length 5–15 μm, diameter 1–2 nm, surface area 712.9 m^2/g) with comparatively higher surface area enables better adsorption capacity 16.63 mg/g. Defluoridation capacity is highly impacted by the pH as it alters the surface charge. At lower pH values positive surface charge is induced on the surface which facilitates negative ion interaction and hence adsorption of fluoride [97].

Impact of surface area and size of CNT is also significant from the studies carried out on aligned CNT (length 200 μ, diameter 20–80 nm) synthesized by catalytic degradation of xylene in the presence of ferrocene as catalyst. The CNT obtained with surface area 74 m^2/g resulted in fluoride uptake capacity of 4.2 mg/g [124].

Modification of CNT is also helpful in achieving improved uptake capacity. MWCNT modified with HA helps in fluoride sorption capacity of 39.22 mg/g [146]. Similar findings were recorded by carbon coated CNT synthesized from coconut shell. The same samples prepared by hand grinding and ball milling resulted in different sorption capacity owing to changes in surface area induced by different techniques, i.e., 78 and 358 m^2/g [95].

Graphite-based nanoadsorbents

Graphene, a single layered carbon allotrope with hexagonal arrangement, and its sister compound Graphite oxide (GO), have emerged as potentially viable materials having widespread applications in development of various kinds of nanocomposites to remediate environmental pollution [180].

GO doped with TiO_2 [137] and Zirconium [143] resulted in fluoride scavenging activity of 342 and 6.12 mg/g, respectively. The one doped with γ-Al_2O_3 decreases defluoridation capacity to 4.68 mg/g [171]. Additional doping with iron enhances the sorption capacity of the nanocomposite that is dependent on the ratio of Al:Fe as well as on the type of iron oxide [117,122,127,133]. The composite prepared with equimolar ratio of both Al and Fe is recorded to have best remediation potential (28.07 mg/g) [133]. Fe_3O_4-based GO nanocomposite displayed uptake capacity of 64.72 mg/g [127] while the one hydrous iron-aluminum oxide enabled removal of 27.75 mg F^-/g [117]. In addition to that experimental data obtained for Graphene oxide-supported sodium acetate, modified goethite (α-FeOOH) NPs (particle size 21.9 nm, surface area 255.24 m^2/g) and akaganeite (β-FeOOH) (particle size 1 nm, surface area 202.60 m^2/g) suggest slightly better scavenging activity of former composite (19.8 mg/g) than that of akaganeite/GO (17.7 mg/g) [122].

In GO-based composites, adsorption is also facilitated through ion exchange mechanism involving -OH groups present on the surface [133] and hampered through presence of counter ions based on their charge to radius ratio. The interfering effect posed by various ions obeys the order $HPO_4^{2-} > SO_4^{2-} > Cl^- > NO_3^-$ [127].

Clay-based nanoadsorbents

Clays and zeolites are hydrated aluminosilicates minerals. These are cheap and have widespread occurrence in nature, and have been extensively used to sequester inorganic pollutants from aqueous solutions through ion exchange mechanism [121,161,165]. The higher affinity of cations, for example, Al^{3+} for fluoride ion has urged the researchers to investigate clay minerals as a potential adsorbent for abatement of fluoride from water. Various types of clay minerals such as bentonite, meixnerite, montmorillonite, pyrophyllite, vermiculite, and kaolinite have been extremely studied for water purification [113,151,160]. The montmorillonite (MM) and bentonite (BT) clay NPs were found very efficient in the removal of fluoride from aqueous media and the maximum defluoridation capacities calculated by Langmuir isotherm were recorded as 11.15 and 9.95 mg/g, respectively [136]. The results of this study are in accordance with the findings of [45] who suggested that hydrated cement is a promising candidate for fluoride abatement from water. Zr(IV)-loaded sodium attapulgite composite has been used as an effective defluoridation agent with the maximum adsorption potential of 24.55 mg/g. The adsorption process followed the pseudo-second-order kinetics and data was well fitted to Langmuir isotherm [178]. Teimouri et al. [159] synthesized a hybrid nanoadsorbent by incorporating chitosan in to the zirconium (IV) oxide-MM clay nano matrix with pore diameter and surface area of 0.923 nm and 62.448 m^2/g, respectively, to adsorb fluoride ions from water. The maximum defluoridation capacity was reported to be 23.0–24.8 mg/g under acidic pH (4.0) and temperature 30 °C. The respective amine group and Zr^{4+} and Al^{3+} cations present in chitosan; metal oxide and clay played the role of potent scavengers to remove fluoride anions from aqueous media. A comparison of fluoride adsorption capacity between *in situ* synthesized neat HA NPs and HA/MM clay nanocomposite was investigated. The fluoride removal capacity of HA/MM nanocomposite was 16.7 mg/g, which is much higher than the adsorption values of 2.44 and 6.7 mg/g reported for individual MM and HA, respectively. The composite exhibited improved adsorption, which is attributed to the exfoliation of MM clay layers, and introduction of HA in between these layers [109]. Iron (III) impregnated MM exploited for its fluoride adsorption and impacts of competitive effect of phosphates on its adsorption. The nanoadsorbent exhibited higher adsorption capacity for fluoride than phosphate in single unit solution, which is attributed to enhanced surface affinity of fluoride as compared to phosphate. The fluoride adsorption decreased with increase in pH. During the first 5 min the increased fluoride adsorption (50%–55%) was observed to happen and over 95% of fluoride adsorption was attained within 3 h. Whereas in binary unit system phosphate interfered the fluoride adsorption process and competed with fluoride for active sites available at nanoadsorbent surface. The HA-modified gismondine type (NaP) zeolite nanoadsorbent was prepared in different molar concentrations of HA/zeolite (1:1, 1:2, 1:4 and 1:6) for detoxification of excessive fluoride from aqueous solution. The nanocomposite prepared with 1:1 (HA/zeolite) molar ratio presented the greatest surface area (45 m^2/g) and had been found most effective in removing fluoride from water with adsorption capacity of 66.60 mg/g under pH (5–8) and temperature 25 °C. The adsorption and ion exchange mechanisms played dominant role in the removal of fluoride and data was best explained by Pseudo-second-order kinetic model [176].

The low fluoride adsorption capacity of HA has limited its practical application alone in fluoride reduction from drinking water. However, it is proven from literature that materials having large surface area and enormous number of functional active sites can be fabricated by tailoring the morphology and pore volume [102]. Moreover, the sorbent with hierarchical meso and macroporous structure allow the quick diffusion of guest molecules inside the channels, thereby promoting their access to available functional sites. Recently it has been discovered that carbon nanotubes can upgrade the photothermal efficiency of HA and in return the HA may improve the poor dispersion of carbonaceous nanotubes. The hybrid adsorbent made up of HA impregnated on carbon nanotubes (300 nm) has been tested for its efficiency to adsorb fluoride from water. The maximum fluoride detoxification potential of nanocomposite was reported to be 11.05 mg/g and adsorption and anion exchange were considered the plausible fluoride removal mechanisms. The adsorption data was best fitted to Freundlich isotherm and pseudo-second-order kinetic models. The maximum adsorption of fluoride was observed in the pH range of 3.0–6.0 followed by a decline in fluoride uptake when the pH was increased from 6.0 to 10.0 [158]. Under weak acidic conditions, the surface of HA and adsorbent become protonated with an increase in the number of functional active sites on the surface which promote the fluoride uptake. Whereas under alkaline conditions the sorbent surface becomes saturated with negative charges which restrict the diffusion of fluoride on the adsorbent and results in the drop of fluoride sorption capacity [138].

Conclusions

Water being the most vital resource for all life on this planet, may be adversely affected qualitatively and quantitatively by different human activities. Access to safe and clean water is an unconditional requirement for health and productive life. The fluoride contamination and its related health problems, is one of significant environmental challenges being faced by most of the countries in the world. Severe health issues arising from drinking water containing higher concentrations of fluoride have urged the scientists to devise more efficient materials and methods to bring down fluoride concentration to the levels that are beneficial for human health. In the current environmental practices, nanotechnology has played an incredible role for detoxification of environmental contaminants. Today the research is primarily focused on inventing low cost and environmentally benign approaches and materials for water purification. Most of the industrial and agricultural by-products are regarded as waste that require proper disposal. However, their beneficial utilization for water treatment can help in reducing the disposal cost, prevention of environmental pollution stemming from the landfill sites and making these sites free for alternative usage. Thus, these materials can better serve the purpose of low cost adsorbents in developing countries like Pakistan where operational cost is a significant parameter while making choice for an adsorbent. In this chapter, numerous low cost metal elements, biopolymers, clay minerals, and carbon-based nanomaterials for fluoride detoxification from water have been summarized. These nanoadsorbents although have exhibited good defluoridation potential; however, they must undergo further research and developmental phases in the laboratories to bring up more improvements in them in terms cost effectiveness, environmental friendliness and enhanced removal efficiencies, and there is an intense need to have their trials in field conditions for ground and drinking water purification.

References

[1] United Nations Environment Programme, Global Water Resource, 2002.
[2] K.D. Brahman, T.G. Kazi, H.I. Afridi, S. Naseem, S.S. Arain, N. Ullah, Evaluation of high levels of fluoride, arsenic species and other physico-chemical parameters in underground water of two sub districts of Tharparkar, Pakistan: a multivariate study, Water Res. 47 (3) (2013) 1005–1020.
[3] A. Farooqi, H. Masuda, N. Firdous, Toxic fluoride and arsenic contaminated groundwater in the Lahore and Kasur districts, Punjab, Pakistan and possible contaminant sources, Environ. Pollut. 145 (3) (2007) 839–849.
[4] T. Rafique, S. Naseem, T.H. Usmani, E. Bashir, F.A. Khan, M.I. Bhanger, Geochemical factors controlling the occurrence of high fluoride groundwater in the Nagar Parkar area, Sindh, Pakistan, J. Hazard. Mater. 171 (1–3) (2009) 424–430.
[5] H. Biglari, A. Chavoshani, N. Javan, A. Hossein Mahvi, Geochemical study of groundwater conditions with special emphasis on fluoride concentration, Iran, Desalin. Water Treat. 57 (47) (2016) 22392–22399.
[6] J. Chen, H. Wu, H. Qian, Y. Gao, Assessing nitrate and fluoride contaminants in drinking water and their health risk of rural residents living in a semiarid region of Northwest China, Expos. Health 9 (3) (2017) 183–195.
[7] A.M. Abdelgawad, K. Watanabe, S. Takeuchi, T. Mizuno, The origin of fluoride-rich groundwater in Mizunami area, Japan—mineralogy and geochemistry implications, Eng. Geol. 108 (1–2) (2009) 76–85.
[8] Ö. Özmen, Ş. Koç, M. Çelik, Evaluation of groundwater quality and contamination around fluorite mineralization, Kaman region, Central Anatolia, Turkey, Geochem. Int. 49 (1) (2011) 76–89.
[9] E.W. Wambu, G. Muthakia, High fluoride water in the Gilgil area of Nakuru County, Kenya, Fluoride 44 (1) (2011) 37.
[10] D. Boyle, M. Chagnon, An incidence of skeletal fluorosis associated with groundwaters of the maritime carboniferous basin, Gaspé region, Quebec, Canada, Environ. Geochem. Health 17 (1) (1995) 5–12.
[11] M.A. Armienta, N. Segovia, Arsenic and fluoride in the groundwater of Mexico, Environ. Geochem. Health 30 (4) (2008) 345–353.
[12] I.E. Nezli, S. Achour, M. Djidel, S. Attalah, Presence and origin of fluoride in the complex terminal water of Ouargla basin (northern Sahara of Algeria), Am. J. Appl. Sci. 6 (5) (2009) 876–881.
[13] B. Frengstad, A.K.M. Skrede, D. Banks, J.R. Krog, U. Siewers, The chemistry of Norwegian groundwaters: III. The distribution of trace elements in 476 crystalline bedrock groundwaters, as analysed by ICP-MS techniques, Sci. Total Environ. 246 (1) (2000) 21–40.
[14] R. Hirata, A. Suhogusoff, A. Fernandes, Groundwater resources in the state of São Paulo (Brazil): the application of indicators, An. Acad. Bras. Cienc. 79 (1) (2007) 141–152.
[15] J.S. Ahn, Geochemical occurrences of arsenic and fluoride in bedrock groundwater: a case study in Geumsan County, Korea, Environ. Geochem. Health 34 (1) (2012) 43–54.
[16] B. Thole, Ground water contamination with fluoride and potential fluoride removal technologies for east and southern Africa, Perspect. Water Pollut. (2013) 66–90.
[17] J. Fawell, K. Bailey, J. Chilton, E. Dahi, Y. Magara, Fluoride in drinking-water, IWA publishing, 2006.
[18] W.H. Organization, Guidelines for Drinking-Water Quality, World Health Organization, Geneva, 2015, p. 2011.
[19] J. Doull, K. Boekelheide, B. Farishian, R. Isaacson, J. Klotz, J. Kumar, N. Reed, Fluoride in Drinking Water: A Scientific Review of EPA's Standards, National Academies, Washington, 2006, pp. 205–223.
[20] Q. Zeng, Y.-S. Cui, L. Zhang, G. Fu, C.-C. Hou, L. Zhao, H.-L. Liu, Studies of fluoride on the thyroid cell apoptosis and mechanism, Zhonghua yu Fang yi xue za zhi [Chinese J. Prev. Med.] 46 (3) (2012) 233–236.

[21] A.L. Choi, Y. Zhang, G. Sun, D.C. Bellinger, K. Wang, X.J. Yang, P. Grandjean, Association of lifetime exposure to fluoride and cognitive functions in Chinese children: a pilot study, Neurotoxicol. Teratol. 47 (2015) 96–101.
[22] M. Diesendorf, J. Colquhoun, B. Spittle, Fluoridation and bones: authors' response to critics, Aust. N. Z. J. Public Health 22 (1) (1998) 165.
[23] F. Liu, J. Ma, H. Zhang, P. Liu, Y.-P. Liu, B. Xing, Y.-H. Dang, Fluoride exposure during development affects both cognition and emotion in mice, Physiol. Behav. 124 (2014) 1–7.
[24] A.J. Malin, C. Lesseur, S.A. Busgang, P. Curtin, R.O. Wright, A.P. Sanders, Fluoride exposure and kidney and liver function among adolescents in the United States: NHANES, 2013–2016, Environ. Int. 132 (2019) 105012.
[25] Y. Ku, H.-M. Chiou, W. Wang, The removal of fluoride ion from aqueous solution by a cation synthetic resin, Sep. Sci. Technol. 37 (1) (2002) 89–103.
[26] T. Ruiz, F. Persin, M. Hichour, J. Sandeaux, Modelisation of fluoride removal in Donnan dialysis, J. Membr. Sci. 212 (1–2) (2003) 113–121.
[27] R.S. Sathish, N. Raju, G. Raju, G. Nageswara Rao, K.A. Kumar, C. Janardhana, Equilibrium and kinetic studies for fluoride adsorption from water on zirconium impregnated coconut shell carbon, Sep. Sci. Technol. 42 (4) (2007) 769–788.
[28] M. Yang, Y. Zhang, B. Shao, R. Qi, H. Myoga, Precipitative removal of fluoride from electronics wastewater, J. Environ. Eng. 127 (10) (2001) 902–907.
[29] S.S. Tripathy, A.M. Raichur, Abatement of fluoride from water using manganese dioxide-coated activated alumina, J. Hazard. Mater. 153 (3) (2008) 1043–1051.
[30] S. Padmavathy, J. Amali, R. Raja, P. Nagarajan, B. Kavitha, A study of fluoride level in potable water of Salem district and an attempt for defluoridation with lignite, Ind. J. Environ. Prot. 23 (2003) 1244–1247.
[31] A. Daifullah, S. Yakout, S. Elreefy, Adsorption of fluoride in aqueous solutions using KMnO4-modified activated carbon derived from steam pyrolysis of rice straw, J. Hazard. Mater. 147 (1–2) (2007) 633–643.
[32] D. Mohan, K.P. Singh, V.K. Singh, Wastewater treatment using low cost activated carbons derived from agricultural byproducts—a case study, J. Hazard. Mater. 152 (3) (2008) 1045–1053.
[33] V. Thergaonkar, W. Nawalakhe, Activated magnesia for fluoride removal, Ind. J. Environ. Health 16 (1971) 241–243.
[34] X. Dou, D. Mohan, C.U. Pittman Jr., S. Yang, Remediating fluoride from water using hydrous zirconium oxide, Chem. Eng. J. 198 (2012) 236–245.
[35] O. Thirunavukkarasu, T. Viraraghavan, K. Subramanian, Arsenic removal from drinking water using iron oxide-coated sand, Water Air Soil Pollut. 142 (1–4) (2003) 95–111.
[36] H. Aloulou, A. Ghorbel, W. Aloulou, R. Ben Amar, S. Khemakhem, Removal of fluoride ions (F−) from aqueous solutions using modified Turkish zeolite with quaternary ammonium, Environ. Technol. (2019) 1–13.
[37] M.S. Onyango, Y. Kojima, O. Aoyi, E.C. Bernardo, H. Matsuda, Adsorption equilibrium modeling and solution chemistry dependence of fluoride removal from water by trivalent-cation-exchanged zeolite F-9, J. Colloid Interface Sci. 279 (2) (2004) 341–350.
[38] S. Samatya, Ü. Yüksel, M. Yüksel, N. Kabay, Removal of fluoride from water by metal ions (Al3+, La3+ and ZrO2+) loaded natural zeolite, Sep. Sci. Technol. 42 (9) (2007) 2033–2047.
[39] M. Islam, R. Patel, Evaluation of removal efficiency of fluoride from aqueous solution using quick lime, J. Hazard. Mater. 143 (1–2) (2007) 303–310.
[40] L. Chen, H.-X. Wu, T.-J. Wang, Y. Jin, Y. Zhang, X.-M. Dou, Granulation of Fe–Al–Ce nano-adsorbent for fluoride removal from drinking water by spray coating on sand in a fluidized bed, Powder Technol. 193 (1) (2009) 59–64.
[41] B.D. Turner, P. Binning, S. Stipp, Fluoride removal by calcite: evidence for fluorite precipitation and surface adsorption, Environ. Sci. Technol. 39 (24) (2005) 9561–9568.
[42] M. YangM, T. Hashimoto, N. Hoshi, H. Myoga, Fluoride removal in a fixed bed packed with granular calcite, Water Res. 33 (16) (1999) 3395–3402.
[43] C. Diaz-Nava, M. Solache-Rios, M. Olguin, Sorption of fluoride ions from aqueous solutions and well drinking water by thermally treated hydrotalcite, Sep. Sci. Technol. 38 (1) (2003) 131–147.
[44] Y. Wang, E.J. Reardon, Activation and regeneration of a soil sorbent for defluoridation of drinking water, Appl. Geochem. 16 (5) (2001) 531–539.
[45] S. Kagne, S. Jagtap, P. Dhawade, S. Kamble, S. Devotta, S. Rayalu, Hydrated cement: a promising adsorbent for the removal of fluoride from aqueous solution, J. Hazard. Mater. 154 (1–3) (2008) 88–95.
[46] R. Singh, S.P. Raghuvanshi, C. Kaushik, Defluoridation of drinking water using brick powder as an adsorbent, Asian J. Chem. 20 (8) (2008) 5818.
[47] R. Tahir, A. Muhammad, K.M. Chadhar, S. Faisal, S.K. Alvi, Defluoridation of water by wasted red brick paving blocks, Desalin. Water Treat. 108 (2018) 207–215.
[48] D. Goswami, A.K. Das, Removal of fluoride from drinking water using a modified fly ash adsorbent, J. Sci. Ind. Res. 65 (1) (2006) 77–79.
[49] L. Lv, Defluorination of drinking water by calcined MgAl-CO3 layered double hydroxides, Desalination 208 (1–3) (2007) 125–133.
[50] L. Lv, J. He, M. Wei, D. Evans, X. Duan, Factors influencing the removal of fluoride from aqueous solution by calcined mg–Al–CO3 layered double hydroxides, J. Hazard. Mater. 133 (1–3) (2006) 119–128.
[51] N. Medellin-Castillo, R. Leyva-Ramos, E. Padilla-Ortega, R.O. Perez, J. Flores-Cano, M. Berber-Mendoza, Adsorption capacity of bone char for removing fluoride from water solution. Role of hydroxyapatite content, adsorption mechanism and competing anions, J. Ind. Eng. Chem. 20 (6) (2014) 4014–4021.
[52] X. Fan, D.J. Parker, M.D. Smith, Adsorption kinetics of fluoride on low cost materials, Water Res. 37 (20) (2003) 4929–4937.
[53] H.S. Parmar, J.B. Patel, P. Sudhakar, V. Koshy, Removal of fluoride from water with powdered corn cobs, J. Environ. Sci. Eng. 48 (2) (2006) 135–138.
[54] R.H. Mckee, Removal of flourides from drinking water, Ind. Eng. Chem. 26 (8) (1934) 849–851.
[55] G. Patel, U. Pal, S. Menon, Removal of fluoride from aqueous solution by CaO nanoparticles, Sep. Sci. Technol. 44 (12) (2009) 2806–2826.
[56] X. Wu, Y. Zhang, X. Dou, M. Yang, Fluoride removal performance of a novel Fe–Al–Ce trimetal oxide adsorbent, Chemosphere 69 (11) (2007) 1758–1764.

[57] A.J. Zarur, J.Y. Ying, Reverse microemulsion synthesis of nanostructured complex oxides for catalytic combustion, Nature 403 (6765) (2000) 65–67.

[58] X. Zhao, Y. Shi, Y. Cai, S. Mou, Cetyltrimethylammonium bromide-coated magnetic nanoparticles for the preconcentration of phenolic compounds from environmental water samples, Environ. Sci. Technol. 42 (4) (2008) 1201–1206.

[59] R.R. Devi, I.M. Umlong, P.K. Raul, B. Das, S. Banerjee, L. Singh, Defluoridation of water using nano-magnesium oxide, J. Exp. Nanosci. 9 (5) (2014) 512–524.

[60] V. Chaudhary, S. Prasad, Rapid removal of fluoride from aqueous media using activated dolomite, Anal. Methods 7 (19) (2015) 8304–8314.

[61] L. Jayarathna, A. Bandara, W. Ng, R. Weerasooriya, Fluoride adsorption on $\gamma - Fe_2O_3$ nanoparticles, J. Environ. Health Sci. Eng. 13 (1) (2015) 54.

[62] B. Pan, J. Xu, B. Wu, Z. Li, X. Liu, Enhanced removal of fluoride by polystyrene anion exchanger supported hydrous zirconium oxide nanoparticles, Environ. Sci. Technol. 47 (16) (2013) 9347–9354.

[63] M. Mohapatra, K. Rout, S. Gupta, P. Singh, S. Anand, B. Mishra, Facile synthesis of additive-assisted nano goethite powder and its application for fluoride remediation, J. Nanopart. Res. 12 (2) (2010) 681–686.

[64] E. Kumar, A. Bhatnagar, U. Kumar, M. Sillanpää, Defluoridation from aqueous solutions by nano-alumina: characterization and sorption studies, J. Hazard. Mater. 186 (2–3) (2011) 1042–1049.

[65] Z. Zhou, Y. Yu, Z. Ding, M. Zuo, C. Jing, Competitive adsorption of arsenic and fluoride on {201} TiO2, Appl. Surf. Sci. 466 (2019) 425–432.

[66] N. Minju, K. Venkat Swaroop, K. Haribabu, V. Sivasubramanian, P. Senthil Kumar, Removal of fluoride from aqueous media by magnesium oxide-coated nanoparticles, Desalin. Water Treat. 53 (11) (2015) 2905–2914.

[67] T. Zhang, Q. Li, Z. Mei, H. Xiao, H. Lu, Y. Zhou, Adsorption of fluoride ions onto non-thermal plasma-modified CeO2/Al2O3 composites, Desalin. Water Treat. 52 (16–18) (2014) 3367–3376.

[68] S. Budyanto, Y.-L. Kuo, J. Liu, Adsorption and precipitation of fluoride on calcite nanoparticles: a spectroscopic study, Sep. Purif. Technol. 150 (2015) 325–331.

[69] M. Chaudhary, A. Maiti, Defluoridation by highly efficient calcium hydroxide nanorods from synthetic and industrial wastewater, Colloids Surf. A Physicochem. Eng. Asp. 561 (2019) 79–88.

[70] A. Dhillon, B. Choudhary, D. Kumar, S. Prasad, Excellent disinfection and fluoride removal using bifunctional nanocomposite, Chem. Eng. J. 337 (2018) 193–200.

[71] M. Zawar, R. Nazir, A. Hamid, E.C. Lima, M.R. Shah, Rapid defluoridation of drinking water by calcium carbonate nanoadsorbent: characterization, adsorption studies and application to real samples' treatment, Water Supply 20 (2) (2020) 667–678.

[72] C.S. Sundaram, N. Viswanathan, S. Meenakshi, Defluoridation of water using magnesia/chitosan composite, J. Hazard. Mater. 163 (2–3) (2009) 618–624.

[73] I.V. Mishakov, A.F. Bedilo, R.M. Richards, V.V. Chesnokov, A.M. Volodin, V.I. Zaikovskii, K.J. Klabunde, Nanocrystalline MgO as a dehydrohalogenation catalyst, J. Catal. 206 (1) (2002) 40–48.

[74] N. Oladoja, S. Hu, J. Drewes, B. Helmreich, Insight into the defluoridation efficiency of nano magnesium oxide in groundwater system contaminated with hexavalent chromium and fluoride, Sep. Purif. Technol. 162 (2016) 195–202.

[75] N. Oladoja, S. Chen, J. Drewes, B. Helmreich, Characterization of granular matrix supported nano magnesium oxide as an adsorbent for defluoridation of groundwater, Chem. Eng. J. 281 (2015) 632–643.

[76] M. Nazari, R. Halladj, Optimization of fluoride adsorption onto a sonochemically synthesized nano-MgO/γ-Al2O3 composite adsorbent through applying the L16 Taguchi orthogonal design, Desalin. Water Treat. 56 (9) (2015) 2464–2476.

[77] Z. Jin, Y. Jia, K.-S. Zhang, L.-T. Kong, B. Sun, W. Shen, J.-H. Liu, Effective removal of fluoride by porous MgO nanoplates and its adsorption mechanism, J. Alloys Compd. 675 (2016) 292–300.

[78] S.M. Maliyekkal, K. Antony, T. Pradeep, High yield combustion synthesis of nanomagnesia and its application for fluoride removal, Sci. Total Environ. 408 (10) (2010) 2273–2282.

[79] K. Sasaki, N. Fukumoto, S. Moriyama, Q. Yu, T. Hirajima, Chemical regeneration of magnesium oxide used as a sorbent for fluoride, Sep. Purif. Technol. 98 (2012) 24–30.

[80] W. Zhang, M. Sun, R. Prins, Multinuclear MAS NMR identification of fluorine species on the surface of fluorinated γ-alumina, J. Phys. Chem. B 106 (45) (2002) 11805–11809.

[81] L. Kong, Y. Tian, Z. Pang, X. Huang, M. Li, N. Li, J. Li, Needle-like Mg-La bimetal oxide nanocomposites derived from periclase and lanthanum for cost-effective phosphate and fluoride removal: characterization, performance and mechanism, Chem. Eng. J. 382 (2020) 122963.

[82] K. Zhang, S. Wu, X. Wang, J. He, B. Sun, Y. Jia, D. Lin, Wide pH range for fluoride removal from water by MHS-MgO/MgCO3 adsorbent: kinetic, thermodynamic and mechanism studies, J. Colloid Interface Sci. 446 (2015) 194–202.

[83] D. Tang, G. Zhang, Efficient removal of fluoride by hierarchical Ce–Fe bimetal oxides adsorbent: thermodynamics, kinetics and mechanism, Chem. Eng. J. 283 (2016) 721–729.

[84] F. Fu, D.D. Dionysiou, H. Liu, The use of zero-valent iron for groundwater remediation and wastewater treatment: a review, J. Hazard. Mater. 267 (2014) 194–205.

[85] C. Zhang, Y. Li, T.-J. Wang, Y. Jiang, J. Fok, Synthesis and properties of a high-capacity iron oxide adsorbent for fluoride removal from drinking water, Appl. Surf. Sci. 425 (2017) 272–281.

[86] M. Hu, H. Shen, S. Ye, Y. Wang, J. Zhang, S. Lv, Facile preparation of a tetraethylenepentamine-functionalized nano magnetic composite material and its adsorption mechanism to anions: competition or cooperation, RSC Adv. 8 (19) (2018) 10686–10697.

[87] T. Prathna, S.K. Sharma, M. Kennedy, Application of iron oxide and iron oxide/alumina nanocomposites for arsenic and fluoride removal: a comparative study, Int. J. Theor. Appl. Nanotechnol. 6 (1) (2018) 1–4.

[88] L. Chai, Y. Wang, N. Zhao, W. Yang, X. You, Sulfate-doped Fe3O4/Al2O3 nanoparticles as a novel adsorbent for fluoride removal from drinking water, Water Res. 47 (12) (2013) 4040–4049.

[89] X. Zhao, J. Wang, F. Wu, T. Wang, Y. Cai, Y. Shi, G. Jiang, Removal of fluoride from aqueous media by Fe3O4@ Al (OH) 3 magnetic nanoparticles, J. Hazard. Mater. 173 (1–3) (2010) 102–109.

[90] Y. Jia, B.-S. Zhu, K.-S. Zhang, Z. Jin, B. Sun, T. Luo, J.-H. Liu, Porous 2-line ferrihydrite/bayerite composites (LFBC): fluoride removal performance and mechanism, Chem. Eng. J. 268 (2015) 325–336.

[91] A. Azari, R.R. Kalantary, G. Ghanizadeh, B. Kakavandi, M. Farzadkia, E. Ahmadi, Iron–silver oxide nanoadsorbent synthesized by co-precipitation process for fluoride removal from aqueous solution and its adsorption mechanism, RSC Adv. 5 (106) (2015) 87377–87391.

[92] I. Abe, S. Iwasaki, T. Tokimoto, N. Kawasaki, T. Nakamura, S. Tanada, Adsorption of fluoride ions onto carbonaceous materials, J. Colloid Interface Sci. 275 (1) (2004) 35–39.

[93] N. Ammavasi, R. Mariappan, Enhanced removal of hazardous fluoride from drinking water by using a smart material: magnetic iron oxide fabricated layered double hydroxide/cellulose composite, J. Environ. Chem. Eng. 6 (4) (2018) 5645–5654.

[94] K. Anand, R. Gengan, A. Phulukdaree, A. Chuturgoon, Agroforestry waste Moringa oleifera petals mediated green synthesis of gold nanoparticles and their anti-cancer and catalytic activity, J. Ind. Eng. Chem. 21 (2015) 1105–1111.

[95] R. Araga, S. Kali, C.S. Sharma, Coconut-shell-derived carbon/carbon nanotube composite for fluoride adsorption from aqueous solution, *CLEAN–Soil, Air*, Water 47 (5) (2019) 1800286.

[96] W. Ayinde, W. Gitari, M. Munkombwe, A. Samie, J.A. Smith, Green synthesis of AgMgOnHaP nanoparticles supported on chitosan matrix: Defluoridation and antibacterial effects in groundwater, J. Environ. Chem. Eng. 104026 (2020).

[97] D. Balarak, Y. Mahdavi, E. Bazrafshan, A.H. Mahvi, Y. Esfandyari, Adsorption of fluoride from aqueous solutions by carbon nanotubes: determination of equilibrium, kinetic, and thermodynamic parameters, Fluoride 49 (1) (2016) 71.

[98] M. Bhaumik, T.Y. Leswifi, A. Maity, V. Srinivasu, M.S. Onyango, Removal of fluoride from aqueous solution by polypyrrole/Fe3O4 magnetic nanocomposite, J. Hazard. Mater. 186 (1) (2011) 150–159.

[99] M.M. Byranvand, A.N. Kharat, One pot green synthesis of gold nanowires using pomegranate juice, Mater. Lett. 134 (2014) 64–66.

[100] P. Cai, H. Zheng, C. Wang, H. Ma, J. Hu, Y. Pu, P. Liang, Competitive adsorption characteristics of fluoride and phosphate on calcined Mg–Al–CO3 layered double hydroxides, J. Hazard. Mater. 213 (2012) 100–108.

[101] M. Chaudhary, S. Rawat, N. Jain, A. Bhatnagar, A. Maiti, Chitosan-Fe-Al-Mn metal oxyhydroxides composite as highly efficient fluoride scavenger for aqueous medium, Carbohydr. Polym. 216 (2019) 140–148.

[102] L. Chen, K.-S. Zhang, J.-Y. He, W.-H. Xu, X.-J. Huang, J.-H. Liu, Enhanced fluoride removal from water by sulfate-doped hydroxyapatite hierarchical hollow microspheres, Chem. Eng. J. 285 (2016) 616–624.

[103] L. Chen, K. Zhang, J. He, X.-G. Cai, W. Xu, J.-H. Liu, Performance and mechanism of hierarchically porous Ce–Zr oxide nanospheres encapsulated calcium alginate beads for fluoride removal from water, RSC Adv. 6 (43) (2016) 36296–36306.

[104] E. Christina, P. Viswanathan, Development of a novel nano-biosorbent for the removal of fluoride from water, Chin. J. Chem. Eng. 23 (6) (2015) 924–933.

[105] K. Das, M. Garcia-Perez, B. Bibens, N. Melear, Slow pyrolysis of poultry litter and pine woody biomass: impact of chars and bio-oils on microbial growth, J. Environ. Sci. Health A 43 (7) (2008) 714–724.

[106] M.H. Dehghani, G.A. Haghighat, K. Yetilmezsoy, G. McKay, B. Heibati, I. Tyagi, V.K. Gupta, Adsorptive removal of fluoride from aqueous solution using single-and multi-walled carbon nanotubes, J. Mol. Liq. 216 (2016) 401–410.

[107] H. Deng, X. Yu, Adsorption of fluoride, arsenate and phosphate in aqueous solution by cerium impregnated fibrous protein, Chem. Eng. J. 184 (2012) 205–212.

[108] N.B. Dewage, A.S. Liyanage, C.U. Pittman Jr., D. Mohan, T. Mlsna, Fast nitrate and fluoride adsorption and magnetic separation from water on α-Fe2O3 and Fe3O4 dispersed on Douglas fir biochar, Bioresour. Technol. 263 (2018) 258–265.

[109] M.S. Fernando, A. Wimalasiri, S. Ratnayake, J. Jayasinghe, G.R. William, D. Dissanayake, R.M. de Silva, Improved nanocomposite of montmorillonite and hydroxyapatite for defluoridation of water, RSC Adv. 9 (61) (2019) 35588–35598.

[110] W. Friess, Collagen–biomaterial for drug delivery, Eur. J. Pharm. Biopharm. 45 (2) (1998) 113–136.

[111] Y. Guo, X. Xu, Y. Shang, B. Gao, Removal of fluoride by carbohydrate-based material embedded with hydrous zirconium oxide nanoparticles, Environ. Sci. Pollut. Res. 25 (28) (2018) 27982–27991.

[112] Y. Guo, Z. Zhu, Y. Qiu, J. Zhao, Enhanced adsorption of acid brown 14 dye on calcined mg/Fe layered double hydroxide with memory effect, Chem. Eng. J. 219 (2013) 69–77.

[113] N. Hamdi, E. Srasra, Removal of fluoride from acidic wastewater by clay mineral: effect of solid–liquid ratios, Desalination 206 (1–3) (2007) 238–244.

[114] J. He, K. Zhang, S. Wu, X. Cai, K. Chen, Y. Li, Z. Jin, Performance of novel hydroxyapatite nanowires in treatment of fluoride contaminated water, J. Hazard. Mater. 303 (2016) 119–130.

[115] M. Iqbal, M. Suleman, S. Ullah, H. Bibi, H. Wahab, Z. Iqbal, M.N. Ahmad, Purification of simulated fluoride-contaminated water using silver nanoparticles synthesized by xanthium stumarium as a bio-template, Fluoride 53 (2020) 112–123.

[116] H. Jiang, X. Li, L. Tian, T. Wang, Q. Wang, P. Niu, X. Luo, Defluoridation investigation of yttrium by laminated Y-Zr-Al tri-metal nanocomposite and analysis of the fluoride sorption mechanism, Sci. Total Environ. 648 (2019) 1342–1353.

[117] S. Kanrar, S. Debnath, P. De, K. Parashar, K. Pillay, P. Sasikumar, U.C. Ghosh, Preparation, characterization and evaluation of fluoride adsorption efficiency from water of iron-aluminium oxide-graphene oxide composite material, Chem. Eng. J. 306 (2016) 269–279.

[118] M. Karthikeyan, K.S. Kumar, K. Elango, Batch sorption studies on the removal of fluoride ions from water using eco-friendly conducting polymer/bio-polymer composites, Desalination 267 (1) (2011) 49–56.
[119] M. Karthikeyan, K. Satheeshkumar, K. Elango, Removal of fluoride ions from aqueous solution by conducting polypyrrole, J. Hazard. Mater. 167 (1–3) (2009) 300–305.
[120] H. Kaygusuz, S. Uzaşçı, F.B. Erim, Removal of fluoride from aqueous solution using aluminum alginate beads, *CLEAN–Soil, Air*, Watermark 43 (5) (2015) 724–730.
[121] G. Kofa, V. Gomdje, C. Telegang, S.N. Koungou, Removal of fluoride from water by adsorption onto fired clay pots: kinetics and equilibrium studies, J. Appl. Chem. 2017 (2017) 1–7.
[122] L. Kuang, Y. Liu, D. Fu, Y. Zhao, FeOOH-graphene oxide nanocomposites for fluoride removal from water: acetate mediated nano FeOOH growth and adsorption mechanism, J. Colloid Interface Sci. 490 (2017) 259–269.
[123] A. Kumar, P. Paul, S.K. Nataraj, Bionanomaterial scaffolds for effective removal of fluoride, chromium, and dye, ACS Sust. Chem. Eng. 5 (1) (2017) 895–903.
[124] Y.-H. Li, S. Wang, X. Zhang, J. Wei, C. Xu, Z. Luan, D. Wu, Adsorption of fluoride from water by aligned carbon nanotubes, Mater. Res. Bull. 38 (3) (2003) 469–476.
[125] Y. Li, S. Wang, X. Zhang, J. Wei, C. Xu, Z. Luan, B. Wei, Removal of fluoride from water by carbon nanotube supported alumina, Environ. Technol. 24 (3) (2003) 391–398.
[126] X.-P. Liao, B. Shi, Adsorption of fluoride on zirconium (IV)-impregnated collagen fiber, Environ. Sci. Technol. 39 (12) (2005) 4628–4632.
[127] L. Liu, Z. Cui, Q. Ma, W. Cui, X. Zhang, One-step synthesis of magnetic iron–aluminum oxide/graphene oxide nanoparticles as a selective adsorbent for fluoride removal from aqueous solution, RSC Adv. 6 (13) (2016) 10783–10791.
[128] X. Liu, M. Wang, S. Zhang, B. Pan, Application potential of carbon nanotubes in water treatment: a review, J. Environ. Sci. 25 (7) (2013) 1263–1280.
[129] J. Ma, Y. Shen, C. Shen, Y. Wen, W. Liu, Al-doping chitosan–Fe (III) hydrogel for the removal of fluoride from aqueous solutions, Chem. Eng. J. 248 (2014) 98–106.
[130] C. Manjunatha, B. Nagabhushana, A. Narayana, S. Pratibha, M. Raghu, Effective and fast adsorptive removal of fluoride on CaAl2O4: Ba nanoparticles: isotherm, kinetics and reusability studies, Mater. Res. Exp. 6 (11) (2019) 115089.
[131] E. Mohammadi, H. Daraei, R. Ghanbari, S.D. Athar, Y. Zandsalimi, A. Ziaee, K. Yetilmezsoy, Synthesis of carboxylated chitosan modified with ferromagnetic nanoparticles for adsorptive removal of fluoride, nitrate, and phosphate anions from aqueous solutions, J. Mol. Liq. 273 (2019) 116–124.
[132] P. Mondal, M.K. Purkait, Green synthesized iron nanoparticles supported on pH responsive polymeric membrane for nitrobenzene reduction and fluoride rejection study: optimization approach, J. Clean. Prod. 170 (2018) 1111–1123.
[133] P. Mondal, M.K. Purkait, Preparation and characterization of novel green synthesized iron–aluminum nanocomposite and studying its efficiency in fluoride removal, Chemosphere 235 (2019) 391–402.
[134] K. Mukunthan, S. Balaji, Cashew apple juice (Anacardium occidentale L.) speeds up the synthesis of silver nanoparticles, Int. J Green Nanotechnol. 4 (2) (2012) 71–79.
[135] A. Mullick, S. Neogi, Acoustic cavitation induced synthesis of zirconium impregnated activated carbon for effective fluoride scavenging from water by adsorption, Ultrason. Sonochem. 45 (2018) 65–77.
[136] A. Naghizadeh, K. Gholami, Bentonite and montmorillonite nanoparticles effectiveness in removal of fluoride from water solutions, J. Water Health 15 (4) (2017) 555–565.
[137] S. Nehra, M. Nair, D. Kumar, Hydrothermally shape-controlled synthesis of TiO2/graphene for fluoride adsorption studies, J. Chem. Eng. Data 64 (12) (2019) 5373–5384.
[138] Y. Nie, C. Hu, C. Kong, Enhanced fluoride adsorption using Al (III) modified calcium hydroxyapatite, J. Hazard. Mater. 233 (2012) 194–199.
[139] K. Pandi, N. Viswanathan, Synthesis of alginate bioencapsulated nano-hydroxyapatite composite for selective fluoride sorption, Carbohydr. Polym. 112 (2014) 662–667.
[140] K. Parashar, N. Ballav, S. Debnath, K. Pillay, A. Maity, Rapid and efficient removal of fluoride ions from aqueous solution using a polypyrrole coated hydrous tin oxide nanocomposite, J. Colloid Interface Sci. 476 (2016) 103–118.
[141] T. Poursaberi, M. Hassanisadi, K. Torkestani, M. Zare, Development of zirconium (IV)-metalloporphyrin grafted Fe3O4 nanoparticles for efficient fluoride removal, Chem. Eng. J. 189 (2012) 117–125.
[142] S.M. Prabhu, S. Meenakshi, Defluoridation of water using synthesized Zr (IV) encapsulated silica gel/chitosan biocomposite: adsorption isotherms and kinetic studies, Desalin. Water Treat. 53 (13) (2015) 3592–3603.
[143] C. Prathibha, A. Biswas, L.A. Chunduri, S.K. Reddy, P. Loganathan, M. Kalaruban, K. Venkatarmaniah, Zr (IV) functionalized graphene oxide anchored sand as potential and economic adsorbent for fluoride removal from water, Diamond Relat. Mater. 109 (2020) 108081.
[144] Z. Qiusheng, L. Xiaoyan, Q. Jin, W. Jing, L. Xuegang, Porous zirconium alginate beads adsorbent for fluoride adsorption from aqueous solutions, RSC Adv. 5 (3) (2015) 2100–2112.
[145] S. Raghav, M. Nair, D. Kumar, Trimetallic oxide entrapped in alginate polymeric matrix employed for adsorption studies of fluoride, Surf. Interfaces 13 (2018) 112–132.
[146] Z. Ruan, Y. Tian, J. Ruan, G. Cui, K. Iqbal, A. Iqbal, S. Yan, Synthesis of hydroxyapatite/multi-walled carbon nanotubes for the removal of fluoride ions from solution, Appl. Surf. Sci. 412 (2017) 578–590.

[147] J. Saikia, S. Sarmah, P. Saikia, R.L. Goswamee, Harmful weed to prospective adsorbent: low-temperature–carbonized Ipomoea carnea stem carbon coated with aluminum oxyhydroxide nanoparticles for defluoridation, Environ. Sci. Pollut. Res. 26 (1) (2019) 721–737.
[148] H. Sehaqui, A. Mautner, U.P. de Larraya, N. Pfenninger, P. Tingaut, T. Zimmermann, Cationic cellulose nanofibers from waste pulp residues and their nitrate, fluoride, sulphate and phosphate adsorption properties, Carbohydr. Polym. 135 (2016) 334–340.
[149] Y. Shang, X. Xu, B. Gao, Q. Yue, Highly selective and efficient removal of fluoride from aqueous solution by ZrLa dual-metal hydroxide anchored bio-sorbents, J. Clean. Prod. 199 (2018) 36–46.
[150] C. Silveira, Q.L. Shimabuku, M. Fernandes Silva, R. Bergamasco, Iron-oxide nanoparticles by the green synthesis method using Moringa oleifera leaf extract for fluoride removal, Environ. Technol. 39 (22) (2018) 2926–2936.
[151] M. Srimurali, A. Pragathi, J. Karthikeyan, A study on removal of fluorides from drinking water by adsorption onto low-cost materials, Environ. Pollut. 99 (2) (1998) 285–289.
[152] C.S. Sundaram, N. Viswanathan, S. Meenakshi, Defluoridation chemistry of synthetic hydroxyapatite at nano scale: equilibrium and kinetic studies, J. Hazard. Mater. 155 (1–2) (2008) 206–215.
[153] C.S. Sundaram, N. Viswanathan, S. Meenakshi, Uptake of fluoride by nano-hydroxyapatite/chitosan, a bioinorganic composite, Bioresour. Technol. 99 (17) (2008) 8226–8230.
[154] C.S. Sundaram, N. Viswanathan, S. Meenakshi, Fluoride sorption by nano-hydroxyapatite/chitin composite, J. Hazard. Mater. 172 (1) (2009) 147–151.
[155] S. Suriyaraj, A. Bhattacharyya, R. Selvakumar, Hybrid Al 2 O 3/bio-TiO 2 nanocomposite impregnated thermoplastic polyurethane (TPU) nanofibrous membrane for fluoride removal from aqueous solutions, RSC Adv. 5 (34) (2015) 26905–26912.
[156] S. Suriyaraj, T. Vijayaraghavan, P. Biji, R. Selvakumar, Adsorption of fluoride from aqueous solution using different phases of microbially synthesized TiO2 nanoparticles, J. Environ. Chem. Eng. 2 (1) (2014) 444–454.
[157] S. Swain, T. Patnaik, R. Dey, Efficient removal of fluoride using new composite material of biopolymer alginate entrapped mixed metal oxide nanomaterials, Desalin. Water Treat. 51 (22–24) (2013) 4368–4378.
[158] Q. Tang, T. Duan, P. Li, P. Zhang, D. Wu, Enhanced defluoridation capacity from aqueous media via hydroxyapatite decorated with carbon nanotube, Front. Chem. 6 (2018) 104.
[159] A. Teimouri, S.G. Nasab, S. Habibollahi, M. Fazel-Najafabadi, A.N. Chermahini, Synthesis and characterization of a chitosan/montmorillonite/ZrO 2 nanocomposite and its application as an adsorbent for removal of fluoride, RSC Adv. 5 (9) (2015) 6771–6781.
[160] D. Thakre, S. Rayalu, R. Kawade, S. Meshram, J. Subrt, N. Labhsetwar, Magnesium incorporated bentonite clay for defluoridation of drinking water, J. Hazard. Mater. 180 (1–3) (2010) 122–130.
[161] M.K. Uddin, A review on the adsorption of heavy metals by clay minerals, with special focus on the past decade, Chem. Eng. J. 308 (2017) 438–462.
[162] L. Van Zwieten, S. Kimber, S. Morris, K. Chan, A. Downie, J. Rust, A. Cowie, Effects of biochar from slow pyrolysis of papermill waste on agronomic performance and soil fertility, Plant and Soil 327 (1–2) (2010) 235–246.
[163] E. Vences-Alvarez, L.H. Velazquez-Jimenez, L.F. Chazaro-Ruiz, P.E. Diaz-Flores, J.R. Rangel-Mendez, Fluoride removal in water by a hybrid adsorbent lanthanum–carbon, J. Colloid Interface Sci. 455 (2015) 194–202.
[164] N. Viswanathan, S. Meenakshi, Development of chitosan supported zirconium (IV) tungstophosphate composite for fluoride removal, J. Hazard. Mater. 176 (1–3) (2010) 459–465.
[165] S. Waghmare, T. Arfin, S. Rayalu, D. Lataye, S. Dubey, S. Tiwari, Adsorption behavior of modified zeolite as novel adsorbents for fluoride removal from drinking water: surface phenomena, kinetics and thermodynamics studies, Int. J. Sci. Eng. Technol. Res. 4 (12) (2015) 4114–4124.
[166] A. Wang, K. Zhou, W. Chen, C. Zhang, X. Liu, Q. Chen, F. Liu, Adsorption of fluoride by the calcium alginate embedded with mg-Al-Ce trimetal oxides, Korean J. Chem. Eng. 35 (8) (2018) 1636–1641.
[167] A. Wang, K. Zhou, X. Liu, F. Liu, Q. Chen, Development of mg–Al–La tri-metal mixed oxide entrapped in alginate for removal of fluoride from wastewater, RSC Adv. 7 (50) (2017) 31221–31229.
[168] J. Wang, X. Lin, X. Luo, Y. Long, A sorbent of carboxymethyl cellulose loaded with zirconium for the removal of fluoride from aqueous solution, Chem. Eng. J. 252 (2014) 415–422.
[169] Y. Xia, X. Huang, W. Li, Y. Zhang, Z. Li, Facile defluoridation of drinking water by forming shell@ fluorapatite nanoarray during boiling egg shell, J. Hazard. Mater. 361 (2019) 321–328.
[170] C. Xu, J. Li, F. He, Y. Cui, C. Huang, H. Jin, S. Hou, Al 2 O 3–Fe 3 O 4–expanded graphite nano-sandwich structure for fluoride removal from aqueous solution, RSC Adv. 6 (99) (2016) 97376–97384.
[171] N. Xu, S. Li, W. Li, Z. Liu, Removal of fluoride by graphene oxide/alumina nanocomposite: adsorbent preparation, characterization, adsorption performance and mechanisms, ChemistrySelect 5 (6) (2020) 1818–1828.
[172] H. Yakun, D. Wenming, X. Huang, X. Jingnian, Z. Menghua, Fluoride removal by lanthanum alginate bead: adsorbent characterization and adsorption mechanism, Chin. J. Chem. Eng. 19 (3) (2011) 365–370.
[173] R. Yao, F. Meng, L. Zhang, D. Ma, M. Wang, Defluoridation of water using neodymium-modified chitosan, J. Hazard. Mater. 165 (1–3) (2009) 454–460.
[174] X. Yu, S. Tong, M. Ge, J. Zuo, Removal of fluoride from drinking water by cellulose@ hydroxyapatite nanocomposites, Carbohydr. Polym. 92 (1) (2013) 269–275.
[175] Y. Yu, C. Wang, X. Guo, J.P. Chen, Modification of carbon derived from sargassum sp. by lanthanum for enhanced adsorption of fluoride, J. Colloid Interface Sci. 441 (2015) 113–120.

[176] M. Zendehdel, B. Shoshtari-Yeganeh, H. Khanmohamadi, G. Cruciani, Removal of fluoride from aqueous solution by adsorption on NaP: HAp nanocomposite using response surface methodology, Process Saf. Environ. Prot. 109 (2017) 172–191.
[177] D. Zhang, H. Luo, L. Zheng, K. Wang, H. Li, Y. Wang, H. Feng, Utilization of waste phosphogypsum to prepare hydroxyapatite nanoparticles and its application towards removal of fluoride from aqueous solution, J. Hazard. Mater. 241 (2012) 418–426.
[178] G. Zhang, Z. He, W. Xu, A low-cost and high efficient zirconium-modified-Na-attapulgite adsorbent for fluoride removal from aqueous solutions, Chem. Eng. J. 183 (2012) 315–324.
[179] J. Zhu, X. Lin, P. Wu, X. Luo, Pectin/Al 2 O 3–ZrO 2 core/shell bead sorbent for fluoride removal from aqueous solution, RSC Adv. 6 (33) (2016) 27738–27749.
[180] Y. Zhu, S. Murali, W. Cai, X. Li, J.W. Suk, J.R. Potts, R.S. Ruoff, Graphene and graphene oxide: synthesis, properties, and applications, Adv. Mater. 22 (35) (2010) 3906–3924.

Chapter 13

Nanofiltration and environmental sustainability

Sukanchan Palit*
Department of Chemical Engineering, University of Petroleum and Energy Studies, Dehradun, India
*Corresponding author: e-mail addresses: sukanchan68@gmail.com, sukanchanp@rediffmail.com, sukanchan92@gmail.com, spalit@ddn.upes.ac.in

Introduction

Nanoscience and nanotechnology are of great importance. Rampant industrialization has caused environmental degradation with some major hazards to humanity. Biological, chemical, and physical hazards are great threats to the society. The basic needs of humans particularly in terms of food, shelter, and proper sanitation are subject to significant disaster. Water treatment and industrial wastewater treatment need immediate attention. Drinking water issues are ravaging nations around the world. Integrated water resource management and wastewater management are dire necessity. Membrane separation processes and other novel separation processes are needed to purify water and wastewater. The development of the Sourirajan-Loeb synthetic membrane in 1960 provided a valuable separation tool in the process industries, but it faced tremendous difficulties and challenges in the early days. Even today the diffusion mechanism is still not fully understood. An improved understanding in the field nanofiltration and nanotechnology applications is necessary. Hopefully, this would lead to a greater effort to tackle global water issues. The crux of this chapter is to review nanofiltration and other membrane separation processes in mitigating industrial pollution control. Energy and environmental sustainability are the immensely vexing issues globally. This chapter discusses issues relating to nanofiltration and sustainability.

The goal of this study

Sustainable development in many forms such as social, economic, energy, and environmental is necessary. The goal of this study is to explore the field of membrane science, particularly nanofiltration, to find sustainable applications. Energy and environmental sustainability are the drivers of global economy today. Today, millions of citizens around the world are without clean drinking water and proper sanitation. The mankind could face gruesome and an unending catastrophe if water remediation needs are not met. Nanofiltration is one of the major and challenging areas of membrane separation processes. The other areas of discussion in this chapter are sustainable development, environmental sustainability, and environmental protection science. A new generation of environmental engineers and scientists are evolving whose main aim is to address global climate change and global warming. Today, United Nations Sustainable Development Goals are targeting numerous issues which include pure drinking water and proper sanitation. This chapter assesses the success of science and engineering in implementation of United Nations Sustainable Development Goals in developed, developing, and poor nations around the world. The development of Sourirajan-Loeb synthetic membrane in 1960 provided a valuable separation technique in chemical process industrial and ushered in a new beginning in the field of membrane science, membrane filtration, and water and wastewater treatment. Today, membranes are more robust, and modules and equipment are better designed. Fouling mechanisms and diffusion processes are less understood today. The author in this chapter addresses these scientific and engineering issues.

Separation Science and Technology, Volume 15. https://doi.org/10.1016/B978-0-323-90763-7.00008-1

Environmental sustainability, scientific sagacity, and the vast vision for the future

Environmental and energy sustainability are the relevant areas of research and development in the global scenario today. The fruits of human civilization today are both energy and environmental sustainability. The vast scientific sagacity, the rendezvous of science, philosophy, and engineering science will all lead a long way in the true emancipation of environmental sustainability and protection of environment. Scientific sagacity, scientific provenance, and scientific revelation in the field of environmental engineering techniques such as nanofiltration are revolutionizing the vast scientific and engineering fabric. Water is very precious to human civilization and human scientific endeavor. For the last two centuries, research and development in water treatment and industrial wastewater treatment are transforming the scientific understanding globally. In this chapter, the author stresses on the recent technologies in membrane science especially nanofiltration.

What is nanofiltration?

Nanofiltration is a relatively recently developed membrane filtration process used most often with low total dissolved solids in water such as surface water and fresh groundwater with the purpose of softening (polyvalent cation removal) and removal of disinfection by-product precursors such as natural organic matter and synthetic organic matter. Nanofiltration is also becoming more widely used in food processing applications such as dairy, for simultaneous concentration and partial mineralization. In Table 1, the author discusses various characteristics of membrane separation processes. Filtration is defined as the separation of two or more components from a fluid stream based primarily on size differences. In conventional usage terminology, it usually refers to the separation of solid immiscible particles from liquid and gaseous streams. In Table 1, the authors discussed osmosis, dialysis, microfiltration, ultrafiltration, nanofiltration, reverse osmosis, electrodialysis, and pervaporation. Membrane filtration extends these applications further to include the separation of dissolved solutes in liquid streams and for separation of gaseous mixtures. Nanofiltration is a membrane filtration based method that uses nanometer sized pores. Nanofiltration membranes have pore sizes from 1 to 10 nm, smaller than that used in ultrafiltration and microfiltration but just larger than that in reverse osmosis. Membranes can be classified by (a) nature of membrane-natural versus synthetic, (b) structure of the membrane-porous versus nonporous, (c) application of the membrane-gaseous phase separations, gas–liquid, liquid–liquid, etc., and (d) mechanism of membrane action-adsorptive versus diffusive, ion-exchange, osmotic, or nonselective membranes. Desalination and water treatment are today interlinked with each other. It has vast and varied engineering applications. Arsenic and heavy metals in groundwater and drinking water contamination are challenging the entire scientific scenario. This chapter attempts to open new windows in the field of environmental protection. The various characteristics of membrane processes such as driving force, retentate, and permeate of different membrane separation processes are elucidated in Table 1.

The challenges of application of nanofiltration and nanotechnology in water treatment

Today application of nanofiltration and nanotechnology in water and wastewater treatment are facing significant scientific and engineering challenges. In drought-prone areas around the world, desalination and other water treatment techniques are the needs of the hour. Millions of people around the world are without clean drinking water and proper sanitation.

TABLE 1 Characteristics of membrane processes (Cheryan [24]).

Process	Driving force	Retentate	Permeate
Osmosis	Chemical potential	Solutes	Water
Dialysis	Concentration difference	Large molecules	Small molecules
Microfiltration	Pressure	Suspended particles	Dissolved solutes
Ultrafiltration	Pressure	Large molecules	Small molecules
Nanofiltration	Pressure	Small molecules	Monovalent ions
Reverse osmosis	Pressure	All solutes, water	Water
Electrodialysis	Voltage/current	Nonionic solutes, water	Ionized solutes
Pervaporation	Pressure	Nonvolatile molecules	Volatile small molecules.

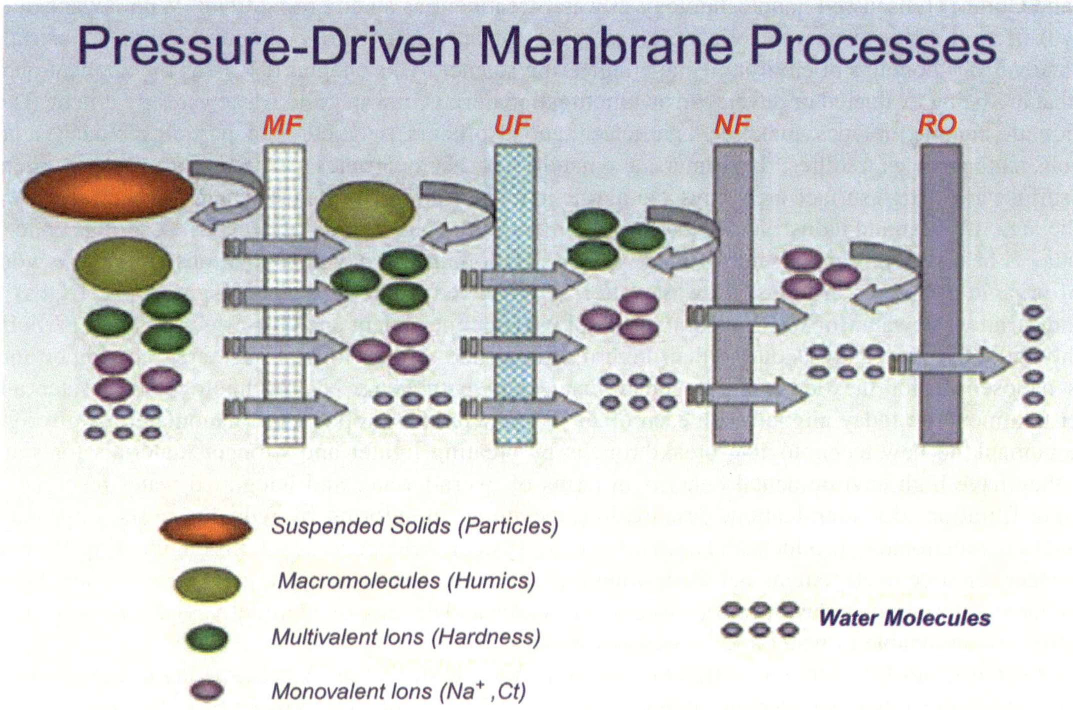

FIG. 1 The technique of pressure-driven membrane processes (www.Wateronline.com).

The situation in the state of West Bengal, India and Bangladesh are extremely grave and thought provoking. Fouling of the nanofiltration membranes is one of the major challenges in membrane separation process. In the domain of nanotechnology, environmental health risk is a major challenge in nanotechnology applications in human society. Pressure driven membrane processes are the ultimate needs of water and wastewater treatment. Fig. 1 discusses the technique of pressure driven membrane processes. Today there are immense challenges in the application of nanofiltration in water and wastewater treatment. Thus the branch of nanofiltration or membrane filtration is a scientific vision and ingenuity in the field of environmental engineering and chemical process engineering.

Environmental pollution control and human society

Environmental pollution control and industrial pollution control are today in the middle of vast scientific introspection. Sustainable development and human society interconnected. Social and economic sustainability are equally important as energy and environmental sustainability. Integrated water resource management and wastewater management are necessary. Humanity is facing immense challenges and tremendous difficulties with regards to water remediation and environmental protection. Scientists, engineers, and civil society are gearing forward for the implementation of 17 United Nations Sustainable Development Goals. The global environmental situation is absolutely grave and is of immense concern to citizens around the world. The applications of United Nations Sustainable Development Goals are of utmost importance for developing, developed and disadvantaged countries around the world. Thus, conventional and nonconventional environmental engineering techniques along with novel separation processes are changing the face of science and engineering in the global scenario. Nanotechnology, nanomaterials, and the whole domain of science of membranes are in the path of newer scientific rejuvenation.

Recent scientific endeavors in the field of nanotechnology

Nanotechnology and nanoengineering are the marvels of science and engineering. Developed, developing and disadvantaged countries around the world are in serious need of application of nanotechnology in water treatment. Nanofiltration and other membrane separation techniques can provide viable solutions. Drinking water and groundwater treatment are the challenges in many developed and developing nations around the globe.

Raval and Gohil [1] discussed nanotechnology in water treatment as an emerging trend. With advancements in nanotechnology, different types of nanomaterials are emerging for application in water purification, industrial wastewater treatment and various domains of environmental engineering science. This chapter discusses the application of nanoscale materials that are being evaluated or developed as functional materials in water and wastewater treatment. The other areas of discussion are nanomembranes, metal nanoparticles, nanoadsorbents, magnetic nanoparticles, bioactive nanoparticles, carbonaceous nanoparticles, zeolites, dendrimers, and nanofibers. Nanoparticles are better for water treatment because of their high surface area. High surface area plays a vital role in membrane performance. There has been significant advancements in the area of water and industrial wastewater treatment in the past couple of decades. A serious concern involving humans today is heavy metals and arsenic groundwater in developing and developed nations around the world. The degradation of organic pollutants in industrial wastewater by advanced catalytic oxidation processes (AOPs) is useful for cleaning industrial wastewater for which existing technologies are inefficient and cost-effective. Nanoparticles are particularly highly attractive for catalysis due to their high reactivity due to their high surface area. An important area of discussion is nanosensor for the detection of contaminants and pathogens. Nanotechnology and water and industrial wastewater treatment are today aligned with each other [5]. The unique properties of nanomaterials offer great promise to provide humankind new technological breakthroughs by creating lighter and stronger materials for water treatment and today they have high environmental benefits in terms of overall water and integrated water resource management by improving filtration, decontamination, desalination, recycling, monitoring by reducing costs, improving efficiency and offering new functionality products and systems. Today systems science and systems engineering are integrated with integrated water resource management and wastewater management. These are the targets of science and humanity today. A variety of nanomaterials are today in various stages of research and development initiatives each possessing unique functionalities that are applicable to water and wastewater remediation.

Argueta and Cifuentes [2] dealt with different areas of nanotechnology such as nanomedicine and nanobiotechnology. Water is the second most precious element of human life and all forms of life. In recent decades, new technologies have emerged worldwide for separation, filtration, purification, and treatment of contaminants in water for example, ultraviolet light, reverse osmosis, membrane technology, activated carbon, polymerization, purification treatment of wastes, and contaminants in water, and more recently the application of nanomembranes technology into the filtration processes and unit operations of chemical engineering. Chemical process engineering, environmental protection science and nanotechnology are today integrated with each other. This chapter covers applications of carbon nanotubes, electrodeposition, carbon nanotubes membranes, and general assembly in water and wastewater treatment.

Bardos et al. [3] dealt with status of nanoremediation and its potential for future deployment and its risk-benefit and benchmarking appraisals. NanoRem (Taking Nanotechnological Remediation Processes from Lab Scale to End User Applications for the restoration of a clean environment) was a comprehensive research project, funded through the European Commission's Seventh Framework Programme, which highly focused on facilitating practical, safe, economic, and exploitable nanotechnology for in-situ remediation of polluted soil and groundwater. Nanoremediation is a huge alternative to soil and water remediation in the global scenario today. Nanoremediation has been most frequently applied to problems of chlorinated solvents and metals such as Chromium VI. *In situ* remediation tools (exploiting biological, chemical, physical stabilization, and thermal processes within the subsurface) are effectively used to avoid excavation of materials or surface treatment of groundwater. Nanoremediation describes the use of nanoparticles in the treatment of contaminated groundwater, drinking water and soil. The authors discussed in detail the status of field-based tests and applications of nanoremediation, risk-benefit appraisal for nanoremediation technologies. Human exposure and hazards from handling these materials are also covered. Nanoremediation can offer significant advantages in some important remediation applications. These scientific benefits are site-specific and imperative rather than representing some kind of over-arching step change in remediation capabilities.

Recent scientific advances in the field of nanofiltration

Nanofiltration and other membrane separation processes are the vexing technologies in modern science and modern human society. Application of nanofiltration in water and industrial wastewater treatment are of significant importance.

Shahmansouri and Bellona [4] discussed nanofiltration technology in water treatment and reuse with its applications and costs. Nanofiltration is a relatively recent development in membrane technology with scientific characteristics that fall between ultrafiltration and reverse osmosis. While reverse osmosis membranes dominate the desalination industry, nanofiltration is highly employed in a variety of water and wastewater treatment and varied industrial applications for the selective removal of ions and organic substances as well as certain seawater desalination applications. Basic economic analyses were also performed to compare the probability of using nanofiltration membranes over alternative processes [1]. The selection of nanofiltration over other treatment technologies is dependent on several factors including pretreatment requirements, influent water quality, treatment facility capacity and treatment goals. The term nanofiltration (NF) appears to have been first used in the mid-1980s to elucidate upon and describe membranes with characteristics that fall between ultrafiltration and the vast domain of reverse osmosis. Because nanofiltration membranes are fundamentally similar to reverse osmosis membranes the designation and description of a membrane as a nanofiltration is based on several scientific and engineering characteristics including effective pore size, molecular weight cut-off and salt rejection. Owing to the remarkable characteristics including selective separation of salts, good organics removal, and relatively low pressure requirements, nanofiltration membranes are being increasingly employed in a wide variety of applications including drinking water and industrial wastewater treatment and in several diverse industries for product purification and treatment (e.g., dairy treatment, chemical, beverage, food, pharmaceutical, pulp and paper, textile, and oil and gas industries) [1]. The authors touched upon methods of economic analysis, nanofiltration in water and wastewater industry, seawater desalination, wastewater treatment in textile industry, and the scientific and engineering perspectives behind applications of membrane science.

According to Farrukh [5], the nanofiltration technique lies between ultrafiltration and reverse osmosis. Nanofiltration is considered a low-cost process and is highly capable of removing pesticides, organic matter, desalination of sea water, oleochemicals process and pollutants from other industrial processes [2]. The important separation mechanism in nanofiltration is attributed to dielectric property and the charged nanofiltration membrane attracts opposite charges to pass through it and repels like charges on the basis of difference in electric potential. Novel separation processes such as membrane science are highly relevant in water and wastewater treatment. The main challenge in the implementation of nanofiltration membrane is its ability toward fouling and low performance at relatively high temperature. This book covers topics from multiple ranges from manufacturing of nanofiltration membranes with their vast and varied applications in industrial wastewater treatment, drinking water treatment, and degradation of pollutants to fouling. Membrane science and other novel separation processes are changing the face of water remediation science today. This book targets effects of high concentrations of organic and inorganic pollutants in wastewater and drinking water on nanofiltration membrane behavior, nanofiltration for the treatment of oil sands produced water, zeolites mixed matrix nanofiltration membranes, micellar enhanced ultrafiltration, fouling of nanofiltration membranes, and electrically and electrochemically assisted nanofiltration [2]. Removal of high concentrations of organic and inorganic matter from aquatic solutions using crossflow spiral wound nanofiltration membranes with semi-industrial pilot plant are discussed in minute details.

Abdel-Fateh [6] discussed nanofiltration systems and applications in wastewater treatment in an extensive review. Nanofiltration membrane is one of the most important activities employed in wastewater treatment domain. In today's scientific world, membrane science mainly nanofiltration has immense scientific as well as engineering importance. It is a relatively recent development in membrane science and technology and it can be aqueous and nonaqueous. Membrane charges play an important role in membrane function, and often nanofiltration membranes have surface negative charges. The primary job of nanofiltration is the selective removal of ions and organic substances and it is used in seawater desalination. The main aim and objective of this review is to illustrate and explore the main applications of nanofiltration process in water reuse. Water reuse stands as a major pillar in the entire domain of water and industrial wastewater treatment. The author discussed in details (1) existing situation of wastewater in Egypt, (2) overview of membrane separation technology, (3) current situation on nanofiltration, (4) list of reverse osmosis and nanofiltration membrane manufacturers, (5) nanofiltration membrane transport models, (6) fouling of nanofiltration membranes, and (7) cost of nanofiltration membranes. Fouling, low flux, instability, and poor durability are technical and engineering barriers to support the viability of nanofiltration membranes. This paper widely enhances the needs of membrane science in integrated water resource management and wastewater management. Air quality management, integrated water resource management and wastewater treatment and management are the highest levels of scientific research pursuit today. Human habitat are today really in the need of a restructuring as regards water science and technology. While Reverse Osmosis and Ultrafiltration usage in water and wastewater treatment are increasingly at a rapid pace, nanofiltration applications are increasing exponentially in water and wastewater treatment and other different applications in the industry like separation of solute or chemical from solution, producing bio-materials, drug industry and flavors. In addition to producing different chemicals using nanofiltration, a recovery of fine chemicals from outlet streams is vastly used in industrial applications. Reverse osmosis and ultrafiltration have been widely used for different applications but their applications are limited. Nanofiltration applications are expanding and replacing other membrane separation processes. A new dawn in the field of membrane science is today emerging fast and soon chemical process engineering and environmental engineering science will usher in a new age of deep scientific contemplation and vision.

Dongre et al. [7] dealt with natural polymer composite membranes for water purification in a well-researched review. The vastly ever-growing human population, environmental pollution, and environmental and ecological degradation cause suffering and immense distress to human race due to chemicals and other water contaminants such as chemicals, heavy metals, pesticides and insecticides. Hence, much attention is made to purify water by advanced technologies employing smart materials and smart eco-materials. Mankind's vast scientific provenance and subtleties as regards environmental protection are in a state of disaster. Thus, the need of natural composite polymer membranes. This review focuses on the existing membrane science and technology involved in water purification and cleaning and vastly portrays molecular-level rationally designed approaches for engineering of biopolymer membranes. Provision of pure drinking water and proper sanitation are burning issues globally. The authors stressed bionanocomposites for biosorption, bioremediation of heavy metals, bionanocomposites for water purification, chitosan bionanocomposites, natural fiber reinforced benign biopolymeric composites, magnetic bionanocomposites, and futuristic nanomaterials. The nanomaterials are highly advantageous over conventional materials due to the special ability to integrate into various multifunctional membranes and a vast array of composites.

Scientific endeavors in the field of environmental sustainability

Nanoscience and nanotechnology are today integrated in the vast domain of sustainable development whether it is social, economic, energy or environmental. Water science and technology are the precious technologies of human civilization today. Environmental sustainability and environmental protection science are the needs of civilization, science and engineering. A new dawn in the field of sustainable development globally is emerging as concerns for global climate change and global warming gets enhanced day by day.

Kuhlman and Farrington [8] discussed the domain of sustainability. Sustainability as a policy concept has its origin in the Brundtland report of 1987. Engineering and technology rapidly moved ahead since then. Scientific understanding and scientific imagination are today transforming the vast domain of sustainability. That document by Dr. Gro Harlem Brundtland, former Prime Minister of Norway, was deeply concerned with the tension between the aspirations of humankind for a better life on the one hand and the vast and important limitations imposed by nature on the other hand. The term sustainability has become immensely popular in policy-oriented research as a target of what public policies ought to achieve and aspire. The authors discussed in minute details history of the concept, people, planet and profit, happiness, well-being and welfare, and resources and the future of sustainability. Weak and strong sustainability are the other pillars of this well researched treatise. In spite of the large body of sustainability, there are lots of work which needs to be done [8]. The practical meaning of sustainability still needs to be fully explored.

United Nations Development Programme Report [9] describes sustainability and inequality in human development. The stance and stature of the science of sustainability needs to be redefined as humanity progresses forward. This report analyzes the theoretical and empirical links between inequality in human development in one hand and sustainable development on the other hand. Many academics and intellectuals have recently voiced immense concern about increasing inequality and its detrimental social impacts slowly destroying the civilization's social fabric. This report specifically looks at causality in both directions. Humankind's immense scientific stance and academic rigor in the field of sustainability are burgeoning phases in human history today. The authors discussed impact of inequality in human development as regards sustainability, effects of within-country, and weak sustainability. The other salient features of this report are strong sustainability and preserving critical forms of natural capital. Effects of between-country inequality are the strong pillars of this report. Another section deals with impact of unsustainability on inequality on human development. The road and the surge for human freedom and human progress lies in the hands of sustainable development. In this paper, the authors deeply elucidated the theoretical and empirical links between inequality in human development and sustainability [9]. In today's scientific world, sustainability, chemical process engineering and environmental engineering science are integrated with each other. The authors of this report mainly target the domains of sustainability and equality in order to get a better understanding of human progress.

Organization of Economic Cooperation and Development Report [10] discussed scientific understanding and scientific perseverance sustainable development linking with economy, society and development. Since the Brundtland Commission published its landmark report in 1987, civilization has come a long way in humankind's reflections on sustainable development. In the last 20 years, significant and important progress has been made. Pro-active businesses and large organizations around the world have brought sustainability to their products and processes. A new generation in the field of technology and engineering are slowly evolving. The authors discussed in minute details the meaning of sustainable development, challenges of a global world, the visionary future, production and consumption, and how to measure sustainability. While people living in the least developed nations often lack basic amenities such as food, water and shelter, but they have access to life of health and quality. The developed world also suffers from the same issues. In the last 200 years, the world's economy has grown six-fold and almost ten-fold in the regions which were first to be industrialized. Standards of living, health, education, shelter and sustainability have drastically improved. These are the areas of scientific introspection in this report.

Environmental sustainability and nanotechnology are the absolute needs of the hour. The vision of industrial pollution control which includes industrial wastewater treatment needs to be more reenvisioned as civilization trudges forward. In this chapter, the author elucidates upon the scientific needs, the scientific evolution and the scientific vision in the application of both environmental sustainability and environmental and water remediation.

Table 2 covers scientific advancements in nanotechnology, nanofiltration, and sustainability. This Table 2 briefly discusses the recent scientific achievements in the field of nanofiltration and water and industrial wastewater treatment. The applications of nanoscale materials for water and wastewater treatment, nanocomposites, and nanomembrane are also described in this table.

TABLE 2 Scientific advancements in the field of nanotechnology, nanofiltration and sustainability.

Authors	Significant research endeavor
Shahmansouri and Bellona [4]	Nanofiltration technology in water and wastewater treatment and its applications and costs. Nanofiltration is a recent development in membrane technology with scientific characteristics between ultrafiltration and reverse osmosis. The authors touched upon economic analysis, seawater desalination and wastewater treatment in textile industry.
Farrukh [5]	The authors discussed in details nanofiltration, ultrafiltration and reverse osmosis tools. This book elucidates nanofiltration of oil sands produced water, zeolites mixed matrix membranes, micellar enhanced ultrafiltration processes, and fouling of nanofiltration membranes.
Abdel-Fateh [6]	A review on nanofiltration applications in industrial wastewater treatment is elucidated. The authors discussed in details (1) the present water situation in Egypt, (2) water reuse alternatives, (3) current situation in the field of nanofiltration, (4) nanofiltration membrane transport models, and (5) fouling of nanofiltration membranes.
Dongre et al. [7]	Application of nanocomposite membranes in water treatment is elucidated in detail. The authors discussed (1) bionanocomposites for biosorption, (2) bioremediation of heavy metals, (3) bionanocomposites for water purification, (4) chitosan bionanocomposites, (5) fiber-reinforced bio nanocomposites, and (6) magnetic bionanocomposites.
Raval and Gohil [1]	The authors discussed in details water treatment as an emerging trend. The authors also discussed (1) application of nanoscale materials for water and wastewater treatment, (2) nanomembranes, (3) metal nanoparticles, (4) metal nanoparticles, (5) nanoadsorbents, (6) magnetic nanoparticles, (7) zeolites, (8) dendrimers, and (9) nanofibers.
Argueta and Cifuentes [2]	This is a lucid review on nanomedicine and nanobiotechnology. The authors discussed (1) carbon nanotubes, (2) electrodeposition, and (3) carbon nanotubes membranes.
Bardos et al. [3]	The authors discussed in details Nanoremediation and its future deployment. Nanoremediation is an alternative to soil and water remediation. The authors discussed the application of nanoparticles in soil and water remediation.
Kuhlman and Farrington [8]	The authors discussed the science of sustainability. Sustainability as a policy concept has its origin in Brundtland Report of 1987. The authors also discussed in details the application of sustainability in people, planet and profit, welfare resources, and the future of sustainability.
United Nations Development Programme Report [9]	The authors covered Sustainability and equality in human development. Also the authors discussed (1) the impact of inequality in human development as regards sustainability and (2) effects of between-country inequality.
Organization of economic cooperation and development [10]	The authors discussed Sustainable development linking with economics, society and development. The authors also discussed (1) the challenges of a global world, (2) production and consumption, and (3) how to measure sustainability.
Lu et al. [11]	The authors covered wastewater treatment for carbon dioxide capture and utilization. The authors also discussed (1) microbial electrolytic carbon capture, (2) microbial electrosynthesis, and (3) simultaneous carbon capture for carbon, nitrogen, and phosphorus removal.
Shannon et al. [12]	The authors discussed Science and technology for purification in the coming decades. The authors also discussed decontamination, disinfection, reuse and reclamation, and the visionary world of desalination science.
Dhingra et al. [13]	The authors covered in details Sustainable nanotechnology. The authors also discussed sustainable nanotechnology through green methods and life-cycle thinking, nanomanufacturing methods and green technology concerns and industrial ecology.

Membrane separation processes, fouling, and cleaning

A major limiting step in membrane science and technology is "fouling" of the membrane. Fouling can be seen as a decline in flux with time of operation. In its strictest scientific sense, the flux decline should occur when all operating parameters, such as, pressure, flow rate, temperature, and feed concentration are kept constant [24]. Fouling problems and its scientific issues were the most important reason for the relatively slow acceptance of membrane technology in the early stage of research and development. Fig. 2 depicts the solute transport mechanisms in membrane separation processes. Scientific vision, vast scientific alacrity and scientific verve and validation are needed to get a deep understanding of fouling and solute transport mechanism. The characteristics of fouling are:

- Changes in membrane properties.
- Changes in feed properties.
- Concentration polarization.

Consequences of fouling

An obvious consequence of fouling is higher capital expense caused by the lower average flux over a process cycle [24]. In addition, depending on the nature and extent of fouling, restoring the flux may require powerful cleaning agents, which may decrease the operating life of the membrane. This continues to be a problem with cellulose acetate membrane, which has a limited pH, temperature, and chlorine tolerance range, although it is of less a problem with other polymeric membranes such as polyethylene, polyethersulfone, polypropylene, and polytetrafluoroethylene and much less with ceramic and other inorganic membranes that can highly tolerate aggressive cleaning procedures.

Global water crisis, nanotechnology, and nanofiltration

Nanotechnology and nanofiltration can be helpful in dealing with global water crisis today. Wastewater engineering and wastewater management will truly open up new windows of innovation and scientific instinct in years to come.

Lu et al. [11] elucidated wastewater treatment for carbon capture and utilization. An intense paradigm shift is underway in wastewater treatment and engineering as the industry heads toward 3% of global electricity consumption and contributes 1.6% of greenhouse gas emission. Although there has been plenty and extensive research in energy efficiency and

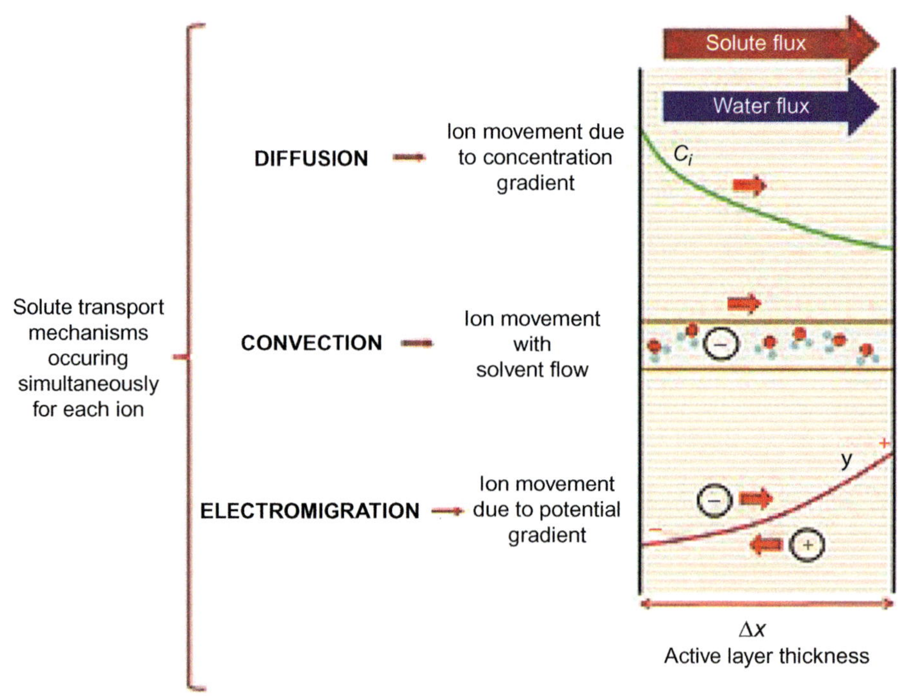

FIG. 2 Solute transport mechanisms in membrane separation processes (Google Images; www.pngegg.org).

renewable energy recovery, studies considering wastewater for carbon capture and utilization are few. This review summarizes alternative wastewater treatment pathways capable of simultaneous carbon dioxide capture and utilization and demonstrates the environmental and economic benefits of microbial electrochemical and phototrophic process. The Paris Agreement on Climate Change aims to hold the global average temperature increase below 2°C, and special efforts are made to limit the increase to 1.5°C by 2100 relative to preindustrial levels. The United Nations Environment Programme recently thoroughly analyzed the major emission gap to achieve and envision this goal. To meet the 2°C target, many studies suggest that it is necessary to balance emission reductions with carbon dioxide level and the massive deployment of negative emission methods because most models indicate it is impossible to reach 1.5°C with a highly 50% chance with a significant negative emissions. Zero-discharge emissions and its abatement are the hallmarks of human scientific progress today. Human mankind's vast scientific progeny, ingenuity and determination are today in the middle of a deep crisis. Carbon capture and industrial wastewater treatment are two opposite sides of the visionary coin today. The authors discussed in minute details approaches for carbon capture unit enabled wastewater treatment, microbial electrolytic carbon capture, microbial electrosynthesis, microalgae cultivation, constructed wetlands, and a vast and varied research pursuit in simultaneous carbon capture unit and carbon/nitrogen/phosphorus removal. Carbon monoxide locations and ambient carbon dioxide capture are the other cornerstones of this well researched treatise. With the emerging and prospective review deeply demonstrates the great potential of using wastewater for carbon capture units, this concept is still at its initial stage. However, a lot of research endeavor is needed and further engineering thrust areas need to be envisioned with the passage of scientific history and time. Carbon capture unit is a promising, emerging, and visionary area of science and engineering. Therefore, technological development and implementation as well as a more detailed technomic, lifecycle, and socioeconomic analyses will be required and envision the potential of such an approach for different industries. Truly a new beginning in the field of carbon capture will emerge and a new vision of mankind will start as time progresses.

Shannon et al. [12] described science and technology needed for water purification in the coming decades. The stance and the challenges of science are today in the middle of vision, prowess, and scientific grit. One of the pervasive problems affecting people around the world is absolute inadequate access of clean water and proper sanitation. A newer beginning is absolute needs of the hour. Problems with water are targeted to grow worse in the coming decades with water scarcity growing globally even in regions where water is considered to be water-rich. Approximately 1.2 billion people lack access to safe drinking water, 2.6 billion have little or no sanitation, millions of people die annually—3900 children a day from diseases transmitted through unsafe water or human excreta. Human scientific research pursuit in the field of environmental protection thus needs to be reframed and reorganized as soon as possible. Water science and technology are at a serious threat and water pollution needs to be mitigated at the utmost. The authors discussed disinfection, decontamination, reuse and reclamation, and desalination science. The work highlighted here plus the tremendous amount of research work done in water and wastewater treatment is veritably sowing the seeds of civilization in water and wastewater treatment. Desalination and membrane science are the areas which are integrated with each other.

Arsenic and heavy metal groundwater remediation and the vast vision for the future

Arsenic and heavy metals contamination of groundwater and drinking water are immense challenges to environmental science and engineering today. The vast vision for the future needs to be aligned with global water research and development initiatives. Membrane science such as nanofiltration and reverse osmosis are the effective and efficient technologies to mitigate water pollution and industrial wastewater treatment. In South Asia particularly India and Bangladesh, drinking water issues are huge human challenges. In such a critical situation, a sound integrated water resource management and wastewater management and engineering should be the hallmarks of human civilization's progress. There are tremendous health effects due to arsenic intake in drinking water. In this paper, the author deeply stresses on certain issues of scientific water innovations such as membrane science. Nanotechnology is a revolutionary area of science and engineering which still needs to be unraveled. Poor and disadvantaged nations around the world are in the middle of an unending water related disaster. The United Nations Development Report reenvisions these water and sanitation issues. Also United Nations Sustainable Development Goals are not properly implemented in many developing countries around the world. Here comes the contribution of scientists, engineers and civil society. A newer beginning will surely emerge in the avenues of human scientific progress if nations around the world take effective steps in mitigating global environmental disasters. There is an immediate need of nonconventional environmental engineering tools such as nanotechnology in heavy metal groundwater remediation. The author in this chapter reiterates and pronounces the need of nanotechnology applications in heavy metal groundwater remediation.

Dhingra et al. [13] discussed sustainable nanotechnology through green methods and life-cycle thinking. Citing the vast and varied applications of nanotechnology, this paper emphasizes the need to conduct life cycle based assessments as early in the development process and also a deep scientific understanding of the potential environmental and health consequences of nanomaterials over the entire life cycle. Technology and engineering are today in the avenues of newer scientific regeneration and sound industrial manufacturing. Sustainable manufacturing is the coin word of global industrialization today. Nanomanufacturing methods often have associated environmental and health effect impacts. These areas need to be envisioned when evaluating nanoproducts for their greenness. The authors discussed in details nanomanufacturing methods and green technology concerns, industrial ecology and life-cycle analysis, and energy intensity of carbon nanofibers and nanoparticles. Proposed solutions for potential health and safety issues in nanomaterials applications in human society are the other areas of research.

Sustainable manufacturing and rapid global industrialization are causing immense scientific and engineering stresses in the rapid progress of human society. A deep scientific introspection is the need of the hour and surely global industrialization will be in a better position as regards health and safety issues of nanomaterials and engineered nanomaterials [14–18]. This is especially true in the field of arsenic and heavy metal groundwater remediation are the immediate needs of the hour. Today United Nations Sustainable Development involves provision of clean drinking water and proper sanitation. In this entire treatise, the author reiterates the need of Sustainable Development Goals in the future emancipation and realization of environmental protection and water and wastewater treatment.

Future perspective

Water science and technology is the futuristic vision of human kind today. In water-scarce countries around the world, desalination and innovative methods are the pillars of research and development initiatives. Future scientific recommendations and future flow of scientific thoughts should be directed toward newer research initiatives in nanofiltration and reverse osmosis. In addition to scarcity of water, water quality is a major issue in many developing and developed nations around the world. The future success of water science lies in the hands of environmental engineers, scientists and global visionaries. In the state of West Bengal, India and the neighboring country Bangladesh, arsenic and fluoride contamination are affecting the population significantly. The application of United Nations Sustainable Development Goals in many developing and developed countries around the world is not properly implemented. So future flow of scientific thoughts should be targeted toward nanotechnology, chemical process engineering, and applied sciences. Zero-discharge norms are the guide words of human progress today. This chapter stresses removal of hazardous contaminants from drinking water, groundwater and industrial wastewater. India as well as many countries around the world is in the grip of an unending global crisis. Lack of pure water, lack of proper sanitation, and illiteracy are the important issues which need to be targeted if the countries around the world truly move forward. So the guideword of today's water science is innovation, vision and purpose [18–25].

Conclusions

Future of civilization today lies in the hands of environmental engineers and scientists. Chemical process engineering, integrated water resource management, wastewater management and systems engineering are today integrated to each other. This chapter describes the status of environmental engineering in mitigating climate change and ecological degradation. Environmental perspectives in today's global scenario need to be overhauled as civilization faces innumerable challenges today. Heavy metal and arsenic groundwater and drinking water contamination are the vexing issues of humankind today. Environmental protection science and environmental sustainability are the absolute needs of the hour. The world of science and engineering is moving at a rapid pace. Today water is a precious item for entire human civilization. Rampant industrialization and enhanced manufacturing are degrading the human environment. Universities and educational institutions around the globe are in search of true integrated water resource management and an innovative wastewater management. This chapter elucidates various conventional and nonconventional environmental engineering techniques for water purification.

Acknowledgment

I would like to thank Dr. Satinder Ahuja for his valuable suggestions and advice.

References

[1] H.D. Raval, J.M. Gohil, Nanotechnology in water treatment: an emerging trend, Int. J. Nucl. Desalin. 4 (2) (2010) 184–188.
[2] C.M. Argueta, J.I. Cifuentes, Nanotechnology: nanomedicine & nanobiotechnology, J. Nanotechnol. Nanomed. Nanobiotechnol. 5 (021) (2018) 1–4.
[3] P. Bardos, C. Merly, P. Kvapil, H.-P. Koschitzky, Status of nanoremediation and its potential for future deployment: risk-benefit and benchmarking appraisals, Remediation 28 (2018) 43–56.
[4] A. Shahmansouri, C. Bellona, Nanofiltration technology in water treatment and reuse: applications and costs, Water Sci. Technol. 71 (3) (2015) 309–319.
[5] M.A. Farrukh, Nanofiltration, Intech Open, 2018.
[6] Abdel-Fateh M.A., (2018), Nanofiltration systems and applications in wastewater treatment, Ain Shams Eng. J., pp 1-16, 2018.
[7] R.S. Dongre, K.K. Sadasivuni, K. Deshmukh, A. Mehta, S. Basu, J.S. Meshram, M.A.A. Al Maadeed, A. Karim, Natural polymer based composite membranes for water purification: a review, Polym. Plast. Technol. Mater. (2019) 1–16, https://doi.org/10.1080/25740881.2018.1563116.
[8] T. Kuhlman, J. Farrington, What is sustainability? Sustain. For. 2 (2010) 3436–3448, https://doi.org/10.3390/su2113436.
[9] United Nations Development Programme Report, (2011), Research Paper, November 2011, Human Development Research Paper, 2011/04, Sustainability and Inequality in Human Development, by Eric Neumayer.
[10] Organization for Economic Co-operation and Development, in: T. Strange, A. Bayley (Eds.), Sustainable Development—Linking Economy, Society and Environment, 2008. Paris, France.
[11] L. Lu, J.S. Guest, C.A. Peters, X. Zhu, G.H. Rau, Z.J. Ren, Wastewater treatment for carbon capture and utilization, Nat. Sustain. 1 (2018) 750–758.
[12] M.A. Shannon, P.W. Bohn, M. Elimelech, J.G. Georgiadis, B.J. Marinas, A.M. Mayes, Science and technology for water purification in the coming decades, Nat. Rev. 453 (2008) 20. Nature Publishing Group.
[13] R. Dhingra, S. Naidu, G. Upreti, R. Sawhney, Sustainable nanotechnology: through green methods and life-cycle thinking, Sustain. For. 2 (2010) 3323–3338, https://doi.org/10.3390/su2103323.
[14] S. Palit, Filtration: Frontiers of the engineering and science of nanofiltration-a far-reaching review, in: U. Ortiz-Mendez, O.V. Kharissova, B.I. Kharisov (Eds.), CRC Concise Encyclopedia of Nanotechnology, Taylor and Francis, 2016, pp. 205–214.
[15] S. Palit, Advanced environmental engineering separation processes, environmental analysis and application of nanotechnology—a far-reaching review, Chapter-14, in: C.M. Hussain, B. Kharisov (Eds.), Advanced Environmental Analysis—Application of Nanomaterials, Vol. 1, The Royal Society of Chemistry, Cambridge, United Kingdom, 2017, pp. 377–416.
[16] C.M. Hussain, B. Kharisov, Advanced Environmental Analysis—Application of Nanomaterials, Vol. 1, The Royal Society of Chemistry, Cambridge, United Kingdom, 2017.
[17] C.M. Hussain, Magnetic nanomaterials for environmental analysis, Chapter 19, in: C.M. Hussain, B. Kharisov (Eds.), Advanced Environmental Analysis—Application of Nanomaterials, Vol. 1, The Royal Society of Chemistry, Cambridge, United Kingdom, 2017, pp. 3–13.
[18] C.M. Hussain, Handbook of Nanomaterials for Industrial Applications, Elsevier, Amsterdam, Netherlands, 2018.
[19] C.M. Hussain, Handbook of Environmental Materials Management, Elsevier, Amsterdam, Netherlands, 2019.
[20] C.M. Hussain, Handbook of Functionalized Nanomaterials for Industrial Applications, Elsevier, Amsterdam, Netherlands, 2020.
[21] C.M. Hussain, Handbook of Manufacturing Applications of Nanomaterials, Elsevier, Amsterdam, Netherlands, 2020.
[22] Hussain C.M. (2020), Handbook of Industrial Applications of Polymer Nanocomposites, Elsevier, Amsterdam, Netherlands.
[23] C.M. Hussain, Handbook of Nanomaterials for Sensing Applications, Elsevier, Amsterdam, Netherlands, 2020.
[24] M. Cheryan, Ultrafiltration and Microfiltration Handbook, Technomic Publishing Company, Inc., Lancaster, Pennsylvania, USA, 1998.
[25] M. Shahid, C. Dumat, N.K. Niazi, S. Khalid, Natasha, Global scale arsenic pollution: increase the scientific knowledge to reduce human exposure, La revue electronique en sciences de l'environnement (2018), https://doi.org/10.4000/vertigo.21331.

Chapter 14

Nanomaterials in polymeric membranes for water treatment applications

C. Ursino and A. Figoli*
Institute on Membrane Technology, Rende, CS, Italy
Corresponding author: e–mail address: a.figoli@itm.cnr.it

Abbreviations

AC	alternating current
Al_2O_3	aluminum
Ag	silver
APTES	aminopropyltriethoxysilane
BSA	bovine serum albumin
BTB	bromothymol blue
CA	cellulose acetate
CAP	chloramphenicol
CNTs	carbon nanotubes
Co	cobalt
CS	chitosan
CR	congo red
CuO	copper oxide
CV	crystal violet
CVD	chemical vapor deposition
DMAC	N,N-dimethylacetamide
DMF	N,N-dimethylformamide
DOPA	dopamine
ECPNC	electrically conductive polymer nanocomposite
EDS	energy dispersive spectroscopy
EMBR	electro-assisted membrane bioreactor
EGCG	epigallocatechin gallate
FO	forward osmosis
FT-IR	Fourier transform infrared spectrophotometer
GA	glutaraldehyde
GN	graphene
GO	graphene oxide
HAP	hydroxyapatite
HBE	hyper branched epoxy
HFs	hollow fibers
IF	inline functionalization
IP	interfacial polymerization
LbL	layer-by-layer
MBR	membrane bioreactor
MB^+	methylene blue
MD	membrane distillation

Separation Science and Technology, Volume 15. https://doi.org/10.1016/B978-0-323-90763-7.00016-0

MF	microfiltration
MMMs	mixed matrix membranes
MOCVD	metal organic vapor deposition
mPEG-SH	poly (ethylene glycol) methyl ether thiol
MWCNTs	multi walled carbon nanotubes
NaX	zeolite
NIPS	nonsolvent-induced phase separation
NMs	nanomaterials
NF	nanofiltration
NFs	nanofibers
NT	nanotubes
NWs	nanowires
NPs	nanoparticles
OD	optical density
PA	polyamide
PAN	polyacrylonitrile
PAM	polyacrylamide
PVA	polyvinyl alcohol
PEI	polyethylenimine
PES	polyethersulfone
PFU/mL	plaque-forming units per milliliter
PI	phase inversion
PMMA	poly (methylmethacrylate)
PPy	polypyrrole
PSf	polysulfone
PTFE	polytetrafluoroethylene
PUs	polyurethane
PV	pervaporation
PVA	polyvinyl alcohol
PVDF	polyvinylidene difluoride
PVDF-HFP	poly (vinylidene fluoride-co-hexafluoropropylene)
RO	reverse osmosis
SA	sodium alginate
SAB	supernatant anaerobic bioreactor
SDS	sodium dodecyl sulfate
SEM	scanning electron microscopy
SiO_2	silicon
SP	sodium polyphosphate
SPECT/CT	single-photon emission computed tomography/computed tomography
SPES	sulfonated polyethersulfone
SPSf	sulfonated polysulfone
SWCNTs	single-walled carbon nanotubes
TA	tannic acid
TiO_2	titanium oxide
TFC	thin film composite
UF	ultrafiltration
UV	ultraviolet
VACNT	vertically aligned carbon nanotube
VIPS	vapor-induced phase separation
VMD	vacuum membrane distillation
WCA	water contact angle
WWTP	wastewater treatment plant
ZnO	zinc oxide
ZrO_2	zirconium

Introduction

The research of the different solutions to overcome the problem of the wastewater treatment shows as recent protagonist the nanomaterials (NMs). This approach based on materials science and nanotechnology opens a new approach to manufacturing higher performing materials. In fact, to date nanotechnology has revolutionized both scientific and technological fields, leading to the production of new materials and devices. In particular, NMs are comprised in the range of 1 to 100 nm and are easily integrated as active components [1]. NMs present unique physicochemical properties, such as high surface to volume ratio, adsorption, and high catalytic activities and reactivity [2,3]. However, NMs present health hazards that differ from those of the native substance [4] and strong concerns regarding environmental contamination, too. One problem using of NMs is their leaching during the water treatment and consequently water contamination [3,5]. Alongside the main features, it is important to emphasize that many NMs possess intrinsic antibacterial properties, and many of them were used precisely for this last peculiarity. For this reason, the research is directing the attention to developing antibacterial NMs highly performing, such as antibiofouling materials and antibacterial coatings [2]. As we know, wastewater treatment represents one of the major problems to overcome. Water pollution is generally due to the presence of bacteria and viruses, heavy metals, oil, soaps and chemicals (e.g., pharmaceutical and personal care products), ions, toxic materials, etc., deriving not only from industrial waste. In this scenario, membrane technology plays a key role in water treatment processes thanks to its important advantages: it does not require chemicals, possesses simple modular design and ease of operation, and again presents advantages in investment and operating costs (low costs). In fact, generally wastewater treatment technologies require (i) high capital investments; (ii) high energy requirements; (iii) high operation and maintenance (O&M) cost; and (iv) large area for the plant. Moreover, the actual methods are unable to completely purify water or reuse the retentate [6]. Membrane processes have been applied with success in water treatment, as microfiltration (MF), ultrafiltration (UF), nanofiltration (NF), reverse osmosis (RO), membrane bioreactor (MBR), forward osmosis (FO), pervaporation (PV), and membrane distillation (MD). However, there are several issues to overcome, e.g., fouling or biofouling phenomenon, that leads to a rapid decline of the membrane performance, permeability, and selectivity. Therefore, in this scenario, the interest in NMs and the related nanocomposite membranes are gaining increasing interest, as evidence of the growing number of publications (Fig. 1) on the topic. Nanocomposite membranes can also be found in the literature with different names as mixed matrix membranes (MMMs) [7], nanoenhanced membranes [8], or hybrid membranes [9]. These new advanced membranes can be prepared by dispersing the nanoparticles (NPs) in the polymer dope solution or introducing them on the membrane's surface by chemical surface modification [10] or coating technique [3]. Improvement of the permeability and selectivity [11,12], reduction of the biofouling [13–15] and increase in the rejection [1,5,16], represent three important features of the NPs loaded into the membranes. In this chapter, an overview of the most common NPs, used in the polymeric membrane preparation for water treatment applications, is described in detail. In particular, for each NPs considered, the properties, the membranes preparation/immobilization and their applications in water treatment, will be treated. The NPs protagonists of this chapter are carbon nanotubes (CNTs), titanium oxide (TiO_2), silver (Ag), and graphene oxide (GO). In addition, the use of other materials, including the 2D materials, will be briefly discussed.

Carbon nanotubes (CNTs)

Structural and functional properties of CNTs

CNTs, constituted by one or more layers of extremely graphitic channels with nanoscale diameter, can be divided into single-walled CNTs (SW-CNT) or multiwalled CNTs (MW-CNT) depending on the arrangement of graphene sheets (single layer or concentric layer of graphene). The resulting CNTs can have three different types of chirality, depending on the rolling angle of the graphene sheets: armchair, zigzag, and chiral one. The type of chirality will affect their transport properties, particularly the electronic properties [17]. Because CNTs are seamless cylinders of graphene (one or more layers, as reported above), the perfect CNTs crystalline structure requires all carbons to be bonded in a hexagonal lattice except at their ends. In fact, often defects can occur during mass production. These defects introduce pentagons, heptagons, and other imperfections in the CNTs structure that reduces the general properties [18]. In the last few years, CNTs have attracted the attention of researchers due to their exceptional properties such as large surface area, higher mechanical strength, thermal stability, attachment capacity (adsorbent), electrical conductivity and partial antibacterial activity, inducing oxidative stress in bacteria, destroying it [3,19,20]. One of the most important peculiarities of the CNTs is the exceptional water transport property. CNTs possess inner selective nanopores that improve transport kinetics, of about 3–4 orders of magnitude higher, respect the expected Hagen–Poiseuille equation, without a decrease in selectivity. This property is due to the fact that the fluid crosses the CNT's channel almost without friction; therefore, the laminar flow has

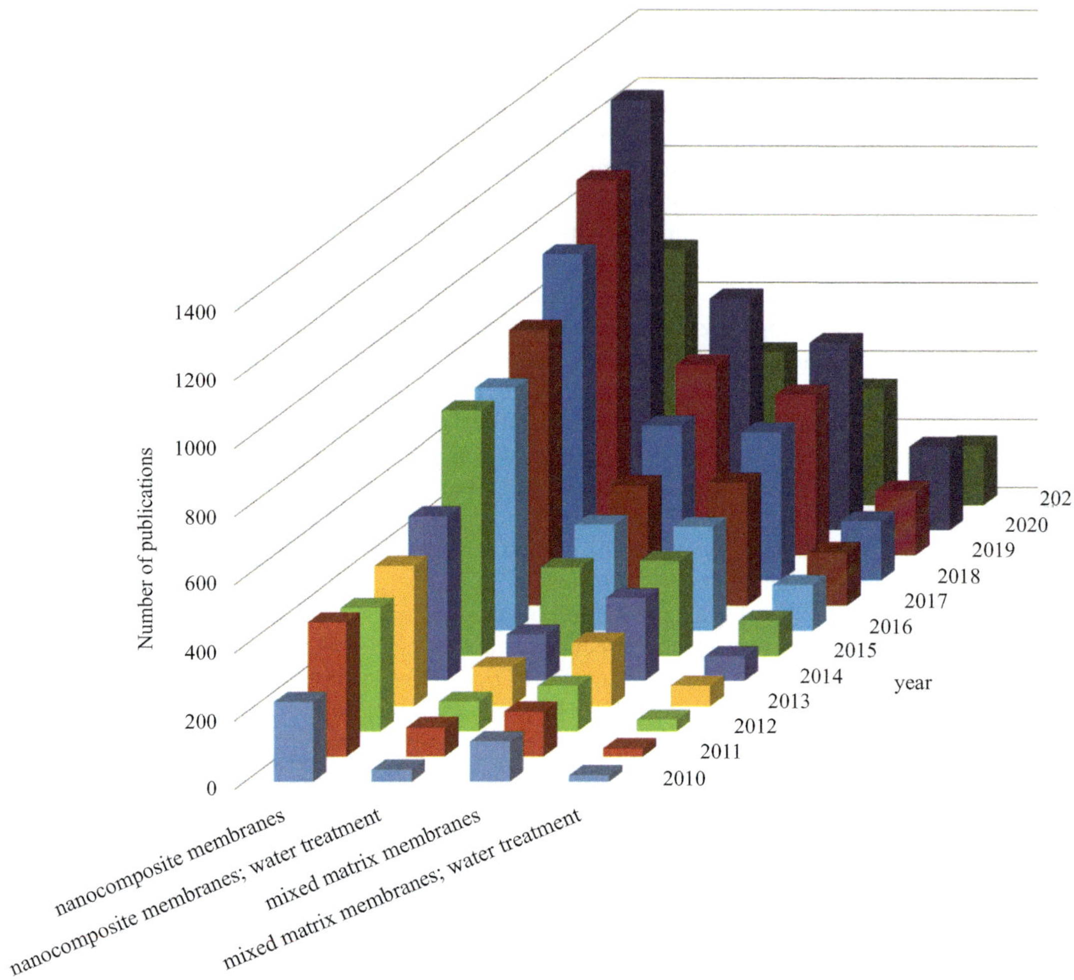

FIG. 1 Numbers of publication related the use of NPs in the membrane technologies for water treatment applications, over the last 10 years. Data acquired from Scopus (www.scopus.com) using the reported keywords (data available on 28/06/2021).

not zero fluid velocity at the pore walls (as predicted by the Hagen–Poiseuille equation) [21,22]. This phenomenon has been widely studied, demonstrating that the interactions between the water molecules are stronger than the interactions of the water molecules with the CNTs wall [23]. Moreover, the water molecules pass through the channel without friction thanks to the interaction between the hydrogen bonding in the chain of water molecules and the hydrophobic core of CNTs and again, between carbon atoms and water molecules [24]; therefore, both hydrophobicity and the roughness [25] as well as curvature of the graphitic surface of the CNTs [24], play key roles.

A disadvantage of their use could be correlated to the difficulty design and achievement of biocompatible CNTs. In fact, some studies have showed that CNTs tend to be water insoluble and are cyto- and ecotoxic [26–28]; however, some functional chemical groups (e.g., hydroxyl, carboxyl, and amine groups) can be used to functionalize CNTs for enhancing their water solubility and biocompatibility [20,28].

CNT membrane preparation

The production of CNTs membranes is not a very simple process. In fact, the size, the distribution, and their uniformity, are very crucial parameters that affected the stability of membranes [29]. Although CNTs can improve membrane selectivity and permeability and simultaneously reduce the phenomenon of fouling, their hydrophobic nature makes their distribution difficult in addition to the aggregation phenomena. Generally, it is possible to distinguish two types of CNTs membranes: (i) freestanding CNTs membranes and (ii) composite CNTs membranes. In the first cases, it is possible to differentiate two alternatives where the CNTs are vertically aligned (VACNT) [30–32] (Fig. 2A) or is possible to have Bucky-paper CNTs membranes [33–35] (Fig. 2B). An important feature of the VACNT is an "ultrafast water transport" through the CNT;

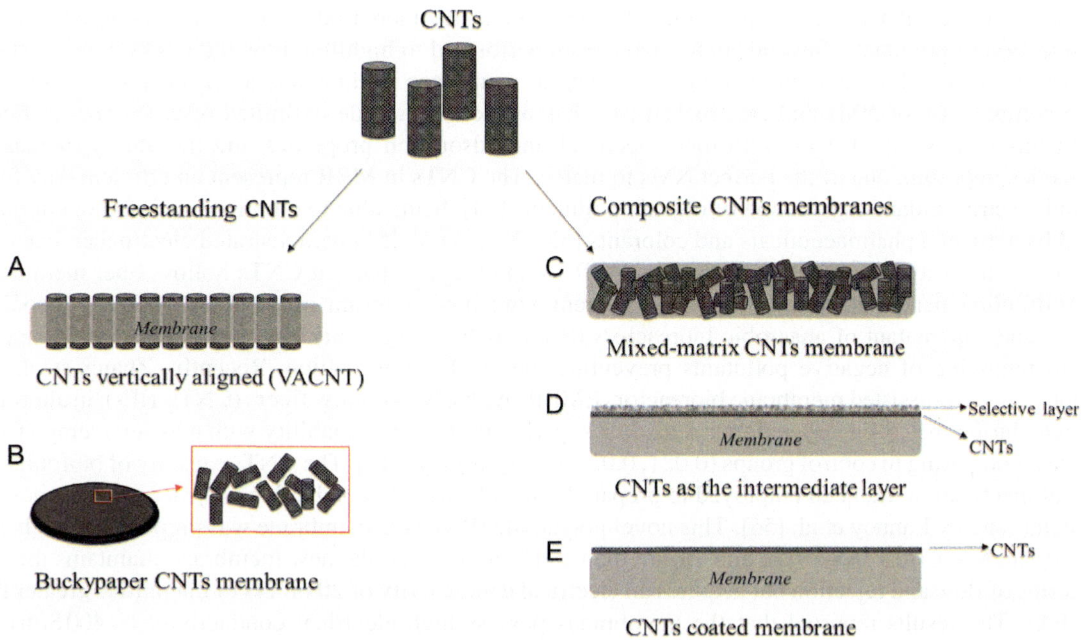

FIG. 2 Types of CNTs membranes. Freestanding CNTs: (A) CNTs vertically aligned (VACNT); (B) Bucky paper CNTs membranes. Composite CNTs membranes: (C) as mixed-matrix; (D) as intermediate layer; (E) coated.

however, the preparation of these membranes is much elaborated with respect to the other techniques. While, the second case, based on disordered CNTs powders, includes three categories of CNTs membranes, depending on whether CNTs were arranged in a disordered manner, or mixed into the polymeric network [36,37]. In this last case, it is possible to differentiate them in Fig. 2C–E.

- Mixed-matrix, where the polymer/CNT compatibility is a crucial factor for membrane performance and stability. The CNTs were dispersed in the dope solution, and the membranes can be prepared via the phase inversion (PI) method [38,39].
- CNTs as the intermediate layer. In this way, it is possible to produce a thin film composite (TFC) membrane [40,41].
- CNTs-coated membranes [42–44].

To overcome the problems of distribution, agglomeration and poor interfacial interaction between CNTs and polymer matrix, during the membranes preparation, or to improve the CNTs membranes performance, functionalized CNTs have been proposed. For this reason, two types of approaches were used to solve the problems mentioned above: mechanical dispersion and functionalization treatment. The mechanical dispersion includes the (i) sonication; (ii) calendaring; (iii) ball milling; (iv) shear mixing; and (v) extrusion. As a mechanical process, the sonication treatment is the most frequently used method for dispersing them in the solution to enhance their distribution [17,37]. However, particular attention should be taken if the sonication treatment is too long and therefore too aggressive. The CNTs can be damaged as the graphene sheet is destroyed [45]. Concerning the functionalization methods, depending on the interaction between the active molecules and carbon atoms on the CNTs, it is possible to distinguish between chemical functionalization and physical methods. The chemical functionalization is due to the covalent bond of functional groups onto the carbon of CNTs. This binding can be performed at the end of the CNTs or on their sidewalls. Generally, it was used to provide hydrophilicity on the CNTs through the attachment of polar groups. It was possible to differentiate the chemical, in direct sidewall or defect functionalization. Compared to chemical functionalization, the physical methods present (noncovalent functionalization) several advantages, such as ease in the method, cheaper and the CNTs are not damaged during the treatment. Physical methods include the using surfactants (nonionic, anionic, and cationic) to improve the CNTs dispersibility and prevent their aggregation phenomena [17].

CNT membrane application for water treatment

CNTs membranes are promising to have superior desalination, disinfection, and filtration properties, and fouling resistance with respect to the common membranes, thanks to their properties reported above. Biofouling phenomenon decreases due to the cytotoxic effect of CNTs, and consequently, membrane life increased. Moreover, functionalized CNTs allow selective rejection of pollutants. Several studies have been performed to highlight how the CNTs improve the membrane performance [36,46,47]. For example, external and internal membrane fouling was a serious problem that affected the MBR. The combination of NMs and electrochemistry has opened up a wide unlimited research area to reduce fouling [48–53]. In this context, the CNTs, with their electrical and adsorption properties and the ability increasing electron transfer kinetics, represent one of the perfect NMs to match. The CNTs in MBR represent an efficient way for adsorptive trapping and electro-oxidant degradation of organic pollutants [54], being able to eliminate various contaminants, such as viruses and bacteria and pharmaceuticals and colorants [55]. Yang et al. [51] demonstrated electrochemical control could provide a promising way to reduce the fouling in MBR using electro-enhanced CNTs hollow fiber membranes (CNTs-HFMs). Antifouling performance was evaluated by removing bovine serum albumin (BSA-92.1%), sodium alginate (SA-87.3%), and supernatant of anaerobic bioreactors (SAB-56.8%). This work shows how the optimal applied voltage is related to removing of negative pollutants preventing the gel layer formation. Recently, Zhang *et al.* reported the combination of electro-assisted membrane bioreactor (EMBR) with CNTs hollow fibers (CNTs-HFs), in alternating current (AC) voltage, during the 73-day operation, results in a superior antifouling capability with a lower average fouling rate of 0.017 bar/day, comparing to control groups (0.021, 0.025, 0.029 bar/day) [48]. The CNTs capacity of biofouling prevention and bacterial inactivation has been employed to prepare highly electrically conductive polymer nanocomposite (ECPNC) tight NF membrane by Lannoy et al. [56]. This novel polyamide (PA) TFC membrane was prepared through cross-linking with carboxylate MWCNTs (10–30 μm and <8 nm diameter). In this way, the new membrane maintains the same performance in terms of flow and rejection but acquires an electrical conductivity of 20 orders of magnitude greater than the base polymer (PA). The results reported that the membranes possess high electrical conductivity (~400 S/m), good NaCl rejection (>95%; 1 g/L NaCl), and good water permeability ($3.2 \pm 0.4\,L/m^2$ h·bar) [56]. CNTs are successfully applied in seawater desalination based on their ability to "high water flux," with high salt rejection and low membrane fouling. Indeed, CNTs are able to reject dissolved ions while enabling water molecules to pass through [36,47]. The selective ions rejection depends on the CNTs diameter, and a mathematical modeling study revealed that a CNT diameter comprised between 0.34 and 0.39 nm (3.4–3.9 Å) allows water passage into the interior and at the same time excludes the bigger ions as Na^+ and Cl^- [57]. However, literature reported several studies where the use of CNTs with larger diameters lead to optimal ion rejection, also thanks to CNTs with chemically modified tips with charged groups (carboxylate or amine) [58]. Furthermore, CNTs functionalized with oleic acid (65 nm and 2 μm, of diameter and length, respectively) were used to coat polytetrafluoroethylene (PTFE) membrane for desalination applications via vacuum membrane distillation (VMD) [59]. Thus obtained PTFE/oleic acid-functionalized CNTs composite membrane showed better performance in terms of water flux ($89\,L/m^2h$) and salt rejection (99.45%) than the same one prepared with carboxylate CNTs. In fact, as reported by the authors the presence of oleic acid with their hydrophobic natures increases hydrophobicity of the produced novel membrane, with a water contact angle (WCA) of 133°, and therefore improves their performance in the VMD process [59]. As the hydrophobicity is crucial parameters in the MD processes, CNTs enhance membranes with high flux and stable performance [34,42,44,46,58]. The use of CNTs composite membranes finds many applications in RO and FO process [30,60–63]. Yang et al. introduced [61] a CNTs interlayer between the dense PA layer and electrospun porous polyacrylonitrile (PAN) support layer in a TFC FO membrane. The introduced CNTs layers allows that the PA layer have a better coverage on the substrate, reducing the defects and improving the selectivity. The osmotic performance was evaluated under AL-FS (active layer facing feed solution) and AL-DS (active layer facing draw solution) mode and in all cases the water flux of the TFC-CNTs-incorporated membranes was higher than unmodified one (~+84%). In addition, the CNTs interlayer enhanced the selection of the TFC membrane during the salty solution experiment (2.0 M NaCl) resulting in a reverse salt flux approximately, 67%–75% lower than the control membrane, in an AL-DS mode. Among all the above applications, CNT composite membranes are also used for removing oil from wastewater [64–66] or heavy metals and dyes [20,67,68] (Table 1).

Titanium oxide (TiO_2)

Functional properties of TiO_2 and photocatalytic activity

Titanium dioxide (TiO_2) is one of the most interesting inorganic materials, thanks to its multiple properties like high photosensitivity, special hydrophilicity, availability, nontoxicity and eco-friendliness [2,3]. Moreover, TiO_2 NPs possess high chemical stability at a low price, and were applied in several water disinfection and decontamination processes [69,70]. In nature, TiO_2 exists in three different crystalline forms: anatase (tetragonal), rutile (tetragonal) and brookite (orthorhombic) (Fig. 3) [3,71]. The anatase phase possesses antibacterial property, finds application in medicine, and it is the preferred

TABLE 1 Brief summary of the most recent CNTs composite membranes for water treatment applications

Polymer material	CNTs type	Membrane application	Main results	Reference
Chitosan	MW-CNTs buckypaper	NF/RO	In correlation to the type of biopolymer modifiers, the best water permeability was obtained for the MWNT/chitosan membrane (0.87 ± 0.03 L/m^2 h·bar), while the MWNT/chitosan-crosslinked membranes showed the highest salt rejection performance (80%–95% for $MgCl_2$, 21%–63% for NaCl, 18%–37% for $MgSO_4$ and 6%–14% for Na_2SO_4)	[34]
PVDF	CNTs coated membrane	MD	The low content of CNTs promote a good dispersion during the coating via spraying method, with good flux (28.4 kg/m^2 h) during 26 h of VMD (3.5 wt% NaCl, T=80°C)	[42]
Cellulose Tri-Acetate	CNTs coated membrane	FO	Membrane fouling was greatly improved via electrochemical operation modes	[60]
PSf	CNTs MMMs	MBR	The addition of CNTs improves the flux, with chemical oxygen demand (COD), volatile fatty acid (VFA), and suspended solids (SS) concentrations of 45, 32 and 15 mg/mL, respectively (85 days of operation)	[50]
PUs	MW-CNTs coated membrane	MBR	CNTs promote the fouling reduction, by adsorbing activated sludge and other contaminants, with an increases of 190% of the overall long-term (90 days of operation in NSMBR mode)	[55]
PTFE/ PVDF	CNTs coated membrane	MD	CNTs functionalized using oleic acid improve the composite membrane performances in terms of WCA (132°), permeation 111.25 L/m^2 h·bar and salt rejection 99.5%	[59]
PAN/PA	CNTs as intermediate layer	FO	The CNTs interlayer improves the FO water flux of ~83.55%, while the reverse salt flux (2.0 NaCl) in the AL-DS mode was 75.58% lower than the virgin membrane	[61]
PSf/PVP	MW-CNTs MMMs	UF	MW-CNTs were functionalized via hydroxylation and carboxylic acid functionalization. Best results was obtained using the carboxylic acid functionalized MCNTs with a BSA (50 ppm) rejection in the range of 63%–72% and a 100% oil rejection (oil/water emulsion, 1000 ppm lube oil)	[65]
-	CNTs hollow fiber membranes (CNTs-HFMs)	EMBR	CNTs-HFMs presented a superior antifouling activity in an EMBR with a lower fouling rate of 0.017 bar/day, comparing to control groups where it was between from 0.021 to 0.029 bar/day, during 73 day of operation	[48]
PVDF	MW-CNTs coated membrane	EMBR	CNTs improves stability, electrical conductivity and wettability, with good antifouling performance against BSA, SA and HA	[52]

phase for photocatalytic processes, while the rutile phase is preferable for use in optoelectronics devices. Depending on the crystalline form, its physical properties and chemical and biological activities change. TiO_2 crystal is an indirect semiconductor in the anatase phase, but it is directly in the rutile phase. Rutile is the most stable and chemically inert form and anatase can be converted in rutile at high temperatures (400–700°C) [72–74].

Generally, TiO_2 is produced in two forms, anatase or rutile, and commercially the most used is TiO_2 Degussa P25. This powder (particle size of 21 nm) is a blend of anatase and rutile (80:20), with 50 m^2/g of the surface area [74]. The photocatalytic activity of TiO_2 is due to the reaction of ultraviolet (UV) radiations and the consequent release of strong oxidants, $^{\bullet}OH$, in aqueous solutions. Photons with wavelengths around 380–388 nm are able to excite the TiO_2 with a band gap energy of 3.2 and 3.0 eV for anatase and rutile phase, respectively [75]. The mechanism of photocatalytic activity of TiO_2 is reported in Fig. 4.

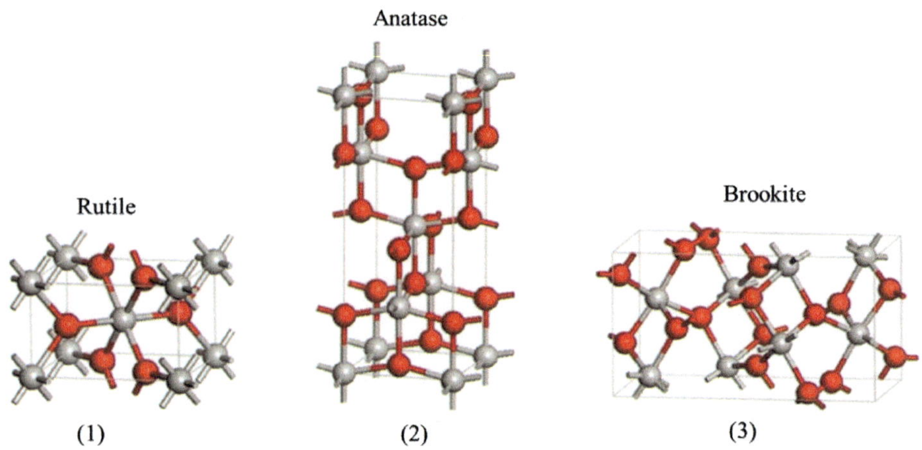

FIG. 3 Representation of TiO_2 crystalline forms of (1) rutile, (2) anatase and (3) brookite (spheres: *red*—02, *gray*—Ti). *Adapted from E. Baranowska-Wójcik, D. Szwajgier, P. Oleszczuk, A. Winiarska-Mieczan, Effects of titanium dioxide nanoparticles exposure on human health—a review, Biol. Trace Elem. Res. 193 (2020) 118–129. doi:10.1007/s12011-019-01706-6.*

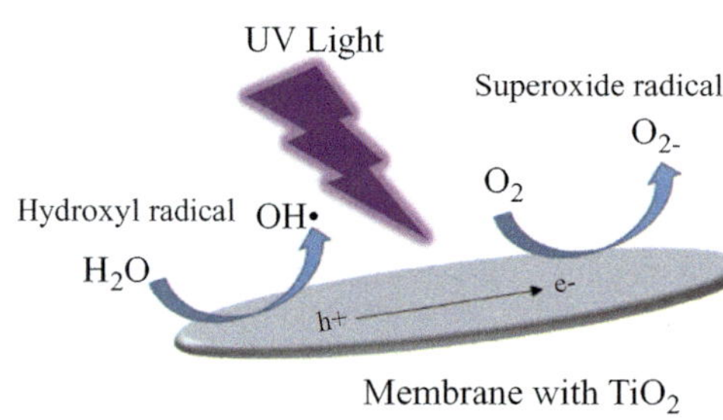

FIG. 4 Mechanism of TiO_2 photocatalytic activity.

After UV irradiation the electron of the photocatalyst becomes excited, producing a pair of a negatively charged free electron and a positively charged electron hole. These electrons and holes react with atmospheric water and oxygen (H_2O and O_2) to yield reactive oxygen species (ROS), such as hydroxyl radicals ($OH^•$), superoxide anions (O_2^-). Superoxide ions can react with H^+ ions deriving from water, thus generating $HO_2^•$ and hydrogen peroxide (H_2O_2) [3,74,75]. These ROS generated after TiO_2 UV irradiation are able to degrade all kinds of contaminants with antimicrobial and disinfection activity. The total efficiency of TiO_2-based photocatalysts depends on three factors: (i) pollutants types and concentration; (ii) UV irradiation in terms of distance and amount absorbed on their surface; and (iii) exposure time of UV irradiation [76]. Finally, after a cycle of the photocatalytic reaction, TiO_2 returns to its original state being ready for another ones. One disadvantage of the use of TiO_2 as a photocatalyst is due to its requirement for UV irradiation; in fact, only 4% can be arrived by the sunlight spectrum. To overcome this problem, new research direction is on the preparation of TiO_2 membranes, which can be directly photo-activated by visible light.

TiO_2 membranes preparation

As reported above, TiO_2 NPs can remove different types of contaminants from wastewater and are comparatively less expensive than other NMs. TiO_2 membranes can be classified in two categories: (i) freestanding-pure TiO_2 photocatalytic membrane reactors with different morphologies such as nanotubes (NTs), nanofibers (NFs), and nanowires (NWs) [76], and (ii) composite TiO_2 photocatalytic membrane. To prepare photocatalytic membranes, TiO_2 NPs can be synthetized using different methods: (i) metal organic chemical vapor deposition (MOCVD); (ii) sol-gel method; (iii) microwave method; (iv) direct oxidation method; and (v) reverse micelle method [74]. Also, in this case, as already reported for CNTs, to avoid agglomeration and improve the stability of TiO_2 NPs (dispersed in solutions), physical and chemical methods are employed. Concerning the membranes preparation, we can differentiate two general TiO_2 NPs introduction, which are TiO_2 suspended and TiO_2 immobilized in/on the support. Obviously, the use of TiO_2 NPs suspended presents several advantages due to the direct contact of the NPs with the pollutants. However, this method requires a posttreatment, where the catalyst is removed.

On the other hand, the possibility of immobilizing the TiO_2 in/on the support requires a single step, but then the catalytic power decreases [77]. In literature, several approaches of TiO_2 immobilization are reported ranging from self-assembly, coating to physical or chemical cross-linking [78]. A comparative study of the effects of mechanical and chemical modification of TiO_2 NPs (Degussa P25; 20nm) for avoiding agglomeration was reported by Razmjou et al. [79]. In this study, two different approaches were examined: (i) dispersion of NPs in di-methylacetamide (DMAC) by sonication in a bath for 15min followed by 10min sonication by a probe (mechanical modification) and (ii) surface pretreatment approaches with aminopropyl-triethoxysilane (APTES) as silane coupling agent (chemical modification). Polyethersulfone (PES) membranes were prepared by immersion precipitation technique using modified TiO_2 NPs at different concentrations (2–4–6wt%). The authors reported that the introduction of TiO_2 NPs improves the membrane performance in each case, but at high TiO_2 concentration the modifications may not be sufficient to prevent agglomerations. Galiano et al. [80] employed two new approaches to enhance TiO_2 NPs (0.5wt%) dispersion in the dope solution via chemical modification using sodium dodecyl sulfate (SDS) combined with mechanical modification using ball milling and ultrasonics. After the functionalization, the TiO_2 NP_S were dispersed in the dope solution, and polyvinylidene difluoride (PVDF) HFs were successfully prepared. TiO_2 NPs can also be just dispersed in the solvent before the addition in the casting solution [81–83]. TiO_2 NPs for removal of (micro)pollutants from the water was reported by Geltmeyer *et al.* [16]. In this work, TiO_2 NPs were immobilized on two types of support: polyamide 6 and silica NFs membranes. Two immobilization techniques were employed, specifically (i) inline functionalization (IF) varying the addition of TiO_2 (0.2–5–35wt%) to the electrospinning solutions, and (ii) dip-coating of the NFs using aqueous TiO_2 NPs suspensions (0.12, 0.34 and 0.5 $mol\,L^{-1}$). The results have been demonstrated that IF was the method preferred for polyamide 6 NFs while dip-coating for silica NFs membranes. Zhang et al. prepared TiO_2 NPs via the sol-gel method, the NPs were added to the solution under stirring at 60°C for 6 h. The solution was then degassed under vacuum at room temperature for 4 h. Organic–inorganic hybrid polysulfone (PSf)-TiO_2 HFs UF via dry–wet spinning method were then obtained [84]. TiO_2/Pebax/PSf-PES TFC membranes were prepared to coat TiO_2 NPs (21 nm, Degussa) on their surface at different concentrations (0–0.01–0.03–0.05 wt%) via dip coating technique. TiO_2 NPs were initially dispersed in distilled water, and after a sonication (1 h) and stirring (1 h), the obtained solution was coated on the TFC membranes. The immersion time was fixed for 1 h and the obtained TiO_2/Pebax/PSf-PES TFC membranes were then washed and dried at room temperature [85]. TiO_2 was grafted onto the surface of PVDF/PAN NFs using a new surface modification method which includes plasma and chemical pretreatments. The authors reported that the crystallinity structure of the obtained TiO_2 NPs can be classified as a polyamorous, with an average particle size larger than pure anatase or rutile phase [86]. The preparation of new environmentally-friendly 3D eggshell membrane (ESM)-supported TiO_2 (TiO_2 anatase, 99.8%) was reported by Li et al. [87]. In this study, a complex of ESM and TiO_2 NPs (TiO_2-polyethyleneimine (PEI)-ESM) was formed by self-assembly approach through electrostatic interaction. ESM samples were treated with PEI and glutaraldehyde (GA) and soaked in TiO_2 NPs aqueous solution for 12 h. In this way, TiO_2-PEI-ESM not only can be used as pollutant adsorbent and photocatalyst but also can be facilely separated from the reacting system and reused.

TiO_2 membranes application for water treatment

As known, membrane fouling is one of the major problems that affect the membrane processes. Generally, membrane fouling and biofouling depends on foulant adhesion/deposition with consequently filtration resistance, pore blocking and gel/cake layer formation. The use of membranes containing TiO_2 presents several advantages due to its photocatalytic property. As reported above, the produced ROS are able to destroy the organic foulant into harmless inorganic substances. Moreover, membranes with TiO_2 NPs possess self-cleaning property, higher hydrophilicity, and antibacterial activity. Ehsani and Aroujalian [88] investigated the effects of TiO_2 NPs concentration (2–10 wt%) on PES/TiO_2 NFs. NFs performance was evaluated in terms for phenol degradation under UVA irradiation. The study highlighted the importance of the TiO_2 NPs concentration during the phenol degradation (phenol concentration 120 mg/L; pH 7), resulting that when the TiO_2 content increased from 4% to 8%, the phenol degradation also increased from 27.0% $\pm$0.7% to 43.0% $\pm$0.3%. When higher TiO_2 NPs concentration was used (10 wt%), the NPs tend to aggregate with consequently reduction of the number of active sites, electron/hole pairs generation rates and OH radicals' formation. TFC PES NF membrane with coated TiO_2 NPs was prepared via interfacial polymerization (IP) using tannic acid (TA) in a single step by Li et al. [89]. Specifically, membrane was prepared at a TiO_2 NPs loading of 0.020 wt%; the membrane was characterized and tested in comparison with a control membrane. The TiO_2 TFC membrane presented a water flux 284% higher than the control TFC membrane and possesses enhanced NaCl and Na_2SO_4 rejections of 57.9% and 94.6%, respectively. The antibacterial activity was evaluated against Escherichia coli (*E. coli*) as model bacteria, at 37°C for 24 h. The TiO_2 TFC membrane presented a significantly higher antibacterial ratio than TFC control membrane both in dark condition (0% vs 77.2% of antibacterial ratio) and after a UV illumination pretreatment (22.5% vs 99.2% of antibacterial ratio). This is a promising study for those applications where real-time UV illumination is not always possible. TiO_2 membranes were also employed for oily wastewater treatment [86,90–94]. For example, PVDF/PAN NFs covered by TiO_2 NPs using plasma and chemical pretreatments were successfully applied for oily wastewater separation [86]. A water/oil mixture (50% v/v) was used during the separation experiments and the results indicated that TiO_2 grafted to the hydrophilic membrane surface prevents direct adhesion of the oil droplets, reducing the fouling phenomenon (oil/water permeability of $\sim 3*10^4$ L/m^2 h·bar after 10 cycle, with respect to $\sim 0.3*10^4$ L/m^2 h·bar after four cycles, for control membrane). Again, PES/sulfonated polysulfone (SPSf)/TiO_2 MMMs were produced using different polymer concentrations (18–28 wt%) and varying the addition of TiO_2 NPs (0.05–0.15 wt%). The TiO_2 NPs size used in this work was 19 nm; the authors reported as the use of smaller NPs (8 nm) leads to aggregation phenomenon due to their larger specific surface area and antiparticle tension, or as larger NPs (up to 25 nm) tend to accumulate on the pore walls due to poor dispersibility. The produced membranes were evaluated in terms of oil rejection performance, with best result for the membrane prepared using 0.075 wt% of TiO_2 NPs. This membrane presents a water permeability of 555 L/m^2 h bar and rejection values of 61%, 72%, and 90% for emulsions containing 100, 300, and 900 ppm oil, respectively [94]. PVDF membranes with immobilized TiO_2 NPs produced using alternative nontoxic solvent [82,95] have been demonstrated an higher Congo Red (CR) and Tartrazine rejection (99%–81% respectively) [95], and a 99% of photocatalytic degradation of methylene blue (MB^+) [82]. Titanium dioxide doped hydroxyapatite (TiO_2-HAP) nanocomposites were synthesized by a hydrothermal method and the prepared PSf/TiO_2-HAP membranes were used for chloramphenicol (CAP) degradation. The experiments results showed that all PSf/TiO_2-HAP membranes were more hydrophilic than the pristine PSf membrane, with a WCA of 60°. The photocatalytic performance was evaluated using 50 mg/L CAP feed solution (pH 5) with and without UV irradiation, with a highest flux of 4.5 L/m^{2}h for the nanocomposite membranes. The CAP degradation rate into various byproducts, increased with increasing TiO_2-HAP concentrations (from 0 to 4 wt%) with a maximum value of 61.59% when 4 wt% of TiO_2 NPs was added [96]. Wu et al. [97] presented PAN-TiO_2 composite microfiber membranes prepared varying the amounts of TiO_2 NPs (0–0.5–0.7 g). This work reported for the first time that TiO2 NPs can be homogeneously distributed on the microfiber surface instead of embedding/agglomerating inside the substrate during the electrospinning. The photocatalytic activity was evaluated by the removal efficiency of Methylene Blue (MB^+) from aqueous solution (50 mL; 5 ppm) under simulated sunlight irradiation (300 W, Ultra Vitalux 230 V E27). Membrane prepared with 0.5 g of TiO_2 NPs (1,1 and 0.05 g of PAN, PVP and sodium polyphosphate (SP), respectively), showed the highest adsorption capacity and the removal of MB^+ achieved 98.9% (Table 2).

Silver (Ag)

Functional properties of Ag and antibacterial activity

Silver (Ag) compounds were widely used in different fields such as biomedical, water and air purification, food production, etc. [3,20,98,99]. Thanks to their broad-spectrum antibacterial activity, nanosilver is the most promising material with respect to the other NPs and is easily available. Like all NPs, also in this case it is possible to have problems of agglomeration, uncontrollable release of Ag^+ or bacteria adhesion; however, the production of composite Ag/membranes for antibacterial applications is very attractive. Ag NPs are able to release silver ions with bactericidal activity, furthermore small Ag NPs (<10 nm in diameter) affect directly viruses although with lower microbicidal effect [99]. Several approaches can be used for Ag NPs production, specifically chemical or physical synthesis. These methods include photo or chemical reduction, seed-mediated growth, templating synthesis, evaporation/vapor condensation, pyrolysis method, etc. [3,99,100]. Recently, near these Ag NPs synthesis, new green synthetic pathways have also been proposed and discussed by Nakamura *et al.* [99]. The antibacterial activity of the Ag NPs is directly correlated with their size, in fact is stronger when the particle size is smaller (<10 nm) but also depends on their shape, and chemistry [101]. As reported by Guo et al. [100], several studies were focused on the possible antimicrobial mechanism of the nanosilver; however, it is still not fully elucidated. It is possible to differentiate between three possible mechanisms:

- Nanosilver-released silver ions; silver ions can interact with the thiol groups of enzymes and proteins necessary for bacterial life, plasma membranes and bacterial DNA, thus affecting cell respiration and destroying the cells.
- Intrinsic antibacterial activity; this mechanism is similar to silver ions, in fact nanosilver can penetrate into bacterial cells, causing several structural changes and cell death.
- Production of ROS as antibacterial agent; ROS production due to oxidative stress, causes impairments in proteins, lipids and DNA, to cause bacteria death [102].

Moreover, Ag NPs possess antiviral action but the mechanism is still not well understood [99].

TABLE 2 Brief summary of the most recent TiO_2 membranes for water treatment applications

Polymer material	TiO_2 type	Membrane application	Main results	Reference
PVDF/ PVP	TiO_2 MMMs	NF	TiO_2 NPs improve the antifouling property of the membranes, in particular the BSA rejection of the TiO_2 membranes was comprised from 78.53% to 95.35%, while the R% of the unmodified membrane was 64.21%	[81]
PVDF/ PMMA	TiO_2 MMMs	UF/MF	The membrane prepared with the highest TiO_2 concentration (0.5 wt%) presented the better photocatalytic degradation of MB (~99%), without PWP changes before and after the exposure to UV light irradiation	[82]
PVDF/ PVP	TiO_2 NFs membranes	NF	The 1.0 wt% of TiO_2 NPs in the NFs improved membrane flux with a copper rejection of 98.18%	[83]
PES	TiO_2 NFs membranes	MF	TiO_2 NPs concentration affected the phenol degradation in synthesized wastewater. Phenol degradation was enhanced by increasing the TiO_2 content until the 8 wt%	[88]
PSf	TiO_2 MMMs	UF	The best HFs performance in terms of hydrophilicity, permeability and anti-fouling properties were obtained when the 2 wt% of TiO_2 NPs was employed	[84]
PES	TiO_2 MMMs	NF	TiO_2 NPs were modified using tannic acid in order to promote the adhesion during the TFC membranes preparation. The produced TFC membranes exhibited a water flux of 28.8 L/m^2 h and good salt rejection (Na_2SO_4, 94.6%)	[89]
PVDF	TiO_2 MMMs	PMR	TiO_2 NPs were used for the production of PMR coupled with a MBR for the treatment of oil refinery wastewater. PMR with TiO_2 improves the fouling resistance without membrane degradation	[91]
PAN	TiO_2 NFs	Adsorption mechanism	TiO_2 NPs were uniformly dispersed on the porous polymer matrix. The new membrane exhibited excellent adsorptive-photocatalytic property to degrade MB (98.9%)	[97]
PES/SPSf	TiO_2 MMMs	UF		[94]
PVDF/ PMMA	TiO_2 MMMs	UF	TiO_2-PVDF/PMMA separation process capacity depends on factors other than the MWCO of the membrane, such as charge. Specifically, the negative charge on the hydrophilic membrane plays a significant role in the rejection process. CR > Tartrazine > Ciprofloxacin	[95]

Nanosilver membranes preparation

Preparation of nanocomposite membranes with Ag NPs, includes several methods based on their preformed/synthesized in situ or directly incorporation during membrane formation. The resulting membrane features will depend on physical and chemical properties of membrane material and NPs incorporation technique. Taurozzi *et al.* [103] presented an interesting study on the effect of two different Ag NPs incorporation routes in the PSf membranes, differing in morphology. In the first approaches, Ag NPs were synthesized ex situ and added to the dope solution as Ag-DMF (N,N-dimethylformamide) organosol. In the second approaches, the NPs were prepared via in situ reduction of ionic Ag^+ by DMF, in the membrane casting solution. The authors reported that in all cases the majority of the observed NPs were located along the internal pore surface of the polymeric matrix with higher density along the top layer for the membranes prepared with the first approaches. Moreover, the ex situ Ag NPs preparation produced a higher NPs density with a relatively higher number of bigger NPs, respect the Ag NPs in situ preparation. Nanocomposite PAN NFs with Ag NPs were produced using two approaches [102]. In the first approaches, the NPs were produced directly in $AgNO_3$/PAN (0.5 wt% of $AgNO_3$) solution using DMF as solvent. In this way, the authors investigated the self-synthesis of Ag NPs within via oxidation reaction. The presence of DMF inhibits the agglomeration of the Ag NPs, according with the reported mechanism:

$$HCONMe_2\, 2\,Ag^+ + H_2O \rightarrow 2Ag^\circ + Me_2NCOOH + 2\,H^+ \quad (1)$$

The second approach involves the UV irradiation (254 nm for 3 and 6 h) of the produced Ag/PAN NFs, in order to promote the decomposition of $AgNO_3$ into Ag NPs. The obtained Ag/PAN NFs were characterized in terms of morphology, energy dispersive spectroscopy (EDS) and Fourier Transform Infrared Spectrophotometer (FT-IR). The results confirmed the presence of the Ag NPs on the NFs in all cases with better antibacterial results for the NPs prepared by UV light. In fact, the UV exposure time (0, 3 and 6 h) affects the $AgNO_3$ conversion into Ag NPs; increasing the exposure time, consequently, increases the NPs concentration and the radial diameter of the bacteria inhibition zone. Ag NPs prepared by in-situ thermal reduction of $AgNO_3$ (0, 12.5, 25, 37.5, 50 and 75 mM) were employed for coated porous chitosan (CS) membrane embedded cotton fabric (Ag NPs-C/CS) [104]. The study reported that when $AgNO_3$ was used at 37.5 mM uniform and small NPs were formed with highest bactericidal effect (4.0 log reduction for *E. coli*). When the concentration increases, Ag NPs tends to aggregate with growing in size. A one step method of Ag NPs synthesis and in situ immobilizing applicable on different substrates was presented by Qi *et al.* [105]. The procedure, reported in Fig. 5 provides to adding dopamine (DOPA) (reducing agent) to $AgNO_3$ solution in the presence of poly (ethylene glycol) methyl ether thiol (mPEG-SH) (stabilizing agent). Synthesis and immobilization of Ag NPs on membranes was performed via a simple dip-coating method.

Varying the reaction time of $AgNO_3$/mPEG-SH/DOPA solution from 1 to 24 h, the Ag NPs diameter increases initially from 58.7 to 141.8 nm (during the first 4 h), up to 295.3 nm when the immersion time increased to 24 h. During the preparation of Ag NPs PES MMMs, the immersion time was fixed at 2 h, obtaining Ag NPs with diameter around 35.2 ± 6.8 nm; the scanning electron microscopy (SEM) images confirmed the NPs generated on all membrane surfaces. Hydrophilization procedure via single-step coating was employed in order to modify PVDF membrane surface using plant-derived epigallocatechin gallate (EGCG) and silver ion (Ag^+) in aqueous solution [106]. The facile procedure involves that PVDF was dipped in ethanol (4 h) and washed in water in order to remove the impurity and subsequently dipped in $AgNO_3$ aqueous solution with EGCG solution (pH 8.0). The obtained membrane presents a WCA of 41° (starting from 110°) and the SEM images evidence that the membrane surface was uniformly decorated with numerous NPs. Ag NPs stability on membrane surface was quantitatively analyzed by Ag leaching test. The results have been demonstrated that the Ag concentration was lower than the contaminant limit of Ag in drinking water (100 μg/L, by WHO), indicating that Ag/EGCG modified PVDF membrane was stable for water treatment application.

Ag membranes application for water treatment

To date, numerous studies have been performed concerning the antibacterial activities of silver/polymer nanocomposites membranes. The membranes preparation containing biocidal NPs is a promising approach to improve the (bio)-fouling resistance [107–111]. In fact, in this case, the NPs on the membrane surface can inactivate bacteria inhibiting their growth and preventing the biofilm formation on the membrane surface (irreversible fouling). An interesting review on the effect of Ag NPs incorporation in different polymeric membranes for fouling mitigation was reported by Bolto et al. [112]. These hybrid membranes can be used for disinfectant applications and antibacterial activities, enhancing the antifouling properties of the membranes, improving the environmental durability and improving porosity, CA and mechanical properties

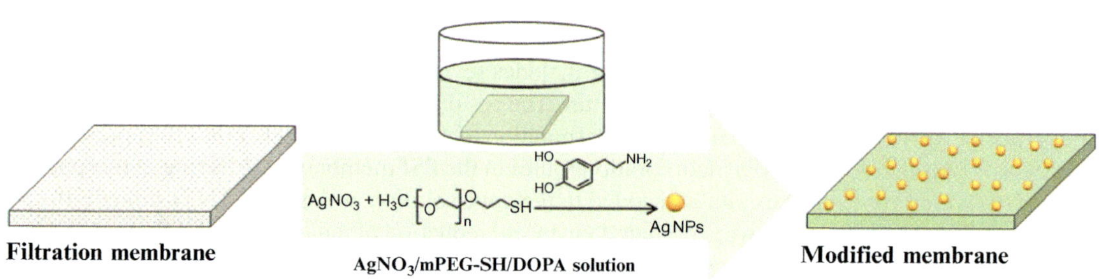

FIG. 5 Synthesis and simultaneous immobilization of Ag NPs on the filtration membrane surfaces in one-step. *Reprinted from L. Qi, Z. Liu, N. Wang, Y. Hu, Facile and efficient in situ synthesis of silver nanoparticles on diverse filtration membrane surfaces for antimicrobial performance, Appl. Surf. Sci. 456 (2018) 95–103. doi:10.1016/j.apsusc.2018.06.066.*

too. The important properties of the nanosilver have been discussed before. In this section, we will be highlighting as the Ag improves the membrane performances for water treatment applications. PAN NFs hybrid composite membranes containing Ag NPs (0.5 wt% of $AgNO_3$ in DMF) for antibacterial activity were reported by Shalaby et al. [102]. Ag/PAN NFs antibacterial performance was evaluated against gram positive *Staphylococcus aureus* (*S. aureus*) and vs Gram negative *E. coli* using agar disk diffusion assay (1×10^5 cells/mL; NFs of 6 mm diameter; 24 h at 37°C). *E. coli* suspension (10^5–10^6 CFU/mL) was also used for the growth kinetics experiment (using 0.1 g PAN NFs). The experiments confirmed that Ag NPs possess similar antibacterial activity vs *E. coli* and *S. aureus* bacteria. In particular, the OD_{600} (optical density at $\lambda = 600$ nm) containing the Ag/PAN NFs presented a value of 0.01 after overnight incubation respect the 0.217 of the control OD. This result indicated that the Ag ions were released from Ag NPs with consequently inhibition of microbial growth. Ag/PAN NFs were an ideal candidate for antibacterial filters for water disinfectant systems. Shim et al. [113], immobilized Ag NPs on cellulose acetate membrane (Ag-CA) for water removal of radioactive iodine species. Ag NPs (30 nm; 0.1 mg) were previously synthesized and the Ag-CA membrane was produced via coating. Ag-CA membrane efficiency was evaluated using NaI aqueous solution (1.0 nM, 3.7 MBq/50 mL) under continuous in-flow conditions (approximately 1.5 mL/s). The amount of radioactivity was determined by using a counter and a single-photon emission computed tomography/computed tomography (SPECT/CT) was employed for the images. SEM images reveal that Ag NPs were incorporated onto the CA membrane without aggregate formation thanks to the hydroxyl and carbonyl groups of the polymeric CA that stabilize the NPs. The iodine removal efficiency was 99.6% and distribution coefficient (K_d) exceeds 10^6 mL/g. The Ag-CA membrane can be successfully used to treat of radioactive iodine wastes, replacing the expensive membrane prepared with gold. PES membranes with Ag NPs (1–70 nm; 0.22 wt%) incorporated via PI were produced and tested against *E. coli, Pseudomonas mendocina* and MS2 bacteriophage (Emesvirus zinderi) by Zodrow et al. [114]. The produced Ag/PES composite membrane was more hydrophilic than the control (68° vs 76°), with an increase in water permeability from 408 to 532 L/m^2 h bar, although the morphology was similar. Antimicrobial properties of membranes were evaluated using a suspension of *E. coli* (10 CFU/mL), observing a 2-log (99%) reduction in E. coli grown on the membrane surface. Biofouling resistance was studied in terms of biofilm formation using *P. mendocina* (10^3 cells/mL), noting only a little growth (if any) on the Ag/PES membrane. Finally, antiviral properties were also evaluated. The experiments results show that incorporating Ag NPs enhanced virus removal by filtration up to $5 \pm 0.2 \times 10^5$ plaque-forming units per milliliter (PFU/mL). The authors referee that the antimicrobial activity was primarily due to the release of Ag^+ ions and the introduction of Ag NPs improved biofouling resistance and virus removal in the nanocomposite membranes [114]. Again, the Ag NPs immobilized on the PES membrane decrease water permeability but increase the BSA rejection rate (1 g/L; pH 7.4) from 42 (membrane without NPs) to 58%. The antimicrobial activity of the Ag/PES membranes was tested against *E. coli* and *S. aureus* as model strains (10^7 CFU/mL). Compared to pristine membrane, the number of viable *E. coli* and *S. aureus* cells attached to the Ag/PES membranes was decreased by 93.9% and 91.6%, respectively [105]. Dolina et al. [115] developed a PES HFs containing Ag NPs formed in situ for biofouling mitigation. Commercial PES HFs were dipped in a solution of $AgNO_3$ (3.5 wt%) and after the diffusion of the ions into the HFs, ions were reduced using ascorbic acid. The modification did not influence the membranes performance in terms of water flux, when demineralized water was used as feed. When a municipal wastewater treatment plant (WWTP) effluent was used as feed (suspended solids were ~15 mg/L), the modified PES HFs presented an approximately 15% higher permeability; this result indicated that the compounds responsible for the biofouling were not fixed to the HFs surface. This study provides an insight into commercial scale use (Table 3).

Graphene oxide (GO)

Structural and functional properties of GO

Graphene (GN) is a two-dimensional NM consisting of a single carbon (C) layer with a sp^2 hybridization. The C atoms arrange themselves to form hexagons with angles of 120° (honeycomb network structure). GN possesses several unique properties such as: ultrathin (0.35 nm in thickness), highly specific surface area (2630 m^2/g^1), excellent thermal and electrical conductivity, photo-absorbent, superior mechanical properties (tensile strength of 130 GPa and Young's modulus of 1060 GPa) and is an excellent charge carrier mobility (1.5×104 cm^2/Vs) [118–120]. GN material dimension can span from <20 nm to μm sizes of nanosheets and meter-scale films and find space in many applications and different fields [121]. Several methods have been reported to produce GN, as chemical vapor deposition (CVD), decomposition of CNTs, liquid-phase or mechanical exfoliation [122]. Graphene oxide (GO) is also a carbon NM, specifically, is an oxidative form of GN which maintains the unique two-dimensional structure (with a GN lamellar structure of 1–30 μm), plus more functional

TABLE 3 Brief summary of the most recent Ag NPs membranes for water treatment applications

Polymer material	Ag type	Membrane application	Main results	Reference
CS	Ag NPs coated membrane	natural water disinfection	The CS membrane reduced the release of silver, with a total Ag in the treated water around 20 mg/L (WHO requirement, <100 mg/L)	[104]
PVDF	Ag NPs coated membrane	MF	PVDF NFs decorated using Ag NPs showed outstanding hydrophilic behavior (0°) and a remarkable 3.9-fold improvement in the separation of water from oily emulsions	[116]
PAN	Ag/GO NFs	MF	The composite membrane presented antibacterial activity against *E. coli* and *S. aureus* with optimal anti-fouling property	[117]
PSf	Ag NPs in situ immobilization	UF	mPEG-SH was employed as ligand, preventing the Ag NPs aggregation. The immobilized silver increased the BSA R% of the 16% and exhibited more antibacterial efficiency against both Gram-negative and Gram-positive bacteria, compared with the pristine membrane	[105]
PVDF	Ag MMMs	MF	Ag NPs were incorporated via thermal treatment, in this way mechanical properties, water flux and antibacterial activity were improved. No Ag leaching was observed during the filtration experiments	[111]
PVDF-HFP	Ag/TiO_2 nanocomposite membrane	MF/UF	Ag-TiO_2/PVDF-HFP nanocomposite membranes possess multifunctional photocatalytic and antimicrobial activities (80.7% norfloxacin degradation efficiencies under visible radiation; 1.3 bacterial log reduction of Gram-negative)	[108]
PAN	Ag NPs coated membrane	MF	100% reduction for Gram-negative bacteria with water flux of 4806 L/m^2h	[110]
PES	Ag NPs in situ formation	MF/UF	The presence of the silver improves the HFs hydrophilicity (WCA from 91° to 86°) and permeability (until the 15%) with only 2.1% of the total amount of silver leached after 8 h of test	[115]
PVC	Ag/TiO_2 nanocomposite membrane	UF	Ag/TiO_2 NPs improved the PWF from 41 to 58.6 L/m^2h. Silver NPs prevented membrane biofouling and the produced membrane exhibited good capability in removal of Lanasol blue 3 R dye from aqueous solution (98.6%)	[107]
PES	Fluorinated Ag NPs	DCMD	Ag NPs are deposited on the PES surface previously functionalized, producing a submicro/nanostructure. Thanks to physical and chemical treatments, WCA increased, and the PES surface became a hydrophobic-hydrophilic surface. The novel membrane showed a stable-high flux (39.37 kg/m^2h) with a high salt rejection (99.99%)	[109]

groups, on the basal plane and at the sheet edges, as hydroxyl, epoxy, and carboxyl groups. The presence of these functional groups makes GO highly hydrophilic and therefore readily dispersible in water [3]. GO presents good chemical stability and adaptability, high mechanical strength, thermal stability and due to the presence of the functional groups is possible to provide a variety of surface-modification reactions, obtaining functionalized GO-based materials [3,118]. Moreover, GO are able to enhance the photocatalytic performance of others NMs, e.g., TiO_2, expanding the light-response range thanks to its semiconductor behavior with an absorption band at about 280 nm. This property is directly correlated with the increase of oxygen content in the GO structure [123–125]. Again, GO possesses antimicrobial properties, which can destroy/inactivate the bacteria with oxidative stress, and extract phospholipids from cell membranes [121]. The introduction of the GO in the membrane matrix, allows to change several features as the morphology, roughness, hydrophilicity, mechanical properties and charge density. A generic structure of GN and GO is reported in Fig. 6.

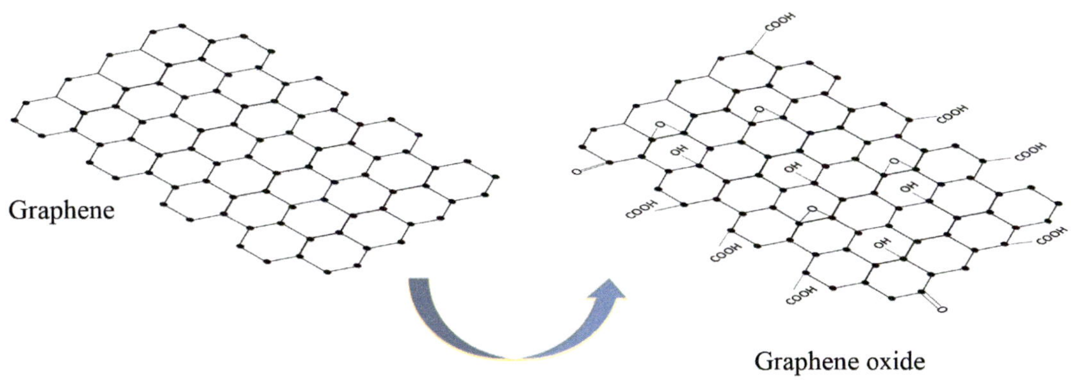

FIG. 6 Generic chemical and physical structures of graphene and graphene oxide.

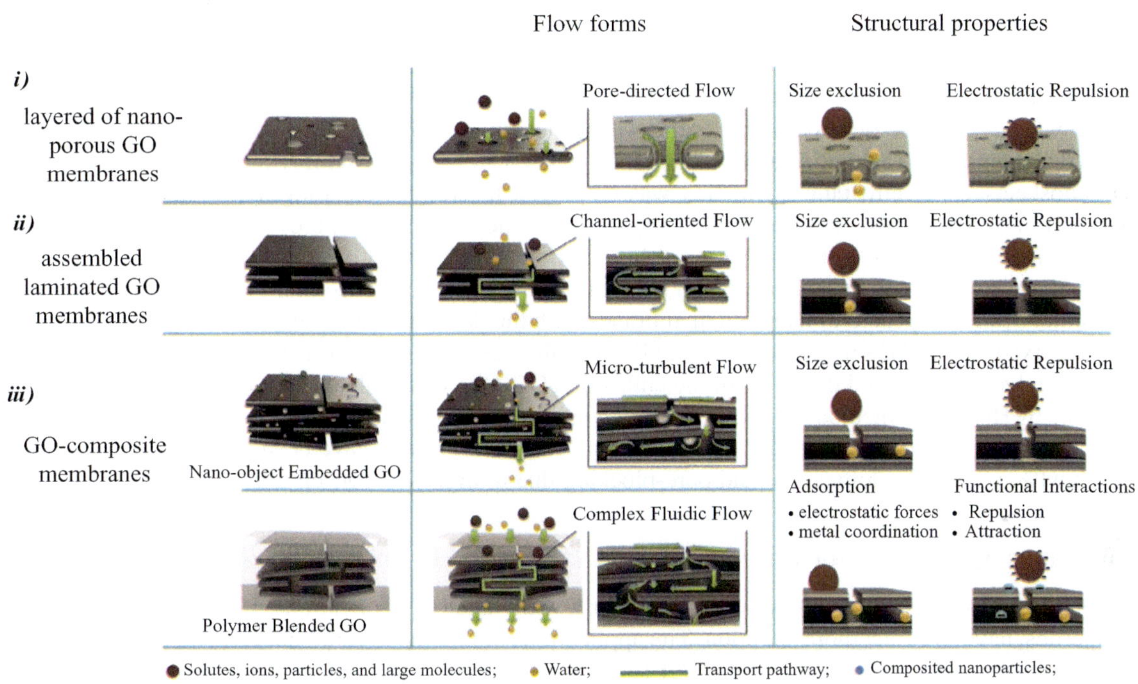

FIG. 7 Main types of GO-based membranes. *Adapted from M. Sun, J. Li, Graphene oxide membranes: functional structures, preparation and environmental applications, Nano Today 20 (2018) 121–137. doi:10.1016/j.nantod.2018.04.007.*

GO membranes preparation

Concerning the membranes containing GO, it is possible to divide them in three different types: (i) layered of nano-porous GO membranes; (ii) assembled laminated GO membranes; (iii) GO-composite membranes [122,126–128], represented in Fig. 7.

In the first case, GO sheets is perforated in order to form a nanopores. This is an ideal separation membrane, with a perfect structure and tunable pore size. Several works were reported in literature about the simulation performance of layer of nano-porous GO membranes [129–132]. The possibility to tune the nanopore size allows improving the membrane selectivity obtaining higher water permeability and salt rejection (approximately three orders of magnitude higher than some types of applied membranes, e.g., RO membranes) [126]. In terms of pore size, permeability and rejection, these properties would be perfect for the RO process; however, as reported by Petukhov et al. [127], to date, no experimental researches were developed to study the transport through this type of membrane. In the second type of GO membranes, the GO nanosheets were assembled in ordered laminated structures. GO membranes are GO paper-like selective and/or functional top layers onto porous supports. Basically, in this case it is possible to produce free-standing or supported membranes. These membranes were fabricated by deposition of a thin GO layer or GO nanocomposites (a few nm to μm) onto relatively thick support membranes (usually >100 μm) [121]. Water flux (or general flux) takes place through the layered laminate GO membrane (interlayers nano-channels). These membranes possess a smaller pore size distribution and higher water permeability, up two orders of magnitude, with respect to the common UF membranes [126]. Several techniques were employed to prepare these membranes, as a different coating (spin-spray-dip-casting coating) [133], vacuum-assisted filtration [134], pressure-assisted methods [135], and casting processes. In the case of laminated structures, membranes performance is directly correlated to the interlayer distance [127]. Finally, the last case includes the MMMs where the GO was used as filler integrated with the polymer. These membranes were developed in order to improve the selectivity and permeability of polymeric material. As reported in literature [121], GO can migrate to the membrane surface during the PI, improving their hydrophilicity, porosity, and rejection.

Abdelkader et al. [136] prepared (GO)/polyacrylamide (PAM)/PES membrane, depositing GO on the PES surface via spin-coating technique and using PAM as an adhesive layer. GO was used at a higher concentration (5.5 mg/mL). Initially, a PAM solution (1.34 mg/mL) was spinning on the PES surface, and then GO solution was deposited over the membrane, increasing the rotation speed. PAM has been employed as a pretreatment step, to filter the divalent ions in desalination application. PA-GO composite membranes were prepared via pressure-assisted ultrafiltration with subsequent IP, using GO powder (0.2 g), by Zhao et al. [137]. PVDF-GO membranes were successfully prepared using a low toxic solvent via Nonsolvent-Induced Phase Separation (NIPS) and Vapor-Induced Phase Separation (VIPS)/NIPS, techniques [124]. GO was initially prepared starting from GN by means of its oxidation, and exfoliation by sonication. The obtained GO sheets possess a lateral size from 0.8 to 3.0 μm (most of them with thickness of 1.0 nm). Several amounts of GO (0–0.125–0.5 wt%) were directly incorporated in the polymeric dope solution and the influence of NIPS/VIPS preparation procedure and GO amounts was studied and evaluated. Layer-by-layer (LbL) assembly of positive chitosan (CS) and negative GO nanosheets via electrostatic interaction was employed to produce blending hydrophilic sulfonated polyethersulfone (SPES) into PES matrix via PI process [138]. GO nanosheets (0.1 wt%), previously prepared from graphene by Hummer's method, were dispersed in an aqueous solution at pH 10. The negatively charged SPES-PES porous substrate was initially firstly immersed in the CS polycation solution (pH 3) and then soaked in the GO solution. The alternative CS and GO treatment can be repeated to fabricate membranes with the desired number of CS/GO layers. Zinadini et al. [139] produced NF mixed matrix PES membranes using GO. Membranes were characterized in terms of morphology and performance on the dye removal, and the effect of the embedded GO NPs was evaluated. GO nanosheets were prepared from natural graphite by Hummer's method and subsequently several amounts were added in the dope solution (0–0.1–0.5–1 wt%). Membranes were then prepared via the PI method; the characterization results have been revealed an optimal GO concentration of 0.5 wt%, without agglomeration and pore-blocking due to high NPs concentration.

GO membranes application for water treatment

Recently, GO-based membranes are increasingly used for a wide range of applications, from MF to RO, and due to their unique properties, have to attract more attention. The applications of GO-membranes cover a large area, from desalination, environmental catalysis decontamination, oil-water separation to dye removal. Several studies reported the successes of incorporating GO into polymeric membranes showing the increase of the water permeability. Again, GO allows to obtain highly performing membranes, improves surface hydrophilicity, with high rejection, stability, photocatalysis action, antifouling properties and mechanical resistance [3,121,126,127,140,141]. Salehi et al. [138] produced SPES-PES FO membranes by LbL assembly of CS and GO as the active layer. Membranes' performance was evaluated using Na_2SO_4 and sucrose solution (1 M) in two different membrane configurations (FS and DS). The authors reported that increasing the LbL deposition, the water permeability dramatically declined (from 65 to 2.95 $L/m^2\,h{\cdot}bar$); however, increasing CS/GO bilayers resulted in more salt rejection (from 0% to 91%); therefore, a compromise between permeability and selectivity was necessary. The experiments have been demonstrated that the LbL CS/GO membranes (bilayers) presented a water flux of 2–4 orders of magnitude higher than the TFC membrane used as control. CS/GO bilayer modified the membrane properties, enhancing the hydrophilicity and improving the water flux and at the same time, it forms a barrier layer preventing water transport. Moreover, LbL CS/GO membrane presented better antifouling properties than the TFC membrane, using SA as a foulant model (200 ppm). In fact, in case of TFC membrane the SA was immediately adsorbed onto the membrane surface, therefore the pore has been blocked. On the contrary, the hydrophilic surface of the LbL membrane allows better anti-fouling performance thanks to the formation an aqueous dense layer that prevents the foulants attachment. Photocatalytic performance of PVDF-GO membranes, evaluated studying the degradation of MB^+ under visible light irradiation ($\lambda > 420$ nm), was reported by Alyarnezhad *et al.* [124]. PVDF-GO membranes, produced using different GO concentration (0–0.125–0.5 wt%), were tested by using MB^+ cationic dye (10 μmol/L) as an organic contaminant. The experiments results show that all the GO-membranes presented higher MB^+ degradation efficiency in comparison of the control membranes. Best photocatalytic results were obtained for the membrane prepared using the 0.125 wt% of GO. Membrane produced using higher concentration of GO (0.5 wt%), presented higher pore size that negatively affected the MB^+ removal efficiency. The preparation of GO-PES membranes for the treatment of industrial dyes was recently reported by Mahmoudian *et al.* [142]. GO was functionalized using hyper branched epoxy (GO-HBE) to improve its compatibility with the polymeric membrane matrix. GO-PES membranes were then prepared dispersing GO (1–7 wt%) in the polymeric dope. Membrane performance was evaluated in terms of water flux, dyes removal (MB^+; methyl orange (MO); bromothymol blue (BTB); crystal violet (CV); acid red 18; acid yellow 36; at 50 ppm) and antifouling properties (BSA, 50 mM, pH 7). The presence of GO-HBE increased surface hydrophilicity, enhancing the water flux (from 45 to 190 L/m^2 h·bar); moreover, leads to the formation of an effective hydration layer on the membranes surface preventing the protein absorption (BSA adsorption was reduced from 78 to 51 ppm/cm^2). When GO-HBE % increases, despite the pore enlargement, the dyes removal rate was improved, due to effective interactions between the additives and dyes groups, obtaining until 85% of dyes removal. GO PSf MMMs via PI method, containing eco-friendly additive as gum Arabic (GA) for sustainable membrane usage, were presented by Chai et al. [143]. GO as nanofiller was previously synthesized via Hummers method; the PSf membranes were prepared adding GA (0–1.5 wt%) and GO nanofiller (0–0.2–0.6–1.0 wt%) in the casting solution. PSf membranes without GA and GO were used as a controlled sample. MMMs were characterized in terms of porosity, pore size, CA and the performance evaluated by water flux and rejection using humic acid (HA, 20 ppm) as model organic foulant. During the phase inversion, the presence of GA and GO leads to a thermodynamic instability that results in a fast exchange rate, and consequently the formation of membranes more porous. Only in case of higher GO % (1.0 wt%), the nanofiller tends to agglomerate, resulting in porosity, pore size and water flux reduction (77%, 19 nm and 54 L/m^2 h·bar). In all cases, the presence of GA and GO improves membranes hydrophilicity (with a WCA from 62° to 56°). Generally, the synergy effect between GO and GA results in a positive enhancement of membranes properties. The porosity increases (from 59% to 78%), as the mean pore size (from 11 to 20 nm). The optimal additive/nanofiller combination was 1.5 and 0.6 wt% of GA and GO, respectively; the produced membrane presented the highest porosity (78%) and the largest mean pore radius (20 nm). Permeability performance was in according to characterization tests, with highest flux at 58.68 L/m^2 h·bar; this indicates that the presence of GA/GO favors the water transport. Antifouling results demonstrated that all GA/GO MMMs have the rejecting ability of more than 95%. Ng and coworkers [144] presented a novel membrane preparation strategy to produce tailored TFC RO membranes for RO applications. PVA was initially employed as adhesive for the hydrophilic GO with different loading (0–0.02 wt%). TFN membranes with two different GO/PVA orientations (coating the PA surface or between PA and substrate (MMMs)) were then produced and characterized. This work highlighted as the presence of GO on the membrane surface enhanced hydrophilicity and membrane properties. An improvement of water flux from 50.94% to 91.93%, good NaCl rejection (98%; 2000 mg/L NaCl) and best antifouling property (100% flux recovery, 5-h filtration using 1000 mg/L organic solutions) were registered for the membrane prepared with GO/PVA a top PA layer. Despite the advantages of the GO as nanofiller, the GO/membranes production and their real full-scale application, e.g., in RO, are still a challenge; in fact, these nanocomposite membranes need to be optimized [145] (Table 4).

Two-dimensional (2D) and other materials

Recently, the attention to 2D nanostructured materials for water purification treatments is growing strongly [7]. These materials can be produced by stacking various kinds of 2D nanosheets, as GN, GO, MXenes [146], layered transition metal dichalcogenides (TMDCs), di-vanadium penta-oxide (V_2O_5), zeolite MCM-22, and MCM-41 silica, layered zeolites, 2D metal organic frameworks (MOFs), 2D covalent organic frameworks (COFs), etc. [3,147]. Such 2D materials possess several properties, as higher mechanical strength, flexible structure and chemical inertness, which to production of highly performing membranes. 2D NMs, also called nanosheets, are characterized by a higher surface areas and an atomic thickness; this structure exposes all of their ions to the surface enabling electronic coupling [7]. The use of 2D NMs allows producing membranes with selective nanochannels for ions or organic molecules, with unique properties that respect the conventional porous membranes. However, their development is still in its infancy, and further research studies are necessary to improve their productions/applications. Concerning water treatment, specifically for desalination and water purification, GN is one of the most 2D materials employed due to their honeycomb structure [3,148]. Others interesting NMs for water purification, desalination or wastewater treatment include: zinc oxide (ZnO) [149,150], copper oxide (CuO) [149], aluminum (Al_2O_3) [151,152], silicon (SiO_2) [153], Iron oxide type II and III (Fe_2O_3 and Fe_3O_4, respectively) [154–156], cobalt (Co) [13], zirconium (ZrO_2) [157], and zeolite (NaX) [158,159]. The induced properties on the membranes, depends on the types of NMs introduced in their matrix. As reported for the mentioned NPs, these properties include: enhanced hydrophilicity [152,158,159], good chemical properties [153,160], higher fluxes and selectivity [3,70], excellent thermal

TABLE 4 Brief summary of the most recent GO membranes for water treatment applications

Polymer material	GO type	Membrane application	Main results	Reference
PA	GO composite membrane	PV	During the PV experiments, GO layer is protected from swelling by PA. Thanks to this synergistic action the membrane possess high desalination performance, with a R% of 99.99% (3.5 wt% NaCl), with a water flux of 24.0 $kg/m^2 h$	[137]
PES	GO composite membrane	UF	GO was functionalized by hyperbranched epoxy in order to improve its dispersion in the polymer matrix. GO increases the hydrophilicity of the membrane, promotes the water flux (until 189 L/m^2 h), reduces the fouling (BSA adsorption 51.1 ppm/cm^2) and enhances the dyes removal (more than 85%)	[142]
PSf	GO MMMs	UF	GO was employed using GA as eco-friendly natural additive. The presence of these two additive, enhances the membrane hydrophilicity, water flux and antifouling activity, due to the presence of more oxygen presents on their functional groups	[143]
PA	GO composite membrane	RO	TFN RO membrane was produced and employed in two orientation where GO/PVA layer was on (i) top PA surface or (ii) between PA layer and microporous substrate. The novel membrane showed similar trend in PWF despite the different GO orientations, with better salt R% and antifouling property for the first (i) orientation mode	[144]
PAM/PES	GO film	NF	GO layer was deposited on the PES substrate surface using PAM as an adhesive layer. The GO-NF membrane enhances the antifouling and antibacterial properties with highest rejection rate for Mg^{2+} and Ca^{2+} ions compared with control membrane	[136]
—	Assembled laminated GO membrane	UF	GO-based lamellar membranes were prepared using ginger plant extractive (GE). The produced membrane shows good separation efficiency of dyes in organic solvent (R% of 98% for RB and MB in MetOH)	[128]
PPy	GO composite membrane	NF	The presence of PPy on the GO membrane results in the R% increase of MB from 60% to 97%. Moreover, the expansion of the nanochannels leads to an increases in PWP of ~14 times	[140]
PES	GO composite membrane	NF	The presence of GO and PVP affects membrane hydrophilicity and mechanical stability. Membrane was applied for arsenic removal from synthetic wastewater, obtaining until the 88.6% of R% (As, feed concentration 55 mg/L)	[141]
PVDF	GO composite membrane	MF/UF	GO improved the mechanical strength of the membranes and the fouling resistance. Moreover, exhibited good photocatalytic performance toward MB^+ dye (removal efficiency of 83.5%)	[124]
—	single-layer GO sheets	water-ethanol separation	The study was performed via molecular dynamics simulations (MD). The results show that in the membrane with a pore size 1.12 nm, the ionization of the carboxyl groups enhances their water-EtOH selectivity coefficient. The work suggests that the ionization of the functional groups can be an effective approach for making high selectivity GO membranes	[129]

and mechanical stabilities [20], magnetic property, antibacterial and antifungal activities [2,154,156,161]. Finally, it is important to underline the use of mixed NPs incorporated in the polymeric membrane matrix. GO, and TiO_2 NPs were often used together to improve the photocatalytic action [162–165]; nanosilver is widely used as bactericide, but their uses are additionally addressed as composites with other NMs, such as GO [100,117,166–168], ZnO [169,170], SiO_2 [171,172], and Fe_3O_4 [172].

Conclusions

In recent years, the properties of NPs have been increasingly investigated in membrane separation processes. NPs possess a large variety of physical and chemical properties, which make them attractive in water applications. Moreover, nanotechnology can be the right solution for energy saving and use of water resources, making wastewater treatment more effective and economical. As reported above, NMs are small and accurate, possess high surface-to volume ratio, high adsorption, and they can be produced by environment-friendly techniques. Furthermore, the use of NMs allows to focus and improve existing methods, enhancing the performance and providing the reuse of the NMs themselves, which reduce the operating costs of the processes [1,6]. This chapter aims to illustrate the most common NPs used for the preparation of polymeric nanocomposite membranes. In particular, four main NPs as CNTs, TiO_2, Ag and GO have been considered. The discussion has been divided in three parts: NPs properties, NPs membranes preparation and their application in water treatments. It is highlighted as the use of NPs into polymeric membranes improves the hydrophilicity, the water flux and at the same time, reduces the (bio)-fouling phenomenon during the water treatment applications. Again, it has been demonstrated as some other properties (mechanical, thermal, and chemical) of these new nanocomposite membranes are enhanced. To foster innovation in this field, the research should be concentrated on overcoming some obstacles related to the use of NPs, such as the adequate adhesion on the polymeric matrix and consequently leaching of NPs during the process. The use of NPs represents an added value in membrane fabrication, important to produce tailored membranes with specific properties for water treatment.

References

[1] P. Jangid, M. Prabhu Inbaraj, Applications of nanomaterials in wastewater treatment, Mater. Today Proc. 43 (2021) 2877–2881, https://doi.org/10.1016/j.matpr.2021.01.126.

[2] W. Huang, F. Tao, F. Li, M. Mortimer, L.H. Guo, Antibacterial nanomaterials for environmental and consumer product applications, NanoImpact. 20 (2020) 100268, https://doi.org/10.1016/j.impact.2020.100268.

[3] C. Ursino, R. Castro-Muñoz, E. Drioli, L. Gzara, M.H. Albeirutty, A. Figoli, Progress of nanocomposite membranes for water treatment, Membranes (Basel) 8 (2018), https://doi.org/10.3390/membranes8020018.

[4] WHO, WHO Guidelines on Protecting Workers From Potential Risks of Manufactured Nanomaterials, World Health Organization, Geneva, Switzerland, 2017, https://doi.org/10.1093/occmed/kqz070.

[5] X. Qu, P.J.J. Alvarez, Q. Li, Applications of nanotechnology in water and wastewater treatment, Water Res. 47 (2013) 3931–3946, https://doi.org/10.1016/j.watres.2012.09.058.

[6] T. Bora, J. Dutta, Applications of nanotechnology in wastewater treatment—a review, J. Nanosci. Nanotechnol. 14 (2014) 613–626, https://doi.org/10.1166/jnn.2014.8898.

[7] L. Loske, K. Nakagawa, T. Yoshioka, H. Matsuyama, 2D nanocomposite membranes: water purification and fouling mitigation, Membranes (Basel) 10 (2020) 1–40, https://doi.org/10.3390/membranes10100295.

[8] N.C. Mueller, B. Van der Bruggen, V. Keuter, P. Luis, T. Melin, W. Pronk, R. Reisewitz, D. Rickerby, G.M. Rios, W. Wennekes, B. Nowack, Nanofiltration and nanostructured membranes—should they be considered nanotechnology or not? J. Hazard. Mater. 211–212 (2012) 275–280, https://doi.org/10.1016/j.jhazmat.2011.10.096.

[9] N. Nasrollahi, L. Ghalamchi, V. Vatanpour, A. Khataee, Photocatalytic-membrane technology: a critical review for membrane fouling mitigation, J. Ind. Eng. Chem. 93 (2021) 101–116, https://doi.org/10.1016/j.jiec.2020.09.031.

[10] L. Upadhyaya, X. Qian, S. Ranil Wickramasinghe, Chemical modification of membrane surface—overview, Curr. Opin. Chem. Eng. 20 (2018) 13–18, https://doi.org/10.1016/j.coche.2018.01.002.

[11] L. Phelane, F.N. Muya, H.L. Richards, P.G.L. Baker, E.I. Iwuoha, Polysulfone nanocomposite membranes with improved hydrophilicity, Electrochim. Acta. 128 (2014) 326–335, https://doi.org/10.1016/j.electacta.2013.11.156.

[12] J. Huang, H. Wang, K. Zhang, Modification of PES membrane with Ag-SiO2: reduction of biofouling and improvement of filtration performance, Desalination 336 (2014) 8–17, https://doi.org/10.1016/j.desal.2013.12.032.

[13] L. Gzara, Z. Ahmad Rehan, S.B. Khan, K.A. Alamry, M.H. Albeirutty, M.S. El-Shahawi, M.I. Rashid, A. Figoli, E. Drioli, A.M. Asiri, Preparation and characterization of PES-cobalt nanocomposite membranes with enhanced anti-fouling properties and performances, J. Taiwan Inst. Chem. Eng. 65 (2016) 405–419, https://doi.org/10.1016/j.jtice.2016.04.012.

[14] A. Alpatova, E.S. Kim, X. Sun, G. Hwang, Y. Liu, M. Gamal El-Din, Fabrication of porous polymeric nanocomposite membranes with enhanced antifouling properties: effect of casting composition, J. Membr. Sci. 444 (2013) 449–460, https://doi.org/10.1016/j.memsci.2013.05.034.
[15] P. Biswas, R. Bandyopadhyaya, Biofouling prevention using silver nanoparticle impregnated polyethersulfone (PES) membrane: E. coli cell-killing in a continuous cross-flow membrane module, J. Colloid Interface Sci. 491 (2017) 13–26, https://doi.org/10.1016/j.jcis.2016.11.060.
[16] J. Geltmeyer, H. Teixido, M. Meire, T. Van Acker, K. Deventer, F. Vanhaecke, S. Van Hulle, K. De Buysser, K. De Clerck, TiO2 functionalized nanofibrous membranes for removal of organic (micro)pollutants from water, Sep. Purif. Technol. 179 (2017) 533–541, https://doi.org/10.1016/j.seppur.2017.02.037.
[17] P.C. Ma, N.A. Siddiqui, G. Marom, J.K. Kim, Dispersion and functionalization of carbon nanotubes for polymer-based nanocomposites: a review, Compos. Part A Appl. Sci. Manuf. 41 (2010) 1345–1367, https://doi.org/10.1016/j.compositesa.2010.07.003.
[18] M.F.L. De Volder, S.H. Tawfick, R.H. Baughman, A.J. Hart, Carbon nanotubes: present and future commercial applications, Science (80-.) 339 (2013) 535–539, https://doi.org/10.1126/science.1222453.
[19] Y. Zhu, L. Li, C. Zhang, G. Casillas, Z. Sun, Z. Yan, G. Ruan, Z. Peng, A.R.O. Raji, C. Kittrell, R.H. Hauge, J.M. Tour, A seamless three-dimensional carbon nanotube graphene hybrid material, Nat. Commun. 3 (2012) 1–7, https://doi.org/10.1038/ncomms2234.
[20] G. Ghasemzadeh, M. Momenpour, F. Omidi, M.R. Hosseini, M. Ahani, A. Barzegari, Applications of nanomaterials in water treatment and environmental remediation, Front. Environ. Sci. Eng. 8 (2014) 471–482, https://doi.org/10.1007/s11783-014-0654-0.
[21] J.K. Holt, H.G. Park, Y. Wang, C.P.G. Stadermann, M.A.B. Artyukhin, Fast mass transport through sub-2-nanometer carbon nanotubes, Science (80-.) 312 (2006) 1034–1037, https://doi.org/10.1126/science.1126298.
[22] M. Majumder, N. Chopra, R. Andrews, B.J. Hinds, Enhanced flow in carbon nanotubes, Nature 438 (2005) 44, https://doi.org/10.1038/438044a.
[23] J.-M.Z. Alexander, I. Kolesnikov, R.O.L. Chun-Keung Loong, P. Thiyagarajan, A.P. Moravsky, C.J. Burnham, Anomalously soft dynamics of water in a nanotube: a revelation of nanoscale confinement, Phys. Rev. Lett. 93 (2004) 035503, https://doi.org/10.1103/PhysRevLett.93.035503.
[24] G. Hummer, J.C. Rasaiah, J.P. Noworyta, Water conduction through the hydrophobic channel of a carbon nanotube, Nature 414 (2001) 188–190, https://doi.org/10.1038/35102535.
[25] S. Joseph, N.R. Aluru, Why are carbon nanotubes fast transporters of water? Nano Lett. 8 (2008) 452–458, https://doi.org/10.1021/nl072385q.
[26] P. Laux, C. Riebeling, A.M. Booth, J.D. Brain, J. Brunner, C. Cerrillo, O. Creutzenberg, I. Estrela-Lopis, T. Gebel, G. Johanson, H. Jungnickel, H. Kock, J. Tentschert, A. Tlili, A. Schäffer, A.J.A.M. Sips, R.A. Yokel, A. Luch, Challenges in characterizing the environmental fate and effects of carbon nanotubes and inorganic nanomaterials in aquatic systems, Environ. Sci. Nano. 5 (2018) 48–63, https://doi.org/10.1039/c7en00594f.
[27] G. Han, P. Ghosh, V.M. Rotello, Advances in experimental medicine and biology, in: Bio-Applications of Nanoparticles, vol. 620, Springer Science+Business Media, USA, 2007.
[28] S.K. Smart, A.I. Cassady, G.Q. Lu, D.J. Martin, The biocompatibility of carbon nanotubes, Carbon N. Y. 44 (2006) 1034–1047, https://doi.org/10.1016/j.carbon.2005.10.011.
[29] D.Y. Koseoglu-Imer, I. Koyuncu, Fabrication and application areas of mixed matrix flat-sheet membranes, in: A. Figoli, J. Hoinkis, S.A. Altinkaya, J. Bundschuh (Eds.), Appl. Nanotechnol. Membr. Water Treat., CRC Press Taylor&Francis Group, London, 2017, pp. 49–65, https://doi.org/10.1201/9781315179070.
[30] K. Li, B. Lee, Y. Kim, High performance reverse osmosis membrane with carbon nanotube support layer, J. Membr. Sci. 592 (2019) 117358, https://doi.org/10.1016/j.memsci.2019.117358.
[31] B.J. Hinds, N. Chopra, T. Rantell, R. Andrews, V. Gavalas, L.G. Bachas, Aligned multiwalled carbon nanotube membranes, Science (80-.) 303 (2004) 62–65, https://doi.org/10.1126/science.1092048.
[32] W. Mi, Y.S. Lin, Y. Li, Vertically aligned carbon nanotube membranes on macroporous alumina supports, J. Membr. Sci. 304 (2007) 1–7, https://doi.org/10.1016/j.memsci.2007.07.021.
[33] E. Elnabawy, I.M.A. Elsherbiny, A.M.A. Abdelsamad, B. Anis, A. Hassan, M. Ulbricht, A.S.G. Khalil, Tailored CNTs buckypaper membranes for the removal of humic acid and separation of oil-in-water emulsions, Membranes (Basel) 10 (2020) 1–16, https://doi.org/10.3390/membranes10050097.
[34] A.A. Alshahrani, H. Al-Zoubi, L.D. Nghiem, M.i.h. Panhuis, Synthesis and characterisation of MWNT/chitosan and MWNT/chitosan-crosslinked buckypaper membranes for desalination, Desalination 418 (2017) 60–70, https://doi.org/10.1016/j.desal.2017.05.031.
[35] X. Yang, J. Lee, L. Yuan, S.R. Chae, V.K. Peterson, A.I. Minett, Y. Yin, A.T. Harris, Removal of natural organic matter in water using functionalised carbon nanotube buckypaper, Carbon N. Y. 59 (2013) 160–166, https://doi.org/10.1016/j.carbon.2013.03.005.
[36] R. Wang, D. Chen, Q. Wang, Y. Ying, W. Gao, L. Xie, Recent advances in applications of carbon nanotubes for desalination: a review, Nanomaterials 10 (2020) 1–28, https://doi.org/10.3390/nano10061203.
[37] M. Sianipar, S.H. Kim, F. Khoiruddin, I.G.W. Iskandar, Functionalized carbon nanotube (CNT) membrane: progress and challenges, RSC Adv. 7 (2017) 51175–51198, https://doi.org/10.1039/c7ra08570b.
[38] J. Yin, G. Zhu, B. Deng, Multi-walled carbon nanotubes (MWNTs)/polysulfone (PSU) mixed matrix hollow fiber membranes for enhanced water treatment, J. Membr. Sci. 437 (2013) 237–248, https://doi.org/10.1016/j.memsci.2013.03.021.
[39] V. Grosso, D. Vuono, M.A. Bahattab, G. Di Profio, E. Curcio, S.A. Al-Jilil, F. Alsubaie, M. Alfife, J.B. Nagy, E. Drioli, E. Fontananova, Polymeric and mixed matrix polyimide membranes, Sep. Purif. Technol. 132 (2014) 684–696, https://doi.org/10.1016/j.seppur.2014.06.023.
[40] M.B. Wu, Y. Lv, H.C. Yang, L.F. Liu, X. Zhang, Z.K. Xu, Thin film composite membranes combining carbon nanotube intermediate layer and microfiltration support for high nanofiltration performances, J. Membr. Sci. 515 (2016) 238–244, https://doi.org/10.1016/j.memsci.2016.05.056.
[41] V. Vatanpour, S.S. Madaeni, R. Moradian, S. Zinadini, B. Astinchap, Fabrication and characterization of novel antifouling nanofiltration membrane prepared from oxidized multiwalled carbon nanotube/polyethersulfone nanocomposite, J. Membr. Sci. 375 (2011) 284–294, https://doi.org/10.1016/j.memsci.2011.03.055.

[42] K.K. Yan, L. Jiao, S. Lin, X. Ji, Y. Lu, L. Zhang, Superhydrophobic electrospun nanofiber membrane coated by carbon nanotubes network for membrane distillation, Desalination 437 (2018) 26–33, https://doi.org/10.1016/j.desal.2018.02.020.

[43] Y. Yang, C. Nie, Y. Deng, C. Cheng, C. He, L. Ma, C. Zhao, Improved antifouling and antimicrobial efficiency of ultrafiltration membranes with functional carbon nanotubes, RSC Adv. 6 (2016) 88265–88276, https://doi.org/10.1039/c6ra18706d.

[44] B. Xie, G. Xu, Y. Jia, L. Gu, Q. Wang, N. Mushtaq, B. Cheng, Y. Hu, Engineering carbon nanotubes enhanced hydrophobic membranes with high performance in membrane distillation by spray coating, J. Membr. Sci. 625 (2021) 118978, https://doi.org/10.1016/j.memsci.2020.118978.

[45] K. Mukhopadhyay, C.D. Dwivedi, G.N. Mathur, Conversion of carbon nanotubes to carbon nanofibers by sonication, Carbon N. Y. 40 (2002) 1373–1376, https://doi.org/10.1016/S0008-6223(02)00074-X.

[46] S. Kar, R.C. Bindal, P.K. Tewari, Carbon nanotube membranes for desalination and water purification: challenges and opportunities, Nano Today 7 (2012) 385–389, https://doi.org/10.1016/j.nantod.2012.09.002.

[47] R. Das, M.E. Ali, S.B.A. Hamid, S. Ramakrishna, Z.Z. Chowdhury, Carbon nanotube membranes for water purification: a bright future in water desalination, Desalination 336 (2014) 97–109, https://doi.org/10.1016/j.desal.2013.12.026.

[48] K. Zhang, Y. Yang, S. Qiao, J. Zhou, X. Quan, Alternating current-enhanced carbon nanotubes hollow fiber membranes for membrane fouling control in novel membrane bioreactors, Chemosphere 277 (2021) 130240, https://doi.org/10.1016/j.chemosphere.2021.130240.

[49] Y. Yang, S. Qiao, J. Zhou, X. Quan, Mitigating membrane fouling based on in situ •OH generation in a novel electro-fenton membrane bioreactor, Environ. Sci. Technol. 54 (2020) 7669–7676, https://doi.org/10.1021/acs.est.0c01428.

[50] J. Mulopo, Bleach plant effluent treatment in anaerobic membrane bioreactor (AMBR) using carbon nanotube/polysulfone nanocomposite membranes, J. Environ. Chem. Eng. 5 (2017) 4381–4387, https://doi.org/10.1016/j.jece.2017.08.027.

[51] Y. Yue, S. Qiao, R. Jin, J. Zhou, X. Quan, Fouling control mechanisms in filtrating natural organic matters by electro-enhanced carbon nanotubes hollow fiber membranes, J. Membr. Sci. 553 (2018) 54–62, https://doi.org/10.1016/j.memsci.2018.02.012.

[52] M. Cao, Y. Zhang, B. Zhang, Z. Liu, X.S. Ma, C. Chen, The preparation of a modified PVDF hollow fiber membrane by coating with multiwalled carbon nanotubes for high antifouling performance, RSC Adv. 10 (2020) 1848–1857, https://doi.org/10.1039/C9RA07542A.

[53] Y. Yue, S. Qiao, R. Jin, J. Zhou, X. Quan, Novel anaerobic electrochemical membrane bioreactor with a CNTs hollow fiber membrane cathode to mitigate membrane fouling and enhance energy recovery, Environ. Sci. Technol. 53 (2019) 1014–1021, https://doi.org/10.1021/acs.est.8b05186.

[54] Y. Yang, S. Qiao, R. Jin, J. Zhou, X. Quan, A novel aerobic electrochemical membrane bioreactor with CNTs hollow fiber membrane by electrochemical oxidation to improve water quality and mitigate membrane fouling, Water Res. 151 (2019) 54–63, https://doi.org/10.1016/j.watres.2018.12.012.

[55] A.M. Almusawy, R.H. Al-Anbari, Q.F. Alsalhy, A.I. Al-Najar, Carbon nanotubes-sponge modified electro membrane bioreactor (EMBR) and their prospects for wastewater treatment applications, Membranes (Basel) 10 (2020) 433, https://doi.org/10.3390/membranes10120433.

[56] C.F. De Lannoy, D. Jassby, K. Gloe, A.D. Gordon, M.R. Wiesner, Aquatic biofouling prevention by electrically charged nanocomposite polymer thin film membranes, in: AMTA/AWWA, Membr. Technol. Conf. Expo, vol. 47, 2013, pp. 1304–1319, https://doi.org/10.1021/es3045168.

[57] Y. Chan, J.M. Hill, Modeling on ion rejection using membranes comprising ultra-small radii carbon nanotubes, Eur. Phys. J. B. 85 (2012) 56, https://doi.org/10.1140/epjb/e2012-21029-0.

[58] B. Corry, Water and ion transport through functionalised carbon nanotubes: implications for desalination technology, Energy Environ. Sci. 4 (2011) 751, https://doi.org/10.1039/c0ee00481b.

[59] Z.A. Pouya, M.A. Tofighy, T. Mohammadi, Synthesis and characterization of polytetrafluoroethylene/oleic acid-functionalized carbon nanotubes composite membrane for desalination by vacuum membrane distillation, Desalination 503 (2021) 114931, https://doi.org/10.1016/j.desal.2021.114931.

[60] Q. Zhang, J. Wang, C.D. Vecitis, Fouling reduction and recovery during forward osmosis of wastewater using an electroactive CNT composite membrane, J. Membr. Sci. 620 (2021), 118803, https://doi.org/10.1016/j.memsci.2020.118803.

[61] Y. Yang, Y. Xu, Z. Liu, H. Huang, X. Fan, Y. Wang, Y. Song, C. Song, Preparation and characterization of high-performance electrospun forward osmosis membrane by introducing a carbon nanotube interlayer, J. Membr. Sci. 616 (2020) 118563, https://doi.org/10.1016/j.memsci.2020.118563.

[62] Y. Wang, R. Ou, Q. Ge, H. Wang, T. Xu, Preparation of polyethersulfone/carbon nanotube substrate for high-performance forward osmosis membrane, Desalination 330 (2013) 70–78, https://doi.org/10.1016/j.desal.2013.09.028.

[63] H.J. Kim, Y. Baek, K. Choi, D.G. Kim, H. Kang, Y.S. Choi, J. Yoon, J.C. Lee, The improvement of antibiofouling properties of a reverse osmosis membrane by oxidized CNTs, RSC Adv. 4 (2014) 32802–32810, https://doi.org/10.1039/c4ra06489e.

[64] Y. Liu, B. Yang, J. Xu, H. Zhao, Y. He, Oil-water separation performance of aligned single walled carbon nanotubes membrane: a reactive molecular dynamics simulation study, J. Mol. Liq. 321 (2021) 114174, https://doi.org/10.1016/j.molliq.2020.114174.

[65] V. Santosh, P.V. Babu, J. Gopinath, N.N.M. Rao, A.V.S. Sainath, A.V.R. Reddy, Development of hydroxyl and carboxylic acid functionalized CNTs–polysulphone nanocomposite fouling-resistant ultrafiltration membranes for oil–water separation, Bull. Mater. Sci. 43 (2020), https://doi.org/10.1007/s12034-020-2079-7.

[66] V. Santosh, J. Gopinath, P.V. Babu, A.V.S. Sainath, A.V.R. Reddy, Acetyl-D-glucopyranoside functionalized carbon nanotubes for the development of high performance ultrafiltration membranes, Sep. Purif. Technol. 191 (2018) 134–143, https://doi.org/10.1016/j.seppur.2017.09.018.

[67] W. Duan, G. Chen, C. Chen, R. Sanghvi, A. Iddya, S. Walker, H. Liu, A. Ronen, D. Jassby, Electrochemical removal of hexavalent chromium using electrically conducting carbon nanotube/polymer composite ultrafiltration membranes, J. Membr. Sci. 531 (2017) 160–171, https://doi.org/10.1016/j.memsci.2017.02.050.

[68] Y. Liu, G. Gao, C.D. Vecitis, Prospects of an electroactive carbon nanotube membrane toward environmental applications, Acc. Chem. Res. 53 (2020) 2892–2902, https://doi.org/10.1021/acs.accounts.0c00544.

[69] S. Iravani, Nanomaterials and nanotechnology for water treatment: recent advances, Inorg. Nano-Metal Chem. 51 (12) (2020) 1615–1645, https://doi.org/10.1080/24701556.2020.1852253.

[70] R. Castro-Muñoz, The role of new inorganic materials in composite membranes for water disinfection, Membranes (Basel) 10 (2020) 1–17, https://doi.org/10.3390/membranes10050101.

[71] E. Baranowska-Wójcik, D. Szwajgier, P. Oleszczuk, A. Winiarska-Mieczan, Effects of titanium dioxide nanoparticles exposure on human health—a review, Biol. Trace Elem. Res. 193 (2020) 118–129, https://doi.org/10.1007/s12011-019-01706-6.

[72] J.-Y. Park, C. Lee, K.-W. Jung, D. Jung, Structure related photocatalytic properties of TiO2, Bull. Korean Chem. Soc. 30 (2009) 402–404.

[73] A. Medvids, P. Onufrijevs, J. Kaupužs, R. Eglitis, J. Padgurskas, A. Zunda, H. Mimura, I. Skadins, S. Varnagiris, Anatase or rutile TiO2 nanolayer formation on Ti substrates by laser radiation: mechanical, photocatalytic and antibacterial properties, Opt. Laser Technol. 138 (2021) 1–8, https://doi.org/10.1016/j.optlastec.2020.106898.

[74] T. Marino, M. Boerrigter, M. Faccini, C. Chaumette, A.F. Lawrence Arockiasamy, J. Bundschuh, Photocatalytic activity and synthesis procedures of TiO2 nanoparticles for potential applications in membranes, in: J.B.A. Figoli, J. Hoinkis, S.A. Altinkaya (Eds.), Appl. Nanotechnol. Membr. Water Treat., CRC Press, Taylor & Francis Group, London, 2020, pp. 127–146, https://doi.org/10.1201/9781315179070.

[75] H. de Lasa, B. Serrano, M. Salaices, Photocatalytic Reaction Engineering, Springer Science+Business Media, New York, 2005, https://doi.org/10.1201/b18533.

[76] S. Riaz, S.J. Park, An overview of TiO2-based photocatalytic membrane reactors for water and wastewater treatments, J. Ind. Eng. Chem. 84 (2020) 23–41, https://doi.org/10.1016/j.jiec.2019.12.021.

[77] N.A.M. Nor, J. Jaafar, A.F. Ismail, M.A. Mohamed, M.A. Rahman, M.H.D. Othman, W.J. Lau, N. Yusof, Preparation and performance of PVDF-based nanocomposite membrane consisting of TiO2 nanofibers for organic pollutant decomposition in wastewater under UV irradiation, Desalination 391 (2016) 89–97, https://doi.org/10.1016/j.desal.2016.01.015.

[78] J. Yin, B. Deng, Polymer-matrix nanocomposite membranes for water treatment, J. Membr. Sci. 479 (2015) 256–275, https://doi.org/10.1016/j.memsci.2014.11.019.

[79] A. Razmjou, J. Mansouri, V. Chen, The effects of mechanical and chemical modification of TiO2 nanoparticles on the surface chemistry, structure and fouling performance of PES ultrafiltration membranes, J. Membr. Sci. 378 (2011) 73–84, https://doi.org/10.1016/j.memsci.2010.10.019.

[80] F. Galiano, X. Song, T. Marino, M. Boerrigter, O. Saoncella, S. Simone, M. Faccini, C. Chaumette, E. Drioli, A. Figoli, Novel photocatalytic PVDF/Nano-TiO2 hollow fibers for environmental remediation, Polymers (Basel) 10 (2018) 1–20, https://doi.org/10.3390/polym10101134.

[81] H. Mahdavi, N. Mazinani, A.A. Heidari, Poly(vinylidene fluoride) (PVDF)/PVDF-g-polyvinylpyrrolidone (PVP)/TiO2 mixed matrix nanofiltration membranes: preparation and characterization, Polym. Int. 69 (2020) 1187–1195, https://doi.org/10.1002/pi.6061.

[82] O. Benhabiles, F. Galiano, T. Marino, H. Mahmoudi, H. Lounici, A. Figoli, Preparation and characterization of TiO2-PVDF/PMMA blend membranes using an alternative non-toxic solvent for UF/MF and photocatalytic application, Molecules 24 (2019) 1–20, https://doi.org/10.3390/molecules24040724.

[83] M.U. Abba, H.C. Man, R.S. Azis, A. Isma Idris, M. Hazwan Hamzah, K.F. Yunos, K.K. Katibi, Novel PVDF-PVP hollow fiber membrane augmented with TiO2 nanoparticles: preparation, characterization and application for copper removal from leachate, Nanomaterials 11 (2021) 399, https://doi.org/10.3390/nano11020399.

[84] H. Zhang, Y. Guo, X. Zhang, X. Hu, C. Wang, Y. Yang, Preparation and characterization of PSF-TiO2 hybrid hollow fiber UF membrane by sol–gel method, J. Polym. Res. 27 (2020) 376, https://doi.org/10.1007/s10965-020-02313-z.

[85] N. Cheshomi, M. Pakizeh, M. Namvar-Mahboub, Preparation and characterization of TiO 2/Pebax/(PSf-PES) thin film nanocomposite membrane for humic acid removal from water, Polym. Adv. Technol. 29 (2018) 1303–1312, https://doi.org/10.1002/pat.4242.

[86] F. Yalcinkaya, A. Siekierka, M. Bryjak, Preparation of fouling-resistant nanofibrous composite membranes for separation of oily wastewater, Polymers (Basel). 9 (2017) 679, https://doi.org/10.3390/polym9120679.

[87] Y. Li, J. Zhou, Y. Fan, Y. Ye, B. Tang, Preparation of environment-friendly 3D eggshell membrane-supported anatase TiO2 as a reusable photocatalyst for degradation of organic dyes, Chem. Phys. Lett. 689 (2017) 142–147, https://doi.org/10.1016/j.cplett.2017.10.019.

[88] M. Ehsani, A. Aroujalian, Fabrication of electrospun polyethersulfone/titanium dioxide (PES/TiO2) composite nanofibers membrane and its application for photocatalytic degradation of phenol in aqueous solution, Polym. Adv. Technol. 31 (2020) 772–785, https://doi.org/10.1002/pat.4813.

[89] T. Li, Y. Xiao, D. Guo, L. Shen, R. Li, Y. Jiao, Y. Xu, H. Lin, In-situ coating TiO2 surface by plant-inspired tannic acid for fabrication of thin film nanocomposite nanofiltration membranes toward enhanced separation and antibacterial performance, J. Colloid Interface Sci. 572 (2020) 114–121, https://doi.org/10.1016/j.jcis.2020.03.087.

[90] Y. Yang, H. Wang, J. Li, B. He, T. Wang, S. Liao, Novel functionalized nano-TiO 2 loading electrocatalytic membrane for oily wastewater treatment, Environ. Sci. Technol. 46 (2012) 6815–6821, https://doi.org/10.1021/es3000504.

[91] C.P.M. de Oliveira, M.M. Viana, G.R. Silva, L.S. Frade Lima, E. Coutinho de Paula, M.C.S. Amaral, Potential use of green TiO2 and recycled membrane in a photocatalytic membrane reactor for oil refinery wastewater polishing, J. Clean. Prod. 257 (2020) 120526, https://doi.org/10.1016/j.jclepro.2020.120526.

[92] C. Chen, L. Chen, S. Chen, Y. Yu, D. Weng, A. Mahmood, G. Wang, J. Wang, Preparation of underwater superoleophobic membranes via TiO 2 electrostatic self-assembly for separation of stratified oil/water mixtures and emulsions, J. Membr. Sci. 602 (2020) 117976, https://doi.org/10.1016/j.memsci.2020.117976.

[93] L. Ma, J. He, J. Wang, Y. Zhou, Y. Zhao, Y. Li, X. Liu, L. Peng, M. Qu, Functionalized superwettable fabric with switchable wettability for efficient oily wastewater purification, in situ chemical reaction system separation, and photocatalysis degradation, ACS Appl. Mater. Interfaces 11 (2019) 43751–43765, https://doi.org/10.1021/acsami.9b15952.

[94] C.N. Matindi, M. Hu, S. Kadanyo, Q.V. Ly, N.N. Gumbi, D.S. Dlamini, J. Li, Y. Hu, Z. Cui, J. Li, Tailoring the morphology of polyethersulfone/sulfonated polysulfone ultrafiltration membranes for highly efficient separation of oil-in-water emulsions using TiO2 nanoparticles, J. Membr. Sci. 620 (2021) 118868, https://doi.org/10.1016/j.memsci.2020.118868.

[95] K.B. Errahmani, O. Benhabiles, S. Bellebia, Z. Bengharez, M. Goosen, H. Mahmoudi, Photocatalytic nanocomposite polymer-TiO2 membranes for pollutant removal from wastewater, Catalysts 11 (2021) 1–15, https://doi.org/10.3390/catal 11030402.

[96] A. Singh, S.K. Ramachandran, M.B. Gumpu, L. Zsuzsanna, G. Veréb, S. Kertész, A. Gangasalam, Titanium dioxide doped hydroxyapatite incorporated photocatalytic membranes for the degradation of chloramphenicol antibiotic in water, J. Chem. Technol. Biotechnol. 96 (2021) 1057–1066, https://doi.org/10.1002/jctb.6617.

[97] J. Wu, S. Yi, Y. Wang, J. Yao, W. Gao, Polymer-based TiO 2 nanocomposite membrane: synthesis and organic pollutant removal, Int. J. Smart Nano Mater. 12 (2021) 129–145, https://doi.org/10.1080/19475411.2021.1901792.

[98] K. Chaloupka, Y. Malam, A.M. Seifalian, Nanosilver as a new generation of nanoproduct in biomedical applications, Trends Biotechnol. 28 (2010) 580–588, https://doi.org/10.1016/j.tibtech.2010.07.006.

[99] S. Nakamura, M. Sato, Y. Sato, N. Ando, T. Takayama, M. Fujita, M. Ishihara, Synthesis and application of silver nanoparticles (Ag nps) for the prevention of infection in healthcare workers, Int. J. Mol. Sci. 20 (15) (2019) 3620–3638, https://doi.org/10.3390/ijms20153620.

[100] L. Guo, W. Yuan, Z. Lu, C.M. Li, Polymer/nanosilver composite coatings for antibacterial applications, Colloids Surfaces A Physicochem. Eng. Asp. 439 (2013) 69–83, https://doi.org/10.1016/j.colsurfa.2012.12.029.

[101] M. López-Heras, I.G. Theodorou, B.F. Leo, M.P. Ryan, A.E. Porter, Towards understanding the antibacterial activity of Ag nanoparticles: electron microscopy in the analysis of the materials-biology interface in the lung, Environ. Sci. Nano 2 (2015) 312–326, https://doi.org/10.1039/C5EN00051C.

[102] T. Shalaby, H. Hamad, E. Ibrahim, O. Mahmoud, A. Al-Oufy, Electrospun nanofibers hybrid composites membranes for highly efficient antibacterial activity, Ecotoxicol. Environ. Saf. 162 (2018) 354–364, https://doi.org/10.1016/j.ecoenv.2018.07.016.

[103] J.S. Taurozzi, H. Arul, V.Z. Bosak, A.F. Burban, T.C. Voice, M.L. Bruening, V.V. Tarabara, Effect of filler incorporation route on the properties of polysulfone-silver nanocomposite membranes of different porosities, J. Membr. Sci. 325 (2008) 58–68, https://doi.org/10.1016/j.memsci.2008.07.010.

[104] S. Jiang, C. Tang, Z. Gong, Z. Zhang, D. Wang, M. Fan, Facile preparation of chitosan coated silver nanoparticles embedded cotton fabric for point-of-use water disinfection, Mater. Lett. 277 (2020), 128256, https://doi.org/10.1016/j.matlet.2020.128256.

[105] L. Qi, Z. Liu, N. Wang, Y. Hu, Facile and efficient in situ synthesis of silver nanoparticles on diverse filtration membrane surfaces for antimicrobial performance, Appl. Surf. Sci. 456 (2018) 95–103, https://doi.org/10.1016/j.apsusc.2018.06.066.

[106] N. Zhang, N. Yang, L. Zhang, B. Jiang, Y. Sun, J. Ma, K. Cheng, F. Peng, Facile hydrophilic modification of PVDF membrane with Ag/EGCG decorated micro/nanostructural surface for efficient oil-in-water emulsion separation, Chem. Eng. J. 402 (2020) 126200, https://doi.org/10.1016/j.cej.2020.126200.

[107] N. Haghighat, V. Vatanpour, M. Sheydaei, Z. Nikjavan, Preparation of a novel polyvinyl chloride (PVC) ultrafiltration membrane modified with Ag/TiO2 nanoparticle with enhanced hydrophilicity and antibacterial activities, Sep. Purif. Technol. 237 (2020) 1163743–1163756, https://doi.org/10.1016/j.seppur.2019.116374.

[108] H. Salazar, P.M. Martins, B. Santos, M.M. Fernandes, A. Reizabal, V. Sebastián, G. Botelho, C.J. Tavares, J.L. Vilas-Vilela, S. Lanceros-Mendez, Photocatalytic and antimicrobial multifunctional nanocomposite membranes for emerging pollutants water treatment applications, Chemosphere 250 (2020), 126299, https://doi.org/10.1016/j.chemosphere.2020.126299.

[109] S. Ghorabi, F.Z. Ashtiani, M. Karimi, A. Fouladitajar, B. Yousefi, F. Dorkalam, Development of a novel dual-bioinspired method for synthesis of a hydrophobic/hydrophilic polyethersulfone coated membrane for membrane distillation, Desalination 517 (2021) 115242, https://doi.org/10.1016/j.desal.2021.115242.

[110] S.A. Parekh, R.N. David, K.K.R. Bannuru, L. Krishnaswamy, A. Baji, Electrospun silver coated polyacrylonitrile membranes for water filtration applications, Membranes (Basel) 8 (2018) 26–29, https://doi.org/10.3390/membranes8030059.

[111] D. Kamarudin, N. Awanis Hashim, B.H. Ong, Y. Kakihana, M. Higa, H. Matsuyama, Multiple effect of thermal treatment approach on PVDF membranes: permeability enhancement and silver nanoparticles immobilization, J. Environ. Chem. Eng. 9 (2021) 105769, https://doi.org/10.1016/j.jece.2021.105769.

[112] B. Bolto, Z. Xie, Recent developments in fouling minimization of membranes modified with silver nanoparticles, J. Membr. Sci. Res. 4 (2018) 111–120, https://doi.org/10.22079/JMSR.2018.79056.1168.

[113] H.E. Shim, J.E. Yang, S.W. Jeong, C.H. Lee, L. Song, S. Mushtaq, D.S. Choi, Y.J. Choi, J. Jeon, Silver nanomaterial-immobilized desalination systems for efficient removal of radioactive iodine species in water, Nanomaterials 8 (2018), https://doi.org/10.3390/nano8090660.

[114] K. Zodrow, L. Brunet, S. Mahendra, D. Li, A. Zhang, Q. Li, P.J.J. Alvarez, Polysulfone ultrafiltration membranes impregnated with silver nanoparticles show improved biofouling resistance and virus removal, Water Res. 43 (2009) 715–723, https://doi.org/10.1016/j.watres.2008.11.014.

[115] J. Dolina, Z. Gončuková, M. Bobák, L. Dvořák, Modification of a hollow-fibre polyethersulfone membrane using silver nanoparticles formed: in situ for biofouling prevention, RSC Adv. 8 (2018) 14552–14560, https://doi.org/10.1039/c8ra02026d.

[116] R. Torres-Mendieta, F. Yalcinkaya, E. Boyraz, O. Havelka, S. Wacławek, J. Maryška, M. Černík, M. Bryjak, PVDF nanofibrous membranes modified via laser-synthesized Ag nanoparticles for a cleaner oily water separation, Appl. Surf. Sci. 526 (2020) 146575, https://doi.org/10.1016/j.apsusc.2020.146575.

[117] W. Jang, J. Yun, Y. Park, I.K. Park, H. Byun, C.H. Lee, Polyacrylonitrile nanofiber membrane modified with ag/go composite for water purification system, Polymers (Basel) 12 (2020) 1–12, https://doi.org/10.3390/polym12112441.

[118] N.G. Khaligh, M.R. Johan, Recent advances in water treatment using graphene-based materials, Mini. Rev. Org. Chem. 17 (2019) 74–90, https://doi.org/10.2174/1570193x16666190516114023.

[119] N. Song, X. Gao, Z. Ma, X. Wang, Y. Wei, C. Gao, A review of graphene-based separation membrane: materials, characteristics, preparation and applications, Desalination 437 (2018) 59–72, https://doi.org/10.1016/j.desal.2018.02.024.

[120] X. Wang, Y. Zhao, E. Tian, J. Li, Y. Ren, Graphene oxide-based polymeric membranes for water treatment, Adv. Mater. Interfaces. 5 (2018) 1–20, https://doi.org/10.1002/admi.201701427.

[121] Y. Jiang, P. Biswas, J.D. Fortner, A review of recent developments in graphene-enabled membranes for water treatment, Environ. Sci. Water Res. Technol. 2 (2016) 915–922, https://doi.org/10.1039/c6ew00187d.

[122] Z.Y. Han, L.J. Huang, H.J. Qu, Y.X. Wang, Z.J. Zhang, Q.L. Rong, Z.Q. Sang, Y. Wang, M.J. Kipper, J.G. Tang, A review of performance improvement strategies for graphene oxide-based and graphene-based membranes in water treatment, J. Mater. Sci. 56 (2021) 9545–9574, https://doi.org/10.1007/s10853-021-05873-7.

[123] Y. Gao, M. Hu, B. Mi, Membrane surface modification with TiO2-graphene oxide for enhanced photocatalytic performance, J. Membr. Sci. 455 (2014) 349–356, https://doi.org/10.1016/j.memsci.2014.01.011.

[124] S. Alyarnezhad, T. Marino, J.B. Parsa, F. Galiano, C. Ursino, H. Garcìa, M. Puche, A. Figoli, Polyvinylidene fluoride-graphene oxide membranes for dye removal under visible light irradiation, Polymers (Basel) 12 (2020) 1–19, https://doi.org/10.3390/polym12071509.

[125] Z. Xu, T. Wu, J. Shi, K. Teng, W. Wang, M. Ma, J. Li, X. Qian, C. Li, J. Fan, Photocatalytic antifouling PVDF ultrafiltration membranes based on synergy of graphene oxide and TiO2 for water treatment, J. Membr. Sci. 520 (2016) 281–293, https://doi.org/10.1016/j.memsci.2016.07.060.

[126] M. Sun, J. Li, Graphene oxide membranes: functional structures, preparation and environmental applications, Nano Today 20 (2018) 121–137, https://doi.org/10.1016/j.nantod.2018.04.007.

[127] D.I. Petukhov, O.O. Kapitanova, E.A. Eremina, E.A. Goodilin, Preparation, chemical features, structure and applications of membrane materials based on graphene oxide, Mendeleev Commun. 31 (2021) 137–148, https://doi.org/10.1016/j.mencom.2021.03.001.

[128] F. Ahmed Janjhi, I. Chandio, A. Ali Memon, Z. Ahmed, K. Hussain Thebo, A.A.A. Pirzado, A.A. Hakro, M. Iqbal, Functionalized graphene oxide based membranes for ultrafast molecular separation, Sep. Purif. Technol. 274 (2021) 117969, https://doi.org/10.1016/j.seppur.2020.117969.

[129] C. Fang, H. Wu, S.-Y. Lee, R.L. Mahajan, R. Qiao, The ionized graphene oxide membranes for water-ethanol separation, Carbon N. Y. 136 (2018) 262–269, https://doi.org/10.1016/j.carbon.2018.04.077.

[130] T. Yu, L. Bai, Z. Xu, X. Yang, Molecular simulation of permeation behaviour of ethanol/water molecules with single-layer graphene oxide membranes, Mol. Simul. 44 (2018) 840–849, https://doi.org/10.1080/08927022.2018.1464161.

[131] Q. Liu, Y. Wu, X. Wang, G. Liu, Y. Zhu, Y. Tu, X. Lu, W. Jin, Molecular dynamics simulation of water-ethanol separation through monolayer graphene oxide membranes: significant role of O/C ratio and pore size, Sep. Purif. Technol. 224 (2019) 219–226, https://doi.org/10.1016/j.seppur.2019.05.030.

[132] C. Sun, M. Liu, B. Bai, Molecular simulations on graphene-based membranes, Carbon N. Y. 153 (2019) 481–494, https://doi.org/10.1016/j.carbon.2019.07.052.

[133] A. Figoli, C. Ursino, F. Galiano, Innovative coating membranes for water treatment, in: E. Drioli, L. Giorno, A. Gugliuzza (Eds.), Funct. Nanostractured Membr., Pan Stanford, Singapore, 2018, p. 604.

[134] B. Liang, W. Zhan, G. Qi, S. Lin, Q. Nan, Y. Liu, B. Cao, K. Pan, High performance graphene oxide/polyacrylonitrile composite pervaporation membranes for desalination applications, J. Mater. Chem. A 3 (2015) 5140–5147, https://doi.org/10.1039/C4TA06573E.

[135] Y. Wei, Y. Zhang, X. Gao, Y. Yuan, B. Su, C. Gao, Declining flux and narrowing nanochannels under wrinkles of compacted graphene oxide nanofiltration membranes, Carbon N. Y. 108 (2016) 568–575, https://doi.org/10.1016/j.carbon.2016.07.056.

[136] B.A. Abdelkader, M.A. Antar, T. Laoui, Z. Khan, Development of graphene oxide-based membrane as a pretreatment for thermal seawater desalination, Desalination 465 (2019) 13–24, https://doi.org/10.1016/j.desal.2019.04.028.

[137] X. Zhao, Z. Tong, X. Liu, J. Wang, B. Zhang, Facile preparation of polyamide-graphene oxide composite membranes for upgrading pervaporation desalination performances of hypersaline solutions, Ind. Eng. Chem. Res. 59 (2020) 12232–12238, https://doi.org/10.1021/acs.iecr.0c01417.

[138] H. Salehi, M. Rastgar, A. Shakeri, Anti-fouling and high water permeable forward osmosis membrane fabricated via layer by layer assembly of chitosan/graphene oxide, Appl. Surf. Sci. 413 (2017) 99–108, https://doi.org/10.1016/j.apsusc.2017.03.271.

[139] S. Zinadini, A. Akbar, M. Rahimi, V. Vatanpour, Preparation of a novel antifouling mixed matrix PES membrane by embedding graphene oxide nanoplates, J. Membr. Sci. 453 (2014) 292–301, https://doi.org/10.1016/j.memsci.2013.10.070.

[140] H. Yu, Y. He, G. Xiao, H. Li, X. Mei, Y. Cheng, F. Zhong, L. Zhou, J.Z. Ou, Intercalation of soft PPy polymeric nanoparticles in graphene oxide membrane for enhancing nanofiltration performances, Sep. Purif. Technol. 272 (2021) 118933, https://doi.org/10.1016/j.seppur.2021.118933.

[141] N.R.N.A. Ghani, S.S. Sulaiman, A. Tahreen, M.S. Jami, Polyether sulfone-graphene oxide- polyvinyl pyrrolidone nanocomposite adsorptive membrane for arsenic removal from wastewater, J. Water Environ. Nanotechnol. 6 (2021) 121–137, https://doi.org/10.22090/JWENT.2021.02.003.

[142] M. Mahmoudian, M.G. Kochameshki, The performance of polyethersulfone nanocomposite membrane in the removal of industrial dyes, Polymer (Guildf) 224 (2021) 1–12, https://doi.org/10.1016/j.polymer.2021.123693.

[143] P.V. Chai, P.Y. Choy, W.C. Teoh, E. Mahmoudi, W.L. Ang, Graphene oxide based mixed matrix membrane in the presence of eco-friendly natural additive gum Arabic, J. Environ. Chem. Eng. 9 (2021) 105638, https://doi.org/10.1016/j.jece.2021.105638.

[144] Z.C. Ng, W.J. Lau, A.F. Ismail, GO/PVA-integrated TFN RO membrane: exploring the effect of orientation switching between PA and GO/PVA and evaluating the GO loading impact, Desalination 496 (2020) 114538, https://doi.org/10.1016/j.desal.2020.114538.

[145] A. Giwa, N. Akther, V. Dufour, S.W. Hasan, A critical review on recent polymeric and nano-enhanced membranes for reverse osmosis, RSC Adv. 6 (2016) 8134–8163, https://doi.org/10.1039/c5ra17221g.

[146] Y.A.J. Al-Hamadani, B.-M. Jun, M. Yoon, N. Taheri-Qazvini, S.A. Snyder, M. Jang, J. Heo, Y. Yoon, Applications of MXene-based membranes in water purification: a review, Chemosphere 254 (2020) 126821, https://doi.org/10.1016/j.chemosphere.2020.126821.

[147] P. Liu, J. Hou, Y. Zhang, L. Li, X. Lu, Z. Tang, Two-dimensional material membranes for critical separations, Inorg. Chem. Front. 7 (2020) 2560–2581, https://doi.org/10.1039/D0QI00307G.

[148] A. Boretti, S. Al-Zubaidy, M. Vaclavikova, M. Al-Abri, S. Castelletto, S. Mikhalovsky, Outlook for graphene-based desalination membranes, NPJ Clean Water 1 (2018) 1–11, https://doi.org/10.1038/s41545-018-0004-z.

[149] N. Nasrollahi, S. Aber, V. Vatanpour, N.M. Mahmoodi, The effect of amine functionalization of CuO and ZnO nanoparticles used as additives on the morphology and the permeation properties of polyethersulfone ultrafiltration nanocomposite membranes, Compos. Part B Eng. 154 (2018) 388–409, https://doi.org/10.1016/j.compositesb.2018.09.027.

[150] I. Munnawar, S.S. Iqbal, M.N. Anwar, M. Batool, S. Tariq, N. Faitma, A.L. Khan, A.U. Khan, U. Nazar, T. Jamil, N.M. Ahmad, Synergistic effect of chitosan-zinc oxide hybrid nanoparticles on antibiofouling and water disinfection of mixed matrix polyethersulfone nanocomposite membranes, Carbohydr. Polym. 175 (2017) 661–670, https://doi.org/10.1016/j.carbpol.2017.08.036.

[151] M. Homayoonfal, M.R. Mehrnia, S. Rahmani, Y.M. Mojtahedi, Journal of Industrial and Engineering Chemistry Fabrication of alumina/polysulfone nanocomposite membranes with biofouling mitigation approach in membrane bioreactors, J. Ind. Eng. Chem. 22 (2014) 357–367, https://doi.org/10.1016/j.jiec.2014.07.031.

[152] B. Ma, C. Hu, X. Wang, Y. Xie, W.A. Jefferson, H. Liu, J. Qu, Effect of aluminum speciation on ultrafiltration membrane fouling by low dose aluminum coagulation with bovine serum albumin (BSA), J. Membr. Sci. 492 (2015) 88–94, https://doi.org/10.1016/j.memsci.2015.05.043.

[153] G.M. Urper-Bayram, N. Bossa, D.M. Warsinger, I. Koyuncu, M. Wiesner, Comparative impact of SiO2 and TiO2 nanofillers on the performance of thin-film nanocomposite membranes, J. Appl. Polym. Sci. 137 (2020) 49382, https://doi.org/10.1002/app.49382.

[154] E. Demirel, B. Zhang, M. Papakyriakou, S. Xia, Y. Chen, Fe2O3 nanocomposite PVC membrane with enhanced properties and separation performance, J. Membr. Sci. 529 (2017) 170–184, https://doi.org/10.1016/j.memsci.2017.01.051.

[155] S. Pan, X. Zhang, Y. Wang, J. Liu, B. Pan, Mesoporous polyacrylonitrile membrane with ultrahigh loading of well-dispersed Fe2O3 nanoparticles: a powerful phosphate scavenger enabling inhibition of microbial regrowth in treated water, J. Membr. Sci. 603 (2020) 118048, https://doi.org/10.1016/j.memsci.2020.118048.

[156] M. Banihashemi, N. Dalali, N. Sehati, B. Farajmand, Decoration of Fe3O4@SiO2@ZnO as a high performance nanosorbent on a stir bar micro-extraction device for preconcentration and determination of cadmium in real water samples, Microchem. J. 154 (2020) 104599, https://doi.org/10.1016/j.microc.2020.104599.

[157] K. Monsef, M. Homayoonfal, F. Davar, Coating carboxylic and sulfate functional groups on ZrO2 nanoparticles: antifouling enhancement of nano-composite membranes during water treatment, React. Funct. Polym. 131 (2018) 299–314, https://doi.org/10.1016/j.reactfunctpolym.2018.08.003.

[158] N. Ma, J. Wei, R. Liao, C.Y. Tang, Zeolite-polyamide thin film nanocomposite membranes: towards enhanced performance for forward osmosis, J. Membr. Sci. 405–406 (2012) 149–157, https://doi.org/10.1016/j.memsci.2012.03.002.

[159] J. Gascon, F. Kapteijn, B. Zornoza, V. Sebastián, C. Casado, J. Coronas, Practical approach to zeolitic membranes and coatings: state of the art, opportunities, barriers, and future perspectives, Chem. Mater. 24 (2012) 2829–2844, https://doi.org/10.1021/cm301435j.

[160] G. Arthanareeswaran, T.K. Sriyamuna Devi, M. Raajenthiren, Effect of silica particles on cellulose acetate blend ultrafiltration membranes: part I, Sep. Purif. Technol. 64 (2008) 38–47, https://doi.org/10.1016/j.seppur.2008.08.010.

[161] S. Kar, M. Subramanian, A.K. Ghosh, R.C. Bindal, S. Prabhakar, J. Nuwad, C.G.S. Pillai, S. Chattopadhyay, P.K. Tewaria, Potential of nanoparticles for water purification: a case study on anti-biofouling behaviour of metal based polymeric nanocomposite membrane, Desalin. Water Treat. 27 (2011) 224–230, https://doi.org/10.5004/dwt.2011.1967.

[162] Y. Liu, Z. Yu, Y. Peng, L. Shao, X. Li, H. Zeng, A novel photocatalytic self-cleaning TiO2 nanorods inserted graphene oxide-based nanofiltration membrane, Chem. Phys. Lett. 749 (2020) 137424, https://doi.org/10.1016/j.cplett.2020.137424.

[163] M.I. Baig, P.G. Ingole, J. Jeon, S.U. Hong, W.K. Choi, H.K. Lee, Water vapor transport properties of interfacially polymerized thin film nanocomposite membranes modified with graphene oxide and GO-TiO2 nanofillers, Chem. Eng. J. 373 (2019) 1190–1202, https://doi.org/10.1016/j.cej.2019.05.122.

[164] H. Xu, M. Ding, W. Chen, Y. Li, K. Wang, Nitrogen–doped GO/TiO2 nanocomposite ultrafiltration membranes for improved photocatalytic performance, Sep. Purif. Technol. 195 (2018) 70–82, https://doi.org/10.1016/j.seppur.2017.12.003.

[165] M. Kumar, Z. Gholamvand, A. Morrissey, K. Nolan, M. Ulbricht, J. Lawler, Preparation and characterization of low fouling novel hybrid ultrafiltration membranes based on the blends of GO-TiO2 nanocomposite and polysulfone for humic acid removal, J. Membr. Sci. 506 (2016) 38–49, https://doi.org/10.1016/j.memsci.2016.02.005.

[166] Y. Mansourpanah, A. Ghanbari, H. Yazdani, A.G. Mohammadi, A. Rahimpour, Silver-polyamidoamine/graphene oxide thin film nanofiltration membrane with improved antifouling and antibacterial properties for water purification and desalination, Desalination 511 (2021), 115109, https://doi.org/10.1016/j.desal.2021.115109.

[167] A.F. Faria, C. Liu, M. Xie, F. Perreault, L.D. Nghiem, J. Ma, M. Elimelech, Thin-film composite forward osmosis membranes functionalized with graphene oxide–silver nanocomposites for biofouling control, J. Membr. Sci. 525 (2017) 146–156, https://doi.org/10.1016/j.memsci.2016.10.040.

[168] X. Sun, J. Qin, P. Xia, B. Guo, C. Yang, C. Song, S. Wang, Graphene oxide–silver nanoparticle membrane for biofouling control and water purification, Chem. Eng. J. 281 (2015) 53–59, https://doi.org/10.1016/j.cej.2015.06.059.

[169] O.T. Mahlangu, R. Nackaerts, J.M. Thwala, B.B. Mamba, A.R.D. Verliefde, Hydrophilic fouling-resistant GO-ZnO/PES membranes for wastewater reclamation, J. Membr. Sci. 524 (2017) 43–55, https://doi.org/10.1016/j.memsci.2016.11.018.

[170] M. Zare, K. Namratha, S. Alghamdi, Y.H.E. Mohammad, A. Hezam, M. Zare, Q.A. Drmosh, K. Byrappa, B.N. Chandrashekar, S. Ramakrishna, X. Zhang, Novel green biomimetic approach for synthesis of ZnO-Ag nanocomposite; antimicrobial activity against food-borne pathogen, biocompatibility and solar photocatalysis, Sci. Rep. 9 (2019) 8303, https://doi.org/10.1038/s41598-019-44309-w.

[171] H. Wu, J. Huang, Y. Liu, Polysulfone ultrafiltration membrane incorporated with Ag-SiO2 nanohybrid for effective fouling control, J. Water Health. 15 (2017) 341–352, https://doi.org/10.2166/wh.2017.234.

[172] A. Nikmah, A. Taufiq, A. Hidayat, H.S. Sunaryono, Excellent antimicrobial activity of Fe 3 O 4 /SiO 2 /Ag nanocomposites, Nano 16 (2021) 2150049, https://doi.org/10.1142/S1793292021500491.

Chapter 15

Applications of two-dimensional nanostructures for water filtration

Iftaykhairul Alam and Indranil Chowdhury*
Department of Civil and Environmental Engineering, Washington State University, Pullman, WA, United States
**Corresponding author: e-mail address: indranil.chowdhury@wsu.edu*

Introduction

Rapid advances in nanotechnology have introduced us to numerous potential nanomaterials with superb properties to be used for environmental technology, and a horde of researchers are putting emphasis on the synthesis, property characterization, modifications, and technological applications of these materials [1]. In general, there are a wide range of 2D layered materials where the atomic layers are weakly bonded together by van der Waals interactions and can be promptly isolated into single or few layer nanosheets with the application of mechanical or liquid exfoliation processes and these nanosheets show excellent electronic properties, exceptional mechanical flexibility, partial optical transparency, and antibacterial properties [1–6]. Graphene has been the heart of research for many years but now the focus has shifted to the application of different 2D nanomaterials and their nanosheets [7]. Chalcogenide, oxide and graphitic layered materials already find applications in batteries as electrodes, thermoelectric devices as semiconductors, catalysts, and lubricants in industry [8]. Due to the van der Waals interactions between neighboring layers of these 2D materials, it is possible to combine different materials without the limitation of lattice matching, each of which has entirely different physical properties [9]. This creates the new opportunities for the researchers to use these materials to apply in developing new technologies. So, the purpose of this review is to delve into the application of 2D nanomaterials (graphene family members, 2D chalcogenides specifically molybdenum disulfide MoS_2, & tungsten disulfide WS_2 and 2D oxides) in the field of environmental treatment and clean technology. With the advancement in the field of nanotechnology, it is highly likely to see the broader applications of these materials in the future much in the same way how we experienced the numerous uses of carbon nanotubes and graphene beyond their initial applications in electronic devices as the researchers and scientists explored more and more with those materials. Application of these 2D nanomaterials in water filtration technology is a growing field in our modern world. Accumulation of foreign substances on membrane surfaces or inside pores hampers the performance of the membranes [10,11]. Fouling can deteriorate membrane performance by lowering water permeability, worsening product water quality, increasing energy consumption, etc. Overgrowth of microorganisms lead to low water quality and pathogen development and this will shorten the membrane life [12]. 2D nanomaterials can be promisingly exploited to make novel membranes due to their unprecedented properties [13,14]. These materials show antifouling and antibacterial properties which may be useful in the path of advancement of membrane technology. Moreover, these materials are one-atom thick ($\sim$1 nm), which can significantly reduce the thickness of the membranes [15].

2D layer materials are ultrathin nanomaterials in which atomic configuration and strength of bond are similar. Electrons in these materials are free to move in 2D plane, but their motion is restricted in 3D plane governed by the quantum mechanics; Nanosheet: stacked assembly or hybrid composite structure formed single to many layers of 2D materials; Graphene is a 2D layer nanomaterial of sp^3 bonded carbon atoms arranged in a hexagonal crystalline structure where every carbon is bonded to a terminal hydrogen, alternatively above and below the layer; Chalcogenide is a chemical compound consisting of at least one chalcogen anion (oxygen (O), sulfur (S), selenium (Se), tellurium (Te), polonium (Po), and at least one more electropositive element molybdenum (Mo), tungsten (W), etc.)

Separation Science and Technology, Volume 15. https://doi.org/10.1016/B978-0-323-90763-7.00014-7

Diversity of 2D materials for water filtration applications and scope of review

The application of 2D nanomaterials in clean technology is still a secrecy specially for the materials other than graphene family members. Monolayers of graphite, hBN, molybdenum disulfide, tungsten disulfide (WS2), tungsten diselenide (WSe2), molybdenum diselenide (MoSe2), etc. have potential to be used for environmental remediation due to their high thermal and chemical stability. Many oxide nanoparticles such as Al_2O_3, ZrO_2, TiO_2, etc. can improve the polymer membrane due to their high hydrophilicity and membrane permeability and also can enhance the antifouling properties of the membrane because of their strong affinity to water molecules [16–18] (Table 1).

That is why our focus is given not only to graphene but also to the other 2D nanomaterials that can be useful for the application in water treatment, membrane technology and other clean technologies. In this review, we discuss the applications and mechanisms involved in antifouling, antimicrobial and desalination properties of 2D graphene family nanomaterials, 2D chalcogenides (MoS_2 & WS_2), and 2D oxides.

Applications graphene family members for water filtration

Antimicrobial properties

Two-dimensional graphene family members have shown strong antimicrobial properties against wide range of microorganisms. Gram-positive and -negative bacterial pathogens, phytopathogens, and biofilm-forming microorganisms have already been tested on GO surfaces to see whether GO can actually kill those microbes [2,4,19–21]. Attachment of bacterial cells to GO-coated surfaces causes the disruption of the cell integrity and loss of cell viability which results in reduction of bacterial growth on the surface [19,21–23]. Physical and chemical interactions during the contact with GO sheets and bacterial cells help the process of antimicrobial activity. Membrane of bacteria gets damaged by the atomically sharp edges of graphene which can penetrate the bacteria cell membrane [19,22,24]. Oxidative stress of GO is also considered as a major component for antimicrobial activity as disturbances in the normal redox states of cells can cause toxic effect resulting in the damage of all components of cells such as proteins, lipids and DNA [4,20]. Graphene nanosheets with higher oxidation degree causes larger scale of irregular membrane perturbation and this tendency increases with the edge length of nanosheets [25]. Adsorption of graphene oxide on the cell membrane was shown to contribute higher antibacterial property when visualized by atomic force microscopy. Cell wrapping may limit bacterial growth by isolating cells from the medium, preventing nutrient absorption, or blocking active sites on the cell surface [26]. Trapping microorganisms within aggregate-reduced graphene sheets (rGO) was also suggested as another mechanism for describing the cytotoxicity of graphene sheets, especially in a suspension [27].

Desalination

Desalination is a process that removes salts and minerals from saline water and it is one of the most promising approaches to supply new fresh water. Though reverse osmosis is considered to be the most effective and energy-efficient desalination technique, nanomaterials have a great deal to offer over desalination [28]. Due to their small size, nanoporous materials can increase the speed of filtration and due to their hydrophobicity they can reject ions and other molecular solutes [28]. Phase transition and molecular alignment of water in confined environments also plays role in increasing the water flux besides offering a smooth, frictionless surface for fast flow of water offered by GO and carbon nanotubes [29,30]. It has been seen on modeling results that nanoporous graphene members can attain water permeability of $400–4000\,L\,m^2\,h^{-1}\,bar^{-1}$ while still rejecting salts, a performance 2 to 3 orders of magnitude higher than current reverse osmosis membranes and can reduce

TABLE 1 2D nanomaterials that have been used for or have potential for membrane applications [13].

Graphene family	2D chalcogenides	2D oxides
Graphene	MoS_2	SiO_2
hBN "white graphene"	WS_2	Fe_3O_4
Graphene oxide	$MoSe_2$	Al_2O_3
Reduced graphene oxide	WSe_2	ZrO_2
Pristine graphene		TiO_2
Functionalized graphene		

required area and energy consumption for desalination [28,31]. Steric effects, hydrodynamic interactions and electrostatic repulsion between charged particles and the pores were found to be involved in rejection of salts [32,33].

For nanoporous graphene membrane, size exclusion and electrostatic interactions play key roles in desalination techniques while selectivity in stacked GO membranes results from adsorption of ions to GO sheets in addition to size exclusion and electrostatic interaction [34]. Also, membranes formed of stacked GO sheets reject both salts and organic molecules due to size exclusion which results in increasing the water flux but decreasing the selectivity of the membrane [35–37].

Applications of transition metal dichalcogenides for water filtration

Emerging 2D nanostructures transition metal dichalcogenides such as molybdenum disulfide (MoS_2) and tungsten disulfide (WS_2) are highly hydrophobic and have extremely low friction. These materials can deter settlement of fouling organisms and provide low adhesion of any fouling that occurs [14,38]. Fouling is one of the major challenges that modern engineers face when operating water filtration membranes. Transition metal dichalcogenides can certainly be the potential candidates in this field. Their unique properties can be utilized for deterring attachment of fouling microorganisms when high concentrations of ions are present in water [39,40]. Electron redistribution between Mo and S in MoS_2 makes the nanopores edge atoms in filter intrinsically charged when nanopores are introduced in MoS_2. These charged pore edges play an important role to alter the ion selectivity and water permeability [28,41]. Unlike graphene-based filters, it has been investigated that with the introduction of mechanical strain, the "open" and "closed" states of the MoS_2 filter can be regulated and the open state of the MoS_2 filter results in high water transparency [41]. The integration of nano-WS_2 well dispersed in the polymers, such as polyether ether ketone (PEEK) has been found to enhance the thermal stability and mechanical strength of the hybrid nanocomposites [42,43]. In one study, blended PES–WS_2 flat-sheet membranes with the incorporation of ultra-low concentrations of nanoparticles (from 0.025% to 0.25%, WS_2/PES ratio) were investigated in terms of permeability, fouling resistance and solute rejection. The optimum permeability of the membrane was found with 0.075%–0.1% WS_2/polymer concentration ratios and fouling resistance was increased by almost 50%. Moreover, solute rejection was significantly enhanced after the incorporation of nanoparticles into the membrane structure [16]. Graphene nanopores with functionalized hydroxyl groups can boost the permeability of the membrane, however, reduces desalination efficiencies. Addition of accurate functional groups to the edges of the nanopores requires complex fabrication [28,44]. MoS_2 is a single-atom-thick membrane with hydrophilic sites which can lead to new water desalination technology (see Fig. 1). Vacancies can be easily introduced into the MoS_2 monolayer which suggests nanoporous MoS_2 can be used in water desalination applications [45]. In one study, it has been found by molecular dynamic simulations, that a nanopore with a single-layer molybdenum disulfide can effectively reject ions and also increase the permeability of the membrane. More than 88% of ions were rejected by membranes having pore areas ranging from 20 to 60 $Å^2$. In that same study, water flux was found to be increased around 2 to 5 times than other nanoporous membranes [38]. Another study reveals that MoS_2 nanopore with "open" and "closed" states can be successfully regulated for water flow and ion filtering. Mechanical stretching of the filter was used to change the size of the nanopore effectively and it has been found that both steric and electrostatic effects contribute to blocking the passage of ions [41].

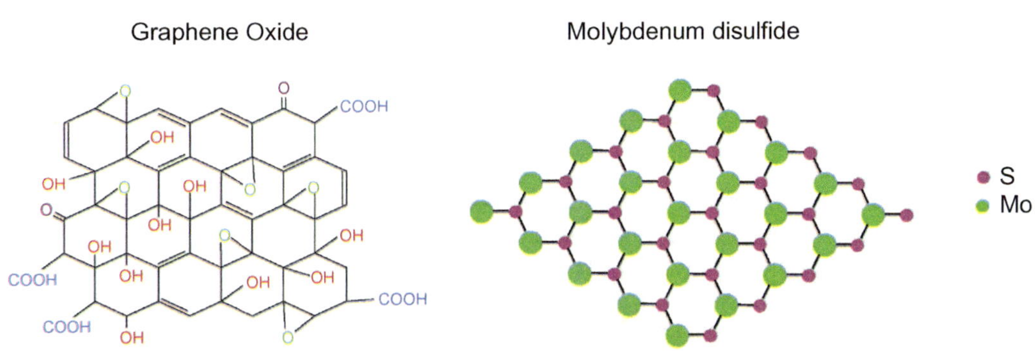

FIG. 1 Introduction of functional groups (i.e., -OH, -COOH, -CO, etc.) makes GO hydrophilic (Left). –COOH is the charged hydrophilic group, and –OH and –CO are the uncharged polar groups on GO structure. There is also presence of hydrophobic π bond domains on the GO structure. MoS_2 consists of weakly coupled layers of S-Mo-S where a Mo atom layer is sandwiched by two S atom layers (Right).

Application of two-dimensional oxides for water filtration

Two-dimensional oxides such as Al_2O_3, ZnO, TiO_2, etc. are being used currently for better performance of water filtration. 2D nano-oxides have shown great potential to improve the water flux or to make the membrane antifouling. Usually these oxides have been used for membrane modification with different polymers. TiO_2 promotes efficient charge separation and charge trapping of ions on the TiO_2 surfaces [46]. Incorporation of TiO_2 into nanocarbon (carbon nano tubes, graphene, etc.) [47–51] is used not only removal of organic pollutants [52,53] but also for decomposition of biological pollutants, removal of contaminants from vapors including NOx, CO, and acetaldehyde [54,55]. Also, TiO_2/nanocarbon composites perform better on self-purification than ozonation and UV irradiation during the application of advanced oxidation technologies for water treatment [56,57]. Many of the polymers such as polyvinylidenefluoride (PVDF) and polyethersulfone (PES) that are used for antifouling membrane preparation are hydrophobic polymers. The hydrophilicity of these polymers can be improved significantly by the addition of nanosized boehmite (aluminum oxide hydroxide (γ-AlOOH) particle, containing hydroxyl groups attached to its surface) [18]. Use of nanoboehmite particles in preparation of mixed matrix membrane also results in reduced fouling in the membranes and this is mainly due to the presence of extra hydroxyl groups on the surface of nanoparticles [58]. It has been found that the 1 wt% nanoboehmite/PES membrane exhibited the highest *FRR* value (96.1%) and the lowest irreversible fouling resistance (*Rir*) value (3.9%) [18]. Multifunctional ZnO also attracted attention due to its outstanding physical, chemical, and efficient antibacterial properties. ZnO nanoparticle as an additive is more economical than TiO_2 and Al_2O_3 [59,60]. In one experiment, the result showed that modified PVDF membrane had photocatalysis self-cleaning capabilities with the addition of nano-ZnO on the inner surface of the membrane. The water permeate flux of this composite membrane was higher than that of a single PVDF membrane [60]. SiO_2 is another common two-dimensional nanoparticle used for membrane modification. PES ultrafiltration membranes blended with SiO_2-Ag composites showed high antibacterial rates. In fact, the antibacterial rates of these hybrid membranes against *E. coli* and *S. aureus* reached 100% [61].

Conclusions and recommendations

During the past decade, significant progresses have been made for understanding the application of two-dimensional nanomaterials to meet up the environmental challenges. These 2D materials have great potential to be used for water treatment and membrane development. With the increasing demand of supplying potable water, using these nanomaterials for desalination, killing bacteria in water and making antifouling membrane will dominate in this field in near future along with convention water treatment processes like reverse osmosis. Nonporous graphene can offer important advantages as the salt rejection performance increases with its narrow pore size distribution. The limited performance of these materials can certainly be overcome by a better understanding of functionalization, modification and combination of the materials. Reducing some functional groups from graphene oxide significantly improves the electrical properties of graphene oxide which is attractive for use in electric devices. But it is one of the challenges to reduce graphene oxide successfully to restore its exceptional electronic and mechanical properties. Combination of the 2D nanomaterials is also a great option for enhancing the favorable properties of these materials. Graphene family members show excellent antibacterial properties when they are used in membranes for water filtration while two-dimensional metal dichalcogenides such MoS_2 have better antifouling properties than graphene materials. The ultra-low concentration range of nano-WS_2 offers great enhancement of antifouling activity in membranes due to their hydrophilicity along with salt rejection properties. Combination of GO/rGO with MoS_2/WS_2 can effectively make any water filtration membrane both antibacterial and antifouling. Combination of these two materials have already been used in electrical devices to overcome the limitation of graphene oxide. 2D oxides are very effective in membrane modification which improves the water flux of the membrane considerably. The membranes modified with 2D nanomaterials will also lower the energy requirements during the desalination processes due to lower operating pressures. Smaller and more modular desalination plants tanks will be possible because of smaller membrane area requirements for 2D nanomaterials. It is also important to highlight the implications of these 2D nanomaterials on environment. Understanding the mechanism of degradation along with the biological and environmental interactions of these materials is an important new field whose results will enable safer and risk free application of these in environmental technologies and clean up techniques. Many of these layer materials can undergo dissolution in biological systems or natural environment while some are likely to be persistent in the environment or in living organisms which may pose the threat of long-latency disease development. It is clear that the 2D nanomaterials are going to the first choice of researchers for the application in the water treatment field. From economical and energy consumption point of view these materials can serve the purposes effectively. But it is also important to understand that research activities for synthesis, fabrication and application of these materials should be done keeping in mind that limits the risk and impact of these materials in natural environment.

References

[1] Z. Wang, W. Zhu, Y. Qiu, X. Yi, A. von dem Bussche, A. Kane, H. Gao, K. Koski, R. Hurt, Biological and environmental interactions of emerging two-dimensional nanomaterials, Chem. Soc. Rev. 45 (6) (2016) 1750–1780.

[2] J. Chen, H. Peng, X. Wang, F. Shao, Z. Yuan, H. Han, Graphene oxide exhibits broad-spectrum antimicrobial activity against bacterial phytopathogens and fungal conidia by intertwining and membrane perturbation, Nanoscale 6 (3) (2014) 1879–1889.

[3] J.N. Coleman, M. Lotya, A. O'Neill, S.D. Bergin, P.J. King, U. Khan, K. Young, A. Gaucher, S. De, R.J. Smith, I.V. Shvets, S.K. Arora, G. Stanton, H.-Y. Kim, K. Lee, G.T. Kim, G.S. Duesberg, T. Hallam, J.J. Boland, J.J. Wang, J.F. Donegan, J.C. Grunlan, G. Moriarty, A. Shmeliov, R.J. Nicholls, J.M. Perkins, E.M. Grieveson, K. Theuwissen, D.W. McComb, P.D. Nellist, V. Nicolosi, Two-dimensional nanosheets produced by liquid exfoliation of layered materials, Science 331 (6017) (2011) 568–571.

[4] S. Gurunathan, J.W. Han, A.A. Dayem, V. Eppakayala, J.-H. Kim, Oxidative stress-mediated antibacterial activity of graphene oxide and reduced graphene oxide in Pseudomonas aeruginosa, Int. J. Nanomedicine 7 (2012) 5901–5914.

[5] U. Halim, C.R. Zheng, Y. Chen, Z. Lin, S. Jiang, R. Cheng, Y. Huang, X. Duan, A rational design of cosolvent exfoliation of layered materials by directly probing liquid–solid interaction, Nat. Commun. 4 (2013).

[6] K.S. Novoselov, D. Jiang, F. Schedin, T.J. Booth, V.V. Khotkevich, S.V. Morozov, A.K. Geim, Two-dimensional atomic crystals, Proc. Natl. Acad. Sci. U. S. A. 102 (30) (2005) 10451–10453.

[7] K.S. Novoselov, V.I. Falko, L. Colombo, P.R. Gellert, M.G. Schwab, K. Kim, Nature 490 (2012) 192.

[8] W. Müller-Warmuth, R. Schöllhorn, Progress in Intercalation Research, Springer, Netherlands, 2012.

[9] X. Duan, C. Wang, A. Pan, R. Yu, X. Duan, Two-dimensional transition metal dichalcogenides as atomically thin semiconductors: opportunities and challenges, Chem. Soc. Rev. 44 (24) (2015) 8859–8876.

[10] Q. Li, M. Elimelech, Organic fouling and chemical cleaning of nanofiltration membranes: measurements and mechanisms, Environ. Sci. Technol. 38 (17) (2004) 4683–4693.

[11] K.R. Zodrow, E. Bar-Zeev, M.J. Giannetto, M. Elimelech, Biofouling and microbial communities in membrane distillation and reverse osmosis, Environ. Sci. Technol. 48 (22) (2014) 13155–13164.

[12] J. Baker, L. Dudley, Desalination 118 (1998) 81.

[13] A.K. Geim, I.V. Grigorieva, Van der Waals heterostructures, Nature 499 (7459) (2013) 419–425.

[14] Q.H. Wang, K. Kalantar-Zadeh, A. Kis, J.N. Coleman, M.S. Strano, Nat. Nanotechnol. 7 (2012) 699.

[15] H. Zhang, Ultrathin two-dimensional nanomaterials, ACS Nano 9 (10) (2015) 9451–9469.

[16] J. Lin, R. Zhang, W. Ye, N. Jullok, A. Sotto, B. Van der Bruggen, Nano-WS2 embedded PES membrane with improved fouling and permselectivity, J. Colloid Interface Sci. 396 (2013) 120–128.

[17] A. Razmjou, A. Resosudarmo, R.L. Holmes, H. Li, J. Mansouri, V. Chen, The effect of modified TiO2 nanoparticles on the polyethersulfone ultrafiltration hollow fiber membranes, Desalination 287 (2012) 271–280.

[18] V. Vatanpour, S.S. Madaeni, L. Rajabi, S. Zinadini, A.A. Derakhshan, Boehmite nanoparticles as a new nanofiller for preparation of antifouling mixed matrix membranes, J. Membr. Sci. 401–402 (2012) 132–143.

[19] O. Akhavan, E. Ghaderi, Toxicity of graphene and graphene oxide nanowalls against bacteria, ACS Nano 4 (10) (2010) 5731.

[20] K. Krishnamoorthy, M. Veerapandian, L.-H. Zhang, K. Yun, S.J. Kim, Antibacterial efficiency of graphene nanosheets against pathogenic bacteria via lipid peroxidation, J. Phys. Chem. C 116 (32) (2012) 17280–17287.

[21] I.E. Mejias Carpio, C.M. Santos, X. Wei, D.F. Rodrigues, Toxicity of a polymer-graphene oxide composite against bacterial planktonic cells, biofilms, and mammalian cells, Nanoscale 4 (15) (2012) 4746–4756.

[22] W. Hu, C. Peng, W. Luo, M. Lv, X. Li, D. Li, Q. Huang, C. Fan, Graphene-based antibacterial paper, ACS Nano 4 (7) (2010) 4317–4323.

[23] F. Perreault, M.E. Tousley, M. Elimelech, Thin-film composite polyamide membranes functionalized with biocidal graphene oxide nanosheets, Environ. Sci. Technol. Lett. 1 (1) (2014) 71–76.

[24] Y. Tu, M. Lv, P. Xiu, T. Huynh, M. Zhang, M. Castelli, Z. Liu, Q. Huang, C. Fan, H. Fang, R. Zhou, Destructive extraction of phospholipids from Escherichia coli membranes by graphene nanosheets, Nat. Nanotechnol. 8 (8) (2013) 594–601.

[25] J. Mao, R. Guo, L.-T. Yan, Simulation and analysis of cellular internalization pathways and membrane perturbation for graphene nanosheets, Biomaterials 35 (23) (2014) 6069–6077.

[26] S. Liu, M. Hu, T.H. Zeng, R. Wu, R. Jiang, J. Wei, L. Wang, J. Kong, Y. Chen, Lateral dimension-dependent antibacterial activity of graphene oxide sheets, Langmuir 28 (33) (2012) 12364–12372.

[27] O. Akhavan, E. Ghaderi, A. Esfandiar, Wrapping bacteria by graphene nanosheets for isolation from environment, reactivation by sonication, and inactivation by near-infrared irradiation, J. Phys. Chem. B 115 (19) (2011) 6279–6288.

[28] D. Cohen-Tanugi, J.C. Grossman, Water desalination across Nanoporous graphene, Nano Lett. 12 (7) (2012) 3602–3608.

[29] A. Alexiadis, S. Kassinos, Molecular simulation of water in carbon nanotubes, Chem. Rev. 108 (12) (2008) 5014–5034.

[30] A. Striolo, The mechanism of water diffusion in narrow carbon nanotubes, Nano Lett. 6 (4) (2006) 633–639.

[31] M. Elimelech, W.A. Phillip, The future of seawater desalination: energy, technology, and the environment, Science 333 (6043) (2011) 712–717.

[32] D. Konatham, J. Yu, T.A. Ho, A. Striolo, Simulation insights for graphene-based water desalination membranes, Langmuir 29 (38) (2013) 11884–11897.

[33] K. Sint, B. Wang, P. Král, Selective ion passage through functionalized graphene nanopores, J. Am. Chem. Soc. 130 (49) (2008) 16448–16449.

[34] F. Perreault, A. Fonseca de Faria, M. Elimelech, Environmental applications of graphene-based nanomaterials, Chem. Soc. Rev. 44 (16) (2015) 5861–5896.

[35] R.K. Joshi, P. Carbone, F.C. Wang, V.G. Kravets, Y. Su, I.V. Grigorieva, H.A. Wu, A.K. Geim, R.R. Nair, Precise and ultrafast molecular sieving through graphene oxide membranes, Science 343 (6172) (2014) 752–754.

[36] R.R. Nair, H.A. Wu, P.N. Jayaram, I.V. Grigorieva, A.K. Geim, Unimpeded permeation of water through helium-leak–tight graphene-based membranes, Science 335 (6067) (2012) 442–444.

[37] P. Sun, F. Zheng, M. Zhu, Z. Song, K. Wang, M. Zhong, D. Wu, R.B. Little, Z. Xu, H. Zhu, Selective trans-membrane transport of alkali and alkaline earth cations through graphene oxide membranes based on cation – Π interactions, ACS Nano 8 (1) (2014) 850–859.

[38] M. Heiranian, A.B. Farimani, N.R. Aluru, Water desalination with a single-layer MoS(2) nanopore, Nat. Commun. 6 (2015) 8616.

[39] P. Dhiman, F. Yavari, X. Mi, H. Gullapalli, Y. Shi, P.M. Ajayan, N. Koratkar, Harvesting energy from water flow over graphene, Nano Lett. 11 (8) (2011) 3123–3127.

[40] J. Yin, X. Li, J. Yu, Z. Zhang, J. Zhou, W. Guo, Generating electricity by moving a droplet of ionic liquid along graphene, Nat. Nanotechnol. 9 (5) (2014) 378–383.

[41] W. Li, Y. Yang, J.K. Weber, G. Zhang, R. Zhou, Tunable, strain-controlled Nanoporous MoS2 filter for water desalination, ACS Nano 10 (2) (2016) 1829–1835.

[42] A.M. Díez-Pascual, M. Naffakh, M.A. Gómez-Fatou, Mechanical and electrical properties of novel poly(ether ether ketone)/carbon nanotube/inorganic fullerene-like WS2 hybrid nanocomposites: experimental measurements and theoretical predictions, Mater. Chem. Phys. 130 (1–2) (2011) 126–133.

[43] A.M. Díez-Pascual, M. Naffakh, C. Marco, G. Ellis, M.A. Gómez-Fatou, High-performance nanocomposites based on polyetherketones, Prog. Mater. Sci. 57 (7) (2012) 1106–1190.

[44] Q. Tang, Z. Zhou, Z. Chen, Graphene-related nanomaterials: tuning properties by functionalization, Nanoscale 5 (11) (2013) 4541–4583.

[45] S. Bertolazzi, J. Brivio, A. Kis, Stretching and breaking of ultrathin MoS2, ACS Nano 5 (12) (2011) 9703–9709.

[46] K. Nagaveni, G. Sivalingam, M.S. Hegde, G. Madras, Photocatalytic degradation of organic compounds over combustion-synthesized Nano-TiO2, Environ. Sci. Technol. 38 (5) (2004) 1600–1604.

[47] E. da Costa, P.P. Zamora, A.J.G. Zarbin, Novel TiO2/C nanocomposites: synthesis, characterization, and application as a photocatalyst for the degradation of organic pollutants, J. Colloid Interface Sci. 368 (1) (2012) 121–127.

[48] T.N. Lambert, C.A. Chavez, B. Hernandez-Sanchez, P. Lu, N.S. Bell, A. Ambrosini, T. Friedman, T.J. Boyle, D.R. Wheeler, D.L. Huber, Synthesis and characterization of titania – graphene nanocomposites, J. Phys. Chem. C 113 (46) (2009) 19812–19823.

[49] Y.J. Lee, G.W. Cao, S.J. Chung, A compact dual-band filtering microstrip antenna with the same polarization planes, in: 2012 Asia Pacific Microwave Conference Proceedings, 2012.

[50] G. Williams, B. Seger, P.V. Kamat, TiO2-graphene nanocomposites. UV-assisted photocatalytic reduction of graphene oxide, ACS Nano 2 (7) (2008) 1487–1491.

[51] X. Zhang, L. Lei, Effect of preparation methods on the structure and catalytic performance of TiO2/AC photocatalysts, J. Hazard. Mater. 153 (1–2) (2008) 827–833.

[52] W. Sigmund, J. Yuh, H. Park, V. Maneeratana, G. Pyrgiotakis, A. Daga, J. Taylor, J.C. Nino, Processing and structure relationships in electrospinning of ceramic fiber systems, J. Am. Ceram. Soc. 89 (2) (2006) 395–407.

[53] K. Woan, G. Pyrgiotakis, W. Sigmund, Photocatalytic carbon-nanotube–TiO2 composites, Adv. Mater. 21 (21) (2009) 2233–2239.

[54] T. Matsunaga, M. Inagaki, Carbon-coated anatase for oxidation of methylene blue and NO, Appl. Catal. Environ. 64 (1–2) (2006) 9–12.

[55] S. Sakthivel, H. Kisch, Daylight photocatalysis by carbon-modified titanium dioxide, Angew. Chem. Int. Ed. 42 (40) (2003) 4908–4911.

[56] K. Ewa, R. Sven, Photoreactors for wastewater treatment: a review, Recent Pat. Eng. 4 (3) (2010) 242–266.

[57] S. Mozia, P. Brozek, J. Przepiorski, B. Tryba, A.W. Morawski, Immobilized TiO2 for phenol degradation in a pilot-scale photocatalytic reactor, J. Nanomater. 2012 (2012) 10.

[58] R. Rinaldi, U. Schuchardt, On the paradox of transition metal-free alumina-catalyzed epoxidation with aqueous hydrogen peroxide, J. Catal. 236 (2) (2005) 335–345.

[59] A. Basile, A. Cassano, N.K. Rastogi, Advances in Membrane Technologies for Water Treatment: Materials, Processes and Applications, Elsevier Science, 2015.

[60] J. Hong, Y. He, Polyvinylidene fluoride ultrafiltration membrane blended with nano-ZnO particle for photo-catalysis self-cleaning, Desalination 332 (1) (2014) 67–75.

[61] H. Yu, Y. Zhang, J. Zhang, H. Zhang, J. Liu, Preparation and antibacterial property of SiO2–ag/PES hybrid ultrafiltration membranes, Des. Water Treat. 51 (16–18) (2013) 3584–3590.

Chapter 16

New nanomaterials for wastewater depollution: Methods using chemometric approaches

Yen T.H. Dang[a,b], Sheeana Gangadoo[b], Vi Khanh Truong[b], Daniel Cozzolino[c], and James Chapman[a,*]
[a]*Advanced Institute of Science and Technology, The University of Da Nang, Da Nang, Viet Nam,* [b]*School of Applied Chemistry and Environmental Sciences, RMIT University, Melbourne, VIC, Australia,* [c]*QAAFI, University of Queensland, QLD, Australia*
**Corresponding author: e–mail address: james.chapman@rmit.edu.au*

Introduction

More severe demands on water systems emerge from rising industry, population increase, and urbanization and our entire water, air, and soil environment [1]. Pollutants are released from industrial operations and residences, wreaking havoc on the ecosystem and human life. Gases such as NO_x, SO_x, CO, and NH_3 are types of pollutants [2], as are organics such as pesticides, herbicides, medicines, biotoxins such as microbes, and heavy metals [2].

Several depollution methods have been developed to manage certain forms of pollution successfully. Many of these technologies operate on the adsorption principle, which is typically regarded as simple, straightforward to operate, and successful in removing the environmental pollutants in issue while not spreading secondary pollution compounds. The uses of solvent extraction, filtration (nano or micro), gravity separation or sedimentation, flotation, coagulation, oxidation, evaporation, precipitation, distillation, reverse osmosis, adsorption, ion exchange [3], or electrolysis are some of the more common and well-established depolluting wastewater techniques [2]. However, the process of adsorption has the broadest range of operational applications and is used in a range of adsorbent nanomaterials. Adsorption and adsorbing nanomaterials have been shown to remove a variety of different organic, inorganic, and biological contaminants, both insoluble and soluble. [2,4]

Many reviews of nanoparticles for water treatment have been published [5–10]. These reviews focus on utilizing nanomaterials and/or adsorbent-specific nanomaterials, metal oxides, nano zero-valent iron, cellulosic materials, or more specialized adsorbate materials (metals, dyes, pharmaceuticals, and personal care products). None, however, examine the use of chemometrics and subsequent machine learning to generate, test, remove, and analyze nanomaterials in the wastewater depollution process.

This chapter seeks to combine some of the past successes on several types of nanomaterials used in the application of water treatment for depollution that have employed chemometrics in their methodologies.

Nanomaterials for water treatment

As the name implies, nanoadsorbents are particles that are nanosized (i.e., in the nanoparticle size range of 100 nm) made of organic or inorganic materials which can adsorb contaminants. The adsorption process occurs due to the high nanoadsorbent's porosity, small size to volume ratio, and active site capability, which can sequester the pollutants, which may have variable molecular size, hydrophobicity, and specification behavior, releasing secondary pollutants if the compounds are broken down. The nanoadsorbent will operate by being both fast and having a high pollutant-binding affinity and capacity. They must, by definition, have chemical regeneration abilities once they have been depleted. For these reasons, researchers have focused on nanotechnology to achieve cleaner water outcomes.

Separation Science and Technology, Volume 15. https://doi.org/10.1016/B978-0-323-90763-7.00013-5

Materials display distinct properties at the nanoscale, primarily determined using their surface area to volume ratio. These properties are important because they boost the ability to increase adsorption capacity. Furthermore, the particle's high surface area to volume ratio demonstrates any catalytic potential and strong reactivity in the water system, resulting in significantly higher adsorption characteristics than traditional materials.

Carbon-based materials

Overall, the adsorbent must have a wide range of surface area, pore volume, and chemical functionality will effectively affect the adsorption processes of the nanoadsorbents. For this reason, several porous materials have been developed, including activated carbon, pillared clays, zeolites [3], mesoporous oxides, polymers, and metal–organic frameworks, with varying water depollution effects. Because of their excellent adsorption capacity and thermal stability, carbon-based adsorbents such as activated carbon, carbon nanotubes, fullerenes, graphene, and graphene oxide have been shown to be very efficient. These carbon-based compounds are highly efficient in eliminating both inorganic and organic pollutants.

Designing carbon-based nanomaterials using chemometrics

When materials are synthesized, data analysis is typically helpful in characterizing them. Indeed, this knowledge may then be used to adjust synthesis settings to create more particular nanomaterial designs. While this process is now iterative and rigorous, it was not at the forefront of many nanomaterials and their first synthetic processes. Complex chemical reaction systems necessitate mass transport, heat transfer, fluid movement, and chemical reaction research to process novel materials. In other words, the creation of new goods, processes, or procedures is multifactorial, and the actual effect of components on the final result is evaluated. Many prior nanotechnology syntheses were created by altering one variable at a time (univariably) [11]. As a result, univariate techniques are defined as the analysis done using only one variable to summarize or explain the given approach, which is applied to a single variable dataset (i.e., temperature, pH, precursor concentration). However, when continuous variables are included in the dataset and are made up of numerous variables, the time required to analyze them is considerable. More complex statistical studies would also be necessary. To overcome this, multivariate analytic approaches have gained traction, revealing the interactions among many factors. Chemometric techniques refer to a collection of multivariate methods that are specially used to analyze such chemical processes. This section will provide examples of how chemometrics may be used to improve the design of carbon-based nanomaterials.

Carbon-based nanomaterials in water depollution

Carbon-based materials have played several roles in human evolution. Carbon exists in a variety of allotropes, including fullerenes, carbon nanotubes, graphene, graphdiyne, carbon nanofoam, and MXene, to name a few. Bankole et al., for example, investigated the effect of the drying time duration, the drying temperature or stirring speed, and mass support on the formation of carbon nanotubes using a bimetallic ($Fe\text{-}Co/CaCO_3$) catalyst. The yield was chosen as the response, and the significance of factors was assessed using the standard plot. At the same time, the regression model was initially tested using ANOVA, a 95% confidence value was obtained, revealing that the stirring speed produced an interaction effect with the mass of support and was significant in the material synthesis process.

The production of activated carbon for use in environmental applications has been extensively researched [12–15]. Using a Doehlert design and response surface approach, for example, atrazine was removed onto an activated carbon-silver material. The following variables were used to optimize the composite during synthesis: $AgNO_3$ (X1), impregnation temperature (X2), and impregnation time (X3). The three variables chosen to optimize atrazine adsorption were temperature (X1), pH (X2), and solution ionic strength (atrazine/$CaCl_2$ ratio; X3). In both situations, the design response inferred the ionic strength of the solution [16].

Heavy metals and dyes are additional dangerous contaminants that are frequently found in wastewater from many sectors. Pollutants have a substantial ecological influence on ecosystems because they may change water's chemical and physical characteristics, affecting flora and wildlife. Dyes, for example, in wastewater, restrict light penetration and can slow photosynthetic rates. These dyes frequently degrade and produce carcinogens that are hazardous to live organisms, including people. Malachite green, a common triphenylmethane dye, has been employed as an antimicrobial agent in aquaculture and the food industry. Malachite green inhibits fungal infections, protozoan infections, and other helminth-related disorders in fish [17]. Malachite green has been shown to have mutagenic and carcinogenic effects, resulting in liver cancer.

Furthermore, metal ions from industrial effluents can find their way into drinking water supplies. Lead (II) is widely utilized in industry, including printing, batteries, pigments, fuels, explosives, and various types of manufacturing. The Environmental Protection Agency has established a maximum Pb^{2+} contamination limit of $15\,L^{-1}$ (EPA). Alipanahpour et al. have studied dye-adsorption and uptake of the dyes and metal ions in a binary system using copper oxide-nanoparticles-activated carbon (CuO-NP-AC). The influence of critical factors in the depollution system was studied and then optimized using artificial neural networks (ANN) and the response surface methodology (RSM). The authors discovered that the most significant simultaneous elimination percentages (>99.0%) were observed at 25 mg L1, 20 mg L1, 0.02 g, 5 min, and 6.0 corresponding to initial Pb^{2+} concentration, initial Malachite Green (MG) concentration, CuO-NP-AC quantity, ultrasonication time, and the pH in the system. The correctness of the RSM equation was validated by the analysis of variance and computation of the correlation coefficient between the predicted and experimental values of ultrasound-assisted simultaneous elimination of analytes, as shown in Fig. 1.

Mahanty et al. (2020) developed a green approach to NP synthesis. The authors used a mycosynthetic technique to create iron oxide nanoparticles. Fig. 2 shows that the particles were ferromagnetic, monodispersed, spherical, and had acceptable crystalline structures in the 73 nm size range.

These particles were then utilized to filter wastewater for the removal of Pb (II), Ni (II), Cu (II), and Zn (II). The initial batch adsorption investigations revealed that 1 g/L of the NPs could remove more than 90% of heavy metals from an aqueous environment at pH 6 at 313 K. These findings were supported by a multifactor chemometric approach, including the RSM and ANN approaches. Adsorption of 9.66 mg/L of heavy metals on 1 g/L of NPs at 308 K and pH 6 removed 98% Pb (II), 97% Ni (II), 92% Cu (II), and 94% Zn (II) ions, according to RSM and ANN analyses [18].

Halboos et al. investigated the use of nano-activated charcoal in the removal of dyes from water systems. The authors employed chemometric separation to extract Gentian violet (GV) from aqueous solutions. Nano-charcoal removed the GV, and the parameters influencing the adsorption process were evaluated concurrently to decrease the number of trials. The researchers discovered that the best clearance effectiveness was 98%. [19]

Azqhandi et al. discovered that carbon nanotubes loaded with zinc and silver as a nanocomposite are effective for dye adsorption. The authors state that methylene blue and basic yellow adsorbed to the nanocomposite were experimentally demonstrated using isotherms and kinetic behaviors, which were then expanded with Freundlich and pseudo-second-order kinetic models [20]. The use of adaptive neuro-fuzzy inference system (ANFIS), generalized regression neural network (GRNN), backpropagation neural network (BPNN), radial basic function neural network (RBFNN), and response surface methodology (RSM) systems were created based on the experimental data supplied by the authors [20]. The methodologies were used to forecast the sorbent's removal performance. This is significant because precise calculations on the degree of depollution determined in the system can be produced. Using the central composite rotatable design of RSM, GRNN, ANFIS, BPNN, and RBFNN, process factors (i.e., dye concentration, adsorbent mass, sonication time) are used on the removal of methylene blue and basic yellow dyes was investigated. The coefficients of determination computed from the validation data for the ANFIS, GRNN, BPNN, RBFNN, and RSM models were 0.9999, 0.9997, 0.9883, 0.9898, and 0.9608 for MB, and 0.9997, 0.9990, 0.9859, 0.9895, and 0.9593 for BY28 dye [20]. When compared to the other models, the ANFIS model was shown to be more exact. The GRNN approach, on the other hand, is significantly simpler

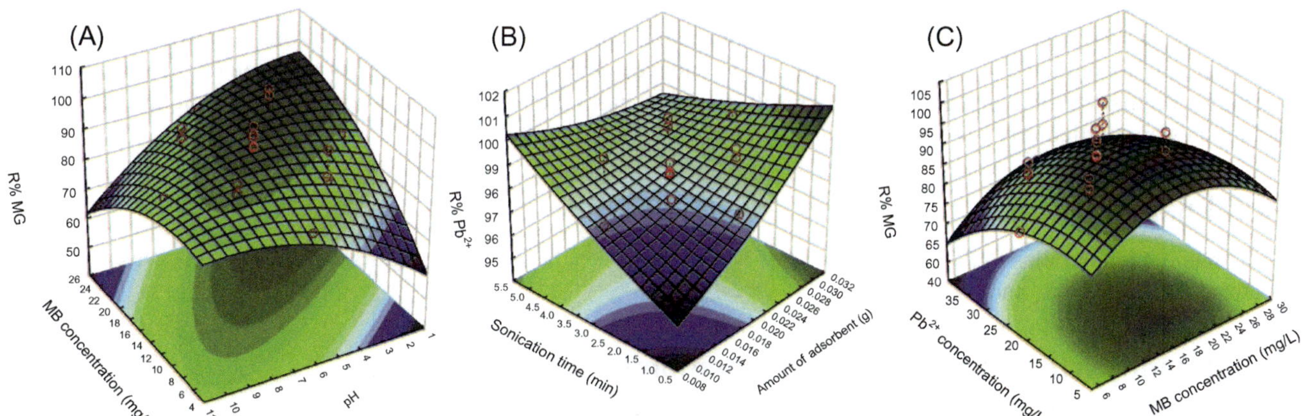

FIG. 1 Response surfaces plot for ultrasound-assisted simultaneous removal of Pb^{2+} ion and MG dye. *Taken from E. A. Dil, M. Ghaedi, A. Asfaram, S. Hajati, F. Mehrabi, A. Goudarzi, Preparation of nanomaterials for the ultrasound-enhanced removal of Pb2+ ions and malachite green dye: chemometric optimisation and modeling. Ultrason. Sonochem. 34 (2017) 677–691.*

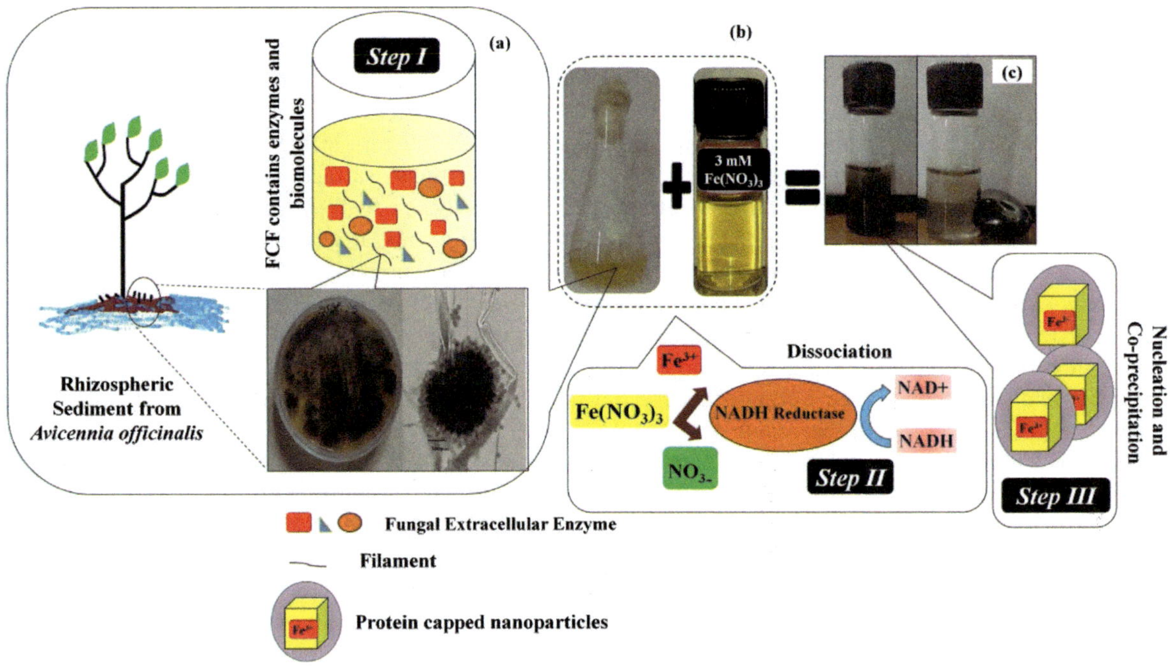

FIG. 2 Shows the mechanism for the synthesis of IO-NPs. (A) Fungal colony with its compound microscopic image and FCF containing enzymes and biomolecules. (B) FCF of fungal isolate STSP 25 (Aspergillus tubingensis), and 3 mM $Fe(NO_3)3$. (C) Reaction mixture showing the visible change in the color from *brown* to the *black* of the reaction mixture after 10 min of incubation at room temperature and suspension of iron oxide nanoparticles (IO-NPs) in the presence of a magnetic field in room temperature. *Taken with permission from S. Mahanty, S. Chatterjee, S. Ghosh, P. Tudu, T. Gaine, M. Bakshi, et al., Synergistic approach towards the sustainable management of heavy metals in wastewater using mycosynthesized iron oxide nanoparticles: biofabrication, adsorptive dynamics and chemometric modeling study. J. Water Process. Eng. 37 (2020) 101426.*

than the ANFIS approach and requires less time for analysis. The authors demonstrate the possible application of chemometrics, and the GRNN method might be used to mimic actual dye systems in wastewater.

Pathogens have also been removed from wastewater using graphene oxide. Rajapaksha et al. employed CuO-NPs and graphene oxide as a pathogenic wastewater bacterial treatment technique in their investigation [4]. The nanomaterial was tested against Escherichia coli ATCC 8739 (E. coli) and Salmonella typhimurium ATCC 14028 (S. typhimurium). The scientists concluded that there was a stage-by-stage killing and eradication process of bacteria when graphene oxide and nanoparticles doped graphene oxide were compared. When compared to the pure GO material, the resultant GO–CuONP nanocomposite was shown to be a highly effective antibacterial nanomaterial, significantly reducing the growth and development of both *E. coli* and *S. typhimurium* bacteria. Fig. 3 shows the low-cost and easy production approach for functionalizing GO with CuONPs, which might be an efficient alternative strategy for eliminating waterborne bacteria [21].

Fig. 4 shows a schematic demonstrating how graphene oxide may remove contaminants from wastewater. Organic pollutants are highly attracted to graphene membranes via noncovalent bonding, hydrogen bonds, van der Waals forces, π–π stacking, and hydrophobic interactions. As a result, the filter has a hydrophobic membrane, demonstrating that graphene membranes are particularly beneficial as nanomaterials for water depollution [2].

Metal nanoparticles

Noble metal nanoparticles (NPs) are being studied for their physicochemical characteristics and possible uses in catalysis, electronics, plasmonics, and biotechnology.

Metal nanoparticle synthesis using chemometrics

Metal nanoparticles of various shapes and sizes have piqued the research community's interest for decades due to their size and form and their optical, electrical, and chemical interaction capabilities. Several strategies for producing nanoparticles have been proposed, with several particular syntheses producing diverse outcomes (i.e., size, shapes, morphology) [23–25]. Top-down and bottom-up techniques have been widely classed as traditional strategies for nanoparticle creation.

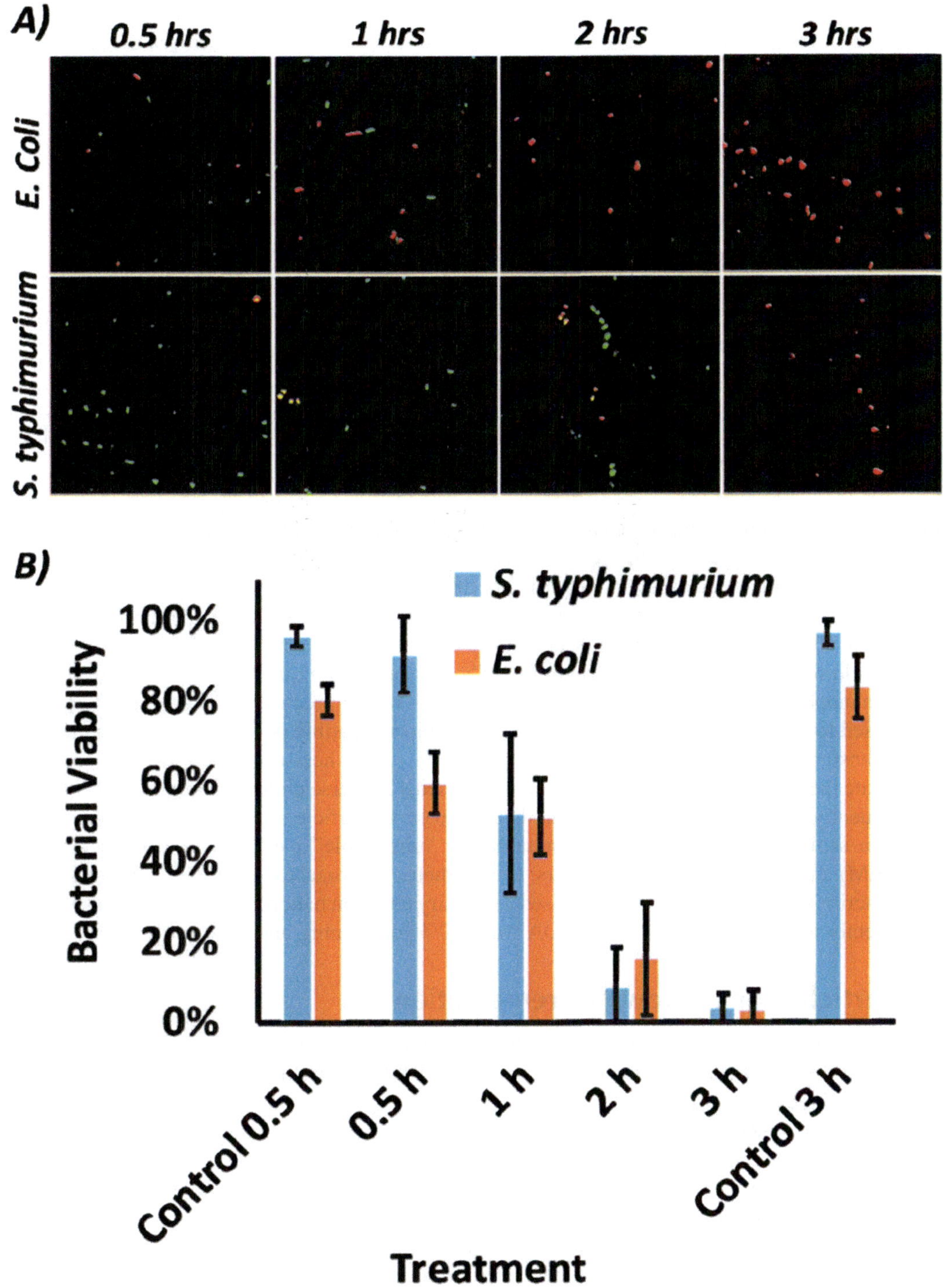

FIG. 3 Bacterial viability as a function of time. (A) Representative CLSM images of cells treated with 3 mg/mL of the GO–CuONP nanocomposite at time intervals of 0.5, 1, 2, and 3 h. Fluorescent staining of the samples with live/dead backlight highlights the live (*green*) and dead (*red*) cells in the CLSM images. (B) Cell viability was quantified as a percentage of live versus dead cells from the CLSM images [22].

These approaches frequently rely on combining various precursors, protective agents, and reducing agents in a single reaction vessel using a single variable experimental design [23]. We mean a limiting reagent or a precursor consumed in the reaction in the desired outcome [26]. For example, while the size of the nanoparticles may be regulated by intuition or the experimentalist during the synthesis process, it has not been possible to quantitatively estimate the size of the nanoparticles based on experimental conditions. The desired size range of the nanoparticles is generally achieved by a trial and

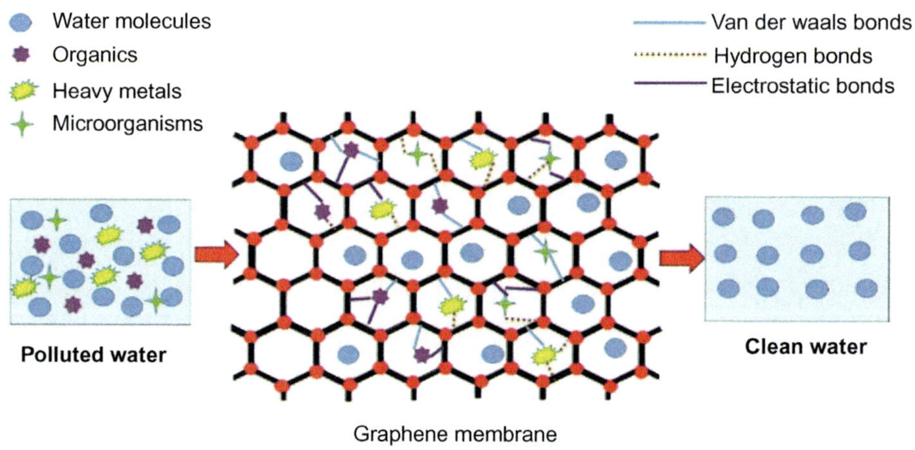

FIG. 4 Schematic showing graphene membrane and the mechanisms for water depollution properties. *Taken with permission from P. Rajapaksha, A. Power, S. Chandra, J. Chapman, Graphene, electrospun membranes and granular activated carbon for eliminating heavy metals, pesticides and bacteria in water and wastewater treatment processes. Analyst 143 (23) (2018) 5629–5645.*

error procedure involving different experimental settings [27]. The ability to anticipate the characteristics of nanoparticles from the synthesis conditions improves our capacity to regulate particle size, shape, morphology, and crystallinity and enables the control of specialized qualities for specific applications, such as wastewater depollution.

Applying a multifactorial statistical design to develop a mathematical model from a series of synthesis experiments, Rondinone et al. have challenged the paradigm of univariate nanoparticle design. This chemometric model quantitatively correlates nanoparticle size with experimental conditions. The author's technique demonstrated the capacity to predict the mean diameter of nanoparticles based on synthesis circumstances with a precision of 1 to 3 nm [27].

Chitosan was used to produce polymeric nanoparticles from biocompatible polymers. Based on the diffuse reflectance infrared Fourier transform (DRIFT) spectroscopic approach, the researchers employed the partial least squares (PLS) regression to estimate the average particle size. The scientists used 42 diverse chitosan samples with varying particle sizes, which PLS subsequently analyzed. The PLS may predict the average particle size on the synthesized sample, according to the authors. The samples were classified as "suitable" or "unsuitable" based on field emission scanning electron microscopy (FESEM) and morphological parameters. Chemometric approaches such as PLS regression, cluster analysis (CA), and linear discriminant analysis (LDA) were used to classify chitosan nanoparticles based on shape successfully. The percentage of correctly identified samples using LDA was 100% for training and 90% for test sets, respectively [28].

Metal nanoparticles for wastewater remediation

These nanoparticles, gold, silver, and copper, have caught the attention of researchers due to their intriguing characteristics. Hatamie et al. adopt a simple approach for the production of Cu NPs for quick colorimetric visual detection of sulfide ions (S^{2-}) in water in their investigation [29].

The Cu NP system functioned as a sensor, detecting sensitive and selective color changes in the presence of micromolar amounts of S^{2}-, which can be seen with the naked eye and tracked using a UV–Vis spectroscopy. In concentration ranges of 12.5 to 50.0×10^{-6} M, the color change quantitatively corresponds with the S^{2-}. The authors of this study spiked tap and river water samples to create a synthetic S^{2-} water sample. The results reveal that the sensor can detect S^{2-} and has strong selectivity, suggesting that it might be utilized to monitor sulfide ions.

Antibiotic treatments in wastewater are also thought to be a hazard. However, several research groups are already seeing success using gold-coated iron oxide nanoparticles, as seen in Fig. 4.

The authors recorded the course of the reaction using a multivariate curve resolution alternative least squares (LS) as a power of the chemometric approach employing UV–Visible spectroscopy and data matrices. For the intermediate species and the end product, the concentrations and removal were fitted to a first-order kinetic profile. The researchers discovered that ampicillin exhibits a faster photocatalytic degradation process and that when exposed to gold-coated iron oxide nanoparticles, it converts to its photodegradation product in less than 30 min [30].

TiO_2 nanoparticles are also utilized as chemical support vehicles for photocatalytic dye decolorization. Textile dyes are a severe water contaminant and must be removed from wastewater systems. During the TiO_2 nanoparticle breakdown, the authors utilized Basic Red, Basic Blue, and Malachite Green concurrently [31]. The authors successfully employed partial

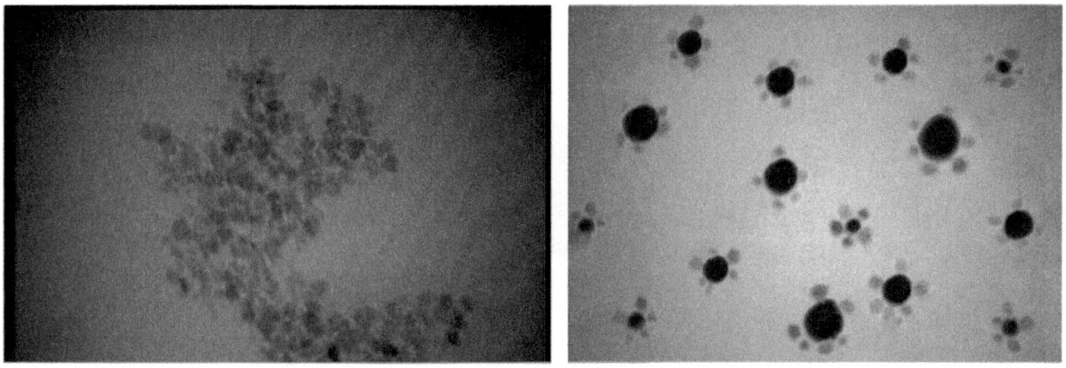

FIG. 5 TEMs of the as-prepared spherical-shaped bare (*left*) and Au-decorated (*right*) magnetic NPs. *The figure was taken with permission from N. Jalili-Jahani, B. Hemmateenejad, M. Shamsipur, Gold-decorated Fe3O4 nanoparticles for efficient photocatalytic degradation of ampicillin: a chemometrics investigation. J. Iran. Chem. Soc. 17 (5) (2020) 1173–1182.*

least squares modeling as a multivariate calibration of the spectrophotometric data. Before and after the photocatalysis process, the scientists tested the dye mixture removal concentrations against an aquatic species, *Spirodela polyrhiza*. The authors investigated the interaction impact of UV-light intensity and reaction time on dye color removal at diverse UV intensities ranging from 12.2 to 47.2 W/m^2 and varied reaction durations, as shown in Fig. 5. The decolorization efficiency rises as the UV-light intensity and response time increases. (See Fig. 6.)

Metal sulfides are recognized to exhibit intriguing electrical characteristics and are used in a variety of technological applications. Nickel sulfide is one of the most well-known metal sulfides. Nickel sulfide (NiS) has gained popularity as a toughening agent for materials, principally employed in the semiconductor sector, catalysis, and rechargeable batteries. The composition of nickel-based nanoparticles is affected by pH, reagent concentration, temperature, and reaction time. Many approaches for synthesizing NiS nanoparticles from simple precursors are accurate "hit or miss."

Multivariate techniques are being utilized to optimize experimental variables while lowering the number of needed experiments significantly. This sort of approach allows for optimal settings and the use of fewer chemicals during laboratory operations. Multivariate processes are both quicker and less expensive than typical univariate methods. Methods can be examined on the influence of several factors at the same time. As previously stated, the building of mathematical models allows for the analysis of the relevance and statistical significance of components and the evaluation of their interaction. Multivariate designs are built on variables and responses (qualitative or quantitative), and some of these are discussed in a study of chemometrics in environmental sensing [32]. The best-operating settings for experimental conditions are found by employing sophisticated experimental designs such as the Doehlert matrix (DM), Central Composite Design (CCD), or Box–Behnken design (BBD) [15,33,34]. Ghaedi et al., for example, employed CCD to investigate the individual and synergistic effects of variables in the synthesis of NiS.

Contact time, pH, MB concentration, SO concentration, and contact time were the variables. The CCD assists in avoiding superfluous tests and in investigating the components' synergy. The scientists created a nickel sulfide nanoparticle loaded activated carbon (NiS-NP-AC) and used it as an adsorbent material to treat water with methylene blue and safranin-O concurrently [35]. The optimal pH, adsorbent dose, methylene blue concentration, safranin-O concentration, and the contact duration were 8.1, 0.022, 17.8 mg/L, 5 mg/L, and 5.46 min, respectively. As a result, when the NiS-NP-AC nanomaterial was employed to adsorb the two dyes concurrently, it did so quickly (within 5 min) and is a potential nanomaterial for adsorption applications and water depollution.

Quantum dots with CdSe starch coating were used as a photocatalyst to decompose methylene green. In a neutral pH range, relatively few levels of CdSe were capable of decolorizing large amounts of methylene green (1000:1) within minutes. To investigate the reaction kinetics, UV–vis was combined with multivariate curve resolution-alternative least squares (MCR-ALS) [36]. Using factor analysis, it was discovered that five chemical variables were recognized for the disintegration of MG. The MCR-ALS was used to resolve the concentration profiles and pure spectra of the identified components. After a few iterations, MCR–ALS convergence was obtained with a deficiency of fit error of 0.49% for the PCA and 0.60% for the model, respectively [37]. Fig. 7 depicts the resulting concentration profiles of the species at various pH profiles combined and compared to the matching pure spectra of each species.

This study makes good use of quantum dots to remove organic pollutants from effluent. Likewise, zeolites have excellent water depollution and adsorption characteristics [3]. A simultaneous elimination of dyes using a nano zeolite was achieved in work by Shojaei et al. To optimize and simulate the dye removal conditions, the authors employed the

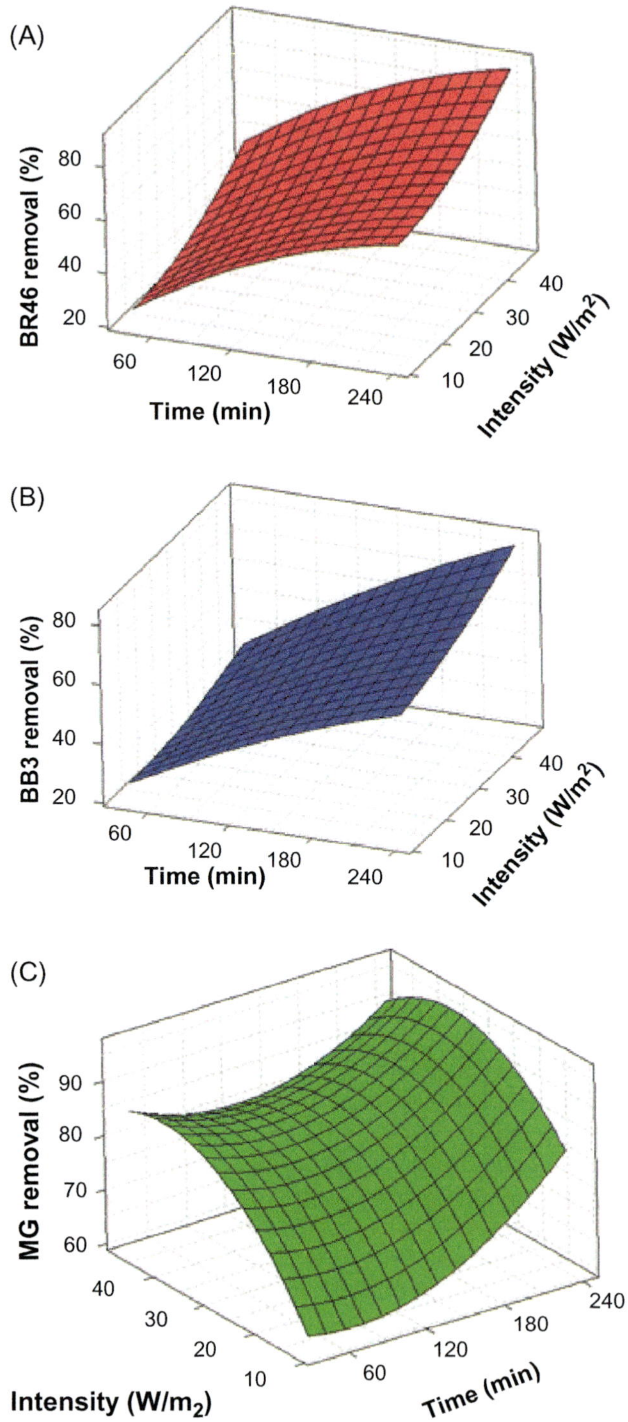

FIG. 6 The response surface plots of the decolorization efficiency (%) as the function of UV light intensity (W/m^2) and reaction time (min) of (A) BR46, (B) BB3, and (C) MG. BR46 CI Basic Red 46, BB3 CI Basic Blue 3, MG Malachite Green. *Image taken with permission from A. R. Khataee, M. Fathinia, A. Naseri, A. Hasanzadeh, F. Vafaei, A. Emami, et al. Modeling and optimisation of simultaneous photocatalysis of three dyes on ceramic-coated TiO2 nanoparticles using chemometrics methods: phytotoxicological assessment during degradation process. Res. Chem. Intermed. 40 (3) (2014), 1283–1302.*

Taguchi technique and RSM. The Taguchi technique was used to assess the quantity of dye removal using a multifactorial approach (dye concentration, the duration of the sonication period, the ionic strength, the adsorbent dose, the temperature, and the pH). The most relevant criteria were identified, and the RSM approach was used to determine the maximum % dye removal [38]. The models were determined to be in corresponding with the experimental data by the authors. These nanozeolites demonstrated that the materials could be reused five times, which is a fantastic outcome for the material application.

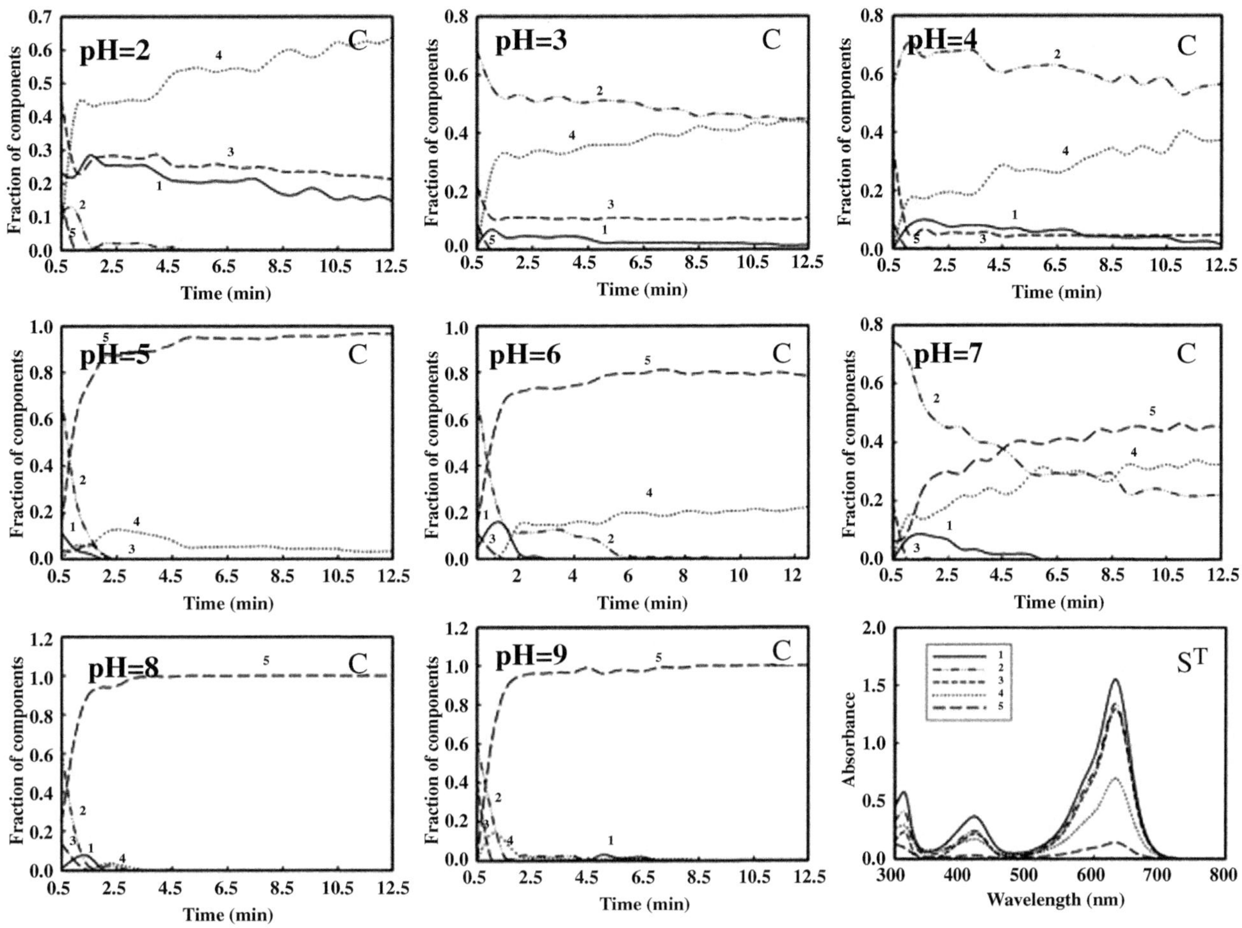

FIG. 7 Concentration profiles resolved by MCR–ALS at different pH ranges. The resolved pure spectra of methyl green and its degradation products are shown in the bottom right image. *The figure was used with permission from B. Hemmateenejad, P. Shadabipour, T. Khosousi, M. Shamsipur, Chemometrics investigation of the light-free degradation of methyl green and malachite green by starch-coated CdSe quantum dots. J. Ind. Eng. Chem. 27 (2015) 384–390.*

Water depollution using machine learning and nanomaterials

Recently, advancements in the design, production, and manipulation of micro/nanoscale robots (MNR) have attracted water depollution and remediation attention. The employment of an MNR results in new adaptability levels due to the variety of motor forms, size, composition, power sources, and other elements. Unfortunately, the fantastic range of nanoparticle shapes makes it difficult and time-consuming to identify critical particle properties for desired applications such as water treatment. However, there is a growing trend of utilizing MNRs for environmental applications, such as heavy metal or organic pollution removal [39]. As we have seen throughout the chapter, direct experiments are labor and resource costly, but computations and simulations are not always practical. For example, the usage of MNR particles might vary in size, making it very difficult to characterize all conceivable dimensions experimentally [40]. As a result, a mechanism for predicting the collective behaviors of MNRs concerning environmental circumstances and the standard nanoparticle properties such as size, shape, reactions, and surroundings should exist. These collective behaviors are precisely what data science and informatics are doing to change the way material science is used in the environment—for example, machine learning (ML), data mining, and ANNs, which enable the prediction of material behavior based on past data rather than direct experimentation or computations.

Zeng and colleagues used ML approaches to forecast the self-propulsion speed and water-cleaning effectiveness of MNRs. The authors successfully identified critical aspects crucial for properly depolluting the Pt-coated MNRs based on particle size, catalyst type, and aspect ratios. The resulting data were compared to the ML predictions, which revealed an accurate forecast with less than 15% error [39]. According to Zeng et al., the gradient boosting algorithm is the most promising method for forecasting MNR environmental decontamination. This research is particularly innovative in the field of environmental sciences.

Although scientific interest in nanoparticle integration into ultrafiltration polymeric membranes is developing, commercial acceptance has yet to occur. A feed forwards ANN model with one hidden layer and a Bayesian regularization training procedure were used to create and forecast the performance of an ultrafiltration nanocomposite [41]. This performance is based on the rejection of solutes, the recovery of fluxes, and clean water flux. These factors allow for the reduction of time and resources spent in the design of nanofiltration membranes. Fetanat et al. utilized experimental data from 200 publications published between 2006 and 2020, totaling 735 samples. The data included ANN model training, validation, and testing. The authors demonstrated that using the 7 variables used for input: polymer concentration, polymer type, filler concentration, average filler size, solvent concentration, solvent type, and contact angle, the best 30 ANN models produce the most accurate estimation of the overall membrane performance (Fig. 8).

Given the originality of this study, it is possible to apply expedited development of new ways of discovery of design process for nanomaterial creation of membranes.

Future directions and conclusions

Nanomaterials have great promise for the next generation of water purification and depollution solutions. Several milestones must be met to accelerate the development of nanomaterials and their applications to water remediation:

(1) Require next-generation nanoadsorbents—nanoparticles can be utilized as an adsorbent or nanoadsorbent. These nanoadsorbents will be highly selective and functional, capable of capturing both polar and nonpolar contaminants in water.

(2) Require Next-Generation Nanomembranes—nanomaterials will require leading and innovative techniques such as electrospun membranes, nanofibers, and graphene nanosheets. These materials will have outstanding size exclusion and pore size properties for water treatment procedures.

(3) Materials should be environmentally friendly and sustainable—nanoparticles employed in producing membranes or nanomaterials must be resilient, reusable, and sustainable. Notably, nanoparticles must be stabilized or permanently attached to membranes or technologies to avoid secondary pollution entering the water system.

(4) The effects of biofouling should be considered, especially when deploying the novel materials for extended periods where nonsterile aqueous environments exist. The antifouling properties of the nanomaterials will generate extension in the lifetime and reuse of the materials for water remediation [42–44].

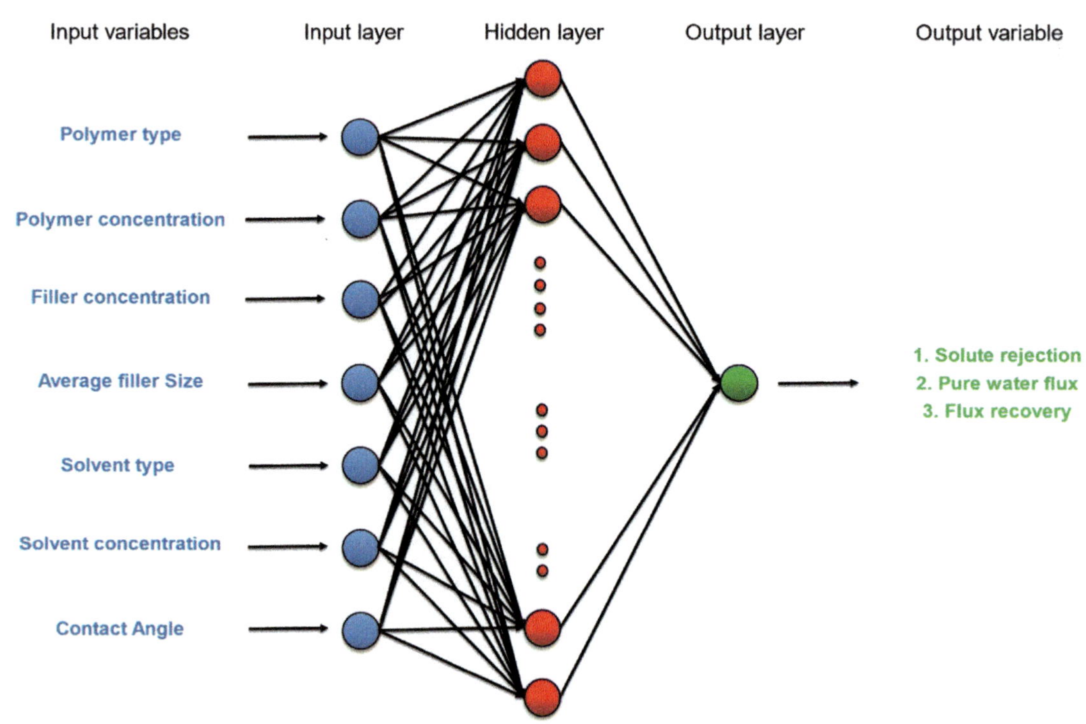

FIG. 8 Architecture of the proposed ANN models for the estimation of (1) solute rejection, (2) pure water flux, and (3) flux recovery using polymer type, polymer concentration, filler concentration, average filler size, solvent type, solvent concentration, and contact angle.

References

[1] L. Jones, F. Regan, J. Chapman, Monitoring of priority substances in waste water effluents, Irish Environmental Protection Agency, 2013.

[2] P. Rajapaksha, A. Power, S. Chandra, J. Chapman, Graphene, electrospun membranes and granular activated carbon for eliminating heavy metals, pesticides and bacteria in water and wastewater treatment processes, Analyst 143 (23) (2018) 5629–5645.

[3] O. Santiago, K. Walsh, B. Kele, E. Gardner, J. Chapman, Novel pre-treatment of zeolite materials for the removal of sodium ions: potential materials for coal seam gas co-produced wastewater, SpringerPlus 5 (1) (2016) 1–16.

[4] P. Rajapaksha, S. Cheeseman, S. Hombsch, B.J. Murdoch, S. Gangadoo, E.W. Blanch, Y.B. Truong, D. Cozzolino, C.F. McConville, R.J. Crawford, Antibacterial properties of graphene oxide-copper oxide nanoparticle nanocomposites, ACS Appl. Bio Mater. 2 (12) (2019) 5687–5696.

[5] R.F. Spalding, M.E. Exner, Occurrence of nitrate in groundwater—a review, J. Environ. Qual. 22 (3) (1993) 392–402.

[6] R.K. Singh, S. Vats, P. Tyagi, Industrial wastewater treatment by biological activated carbon-a review, Res. J. Pharm. Biol. Chem. Sci. 2 (4) (2011) 1053–1058.

[7] M. Kumar, A. Puri, A review of permissible limits of drinking water, Indian J. Occup. Environ. Med. 16 (1) (2012) 40.

[8] S. Wang, H. Sun, H.M. Ang, M.O. Tadé, Adsorptive remediation of environmental pollutants using novel graphene-based nanomaterials, Chem. Eng. J. 226 (2013) 336–347.

[9] P. Suja, C. Reshmi, P. Sagitha, A. Sujith, Electrospun nanofibrous membranes for water purification, Poly. Rev. 57 (3) (2017) 467–504.

[10] S.N. Zulkifli, H.A. Rahim, W.-J. Lau, Detection of contaminants in water supply: a review on state-of-the-art monitoring technologies and their applications, Sens. Actuators B 255 (2018) 2657–2689.

[11] D. Cozzolino, J. Roberts, Applications and developments on the use of vibrational spectroscopy imaging for the analysis, monitoring and characterisation of crops and plants, Molecules 21 (6) (2016) 755.

[12] T.W. Lambert, C.F. Holmes, S.E. Hrudey, Adsorption of microcystin-LR by activated carbon and removal in full scale water treatment, Water Res. 30 (6) (1996) 1411–1422.

[13] B.E. Reed, M. Jamil, B. Thomas, Effect of pH, empty bed contact time and hydraulic loading rate on lead removal by granular activated carbon columns, Water Environ. Res. 68 (5) (1996) 877–882.

[14] J. Xia, K. Noda, S. Kagawa, N. Wakao, Production of activated carbon from bagasse (waste) of sugarcane grown in Brazil, J. Chem. Eng. Japan 31 (6) (1998) 987–990.

[15] K. Nakagawa, A. Namba, S.R. Mukai, H. Tamon, P. Ariyadejwanich, W. Tanthapanichakoon, Adsorption of phenol and reactive dye from aqueous solution on activated carbons derived from solid wastes, Water Res. 38 (7) (2004) 1791–1798.

[16] A.L. Cazetta, L. Spessato, V.C. Almeida, The use of chemometric tools for screening and optimisation of variables in the preparation and application of carbon-based materials, J. Taiwan Inst. Chem. Eng. 121 (2021) 321–336.

[17] E.A. Dil, M. Ghaedi, A. Asfaram, S. Hajati, F. Mehrabi, A. Goudarzi, Preparation of nanomaterials for the ultrasound-enhanced removal of Pb2+ ions and malachite green dye: chemometric optimisation and modeling, Ultrason. Sonochem. 34 (2017) 677–691.

[18] S. Mahanty, S. Chatterjee, S. Ghosh, P. Tudu, T. Gaine, M. Bakshi, S. Das, P. Das, S. Bhattacharyya, S. Bandyopadhyay, Synergistic approach towards the sustainable management of heavy metals in wastewater using mycosynthesized iron oxide nanoparticles: biofabrication, adsorptive dynamics and chemometric modeling study, J. Water Process. Eng. 37 (2020) 101426.

[19] M.H. Halboos, B.J. Hussein, A.A. Sayhood, Used central composite chemometric method to adsorption gentian violet dye by nano activated charcoal, J. Phys. Conf. Ser. 1853 (1) (2021) 012027.

[20] M.H. Ahmadi Azqhandi, M. Shekari, B. Ghalami-Choobar, Synthesis of carbon nanotube-based nanocomposite and application for wastewater treatment by ultrasonicated adsorption process, Appl. Organomet. Chem. 32 (8) (2018) e4410.

[21] P. Rajapaksha, S. Cheeseman, S. Hombsch, B.J. Murdoch, S. Gangadoo, E.W. Blanch, Y. Truong, D. Cozzolino, C.F. McConville, R.J. Crawford, V.K. Truong, A. Elbourne, J. Chapman, Antibacterial properties of graphene oxide–copper oxide nanoparticle nanocomposites, ACS Appl. Bio Mater. 2 (12) (2019) 5687–5696.

[22] P. Rajapaksha, S. Cheeseman, S. Hombsch, B.J. Murdoch, S. Gangadoo, E.W. Blanch, Y. Truong, D. Cozzolino, C.F. McConville, R.J. Crawford, Antibacterial properties of graphene oxide–copper oxide nanoparticle nanocomposites, ACS Appl. Bio Mater. 2 (12) (2019) 5687–5696.

[23] J. Chapman, E. Weir, F. Regan, Period four metal nanoparticles on the inhibition of biofouling, Colloids Surf. B Biointerfaces 78 (2) (2010) 208–216.

[24] J. Chapman, T. Sullivan, F. Regan, Nanoparticles in Anti-Microbial Materials: Use and Characterisation, Royal Society of Chemistry, 2012, pp. 1–225.

[25] S. Gangadoo, D. Stanley, R.J. Hughes, R.J. Moore, J. Chapman, The synthesis and characterisation of highly stable and reproducible selenium nanoparticles, Inorg. Nano-Met. Chem. (2017) 0–0.

[26] H. Duan, D. Wang, Y. Li, Green chemistry for nanoparticle synthesis, Chem. Soc. Rev. 44 (16) (2015) 5778–5792.

[27] A.J. Rondinone, A.C.S. Samia, Z.J. Zhang, A Chemometric approach for predicting the size of magnetic spinel ferrite nanoparticles from the synthesis conditions, J. Phys. Chem. B 104 (33) (2000) 7919–7922.

[28] M. Khanmohammadi, H. Elmizadeh, K. Ghasemi, Investigation of size and morphology of chitosan nanoparticles used in drug delivery system employing Chemometric technique, Iran J. Pharm. Res. 14 (3) (2015) 665–675.

[29] A. Hatamie, B. Zargar, A. Jalali, Copper nanoparticles: a new colorimetric probe for quick, naked-eye detection of sulfide ions in water samples, Talanta 121 (2014) 234–238.

[30] N. Jalili-Jahani, B. Hemmateenejad, M. Shamsipur, Gold-decorated Fe3O4 nanoparticles for efficient photocatalytic degradation of ampicillin: a chemometrics investigation, J. Iran. Chem. Soc. 17 (5) (2020) 1173–1182.

[31] A.R. Khataee, M. Fathinia, A. Naseri, A. Hasanzadeh, F. Vafaei, A. Emami, Y. Hanifehpour, S.W. Joo, Modeling and optimisation of simultaneous photocatalysis of three dyes on ceramic-coated TiO_2 nanoparticles using chemometrics methods: phytotoxicological assessment during degradation process, Res. Chem. Intermed. 40 (3) (2014) 1283–1302.

[32] J. Chapman, V.K. Truong, A. Elbourne, S. Gangadoo, S. Cheeseman, P. Rajapaksha, K. Latham, R.J. Crawford, D. Cozzolino, Combining Chemometrics and sensors: toward new applications in monitoring and environmental analysis, Chem. Rev. (2020).

[33] G.E. Box, W.H. Hunter, S. Hunter, Statistics for experimenters, Vol. 664, John Wiley and Sons, New York, 1978.

[34] D. Massart, B. Vandeginste, L. Buydens, S. De Jong, P. Lewi, J. Smeyers-Verbeke, Handbook of Chemometrics and Qualimetrics: Part A, Elservier Science B. V, 1997.

[35] M. Ghaedi, M. Pakniat, Z. Mahmoudi, S. Hajati, R. Sahraei, A. Daneshfar, Synthesis of nickel sulfide nanoparticles loaded on activated carbon as a novel adsorbent for the competitive removal of methylene blue and safranin-O, Spectrochim. Acta A Mol. Biomol. Spectrosc. 123 (2014) 402–409.

[36] K. Ahmadi, Y. Abdollahzadeh, M. Asadollahzadeh, A. Hemmati, H. Tavakoli, R. Torkaman, Chemometric assisted ultrasound leaching-solid phase extraction followed by dispersive-solidification liquid–liquid microextraction for determination of organophosphorus pesticides in soil samples, Talanta 137 (2015) 167–173.

[37] B. Hemmateenejad, P. Shadabipour, T. Khosousi, M. Shamsipur, Chemometrics investigation of the light-free degradation of methyl green and malachite green by starch-coated CdSe quantum dots, J. Ind. Eng. Chem. 27 (2015) 384–390.

[38] S. Shojaei, S. Shojaei, S.S. Band, Application of chemometrics into removal of dyes by NaX Nanozeolites, Simultaneous Model (2021), https://doi.org/10.2139/ssrn.3809612.

[39] M. Zeng, S. Yuan, D. Huang, Z. Cheng, Accelerated design of catalytic water-cleaning nanomotors via machine learning, ACS Appl. Mater. Interfaces 11 (43) (2019) 40099–40106.

[40] L. Soler, V. Magdanz, V.M. Fomin, S. Sanchez, O.G. Schmidt, Self-propelled micromotors for cleaning polluted water, ACS Nano 7 (11) (2013) 9611–9620.

[41] M. Fetanat, M. Keshtiara, Z.-X. Low, R. Keyikoglu, A. Khataee, Y. Orooji, V. Chen, G. Leslie, A. Razmjou, Machine learning for advanced design of nanocomposite ultrafiltration membranes, Ind. Eng. Chem. Res. 60 (14) (2021) 5236–5250.

[42] J. Chapman, F. Regan, Nanofunctionalized superhydrophobic antifouling coatings for environmental sensor applications-advancing deployment with answers from nature, Adv. Eng. Mater. 14 (4) (2012).

[43] J. Chapman, L. Le Nor, R. Brown, E. Kitteringham, S. Russell, T. Sullivan, F. Regan, Antifouling performances of macro- to micro- to nano-copper materials for the inhibition of biofouling in its early stages, J. Mater. Chem. B 1 (45) (2013) 6194–6200.

[44] J. Chapman, C. Hellio, T. Sullivan, R. Brown, S. Russell, E. Kiterringham, L. Le Nor, F. Regan, Bioinspired synthetic macroalgae: examples from nature for antifouling applications, Int. Biodeter. Biodegr. 86 (2014) 6–13.

Index

Note: Page numbers followed by "*f*" indicate figures and "*t*" indicate tables.

X

Z

9780323907637